Introduction to Feedback Control

Introduction to Feedback Control

Li Qiu
Hong Kong University of Science and Technology
Kemin Zhou
Louisiana State University

Prentice Hall
Upper Saddle River Boston Columbus San Francisco
New York Indianapolis London Toronto Sydney Singapore
Tokyo Montreal Madrid Hong Kong Mexico City Munich
Paris Amsterdam Cape Town

Vice President and Editorial Director, ECS: Marcia J. Horton
Associate Editor: Alice Dworkin
Editorial Assistant: William Opaluch
Director of Team-Based Project Management: Vince O'Brien
Senior Managing Editor: Scott Disanno
Production Liaison: Irwin Zucker
Production Editor: Haseen Khan, Laserwords
Senior Operations Specialist: Alan Fischer
Operations Specialist: Lisa McDowell
Marketing Manager: Tim Galligan
Marketing Assistant: Mack Patterson
Creative Director: Jayne Conte
Cover Designer: Bruce Kenselaar
Cover Images: Duncan Babbage, Jean-Joseph Renucci, Mark Evans, Andrey Volodin//iStockphoto
Art Editor: Gregory Dulles
Composition/Full-Service Project Management: Laserwords, Inc.

Library of Congress Cataloging-in-Publication Data

Qiu, Li, 1961-
Introduction to feedback control / Li Qiu, Kemin Zhou.
p. cm.
Includes bibliographical references and index.
ISBN 0-13-235396-2 (alk. paper)
1. Feedback control systems. I. Zhou, Kemin. II. Title.
TJ216.Q23 2009
629.8′3–dc22

2009004028

Prentice Hall
is an imprint of
PEARSON

www.pearsonhighered.com

10 9 8 7 6 5 4 3 2 1

ISBN 13: 978-0-13-235396-0
ISBN 10: 0-13-235396-2

Contents

Preface

Do we need another textbook on classical control? Who needs another textbook on classical control when there are a few dozen existing ones that we have learned or taught from, such as *Modern Control Systems* (11th Edition) by Dorf and Bishop, *Feedback Control of Dynamic Systems* (5th Edition) by Franklin, Powell, and Emami-Naeini, *Modern Control Engineering* (4th Edition) by Ogata? No, we by no means believe that we can do a significantly better job in presenting the classical control theory to justify the writing of another textbook on the subject. However, we believe that the development of modern optimal and robust control theory in the last 30 years now calls for a significant change in the teaching of classical control. It is our goal to integrate the modern optimal and robust control theory into the classical control theory using tools already available from the context of classical control. We hope this objective has been achieved in our book.

Obviously, we still include a significant portion of the well-known classical control material in our work, albeit with some twists whenever appropriate in consideration of recent developments and the available modern computational tools. For example, we completely take out the material on the signal flow graph covered in many classical control textbooks, provide significant coverage on two-degree-of-freedom control, add Kharitonov robust stability results of polynomials, discuss in detail the effects of nonminimum phase zeros on the system performance and their relationship with overshoot and undershoot, and introduce a Routh table method for computing the 2-norm of a signal. Instead of introducing the detailed techniques of drawing an accurate root locus, we emphasize on how to quickly sketch a root locus with minimum effort indicating the trend of the root loci to help the analysis and design of a control system, leaving the detailed work to modern computational tools. In the frequency-domain analysis, we introduce a completely new method to visualize frequency responses from the Riemann sphere in addition to the classical Bode diagram, Nyquist plot, and Nichols chart. It turns out that this representation of a frequency response on the Riemann sphere is arguably the most natural method of considering the robustness issue of dynamical systems.

We add considerable material on modern optimal and robust control without introducing undue advanced tools. Consequently, we limit our presentation to single-input and single-output (SISO) systems with rational transfer functions. We do this to avoid the introduction of state space techniques which are widely used in modern optimal and robust control theory in dealing with multi-input and multi-output (MIMO) systems, but which require advanced linear algebra tools. On the other hand, we have tried to make the book self-contained and we have tinted parts of the text containing sophisticated mathematical reasoning to indicate that these parts may be skipped without affecting the basic understanding of the book.

We have intentionally tried to keep the book as short as possible so that most of the materials in the book can be covered in a one-semester course. Hence, we have faced many tough choices and ultimately this book reflects our own preference. We do intend to expand our presentation and coverage beyond this book in the future through a web site. We expect to update and improve our presentation continuously as we receive more feedbacks from readers. Hence a web site is maintained (**http://www.ee.ust.hk/~eeqiu/minisites/ifc**), where readers can obtain updates, corrections, and additional materials related to the book and post their comments and feedback to the authors.

We would like to express our sincere gratitude to many colleagues and friends for their help, encouragement, and support during the process of writing this book. In particular, we wish to thank Professors Ben Chen (National University of Singapore), Jie Chen (University of California, Riverside), Tongwen Chen (University of Alberta), Xiang Chen (University of Windsor), Peilin Fu (National University, San Diego), Huijun Gao (Harbin Institute of Technology), Tryphone Georgiou (University of Minnesota, Minneapolis), Fathi H. Gorbel (Rice University, Houston), Guoxiang Gu (Louisiana State University, Baton Rouge), Robert G. Landers (Missouri University of Science and Technology), Zexiang Li (Hong Kong University of Science and Technology), Derong Liu (University of Illinois, Chicago), Andrea Serrani (Ohio State University, Columbus), Weizhou Su (South China University of Technology), and Lihua Xie (Nanyang Technological University, Singapore) for their detailed reviews and constructive comments of various versions of this manuscript. We would also like to thank Professors Xiren Cao (Hong Kong University of Science and Technology), Hanfu Chen (Chinese Academy of Sciences), Lei Guo (Chinese Academy of Sciences), and Youxian Sun (Zhejiang University) for their encouragement and support.

In addition, we wish to thank Jingjing Li, Lili Kong, and Laurentiu Dan Marinovici of Louisiana State University, Dr. Wai-Chuen Gan of ASM Assembly Automation Ltd, Dr. Yiu Kuen Yiu of Hong Kong University of Science and Technology, the students in the ELEC271 classes in Hong Kong University of Science and Technology, and the students in the 2004 and 2005 classes of the Control and Mechatronics Division, Shenzhen Graduate School of the Harbin Institute of Technology, for reading and commenting on parts of this manuscript.

Our special thanks go to Yongxin Pang of China Southern Power Grid and Yu Liang of Hong Kong University of Science and Technology who did most of the art work and MATLAB graphics, to Jacqueline Wah of KGSupport who helped in the English editing, and to Enzhe Zhang of the University of Cambridge who created the cover design. We are also very grateful to Alice Dworkin of Prentice Hall for her assistance throughout the preparation of the manuscript.

We would also like to thank Hong Kong Research Grants Council for its continuous support during the writing of this manuscript.

This book is written for the next generation of engineers, researchers, and practitioners interested in feedback control. We dedicate this book to the next generation of our families, Luna Qiu, Celina Qiu, Eric Zhou, Catherine Zhou, and Albert Zhou.

Li Qiu eeqiu@ust.hk

Kemin Zhou kemin@ece.lsu.edu

CHAPTER 1

Overview

This book discusses feedback systems and feedback control. A system is called a **feedback system** if it feeds back the knowledge of the system dynamical behavior for the control of the system. Feedback systems are everywhere; for example, a temperature control system such as an air conditioner is a feedback system since it measures the temperature to turn on or off the compressor motor. A human being is a very complex feedback system, who uses vision, touch, smell, etc., to coordinate many complex activities.

1.1 INTRODUCTION

Let us begin with an example. Figure 1.1 shows a car running along a straight road. Assume that the objective of the operation is to keep the car as close to the center line as possible. If there is no human or machine driving the car, it is likely to deviate further and further away from the center line, possibly because of imperfection of the car, imperfection of the road, wind gusts, or even earthquakes. Such a system is said to be unstable. One of the purposes of a driver is to make the system stable, i.e., to keep the car in the neighborhood of the center line. Here the car is the object of control, which is often called a **plant**. The driver plays the role of a **controller**. The task of making the combined system consisting of the plant and controller stable is called **stabilization**.

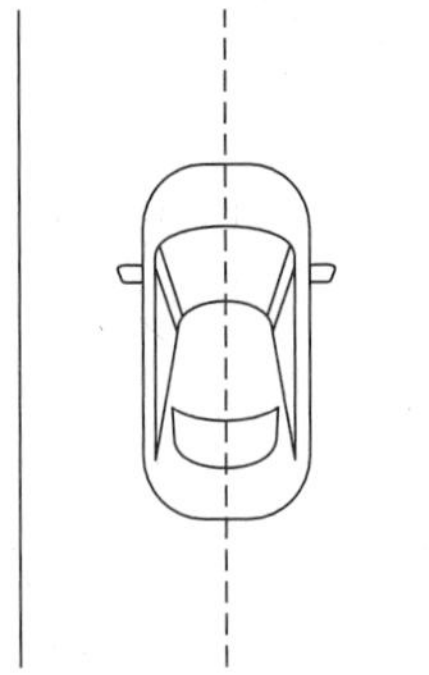

FIGURE 1.1: A car running in a straight road.

There are several conceivable schemes for achieving stabilization:

- **Open-loop control**: driving the car with eyes closed. This is clearly not going to work well, since the factors that affect the car position, such as wind, road conditions, etc., cannot be predicted. It is impossible to have a predetermined way to drive the car without observing the car position in real time. In general, one can never stabilize an unstable plant by open-loop control. It is possible to improve the performance of a stable plant by open-loop control, but even this is not very good.

- **Closed-loop control**: turning the wheel according to the deviation of the car position. Common sense indicates that as long as the wheels are turned in a proper way, the system can indeed be stabilized. Since the driver uses the information on the output to adjust the input, closed-loop control is also called **feedback control**. In general, feedback is essential to almost all control systems.

Automatic control is simply to use machines as controllers, to replace human beings.

Let us try to extract some essence of the plant from the car example we have just presented. First, a plant is a physical process that can be influenced from outside. The medium that can be used to influence the plant is called the **input**. In the example, the input is the wheel angle. The plant also produces some result, which is our concern. This result is called the **output**. In the example, the output is the deviation of the car from the center line. In addition to the input that can be manipulated, there are factors that influence the behavior of the plant but cannot be controlled, such as wind gust and road condition in the example. Such factors are called **disturbances**. In order to know and predict the behavior of a plant, and especially to know if the output is and will be satisfactory, we need to extract information from the plant. The information may or may not be the output itself, depending on the convenience and the technology available. Such information is called **measurement**. A plant can often be represented by a block diagram as shown in Figure 1.2. Here, u is the input, z is the output, d is the disturbance, and y is the measurement.

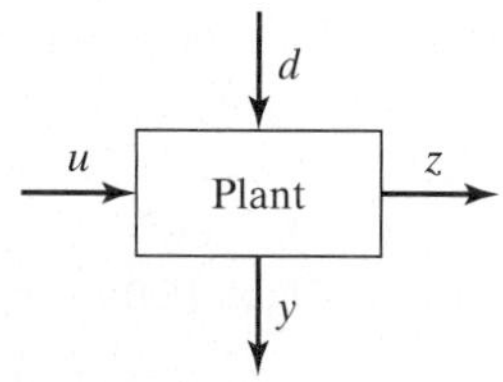

FIGURE 1.2: A plant.

If we take a closer look at a human feedback system, we can see that the human actually has to perform several subtasks. He/she has to observe the car position, then make some decisions regarding what actions to take, and finally exercise his/her decision and influence the motion of the car. Normally these three stages are called **sensing**, **control**, and **actuation**, respectively. If we are to replace a human driver by a machine, we need to build a sensor, a controller, and an actuator. A sensor measures a physical variable that can be used to deduce the behavior of the concerned output, such as the deviation of the car position, and turn it into a signal, usually an electrical signal, that the controller can read. The controller, often a computer or an electric circuit, takes the reading from the sensor, determines the action needed to correct the car position, and sends the decision to an actuator. The actuator then generates the quantity which influences the plant.

Therefore, a stabilizing closed-loop system can be represented by a block diagram as in Figure 1.3.

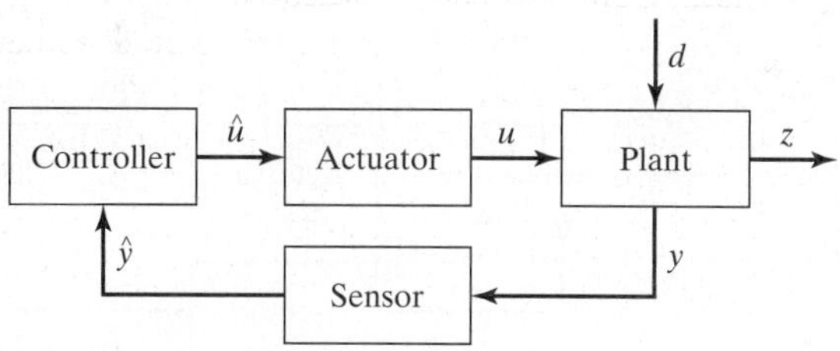

FIGURE 1.3: Structure of a feedback system for stabilization purpose.

Let us now look at another example in Figure 1.4. The purpose in this example is to have a car running along a hilly road against a persistent wind, so that its speed follows an external command, such as the maximum and minimum speed limits in certain segments of the road. There are two subtasks in this problem. The first is speed following even without the uphill or downhill slopes, or the head or tail wind. This problem is called **tracking**. The second is the reduction or complete elimination of the effect of the slopes and wind

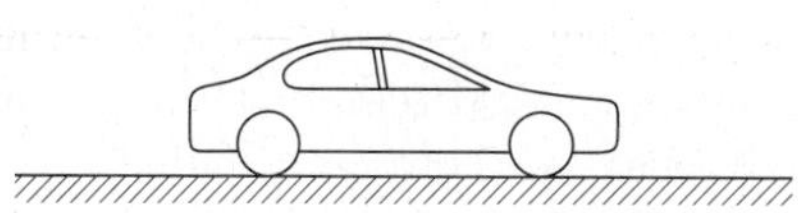

FIGURE 1.4: Speed control of a moving car.

on the car speed. This problem is called **disturbance rejection**. The overall problem, encompassing tracking and disturbance rejection, is called **regulation**. If the speed command is known to be a piecewise constant function, the tracking problem is called **set-point tracking**. Suppose that the acceleration of the car can be controlled by the gas pedal position. Again we have two possible schemes:

- **Open-loop control**: setting the gas pedal movement according to a computed position profile derived from the speed command, the road conditions in different segments of the road, and the wind speed obtained from an accurate weather forecast. One can imagine that the scheme will not work well since any error in the computed position profile, the road conditions, and the weather forecast will cause the speed to settle at the wrong value or not to settle at all because of the integration effect (the speed is proportional to the integral of the gas pedal position and hence small errors in the position may accumulate into large errors in speed).
- **Closed-loop control**: adjusting the gas pedal position according to the actual speed of the car. Since we can accelerate or decelerate the car in real time, according to the speed measurement, we should be able to control the speed of the car within a small neighborhood of the command as long as we take correct actions. Whether or not we know the road conditions or the wind speed is not important.

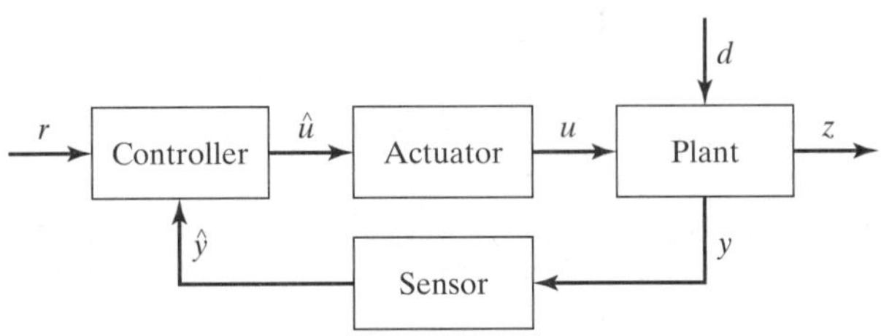

FIGURE 1.5: Structure of a feedback system for regulation purpose.

One can see that the major difference between regulation and stabilization is that there is a command signal, also called a **reference signal**, in the regulation situation. The controller needs to process this reference signal, in addition to other signals processed by the stabilizing controller. The structure of a feedback system for a regulation purpose is shown in Figure 1.5. Another difference between regulation and stabilization is that in the regulation problem, the disturbance is often assumed to be persistent and has some known features, such as being piecewise constant or piecewise sinusoids, whereas in the stabilization problem, the disturbance is assumed to be unknown and temporary in nature. The study of stabilization is important not only because there are genuine stabilization problems, such as suppressing vibration, balancing a pendulum, etc., but also because it is the key step in achieving regulation.

1.2 BASIC CONCEPTS

A **signal** $x(t)$ is a real-valued function of time. Conceptually, the time axis is the whole real axis $\mathbb{R}$ from $-\infty$ to ∞. Hence, a signal is a function from $\mathbb{R}$ to $\mathbb{R}$. However, we will mostly deal with unilateral or one-sided signals, i.e., signals $x(t)$ with $x(t) = 0$ for all $t < 0$. One such typical signal is the **unit step** signal, as shown in Figure 1.6. We use a special notation $\sigma(t)$ to denote the unit step signal

$$\sigma(t) = \begin{cases} 0 & t < 0 \\ 1 & t \geq 0. \end{cases}$$

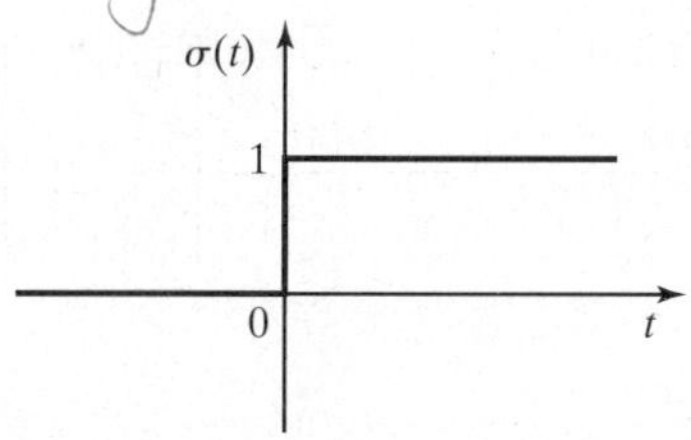

FIGURE 1.6: The unit step signal.

The unit step signal also comes handy in describing other unilateral signals, e.g., unilateral sinusoidal (function) signal $\sin \omega t \, \sigma(t)$ and unilateral exponential (function) signal $e^{\lambda t}\sigma(t)$. Another special signal that we often need to use is the so-called Dirac **unit impulse** signal $\delta(t)$, which is defined by the property:

$$\begin{cases} \delta(t) = 0, & t \neq 0 \\ \int_{-\infty}^{t} \delta(\tau) d\tau = \sigma(t), & \end{cases} \tag{1.1}$$

or by the property

$$\delta(t) = \frac{d\sigma(t)}{dt}. \tag{1.2}$$

The unit impulse function is not a function in the strict sense. It is a distribution and sometimes called a **singular function** or a **generalized function**. If we imagine a function as the distribution of mass in a long, straight string, then the unit impulse function means that there is a point mass of one unit at the middle of the string. This point mass occupies zero length in the string and hence has infinite density at the point, whereas the usual functions correspond to distributions with finite density. The unit impulse function $\delta(t)$ can be approximated by a narrow square pulse as shown in Figure 1.7(a). We also have a corresponding approximation of the unit step function as shown in Figure 1.7(b), which makes properties (1.1) and (1.2) consistent.

A **system** $\boldsymbol{S}$ is a map that transforms a signal vector

$$\boldsymbol{u}(t) = \begin{bmatrix} u_1(t) \\ \vdots \\ u_m(t) \end{bmatrix},$$

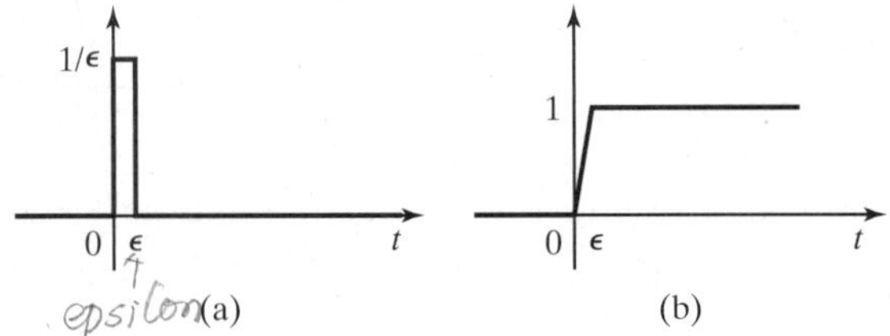

FIGURE 1.7: Approximate (a) unit impulse and (b) unit step.

called the **input**, to another signal vector

$$\boldsymbol{y}(t) = \begin{bmatrix} y_1(t) \\ \vdots \\ y_p(t) \end{bmatrix},$$

called the **output**.

We write

$$\boldsymbol{y}(t) = \boldsymbol{S}[\boldsymbol{u}(t)] \quad \text{or} \quad \boldsymbol{y} = \boldsymbol{S}\boldsymbol{u}$$

to mean that $\boldsymbol{S}$ is a system with input $\boldsymbol{u}(t)$ and output $\boldsymbol{y}(t)$. We also use the block diagram as in Figure 1.8 to represent a system graphically. For example, plants, controllers, sensors, and actuators are all systems.

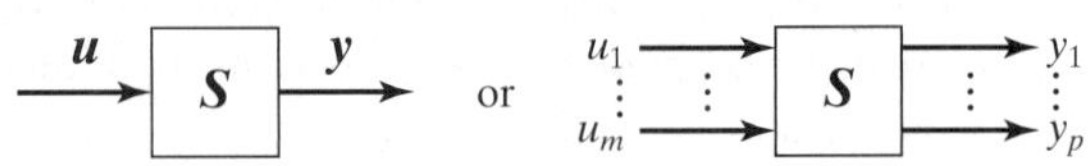

FIGURE 1.8: Graphic representation of a system.

A system is said to be a single-input–single-output (SISO) system if it has only one input and one output. If it has more than one input or more than one output, it is said to be a multi-input–single-output (MISO), single-input–multi-output (SIMO), or multi-input–multi-output (MIMO) system.

For technical reasons, we will mostly study SISO control systems in this book. Only occasionally will we deal with MISO and SIMO systems. The study of general MIMO systems requires more sophisticated tools and will be covered in more advanced books. For a clearer distinction, SISO systems will be denoted by normal upper case letters such as S, P, C, and MIMO (including MISO and SIMO) systems will be denoted by bold upper case letters such as $\boldsymbol{S}, \boldsymbol{P}, \boldsymbol{C}$. Similarly, scalar signals will be denoted by normal lower case letters such as x, u, y, and vector signals will be denoted by bold lower case letters such as $\boldsymbol{x}, \boldsymbol{u}, \boldsymbol{y}$.

A system is said to be **static** or **memoryless** if the value of the output $y(t_0)$ at some time instance t_0 depends only on the value of the input $u(t_0)$ at t_0. Such systems are easy to deal with, but are very limited. We will be more interested in **dynamic systems** whose output $y(t_0)$ at time t_0 also depends on the value of the input $u(t)$ at other time instances $t \neq t_0$. One might feel awkward if the output $y(t_0)$ depends on a future input $u(t)$ with $t > t_0$. Indeed, this is not what we usually have. What we usually have are systems whose output at time t_0 depends only on the input $u(t)$ in the past or current time instances $t \leq t_0$. Such systems are called **causal systems**. All real-time physical systems are causal systems. In theoretical studies, we occasionally have to deal with noncausal systems. They are theoretical abstractions. From the applications point of view, they can only occur in non-real-time situations.

Finally in this section, let us pay attention to two very special MISO and SIMO systems that appear in almost all interconnected systems as will be seen throughout this book. They are too simple to be called systems. Instead, they are called a **summing point** and a **pickoff point**, respectively, as shown in Figure 1.9.

Summing point: $y = u_1 \pm u_2$ Pickoff point: $y_1 = u, y_2 = u$

FIGURE 1.9: Summing point and pickoff point.

1.3 BASIC STRUCTURES OF FEEDBACK SYSTEMS

The purpose of this book is to study the design of a controller to satisfy given specifications in terms of stability and performance. As we have seen, there are two typical control tasks: stabilization and regulation. The control system structure for stabilization is given by Figure 1.3. However, the structure is not very convenient for a theory. First, there is no theory for the selection of actuators and sensors. Second, the effect of disturbance is very difficult to model. Third, in most of the applications, taking the measurement the same as the output has important advantages. For these reasons, we usually absorb the sensors and actuators into the plant, simplify the way the disturbance enters the system so that there is only an input disturbance and an output disturbance, and assume that the measurement and the output are the same. The general structure in Figure 1.3 then becomes a simpler yet more abstract structure as shown in Figure 1.10. It is this structure that we will use in our theory development. The stabilization problem then becomes the following mathematical problem: Given plant P, design controller C so that the system shown in Figure 1.10 has "good" stability.

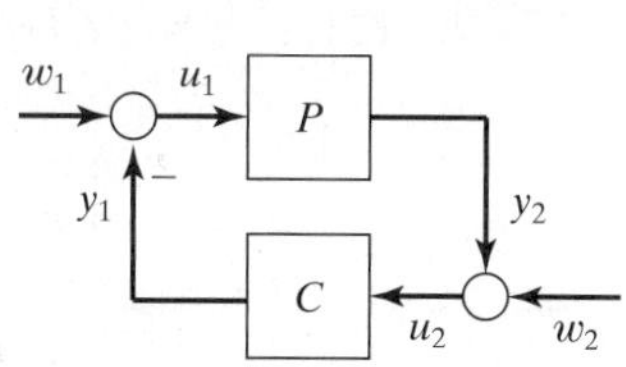

FIGURE 1.10: Feedback system for stabilization.

One may have noticed that in Figure 1.10 there is a minus sign attached to the signal y_1. This means that u_1 is w_1 minus y_1. This usage is mostly customary because the controller usually has a negative effect, in the sense that when y_2 is too big, the controller tries to reduce u_1, and vice versa. This type of feedback is called **negative feedback**. One may argue that if we replace C by $-C$, then the minus sign in Figure 1.10 can be removed. This is indeed a valid argument and is also an accepted practice. However, the tradition of keeping the negative sign there has strong reasons and we will follow this tradition throughout the book.

Similarly, in the regulation problem, we can also absorb the sensor and actuator to the plant and group the disturbances and noises into two groups: input disturbances and output noises. After doing these, Figure 1.5 becomes a simpler yet more abstract structure as shown in Figure 1.11. The regulation problem can be formulated into the following mathematical problem: Given plant P, design controller $\boldsymbol{C}$ so that good performance in tracking is achieved. Notice that the controller here is a MISO system. It has two inputs and one output. Such a controller is called a **two-degree-of-freedom (2DOF) controller**. Also, notice that there is a minus sign attached to the feedback signal y. This reflects the usual practice that the controller $\boldsymbol{C}$ generates the control signal u somehow from the difference between reference r and feedback y.

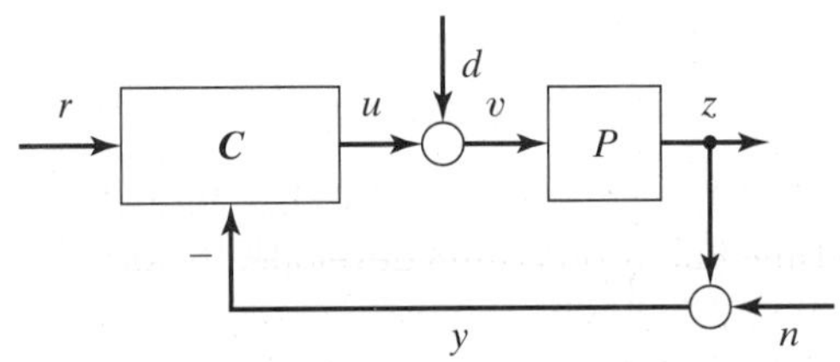

FIGURE 1.11: Feedback system for regulation.

In the early years of feedback control, the regulation problem was mostly solved by a more special feedback structure shown in Figure 1.12, which is called the **one-degree-of-freedom (1DOF) control** or **unity feedback**. In this structure, instead of taking the information of the reference and the measurement independently, the controller is driven by the difference between the reference and the measurement. Such a structure is simpler than the 2DOF structure since the controller is a SISO system and is actually a special case of the 2DOF structure by setting

$$\boldsymbol{C}\begin{bmatrix} r \\ -y \end{bmatrix} = C(r-y).$$

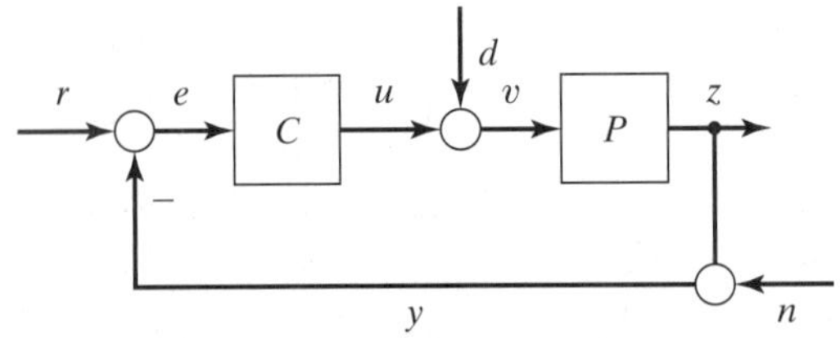

FIGURE 1.12: Unity feedback system.

This simple, or primitive, structure is still widely used today and is often considered as the default choice by many practitioners. However, as we will see in later parts of this book, the extra design freedom in 2DOF control over the 1DOF control provides significant advantages in achieving better performance and in trading off the often conflicting tracking and disturbance rejection requirements. One strong message of this book is that when we have a regulation problem, or even a pure tracking problem in the case when the disturbance is not present, the use of a 2DOF controller should be considered whenever possible.

1.4 ABOUT THIS BOOK

The main features of this book are as follows:

- It is a blend of classical (or even preclassical) and modern (or even postmodern) approaches.

  Most control textbooks treat classical control theory and modern control theory separately, with the implied message that the classical control theory is more elementary and more accessible to beginners, while the modern control theory is more advanced and more sophisticated. One common division of the classical and modern approaches is that the classical approach is based on transfer functions, whereas the modern approach is based on state space descriptions. Another division of the classical and modern approaches is based on the time line: everything developed before the 1950s is classical, everything developed after the 1950s is modern, and those developed during the 1950s are in the gray zone. We believe that all these separations, divisions, and segregations only do more harm than good.

  In this book, we try to break the fine line between the classical and modern approaches, and integrate control theory development in different stages into a unified theory for SISO system analysis and design. As in classical control, we mainly use the transfer function as a system model, and try to design simple controllers using intuitive techniques. As in modern control, we emphasize quantitative analysis and analytical design, and try to design optimal controllers and understand fundamental design limitations, i.e., what feedback control can or cannot do. We attempt not to sacrifice mathematical rigor and attempt to make connections to computer-aided analysis and design.

- The use of 2DOF controllers in regulation problems is emphasized.

  2DOF controllers are not new. They appeared in the earlier days when feedback control first became a widely used practice. Many *ad hoc* control schemes in animals and machines are 2DOF by nature.

  However, history took a sharp turn when many popular textbooks and other publications defined a feedback control system as "a system that maintains a prescribed relationship between the output and some reference input by comparing them and using the difference as a means of control" (Ogata, 2008), or a system in which "the controlled signal $c(t)$ should be fed back and compared with the reference input, and an actuating signal proportional to the difference of the input and the output must be sent through the system to correct the error" (Kuo and Golnaraghi, 2002).

Such a definition, though covering a good number of situations when it is indeed the difference of the reference and the outcome that is used in the decision-making process of the controller, misses many other practices where the controllers process the reference and outcome independently.

A more accurate definition was given in Dorf and Bishop (2008): "A feedback control system is a control system that tends to maintain a prescribed relationship of one system variable to another by comparing functions of these variables and using the difference as a means of control." The difference between this definition and the one in Ogata (2008) is the word "functions". In general, the function of the reference and that of the output are allowed to be completely different, giving rise to two complete degrees of freedom.

In this book, we take the unprecedented step in introducing the notation $\sigma(t)$ to denote the unit step function. The unit step function has wide applications in diverse disciplines. However, it has not had a standard notation as its sister, the unit impulse signal $\delta(t)$, does. It has sometimes been denoted by $u(t)$, $\mathbf{1}(t)$, or other variations. The unit step is the integral of the unit impulse and the unit impulse is the differentiation of the unit step; as such, considering the meanings of Greek letters Σ and Δ as sum and difference, respectively, we now simply have the Σ of $\delta(t)$ as $\sigma(t)$ and the Δ of $\sigma(t)$ as $\delta(t)$.

We attempt to make this book self-contained so that it not only tells us "how" but also "why". We do not share the argument that in undergraduate textbooks rigorous reasonings should give way to intuitions and design recipes. We strongly believe that learning a few "whys" is better than learning many "hows". In this book, we provide as much as possible, the reasons and justifications behind theorems, design procedures, algorithms, etc.

Such reasons and justifications might be difficult to digest for an average reader, but these insights definitely provide a source of information for instructors and students with an interest in in-depth exploration. We tint the parts of the text containing sophisticated mathematical reasoning to indicate that these parts may be skipped without affecting the basic understanding of the book.

While we pay attention to the theoretic soundness of the theory, we also pay attention to the illustrative examples, which are usually quite simple, yet informative, and to case studies, which are nontrivial but are commonly accessible in undergraduate laboratories. We strongly suggest that the use of this book be accompanied by control experiments of real physical systems, starting with system modeling, going through controller design, analysis, and redesign, and ending with controller implementation and hardware-in-the-loop simulations.

In this textbook, computer-aided analysis and design are integrated into the presentation of theory and examples. MATLAB, together with SIMULINK, is used as the programming platform. Analysis and design procedures are stated in the form of algorithms so that they can be programmed easily. Some of the exercise problems are specifically labeled as MATLAB problems to give ample opportunities for students to practice their MATLAB program skills and to strengthen their understanding of theory by translating the algorithms into programs. If all MATLAB problems in the book are completed, enough number of programs can be generated to form a MATLAB toolbox of SISO system analysis and design.

Finally, in this textbook, we attempt to create opportunities to motivate the student into further exploration of the deeper and wider space of feedback control theory. We do this by pointing out the insufficiencies of the content of this book, by referring to sources in the literature for materials beyond the coverage of this book, and by giving several extra credit problems which require a fair amount of extra reading and thinking.

PROBLEMS

1.1. Give several examples of stabilization control.

1.2. Give several examples of regulation control.

1.3. Suppose that you are a meticulously lawful driver who always follows the speed limit closely, never overspeed and seldom underspeed. Do you think that you are a unity feedback controller or a 2DOF controller?

1.4. Find the derivatives of unilateral functions $(\sin\ \omega t)\sigma(t)$, $(\cos\ \omega t)\sigma(t)$, and $e^{\lambda t}\sigma(t)$.

NOTES AND REFERENCES

A prerequisite of this course is a basic knowledge on signals and systems. An excellent textbook on this is

A. V. Oppenheim and A. S. Willsky with S. H. Newab, *Signals and Systems*, 2nd Edition, Prentice Hall, Upper Saddle River, New Jersey, 1996.

Other good references are

H. Kwakernaak and R. Sivan, *Modern Signals and Systems*, Prentice Hall, Englewood Cliffs, New Jersey, 1991.

E. A. Lee and P. Varaiya, *Structure and Interpretation of Signals and Systems*, Addison Wesley, Boston, 2003.

There are several commonly used textbooks on control systems, some essentially written over 30 years ago, such as

R. C. Dorf and R. H. Bishop, *Modern Control Systems*, 11th Edition, Pearson Prentice Hall, Upper Saddle River, New Jersey, 2008.

B. C. Kuo and F. Golnaraghi, *Automatic Control Systems*, 8th Edition, Prentice Hall, Englewood Cliffs, New Jersey, 2002.

K. Ogata, *Modern Control Engineering*, 5th Edition, Pearson Prentice Hall, Upper Saddle River, New Jersey, 2008.

and some are more recent, such as

G. F. Franklin, J. D. Powell, and A. Emami-Naeini, *Feedback Control of Dynamic Systems*, 5th Edition, Pearson Prentice Hall, Upper Saddle River, New Jersey, 2006.

G. C. Goodwin, S. F. Graebe, and M. E. Salgado, *Control Systems Design*, Prentice Hall, Upper Saddle River, New Jersey, 2001.

An early book introducing 2DOF controllers is

I. Horowitz, *Synthesis of Feedback Systems*, Academic Press, London, 1963.

A recent textbook advocating the use of 2DOF controllers is

W. A. Wolovich, *Automatic Control Systems*, Sauders College Publishing, Fort Worth, 1994.

Two textbooks with the aim of integrating classical and modern approaches are

J. C. Doyle, B. A. Francis, and A. R. Tannenbaum, *Feedback Control Theory*, Macmillan Publishing Company, New York, 1992.

J. W. Helton and O. Merino, *Classical Control Using H^∞ Methods: Theory, Optimization, and Design*, SIAM, Philadelphia, 1998.

CHAPTER 2

Modeling and Simulation

The purpose of modeling is to understand the relationships among various physical variables in a system, in particular that between the input and the output of the system. Such a relationship is called a **mathematical model**. There are several ways to obtain a mathematical model. The first is to write down a complete set of equations relating the different variables in the system based on physical laws. The equations are usually differential equations. This method is called **first principle modeling**. The second is to find out the relationship between input and output, as well as other signals, from a set of experimental data. This method is called **system identification**. In most cases, modeling is done by combining the above two methods. It happens quite often that the structure of the model is determined from physical principles and the parameters are obtained from experiments.

In general, modeling is time consuming, problem dependent, and theoretically underdeveloped, but a very important step toward controller design.

In this chapter, only first principle modeling will be covered. We will discuss, using examples, the modeling of electrical, mechanical, and electromechanical systems. These are the systems that we are most likely to meet in applications. However, by no means do they exhaust all possible systems. Mathematical models may take different forms. We will see how we can process or simplify a model to make it easy to use and how to convert a model from one form to another.

2.1 MODELING BASED ON FIRST PRINCIPLES

In this section, we will discuss the modeling of various types of systems using examples.

2.1.1 Electrical systems

We will basically consider two types of circuits: one consisting of active RLC circuits containing possibly lumped resistors, inductors, capacitors, ideal sources, and controlled sources; and the other one consisting of circuits containing operational amplifiers.

First let us consider a simple RLC circuit as shown in Figure 2.1, where the diamond symbol labeled gv_1 means a voltage-controlled current source whose current is proportional to the voltage across the capacitor C_1. The input and output of the system are $v_i(t)$ and $v_o(t)$, respectively.

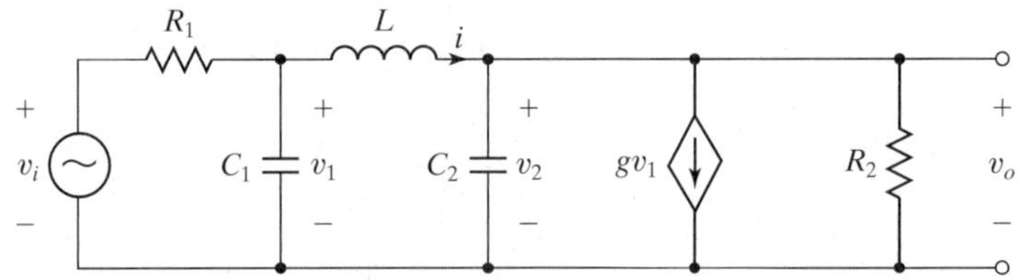

FIGURE 2.1: A simple RLC circuit.

The use of intermediate variables in addition to the input and output greatly facilitates the generation of the differential equation model. In an RLC circuit, we usually choose capacitor voltages and inductor currents as the intermediate variables. Let the number of such intermediate variables be n. Then we can write down n independent differential equations using Kirchhoff's current law (KCL) and Kirchhoff's voltage law (KVL), applied to nodes and loops involving inductors and capacitors. Here, by independent equations we mean that any one of them cannot be generated from the others among the equations obtained.

Referring back to the RLC circuit shown in Figure 2.1, we choose $v_1(t), v_2(t)$, and $i(t)$ as intermediate variables. Then applying KCL to the node connecting R_1, C_1, and L gives

$$C_1 \frac{dv_1(t)}{dt} = \frac{v_i(t) - v_1(t)}{R_1} - i(t). \tag{2.1}$$

Applying KCL to the node connecting to L, C_2, R_2, and the controlled source gives

$$C_2 \frac{dv_2(t)}{dt} = i(t) - gv_1(t) - \frac{v_2(t)}{R_2}. \tag{2.2}$$

Applying KVL to the loop containing C_1, L, and C_2 gives

$$L \frac{di(t)}{dt} = v_1(t) - v_2(t). \tag{2.3}$$

The output should be expressed as a function of the intermediate variables and the input variable. For the RLC circuit at hand, the output $v_o(t)$ is simply one of the intermediate variables,

$$v_o(t) = v_2(t). \tag{2.4}$$

The differential equations (2.1), (2.2), and (2.3), as well as the algebraic equation (2.4), give a model of the RLC circuit.

The model can also be organized in the following matrix form

$$\begin{bmatrix} C_1 & 0 & 0 \\ 0 & C_2 & 0 \\ 0 & 0 & L \end{bmatrix} \begin{bmatrix} \dot{v}_1(t) \\ \dot{v}_2(t) \\ \dot{i}(t) \end{bmatrix} = \begin{bmatrix} -1/R_1 & 0 & -1 \\ -g & -1/R_2 & 1 \\ 1 & -1 & 0 \end{bmatrix} \begin{bmatrix} v_1(t) \\ v_2(t) \\ i(t) \end{bmatrix} + \begin{bmatrix} 1/R_1 \\ 0 \\ 0 \end{bmatrix} v_i(t) \tag{2.5}$$

$$v_o(t) = \begin{bmatrix} 0 & 1 & 0 \end{bmatrix} \begin{bmatrix} v_1(t) \\ v_2(t) \\ i(t) \end{bmatrix}. \tag{2.6}$$

Note that we have used the alternative notation $\dot{x}(t)$ to denote the derivative $\dfrac{dx(t)}{dt}$. One may premultiply both sides of (2.5) by the inverse of the matrix on the left-hand side of (2.5). We then obtain

$$\begin{bmatrix} \dot{v}_1(t) \\ \dot{v}_2(t) \\ \dot{i}(t) \end{bmatrix} = \begin{bmatrix} -\dfrac{1}{R_1C_1} & 0 & -\dfrac{1}{C_1} \\ -\dfrac{g}{C_2} & -\dfrac{1}{R_2C_2} & \dfrac{1}{C_2} \\ \dfrac{1}{L} & -\dfrac{1}{L} & 0 \end{bmatrix} \begin{bmatrix} v_1(t) \\ v_2(t) \\ i(t) \end{bmatrix} + \begin{bmatrix} \dfrac{1}{R_1C_1} \\ 0 \\ 0 \end{bmatrix} v_i(t) \tag{2.7}$$

$$v_o(t) = \begin{bmatrix} 0 & 1 & 0 \end{bmatrix} \begin{bmatrix} v_1(t) \\ v_2(t) \\ i(t) \end{bmatrix}. \tag{2.8}$$

As we will see later, the model of this form is called a **state space model**.

Next let us consider circuits containing operational amplifiers (op-amps), which are simply called op-amp circuits. Such circuits have wide applications. We often use such a circuit to realize a set of differential equations or a system.

An ideal op-amp, as shown in Figure 2.2, is an amplifier with infinite gain, infinite input impedance, and zero output impedance.

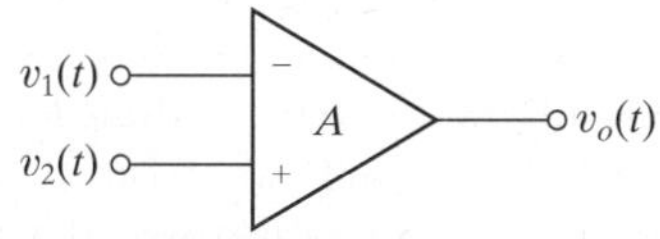

FIGURE 2.2: An operational amplifier.

Consider the circuit shown in Figure 2.3. This is a system with m inputs $u_1(t), \ldots, u_m(t)$ and one output $y(t)$. Owing to the infinite gain of the op-amp, the

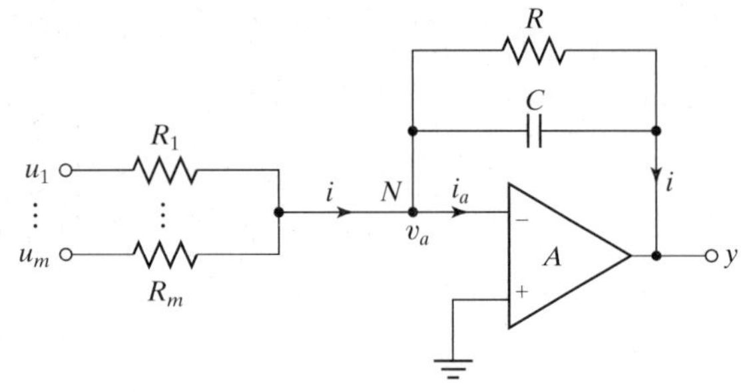

FIGURE 2.3: An op-amp circuit.

node N is a virtual ground, i.e., $v_a(t) = 0$. Owing to the infinite input impedance of the op-amp, the current flowing into the op-amp is zero, i.e., $i_a(t) = 0$. Thus we have

$$i(t) = \frac{u_1(t)}{R_1} + \frac{u_2(t)}{R_2} + \cdots + \frac{u_m(t)}{R_m}$$

and

$$i(t) = -\frac{1}{R}y(t) - C\frac{dy(t)}{dt}.$$

Hence

$$RC\frac{dy(t)}{dt} + y(t) = -\left[\frac{R}{R_1}u_1(t) + \cdots + \frac{R}{R_m}u_m(t)\right].$$

If $C = 0$, i.e., the capacitor is disconnected, then

$$y(t) = -\left[\frac{R}{R_1}u_1(t) + \cdots + \frac{R}{R_m}u_m(t)\right],$$

i.e., the output is a negatively weighted sum of the inputs. In this case, the circuit is called an (inverting) summing amplifier. If further, $m = 1$, the circuit is called an (inverting) amplifier.

If $R = \infty$, i.e., the resistor is disconnected, then

$$y(t) = -\left[\frac{1}{R_1 C}\int_0^t u_1(\tau)d\tau + \cdots + \frac{1}{R_m C}\int_0^t u_m(\tau)d\tau\right],$$

i.e., the output is a negatively weighted sum of the integrals of the inputs. If further, $m = 1$, then the output is proportional to the integral of the input. In this case, the circuit is called an (inverting) integrator.

Operational amplifiers are the building blocks of most electronic systems existing today. In an abstract level, we do not wish to keep track of the inverting effect and the coefficient of the integrators. We take pure integrators and weighted summing amplifiers as the basic components. We schematically represent them by diagrams shown in Figure 2.4. These two types of components can be used as the building blocks of sophisticated op-amp circuits that can be used in system simulation and implementation, as will be discussed in Section 2.7.

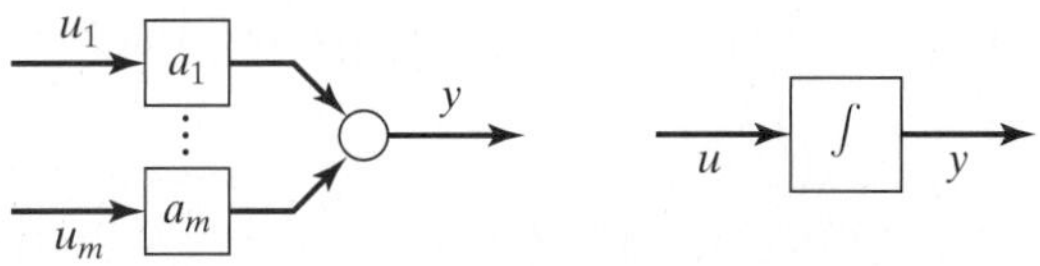

FIGURE 2.4: Schematic symbols for a summing amplifier and an integrator.

2.1.2 Mechanical systems

Typical mechanical systems may involve two kinds of motion: linear motion and rotational motion. In this section, we will look at two examples of mechanical systems. The first example is one with linear motion. The second example is one with rotational motion. Of course, a general mechanical system may have both linear and rotational motion, as will be seen in Section 2.10.

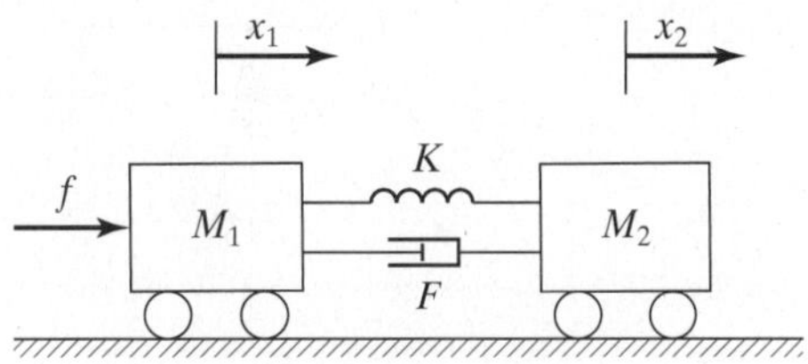

FIGURE 2.5: A two-cart system.

Consider two carts moving on a table, as shown in Figure 2.5. The carts, assumed to have masses M_1 and M_2, respectively, are connected by a spring. A force $f(t)$ is applied to cart M_1 and we wish to control the position of cart M_2. An ideal spring generates a force that is proportional to the relative displacement of its two ends. The spring in this system is assumed to be nonideal and its effect is equivalent to the sum of an ideal spring, depicted by the symbol in the upper part of the connection, and a friction device, depicted by the symbol in the lower part of the connection, which generates a force proportional to the relative velocity of its two ends. Assume that the reference points for the positions of the two carts are chosen so that the spring is at rest when $x_1(t) = x_2(t)$.

To model a mechanical system like this, we usually choose the positions of all independent rigid bodies as the internal variables and write down the equations according to Newton's second law. Applying Newton's second law to the first body, we obtain

$$M_1 \frac{d^2x_1(t)}{dt^2} = f(t) - K(x_1(t) - x_2(t)) - F\left(\frac{dx_1(t)}{dt} - \frac{dx_2(t)}{dt}\right). \tag{2.9}$$

Applying Newton's second law to the second cart, we obtain

$$M_2 \frac{d^2x_2(t)}{dt^2} = K(x_1(t) - x_2(t)) + F\left(\frac{dx_1(t)}{dt} - \frac{dx_2(t)}{dt}\right). \tag{2.10}$$

Finally, we identify the output and relate it to the internal variables and the input. In this case, the output is simply $x_2(t)$.

We can also reorganize the equations into the following matrix form

$$\begin{bmatrix} M_1 & 0 \\ 0 & M_2 \end{bmatrix}\begin{bmatrix} \ddot{x}_1(t) \\ \ddot{x}_2(t) \end{bmatrix} + \begin{bmatrix} F & -F \\ -F & F \end{bmatrix}\begin{bmatrix} \dot{x}_1(t) \\ \dot{x}_2(t) \end{bmatrix} + \begin{bmatrix} K & -K \\ -K & K \end{bmatrix}\begin{bmatrix} x_1(t) \\ x_2(t) \end{bmatrix} = \begin{bmatrix} 1 \\ 0 \end{bmatrix} f(t). \quad (2.11)$$

Models of the form (2.11) are very typical for mechanical systems. Such a model is called a **second-order model.**

Next let us consider a pendulum shown in Figure 2.6. Here a torque $\tau_i(t)$ can be applied around the pivot point and we are concerned with the angle $\theta(t)$ between the pendulum and the vertical downward direction. The length of the pendulum is L and the mass M of the pendulum is concentrated at its tip.

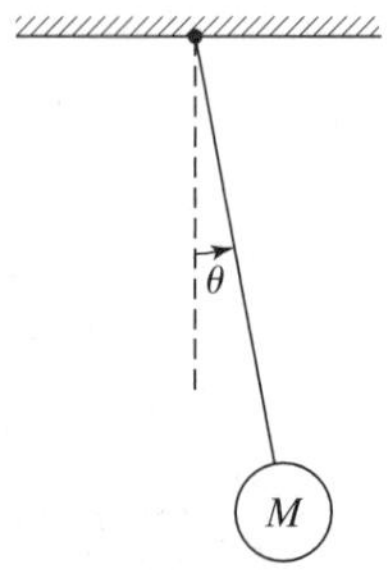

FIGURE 2.6: A pendulum.

In a rotational motion, Newton's second law takes the form

$$J\frac{d^2\theta(t)}{dt^2} = \tau(t)$$

where J is called the moment of inertia, $\theta(t)$ is the angular displacement, and $\tau(t)$ is the total torque applied.

Applying this to the pendulum system, we know that the moment of inertia is $J = ML^2$ and there are two torques applied to the system: the externally applied torque $\tau_i(t)$, and the torque due to the gravity of the mass which is $MgL\sin\theta(t)$. Therefore, the equation governing the motion is given by

$$ML^2\frac{d^2\theta(t)}{dt^2} = \tau_i(t) - MgL\sin\theta(t).$$

This is a second-order differential equation. Here the input is the torque $\tau_i(t)$ and the output is the angle $\theta(t)$.

2.1.3 Electromechanical systems

A simple electromechanical system is an armature-controlled direct current (DC) motor with a load, shown in Figure 2.7.

A DC motor has two sets of windings. One set is mounted on the stator and is used to generate the magnetic field. In an armature-controlled DC motor,

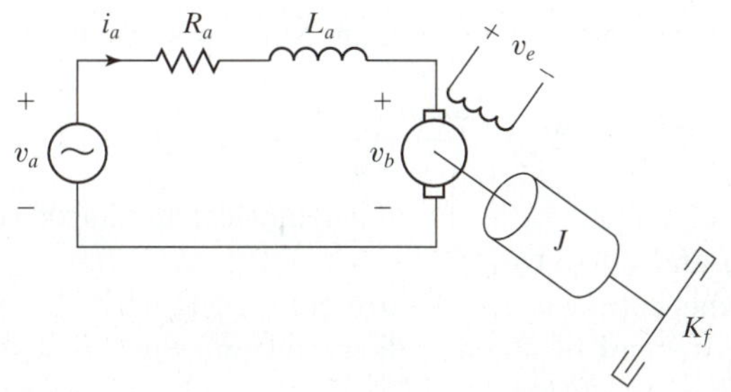

FIGURE 2.7: An armature-controlled DC motor system.

the current to this set of windings is set to be constant so that the magnetic field in the motor is constant. The other set is mounted on the rotor and is used to generate the torque through the magnetic force. The current through this winding is controllable so a controlled torque can be obtained. When the motor shaft turns, the magnetic field also generates a potential in the rotor winding as a result of the Faraday induction. This potential is called the back electro-motive force (back emf). There are two basic relations in a DC motor. One is that the torque in the motor shaft is proportional to the armature current via the torque constant K_t, i.e.,

$$\tau_m(t) = K_t i_a(t).$$

The other is that the back emf $v_b(t)$ is proportional to the motor velocity $\omega(t)$ via the back emf constant K_b, i.e.,

$$v_b(t) = K_b \omega(t).$$

The torque in the motor shaft then drives the load, which consists of a mass with a moment of inertia J and a counteractive friction torque proportional to the motor velocity via the friction coefficient K_f. Therefore, the whole DC motor system including the armature circuit and the mechanical load can be described by the following parameters and variables:

R_a:	armature resistance
L_a:	armature inductance
J:	moment of inertia of the load
K_f:	friction coefficient
K_t:	torque constant
K_b:	back emf constant
$v_a(t)$:	armature voltage
$i_a(t)$:	armature current
$v_b(t)$:	back electro-motive force (back emf)
$\tau_m(t)$:	motor torque
$\theta(t)$:	angular position of the motor shaft
$\omega(t)$:	angular velocity of the motor shaft ($= \dot{\theta}(t)$)

Applying KVL to the armature circuit, we obtain

$$R_a i_a(t) + L_a \frac{di_a(t)}{dt} + K_b \frac{d\theta(t)}{dt} = v_a(t). \tag{2.12}$$

Applying the rotational version of Newton's second law to the motor shaft, we obtain

$$J\frac{d^2\theta(t)}{dt^2} + K_f\frac{d\theta(t)}{dt} = K_t i_a(t). \quad (2.13)$$

These two equations give the mathematical model of the DC motor system with input $v_a(t)$ and output $\theta(t)$.

In some applications, we are concerned with the angular velocity (speed) of the motor, instead of the angular position. Such cases are called **speed control cases**. Replacing $\frac{d\theta(t)}{dt}$ by $\omega(t)$ in (2.12) and (2.13), we get the differential equation model of a DC motor system in the speed control case.

$$R_a i_a(t) + L_a\frac{di_a(t)}{dt} + K_b\omega(t) = v_a(t) \quad (2.14)$$

$$J\frac{d\omega(t)}{dt} + K_f\omega(t) = K_t i_a(t). \quad (2.15)$$

These two equations give the mathematical model of the DC motor system with input $v_a(t)$ and output $\omega(t)$.

Another interesting electromechanical system is a magnetic-ball suspension system shown in Figure 2.8. The coil at the top, after being fed with current, produces a magnetic field. The magnetic field generates an attracting force on the steel ball.

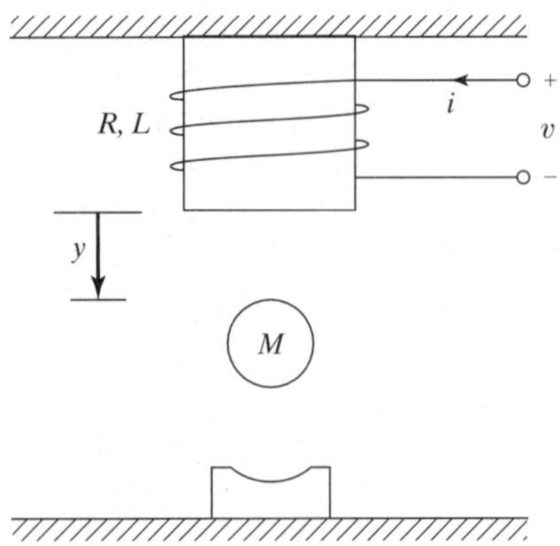

FIGURE 2.8: A magnetic-ball suspension system.

Here, the voltage applied to the coil $v(t)$ is the input, the distance from the ball to the coil $y(t)$ is the output, and the lifting force generated by the magnetic field on the ball is approximately given by $f(t) = K\frac{i^2(t)}{y(t)}$. The other parameters are mass of the ball M, winding resistance R, and winding inductance L.

Applying KVL to the coil, we obtain

$$Ri(t) + L\frac{di(t)}{dt} = v(t).$$

Applying Newton's second law to the ball, we obtain

$$M\frac{d^2y(t)}{dt^2} = -K\frac{i^2(t)}{y(t)} + Mg.$$

These two equations then give the mathematical model of the system. It is noted that the variables $y(t)$ and $i(t)$ are involved nonlinearly in the equations and this system is called a **nonlinear system**.

2.2 STATE SPACE MODEL AND LINEARIZATION

The mathematical models obtained in the previous sections consist of sets of differential equations with different orders and these equations involve variables other than the inputs and outputs. For the sake of systematic study, we need to put them in standard forms. One of the commonly used standard forms is the state space form.

Definition 2.1. The **state variables** of a system are a set of independent variables whose values at time t_0, together with input for all $t \geq t_0$, determine the behavior of the system for all $t \geq t_0$.

This definition looks very abstract but in many situations the state variables can be chosen intuitively. For electrical circuits, we can always choose the voltages across independent capacitors and the currents through independent inductors as state variables. For mechanical systems, we can always choose the positions and velocities of independent rigid bodies as state variables. Suppose that a differential equation model of a system is already obtained, the variables in the differential equations are the input $u(t)$ and the internal variables $v_1(t), \ldots, v_p(t)$, and the highest order of the derivatives of $v_i(t)$ in the differential equations is q_i. Then we can choose

$$v_i(t), \dot{v}_i(t), \ddot{v}_i(t), \ldots, v_i^{(q_i-1)}(t) \qquad\qquad i = 1, 2, \ldots, p$$

as the state variables. In this case, the total number of state variables is $\sum_{i=1}^{p} q_i$.

After the state variables are chosen, usually named $x_1(t), x_2(t), \ldots, x_n(t)$, we put them into a vector

$$\boldsymbol{x}(t) = \begin{bmatrix} x_1(t) \\ x_2(t) \\ \vdots \\ x_n(t) \end{bmatrix} \in \mathbb{R}^n.$$

This vector is called a **state vector**. Here and in the sequel, we use bold font letters to denote vectors (or matrices) and vector-valued (or matrix-valued) functions, whereas we use normal font letters to denote scalars and scalar-valued functions. Then the set of mixed ordered differential equations can be converted into a set of first-order differential equations plus an algebraic equation

$$\dot{\boldsymbol{x}}(t) = \boldsymbol{f}[\boldsymbol{x}(t), u(t), t] \tag{2.16}$$

$$y(t) = g[\boldsymbol{x}(t), u(t), t] \tag{2.17}$$

where $u(t) \in \mathbb{R}$ is the input, $y(t) \in \mathbb{R}$ is the output,

$$\boldsymbol{f}[\boldsymbol{x}(t), u(t), t] = \begin{bmatrix} f_1[\boldsymbol{x}(t), u(t), t] \\ f_2[\boldsymbol{x}(t), u(t), t] \\ \vdots \\ f_n[\boldsymbol{x}(t), u(t), t] \end{bmatrix} : \mathbb{R}^n \times \mathbb{R} \times \mathbb{R} \to \mathbb{R}^n$$

is a vector-valued function, and

$$g[\boldsymbol{x}(t), u(t), t] : \mathbb{R}^n \times \mathbb{R} \times \mathbb{R} \to \mathbb{R}$$

is a scalar-valued function. The number of state variables n is called the **order** of the system. The set of first-order differential equations (2.16) is called the **state equation** of the system. The algebraic equation (2.17) is called the **output equation** of the system. Together they form the state space model of the system.

We always assume that the system starts operation at time $t = 0$; namely, we assume that the input $u(t)$ is a unilateral signal whose value before the initial time is zero. To determine the state vector $\boldsymbol{x}(t)$ from the differential equation (2.16), the input $u(t)$ alone is not sufficient. According to Definition 2.1, the initial value of the state $\boldsymbol{x}(0)$ is also needed. This initial value is called the **initial condition**. To conform to our standard mathematical treatment of signals, we also view $\boldsymbol{x}(t)$ as a unilateral function. If the initial condition is nonzero, then $\boldsymbol{x}(t)$ has a jump discontinuity at $t = 0$ and its derivative $\dot{\boldsymbol{x}}(t)$ contains impulse functions.

Among the models of the systems discussed in Section 2.1, the one for the active RLC circuit has already been put in the state space form.

EXAMPLE 2.2

For the pendulum system introduced in the last section, the differential equation model directly obtained from Newton's second law is

$$ML^2\ddot{\theta}(t) = \tau_i(t) - MgL\sin\theta(t)$$

with input $\tau_i(t)$ and output $\theta(t)$. Renaming $x_1(t) = \theta(t), x_2(t) = \dot{\theta}(t)$, $u(t) = \tau_i(t)$, $y(t) = \theta(t)$, we obtain the state space model

$$\begin{bmatrix} \dot{x}_1(t) \\ \dot{x}_2(t) \end{bmatrix} = \begin{bmatrix} x_2(t) \\ \dfrac{-MgL\sin x_1(t) + u(t)}{ML^2} \end{bmatrix}$$

$$y(t) = x_1(t)$$

with

$$\begin{bmatrix} x_1(0) \\ x_2(0) \end{bmatrix} = \begin{bmatrix} \theta(0) \\ \dot{\theta}(0) \end{bmatrix}.$$

EXAMPLE 2.3

The magnetic suspension system has the differential equation model

$$Ri(t) + L\frac{di(t)}{dt} = v(t)$$

$$M\frac{d^2y(t)}{dt^2} = -K\frac{i^2(t)}{y(t)} + Mg.$$

Choose $x_1(t) = i(t), x_2(t) = y(t), x_3(t) = \dot{y}(t), u(t) = v(t)$, and we get the state space model

$$\begin{bmatrix} \dot{x}_1(t) \\ \dot{x}_2(t) \\ \dot{x}_3(t) \end{bmatrix} = \begin{bmatrix} \dfrac{u(t) - Rx_1(t)}{L} \\ x_3(t) \\ -\dfrac{Kx_1^2(t)}{Mx_2(t)} + g \end{bmatrix}$$

$$y(t) = x_2(t)$$

with

$$\begin{bmatrix} x_1(0) \\ x_2(0) \\ x_3(0) \end{bmatrix} = \begin{bmatrix} i(0) \\ y(0) \\ \dot{y}(0) \end{bmatrix}.$$

Definition 2.4. A system is said to be **linear** if it can be described by linear differential equations, in particular, if the functions $\boldsymbol{f}$ and g in its state space model are linear functions of $\boldsymbol{x}(t)$ and $u(t)$.

For a linear system, the state space model takes the following matrix form:

$$\dot{\boldsymbol{x}}(t) = \boldsymbol{A}(t)\boldsymbol{x}(t) + \boldsymbol{b}(t)u(t)$$

$$y(t) = \mathbf{c}(t)\boldsymbol{x}(t) + d(t)u(t)$$

where $\boldsymbol{A}(t) \in \mathbb{R}^{n\times n}$ is an $n \times n$ matrix, possibly depending on time t, and $\boldsymbol{b}(t) \in \mathbb{R}^{n\times 1}$ and $\mathbf{c}(t) \in \mathbb{R}^{1\times n}$ are, respectively, column and row vectors depending possibly on time t. For example, the RLC circuit in Section 2.1 is a linear system and its state space equations were already in the matrix form as in (2.7) and (2.8)

Theorem 2.5 (Superposition Principle). Assume that a linear system has zero initial condition. If input $u_1(t)$ produces output $y_1(t)$ and input $u_2(t)$ produces output $y_2(t)$, then input $\alpha_1 u_1(t) + \alpha_2 u_2(t)$ produces output $\alpha_1 y_1(t) + \alpha_2 y_2(t)$ for all $\alpha_1, \alpha_2 \in \mathbb{R}$.

Proof. With zero initial condition, if input $u_1(t)$ produces output $y_1(t)$ and input $u_2(t)$ produces output $y_2(t)$, then there are $\boldsymbol{x}_1(t)$ and $\boldsymbol{x}_2(t)$ with $\boldsymbol{x}_1(0) = \mathbf{0}$ and $\boldsymbol{x}_2(0) = \mathbf{0}$ satisfying

$$\dot{\boldsymbol{x}}_1(t) = \boldsymbol{A}(t)\boldsymbol{x}_1(t) + \boldsymbol{b}(t)u_1(t)$$

$$y_1(t) = \mathbf{c}(t)\boldsymbol{x}_1(t) + d(t)u_1(t)$$

and

$$\dot{\boldsymbol{x}}_2(t) = \boldsymbol{A}(t)\boldsymbol{x}_2(t) + \boldsymbol{b}(t)u_2(t)$$
$$y_2(t) = \boldsymbol{c}(t)\boldsymbol{x}_2(t) + d(t)u_2(t).$$

If we add the two state equations and the two output equations, respectively, and define $u(t) = \alpha_1 u_1(t) + \alpha_2 u_2(t)$, $\boldsymbol{x}(t) = \alpha_1 \boldsymbol{x}_1(t) + \alpha_2 \boldsymbol{x}_2(t)$, and $y(t) = \alpha_1 y_1(t) + \alpha_2 y_2(t)$, then we obtain $\boldsymbol{x}(0) = \boldsymbol{0}$ and

$$\dot{\boldsymbol{x}}(t) = \boldsymbol{A}(t)\boldsymbol{x}(t) + \boldsymbol{b}(t)u(t)$$
$$y(t) = \boldsymbol{c}(t)\boldsymbol{x}(t) + d(t)u(t).$$

This implies that $y(t)$ is the output of the system with zero initial condition and input $u(t)$. □

Definition 2.6. A system is said to be **time-invariant** if it can be described by differential equations with constant coefficients, in particular, if the functions $\boldsymbol{f}$ and g in its state space model do not depend on the time t explicitly.

All examples of real physical systems considered so far are time-invariant systems.

Theorem 2.7. Assume that a time-invariant system has zero initial condition. Also assume that zero input generates zero output. If input $u(t)$ produces output $y(t)$, then input $u(t-\tau)$ produces output $y(t-\tau)$ for all $\tau \geq 0$.

A linear time-invariant (LTI) system has the following form of state space model:

$$\dot{\boldsymbol{x}}(t) = \boldsymbol{A}\boldsymbol{x}(t) + \boldsymbol{b}u(t)$$
$$y(t) = \boldsymbol{c}\boldsymbol{x}(t) + du(t)$$

where $\boldsymbol{A} \in \mathbb{R}^{n\times n}$, $\boldsymbol{b} \in \mathbb{R}^{n\times 1}$, $\boldsymbol{c} \in \mathbb{R}^{1\times n}$, and $d \in \mathbb{R}$ are constant matrices.

EXAMPLE 2.8

The DC motor system is an LTI system. Indeed, in the position control case, if we choose the state variables, input and output variables as $x_1(t) = i_a(t), x_2(t) = \theta(t), x_3(t) = \omega(t), u(t) = v_a(t), y(t) = \theta(t)$, then the state space equation can be written as

$$\begin{bmatrix} \dot{x}_1(t) \\ \dot{x}_2(t) \\ \dot{x}_3(t) \end{bmatrix} = \begin{bmatrix} -\dfrac{R_a}{L_a} & 0 & -\dfrac{K_b}{L_a} \\ 0 & 0 & 1 \\ \dfrac{K_t}{J} & 0 & -\dfrac{K_f}{J} \end{bmatrix} \begin{bmatrix} x_1(t) \\ x_2(t) \\ x_3(t) \end{bmatrix} + \begin{bmatrix} \dfrac{1}{L_a} \\ 0 \\ 0 \end{bmatrix} u(t)$$

$$y = \begin{bmatrix} 0 & 1 & 0 \end{bmatrix} \begin{bmatrix} x_1(t) \\ x_2(t) \\ x_3(t) \end{bmatrix}.$$

In the speed control case, if we choose $x_1(t) = i_a(t), x_2(t) = \omega(t), u(t) = v_a(t)$, $y(t) = \omega(t)$, then we obtain

$$\begin{bmatrix} \dot{x}_1(t) \\ \dot{x}_2(t) \end{bmatrix} = \begin{bmatrix} -\dfrac{R_a}{L_a} & -\dfrac{K_b}{L_a} \\ \dfrac{K_t}{J} & -\dfrac{K_f}{J} \end{bmatrix} \begin{bmatrix} x_1(t) \\ x_2(t) \end{bmatrix} + \begin{bmatrix} \dfrac{1}{L_a} \\ 0 \end{bmatrix} u(t)$$

$$y(t) = \begin{bmatrix} 0 & 1 \end{bmatrix} \begin{bmatrix} x_1(t) \\ x_2(t) \end{bmatrix}.$$

In the rest of this section, we will study how to approximate a nonlinear system by a linear one. Such a process is called **linearization**.

We will deal only with time-invariant systems. Assume that a system is described by a state space model.

$$\dot{\boldsymbol{x}}(t) = \boldsymbol{f}[\boldsymbol{x}(t), u(t)]$$
$$y(t) = g[\boldsymbol{x}(t), u(t)]$$

and $\boldsymbol{f}$ and g are continuously differentiable functions, i.e., $\boldsymbol{f}$ and g are sufficiently smooth functions.

Definition 2.9. A triple of constant vectors $(u_0, \boldsymbol{x}_0, y_0) \in \mathbb{R} \times \mathbb{R}^n \times \mathbb{R}$ is said to be an **operating point** of the system if

$$0 = \boldsymbol{f}(\boldsymbol{x}_0, u_0)$$
$$y_0 = g(\boldsymbol{x}_0, u_0).$$

The physical meaning of an operating point is that if the system has initial condition $\boldsymbol{x}_0$ and a constant input u_0 is applied, then the state and output will stay at constant values $\boldsymbol{x}_0$ and y_0, respectively, for all time, i.e.,

$$u(t) = u_0, \boldsymbol{x}(0) = \boldsymbol{x}_0 \Longrightarrow \boldsymbol{x}(t) = \boldsymbol{x}_0, y(t) = y_0.$$

Since $\boldsymbol{f}$ and g are sufficiently smooth, we can conclude that

$$u(t) - u_0 \text{ and } \boldsymbol{x}(0) - \boldsymbol{x}_0 \text{ are small} \Longrightarrow \boldsymbol{x}(t) - \boldsymbol{x}_0 \text{ and } y(t) - y_0 \text{ are small.}$$

Denote

$$\tilde{u}(t) = u(t) - u_0$$
$$\tilde{\boldsymbol{x}}(t) = \boldsymbol{x}(t) - \boldsymbol{x}_0$$
$$\tilde{y}(t) = y(t) - y_0.$$

Replace $\boldsymbol{f}$ and g by their differentials:

$$\dot{\tilde{\boldsymbol{x}}}(t) = \left.\frac{\partial \boldsymbol{f}}{\partial \boldsymbol{x}}\right|_{\substack{\boldsymbol{x}=\boldsymbol{x}_0 \\ u=u_0}} \tilde{\boldsymbol{x}}(t) + \left.\frac{\partial \boldsymbol{f}}{\partial u}\right|_{\substack{\boldsymbol{x}=\boldsymbol{x}_0 \\ u=u_0}} \tilde{u}(t) + \text{high-order terms}$$

$$\tilde{y}(t) = \left.\frac{\partial g}{\partial \boldsymbol{x}}\right|_{\substack{\boldsymbol{x}=\boldsymbol{x}_0 \\ u=u_0}} \tilde{\boldsymbol{x}}(t) + \left.\frac{\partial g}{\partial u}\right|_{\substack{\boldsymbol{x}=\boldsymbol{x}_0 \\ u=u_0}} \tilde{u}(t) + \text{high-order terms}$$

where

$$\frac{\partial \boldsymbol{f}}{\partial \boldsymbol{x}} = \begin{bmatrix} \frac{\partial f_1}{\partial x_1} & \cdots & \frac{\partial f_1}{\partial x_n} \\ \vdots & & \vdots \\ \frac{\partial f_n}{\partial x_1} & \cdots & \frac{\partial f_n}{\partial x_n} \end{bmatrix}, \quad \frac{\partial \boldsymbol{f}}{\partial u} = \begin{bmatrix} \frac{\partial f_1}{\partial u} \\ \vdots \\ \frac{\partial f_n}{\partial u} \end{bmatrix}, \quad \frac{\partial g}{\partial \boldsymbol{x}} = \begin{bmatrix} \frac{\partial g}{\partial x_1} & \cdots & \frac{\partial g}{\partial x_n} \end{bmatrix}.$$

Since $\tilde{u}(t), \tilde{\boldsymbol{x}}(t), \tilde{y}(t)$ are small, we can neglect the high-order terms and approximate the original system by the following linear system:

$$\dot{\tilde{\boldsymbol{x}}}(t) = \boldsymbol{A}\tilde{\boldsymbol{x}}(t) + \boldsymbol{b}\tilde{u}(t)$$
$$\tilde{y}(t) = \boldsymbol{c}\tilde{\boldsymbol{x}}(t) + d\tilde{u}(t)$$

where

$$\boldsymbol{A} = \left.\frac{\partial \boldsymbol{f}}{\partial \boldsymbol{x}}\right|_{\substack{\boldsymbol{x}=\boldsymbol{x}_0 \\ u=u_0}}, \qquad \boldsymbol{b} = \left.\frac{\partial \boldsymbol{f}}{\partial u}\right|_{\substack{\boldsymbol{x}=\boldsymbol{x}_0 \\ u=u_0}},$$
$$\boldsymbol{c} = \left.\frac{\partial g}{\partial \boldsymbol{x}}\right|_{\substack{\boldsymbol{x}=\boldsymbol{x}_0 \\ u=u_0}}, \qquad d = \left.\frac{\partial g}{\partial u}\right|_{\substack{\boldsymbol{x}=\boldsymbol{x}_0 \\ u=u_0}}.$$

This linear system is called a **linearized system** of the original nonlinear system.

EXAMPLE 2.10

The magnetic suspension system is a nonlinear system described by state space equation

$$\begin{bmatrix} \dot{x}_1(t) \\ \dot{x}_2(t) \\ \dot{x}_3(t) \end{bmatrix} = \begin{bmatrix} \frac{u(t) - Rx_1(t)}{L} \\ x_3(t) \\ -\frac{x_1^2(t)}{Mx_2(t)} + g \end{bmatrix}$$
$$y(t) = x_2(t).$$

A usual control problem is to lift the ball to a certain height and suspend it at that height. Hence we wish to linearize it around an operating point with $y(t) = y_0$.

To get the operating point, solve equations

$$0 = \frac{u_0 - Rx_{10}}{L}$$
$$0 = x_{30}$$
$$0 = -\frac{x_{10}^2}{Mx_{20}} + g$$
$$y_0 = x_{20}.$$

This gives

$$u_0 = R\sqrt{Mgy_0}, \qquad \begin{bmatrix} x_{10} \\ x_{20} \\ x_{30} \end{bmatrix} = \begin{bmatrix} \sqrt{Mgy_0} \\ y_0 \\ 0 \end{bmatrix}, \qquad y_0 = y_0$$

which means that to suspend the ball at height y_0 in the steady state, one needs to apply a constant voltage $u(t) = R\sqrt{Mgy_0}$ to the coil. Denote the deviations of the input, state, and output variables from the operating point by

$$\begin{aligned} \tilde{u}(t) &= u(t) - u_0 = u(t) - R\sqrt{Mgy_0} \\ \tilde{\boldsymbol{x}}(t) &= \boldsymbol{x}(t) - \boldsymbol{x}_0 = \begin{bmatrix} x_1(t) - \sqrt{Mgy_0} \\ x_2(t) - y_0 \\ x_3(t) \end{bmatrix} \\ \tilde{y}(t) &= y(t) - y_0. \end{aligned}$$

Now, the linearized model of the deviation variables is

$$\begin{aligned} \dot{\tilde{\boldsymbol{x}}}(t) &= \boldsymbol{A}\tilde{\boldsymbol{x}}(t) + \boldsymbol{b}\tilde{u}(t) \\ \tilde{y}(t) &= \boldsymbol{c}\tilde{\boldsymbol{x}}(t) + d\tilde{u}(t) \end{aligned}$$

where

$$\boldsymbol{A} = \left.\frac{\partial \boldsymbol{f}}{\partial \boldsymbol{x}}\right|_{\substack{\boldsymbol{x}=\boldsymbol{x}_0 \\ u=u_0}} = \begin{bmatrix} -\dfrac{R}{L} & 0 & 0 \\ 0 & 0 & 1 \\ -2\sqrt{\dfrac{g}{My_0}} & \dfrac{g}{y_0} & 0 \end{bmatrix}, \quad \boldsymbol{b} = \left.\frac{\partial \boldsymbol{f}}{\partial u}\right|_{\substack{\boldsymbol{x}=\boldsymbol{x}_0 \\ u=u_0}} = \begin{bmatrix} \dfrac{1}{L} \\ 0 \\ 0 \end{bmatrix},$$

$$\boldsymbol{c} = \left.\frac{\partial g}{\partial \boldsymbol{x}}\right|_{\substack{\boldsymbol{x}=\boldsymbol{x}_0 \\ u=u_0}} = \begin{bmatrix} 0 & 1 & 0 \end{bmatrix}, \qquad d = \left.\frac{\partial g}{\partial u}\right|_{\substack{\boldsymbol{x}=\boldsymbol{x}_0 \\ u=u_0}} = 0.$$

2.3 TRANSFER FUNCTIONS AND IMPULSE RESPONSES

Consider an LTI system described by state space equation

$$\begin{aligned} \dot{\boldsymbol{x}}(t) &= \boldsymbol{A}\boldsymbol{x}(t) + \boldsymbol{b}u(t) \\ y(t) &= \boldsymbol{c}\boldsymbol{x}(t) + du(t). \end{aligned}$$

Take the Laplace transform with zero initial conditions:

$$s\boldsymbol{X}(s) = \boldsymbol{A}\boldsymbol{X}(s) + \boldsymbol{b}U(s) \tag{2.18}$$

$$Y(s) = \boldsymbol{c}\boldsymbol{X}(s) + dU(s). \tag{2.19}$$

Now a set of differential equations in the time domain becomes a set of algebraic equations in the frequency domain. There are a total of $n+1$ equations in (2.18)–(2.19) and we can use them to eliminate the n variables in $\boldsymbol{X}(s)$ to obtain an equation relating the input $U(s)$ and the output $Y(s)$. Linear algebra now gives a formal method of capturing this process. From (2.18), we get

$$\boldsymbol{X}(s) = (s\boldsymbol{I} - \boldsymbol{A})^{-1}\boldsymbol{b}U(s).$$

Plugging it into (2.19), we obtain

$$\boxed{\frac{Y(s)}{U(s)} = \boldsymbol{c}(s\boldsymbol{I} - \boldsymbol{A})^{-1}\boldsymbol{b} + d,}$$

i.e., the ratio of the Laplace transform of the output over that of the input is a fixed function independent of the input.

Definition 2.11. The **transfer function** of an LTI system is the ratio of the Laplace transform of the output over that of the input when the initial condition is zero, i.e.,

$$G(s) = \frac{Y(s)}{U(s)}.$$

EXAMPLE 2.12

Let us continue to consider the DC motor system in Example 2.8. In the position control case, the transfer function is

$$\begin{aligned}
G(s) &= \begin{bmatrix} 0 & 1 & 0 \end{bmatrix} \begin{bmatrix} s + \dfrac{R_a}{L_a} & 0 & \dfrac{K_b}{L_a} \\ 0 & s & -1 \\ -\dfrac{K_t}{J} & 0 & s + \dfrac{K_f}{J} \end{bmatrix}^{-1} \begin{bmatrix} \dfrac{1}{L_a} \\ 0 \\ 0 \end{bmatrix} \\
&= \frac{K_t/(L_a J)}{s(s + R_a/L_a)(s + K_f/J) + K_t K_b/(L_a J)s} \\
&= \frac{K_t}{L_a J s^3 + (R_a J + K_f L_a)s^2 + (R_a K_f + K_t K_b)s}.
\end{aligned}$$

One may feel that the inverse of the 3×3 matrix is hard to compute, but the computation can be significantly simplified if one notices that only the element in the second row and the first column of the inverse is needed since all other elements will be multiplied by zero when forming the transfer function. Computing only one element is, of course, much simpler than computing all elements in the inverse. In the speed control case, the transfer function is

$$\begin{aligned}
G(s) &= \begin{bmatrix} 0 & 1 \end{bmatrix} \begin{bmatrix} s + \dfrac{R_a}{L_a} & \dfrac{K_b}{L_a} \\ -\dfrac{K_t}{J} & s + \dfrac{K_f}{J} \end{bmatrix}^{-1} \begin{bmatrix} \dfrac{1}{L_a} \\ 0 \end{bmatrix} \\
&= \frac{K_t/(L_a J)}{(s + R_a/L_a)(s + K_f/J) + K_t K_b/(L_a J)} \\
&= \frac{K_t}{L_a J s^2 + (R_a J + K_f L_a)s + (R_a K_f + K_t K_b)}.
\end{aligned}$$

For systems with a small number of state variables, it is probably more convenient to obtain the transfer function by directly manipulating the Laplace transform of the state space model (2.18) and (2.19). For example, in the speed control case, the Laplace transform of the state space model is

$$\begin{aligned} sX_1(s) &= -\frac{R_a}{L_a}X_1(s) - \frac{K_b}{L_a}X_2(s) + \frac{1}{L_a}U(s) \\ sX_2(s) &= \frac{K_t}{J}X_1(s) - \frac{K_f}{J}X_2(s) \\ Y(s) &= X_2(s). \end{aligned}$$

Substitute $X_1(s)$ from the second equation into the first equation and note that $Y(s) = X_2(s)$ from the third equation. We then get

$$\left[\frac{J}{K_t}\left(s + \frac{R_a}{L_a}\right)\left(s + \frac{K_f}{J}\right) + \frac{K_b}{L_a}\right]Y(s) = \frac{1}{L_a}U(s).$$

Consequently,

$$G(s) = \frac{Y(s)}{U(s)} = \frac{K_t}{L_aJs^2 + (R_aJ + K_fL_a)s + (R_aK_f + K_tK_b)},$$

the same result as the one obtained by matrix inversion.

EXAMPLE 2.13

Let us continue with Example 2.10, the magnetic suspension system. The transfer function of the linearized model is

$$\begin{aligned} G(s) &= \begin{bmatrix} 0 & 1 & 0 \end{bmatrix} \begin{bmatrix} s + \frac{R}{L} & 0 & 0 \\ 0 & s & -1 \\ 2\sqrt{\frac{g}{My_0}} & -\frac{g}{y_0} & s \end{bmatrix}^{-1} \begin{bmatrix} \frac{1}{L} \\ 0 \\ 0 \end{bmatrix} \\ &= \frac{-2\sqrt{\frac{g}{My_0}}\frac{1}{L}}{\left(s + \frac{R}{L}\right)\left(s^2 - \frac{g}{y_0}\right)} = \frac{-2\sqrt{gy_0}}{\sqrt{M}(Ls + R)(y_0s^2 - g)}. \end{aligned}$$

The transfer function of an LTI system with a state space model is always a ratio of two polynomials

$$G(s) = \frac{b(s)}{a(s)}$$

where $b(s)$ is called the numerator polynomial and $a(s)$ is called the denominator polynomial. We assume that polynomials $b(s)$ and $a(s)$ are **coprime**, i.e., they do not have common factors.

The ratio of two polynomials is also called a **rational function**. So the transfer function of an LTI system with a state space model is a rational function. We often denote a transfer function, or any rational function, in either of the following forms

$$G(s) = \frac{b_0 s^m + b_1 s^{m-1} + \cdots + b_m}{a_0 s^n + a_1 s^{n-1} + \cdots + a_n}$$

or

$$G(s) = K \frac{(s - z_1)(s - z_2) \cdots (s - z_m)}{(s - p_1)(s - p_2) \cdots (s - p_n)}.$$

The first form is called the **unfactored form** and the second form is called the **factored form**. Here $z_1, z_2, \ldots, z_m$, the roots of $b(s)$, are called the **zeros** of $G(s)$ and $p_1, p_2, \ldots, p_n$, the roots of $a(s)$, are called the **poles** of $G(s)$. K is called the **(high frequency) gain** of $G(s)$. The factored form is also called the **zero-pole-gain form**.

Several additional definitions will be needed. A transfer function or system $G(s)$ is said to be **proper** if $\deg b(s) \le \deg a(s)$, or equivalently $|G(\infty)| < \infty$. It is said to be **strictly proper** if $\deg b(s) < \deg a(s)$, or equivalently $G(\infty) = 0$. It is said to be **bi-proper** if $\deg b(s) = \deg a(s)$, or equivalently $0 \ne |G(\infty)| \ne \infty$. Transfer functions obtained from state space models are always proper, but occasionally nonproper transfer functions appear in abnormal cases. For a proper transfer function, the difference $\deg a(s) - \deg b(s)$ is called the **relative degree** of $G(s)$, and $\deg a(s)$ is called the **order** or **degree** of $G(s)$. $G(0)$ is called the **DC gain** of $G(s)$. Notice the difference between (high frequency) gain and DC gain.

In the transfer function of a state space model

$$G(s) = \boldsymbol{c}(s\boldsymbol{I} - \boldsymbol{A})^{-1}\boldsymbol{b} + d = \frac{\boldsymbol{c}\,\mathrm{adj}(s\boldsymbol{I} - \boldsymbol{A})\,\boldsymbol{b}}{\det(s\boldsymbol{I} - \boldsymbol{A})} + d$$

where adj means the adjugate of a matrix (see Appendix B), if there is no common factor on the above denominator and numerator, then $a(s) = \det(s\boldsymbol{I} - \boldsymbol{A})$, i.e., $a(s)$ is the characteristic polynomial of matrix $\boldsymbol{A}$, and the poles of the system are the eigenvalues of $\boldsymbol{A}$. If there are common factors on the above denominator and numerator, then $a(s)$ is only a factor of $\det(s\boldsymbol{I} - \boldsymbol{A})$ and the poles of the system are part of the eigenvalues of $\boldsymbol{A}$. In this case, some of the eigenvalues of $\boldsymbol{A}$ do not appear as the poles of the transfer function $G(s)$. They become hidden from the transfer function and hence are called the **hidden poles**. This again is an abnormal phenomenon and is prone to trouble. Extra care needs to be taken in this case. We will assume that this does not happen in our development.

In MATLAB, one can represent a polynomial

$$p(s) = p_0 s^n + p_1 s^{n-1} + \cdots + p_n$$

by a vector

$$\boldsymbol{p} = \begin{bmatrix} p_0 & p_1 & \cdots & p_n \end{bmatrix}.$$

To find the roots of $p(s)$, we can type

```
>> roots(p)
```

One often needs to compute the sums and products of polynomials. Let

$$p(s) = s^3 + 2s^2 + 3s + 4, \quad q(s) = 5s + 6.$$

Represent them in MATLAB by

```
>> p=[1 2 3 4];
>> q=[5 6];
```

However, neither of the following commands

```
>> p+q
>> p*q
```

would give you what you want since the first requires the two vectors to have the same dimension and the second is, in general, not defined at all. For polynomial addition, one has to augment either p or q with enough zeros so that they have the same dimension. For the example above, we should do

```
>> p+[0 0 q];
```

For polynomial multiplication, one has to use the command “conv”:

```
>> conv(p,q)
```

One may find these inconvenient and counterintuitive. In Section 2.6, we will present an alternative way of representing and operating polynomials in MATLAB.

Since the transfer function $G(s)$ of an LTI system is the ratio of the output Laplace transform $Y(s)$ and the input Laplace transform $U(s)$, if $U(s) = 1$, i.e., $u(t) = \delta(t)$, then $Y(s) = G(s)$ and $y(t) = \mathcal{L}^{-1}[G(s)] = g(t)$. Hence the inverse Laplace transform of $G(s)$, denoted by $g(t)$, is called the **impulse response**.

2.4 SIMPLIFYING BLOCK DIAGRAMS

Interconnected systems are often conveniently represented by block diagrams. For example, the system in the last section $Y(s) = G(s)U(s)$ can be represented using the block diagram in Figure 2.9.

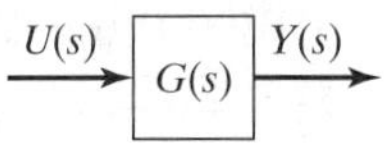

FIGURE 2.9: Block diagram representation.

Block diagrams are particularly useful when dealing with complex systems consisting of collections of interconnected subsystems. They can be simplified using the equivalence relationships in Table 2.1.

We shall now see how a complex system block diagram can be simplified.

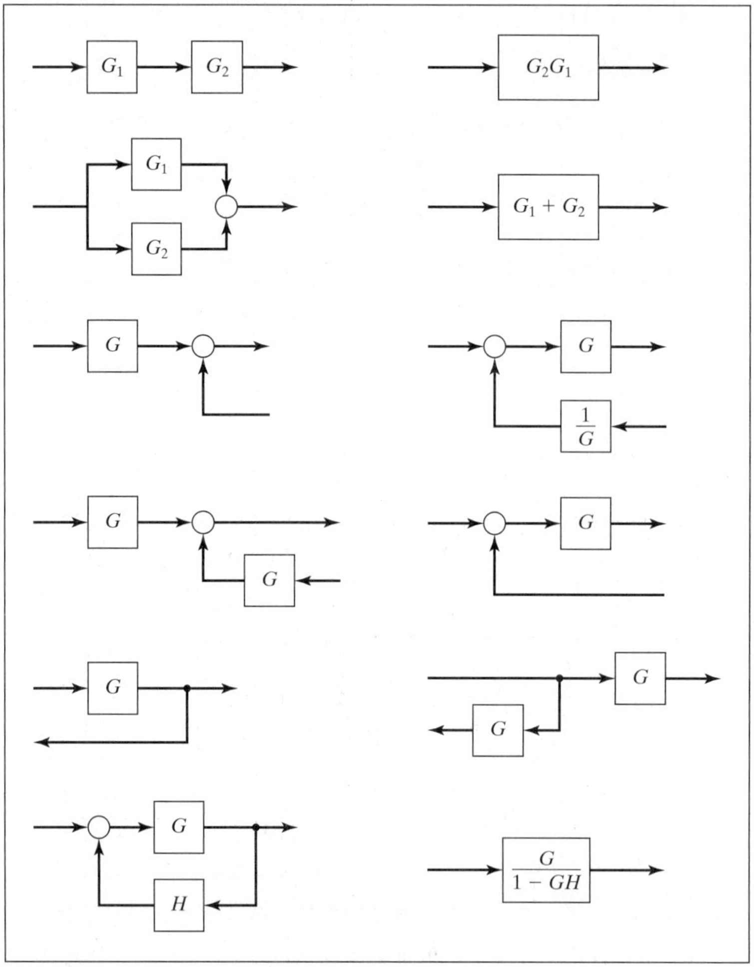

TABLE 2.1: Equivalent block diagrams.

EXAMPLE 2.14

Consider the simple feedback system shown in Figure 2.10. To find the relationship between r and y, we shall write down all the equations:

$$Y(s) = P(s)U(s), \quad U(s) = C(s)E(s), \quad E(s) = F(s)R(s) - H(s)Y(s).$$

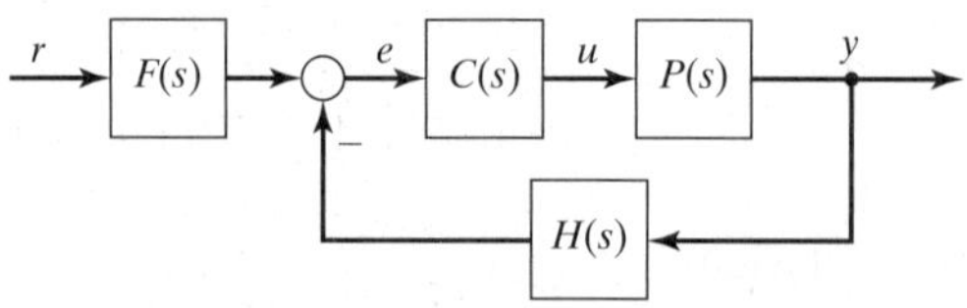

FIGURE 2.10: A simple feedback system.

We shall now eliminate the intermediate variables, $E(s)$ and $U(s)$, to get

$$Y(s) = P(s)C(s)\left[F(s)R(s) - H(s)Y(s)\right].$$

Thus solving $Y(s)$, we get

$$Y(s) = \frac{P(s)C(s)F(s)}{1 + P(s)C(s)H(s)} R(s)$$

i.e., the transfer function from $R(s)$ to $Y(s)$ is given by

$$\frac{Y(s)}{R(s)} = \frac{P(s)C(s)F(s)}{1 + P(s)C(s)H(s)}.$$

EXAMPLE 2.15

Consider the feedback control system shown in Figure 2.11. Here we omitted the variable s in the transfer function notation to make the expressions more compact.

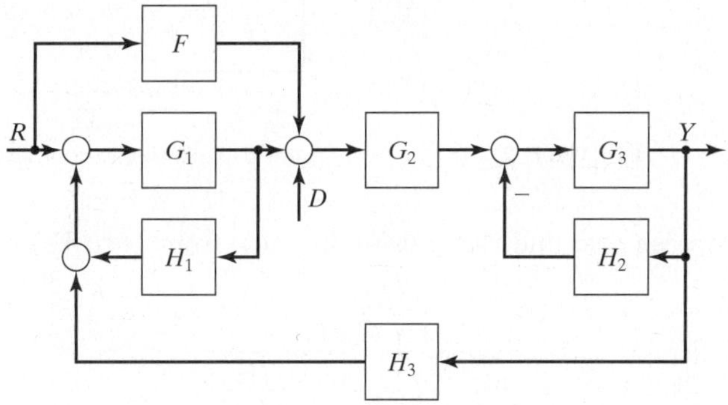

FIGURE 2.11: Original diagram of Example 2.15.

We shall compute the transfer function from R to Y. Thus, we shall assume $D = 0$ and we can simplify the block diagram by first closing the two inner loops: $G_1 \to H_1 \to G_1$ loop and $G_3 \to H_2 \to G_3$ loop, which results in the block diagram in Figure 2.12. We then move the first summing junction to the place of the second summing junction to get the block diagram in Figure 2.13. Finally, closing the loop, we have

$$\begin{aligned}\frac{Y}{R} &= \left(F + \frac{G_1}{1 - G_1H_1}\right) \frac{\dfrac{G_2G_3}{1 + G_3H_2}}{1 - \dfrac{G_2G_3}{1 + G_3H_2}\dfrac{G_1H_3}{1 - G_1H_1}} \\ &= \frac{G_2G_3(F - FG_1H_1 + G_1)}{1 - G_1H_1 + G_3H_2 - G_1G_3H_1H_2 - G_1G_2G_3H_3}.\end{aligned}$$

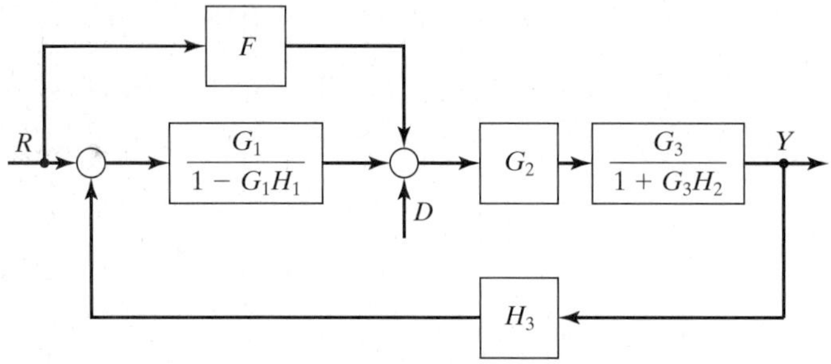

FIGURE 2.12: Block diagram of Example 2.15 with inner loops closed.

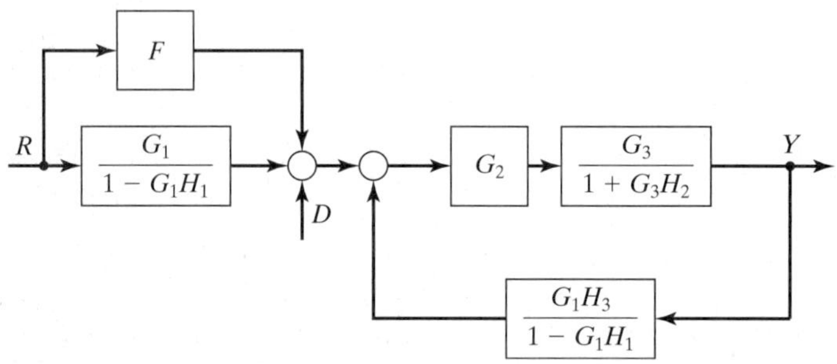

FIGURE 2.13: Further simplified block diagram of Example 2.15.

We can also find the transfer function from D to Y as

$$\frac{Y}{D} = \frac{\dfrac{G_2G_3}{1+G_3H_2}}{1 - \dfrac{G_2G_3}{1+G_3H_2}\,\dfrac{G_1H_3}{1-G_1H_1}}$$

$$= \frac{G_2G_3(1-G_1H_1)}{1-G_1H_1+G_3H_2-G_1G_3H_1H_2-G_1G_2G_3H_3}.$$

In general, block diagrams can always be simplified by using the block diagram algebra as shown in Table 2.1.

2.5 TRANSFER FUNCTION MODELING

The modeling of complicated interconnected LTI systems can be done in the Laplace transform domain using transfer functions and block diagrams in cases when the transfer functions of the subsystems are known.

Let us use an armature-controlled DC motor with a load torque, shown in Figure 2.7, as an example to see how this can be done. Let us consider the case when the load not only contains an inertia torque proportional to the angular acceleration and a friction torque proportional to the angular velocity, but also includes a nonzero possibly time-varying torque $\tau_d(t)$ independent of the angular position, velocity, and acceleration. Such a system can be considered as an interconnected system with electrical, magnetic, and mechanical subsystems.

First notice that in the electrical part, we have

$$I_a(s) = \frac{1}{L_a s + R_a}[V_a(s) - V_b(s)]. \tag{2.20}$$

Then the torque generated by the motor is given by

$$T_m(s) = K_t I_a(s). \tag{2.21}$$

We also know that the back emf voltage is given by

$$V_b(s) = K_b \Omega(s). \tag{2.22}$$

The mechanical part has relations

$$\Omega(s) = \frac{1}{Js + K_f}[T_m(s) - T_d(s)] \tag{2.23}$$

and

$$\Theta(s) = \frac{1}{s}\Omega(s)$$

where $T_d(s)$ is the Laplace transform of a possible load torque. Combining all of these equations, we can see that the block diagram of the whole system is as in Figure 2.14. Simplifying the block diagram gives

$$\Theta(s) = \frac{1}{s\left[(L_a s + R_a)(Js + K_f) + K_t K_b\right]}\begin{bmatrix} K_t & -(L_a s + R_a) \end{bmatrix}\begin{bmatrix} V_a(s) \\ T_d(s) \end{bmatrix}.$$

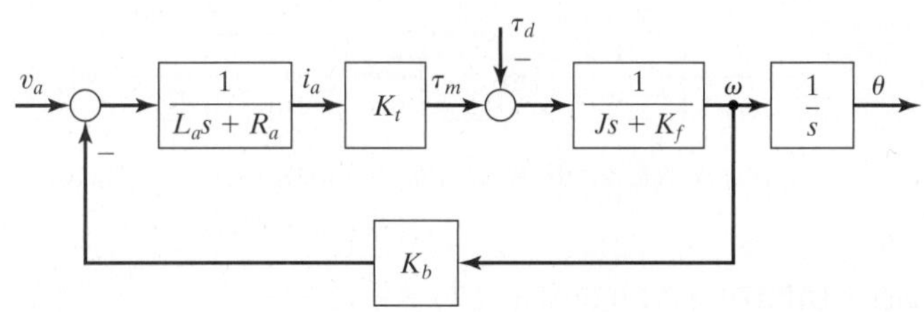

FIGURE 2.14: Block diagram of an armature-controlled DC motor.

Another illustrative example is a field-controlled DC motor shown in Figure 2.15. A field-controlled DC motor has the same structure as an armature-controlled DC motor. The difference is that in the field-controlled case, the armature current $i_a(t)$ is set to be constant but the field circuit is used to control the varying torque. In this case, the torque is related to the field current as

$$\tau_m(t) = K_t i_f(t) \tag{2.24}$$

where $i_f(t)$ is the field current. In the Laplace domain, we have

$$T_m(s) = K_t I_f(s) \tag{2.25}$$

where $I_f(s)$ is the Laplace transform of $i_f(t)$.

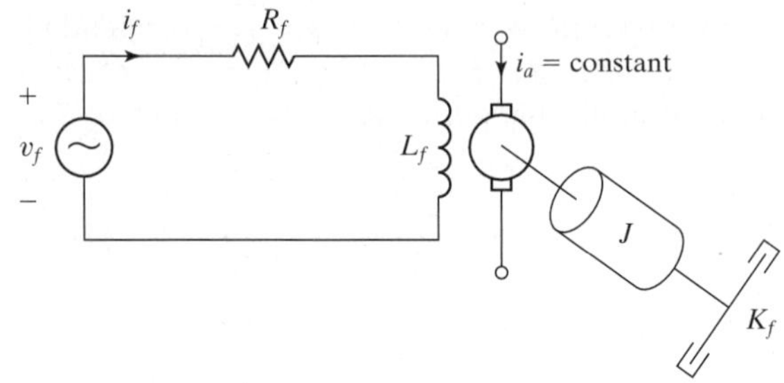

FIGURE 2.15: A field-controlled DC motor system.

The field current is generated by a field voltage through a field circuit and satisfies the following equation:

$$v_f(t) = L_f \frac{di_f(t)}{dt} + R_f i_f(t) \tag{2.26}$$

which, in terms of Laplace transforms, gives

$$\frac{I_f(s)}{V_f(s)} = \frac{1}{L_f s + R_f}. \tag{2.27}$$

Combining equations (2.23), (2.25), and (2.27), we get

$$\Theta(s) = \frac{1}{(L_f s + R_f)(Js + K_f)s} \begin{bmatrix} K_t & -(L_f s + R_f) \end{bmatrix} \begin{bmatrix} V_f(s) \\ T_d(s) \end{bmatrix}.$$

An interconnection block diagram for the system is shown in Figure 2.16.

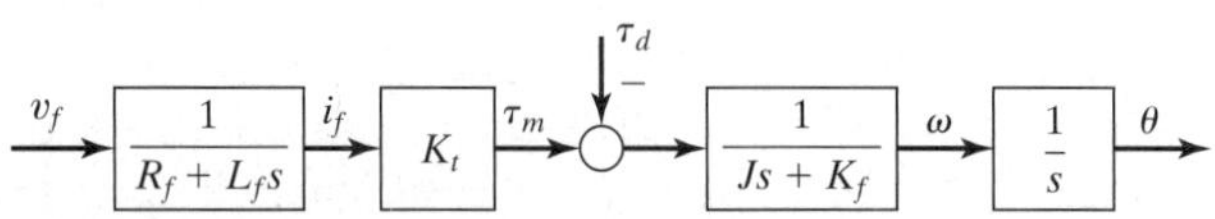

FIGURE 2.16: Block diagram of a field-controlled DC motor.

2.6 MATLAB MANIPULATION OF LTI SYSTEMS

In the MATLAB Control Systems Toolbox, a system can be represented by a single variable, no matter whether it is described by a state space model or by transfer function model. Suppose that we have a system described by its state space model:

$$\dot{\boldsymbol{x}}(t) = \begin{bmatrix} 0 & 1 \\ -1 & -2 \end{bmatrix} \boldsymbol{x}(t) + \begin{bmatrix} 0 \\ 1 \end{bmatrix} u(t)$$
$$y(t) = \begin{bmatrix} 2 & 1 \end{bmatrix} \boldsymbol{x}(t)$$

and we wish to name it as F. Then the following sequence of commands assigns the variable F with its state space description:

```
>> A=[0 1; -1 -2];
>> B=[0 ; 1];
```

```
>> C=[2 1];
>> D=0;
>> F=ss(A,B,C,D);
```

Suppose we now have a system described by its transfer function model

$$\frac{s+2}{s^2+2s+1}$$

and we wish to name it as G. The following sequence of commands assigns the variable G with its transfer function description:

```
>> num=[1 2];
>> den=[1 2 1];
>> G=tf(num,den);
```

The different descriptions of a system can be easily converted from one to another. The command

```
>> F=tf(F);
```

converts the description of F from state space to transfer function. By doing this we can find that F and G are actually the same system since they have the same transfer function. Also the command

```
>> G=ss(G);
```

converts the description of G from transfer function to state space. However, we do not necessarily get exactly the same state space description as the original F although G and F have the same transfer function. This is because a system may have different state space descriptions. Now F is in transfer function form. Let us run

```
>> F=ss(F);
```

We will observe that this state space description of F is different from the original state space description of F, but is actually the same as the state space description of G obtained from the conversion. This is because by running the transfer function to state space conversion, the computer program chooses, among many possibilities, a particular canonical form of the state space description. If the very original state space description is not of this canonical form, then a state space to transfer function to state space conversion will not give the same thing back. The original state space description of F is forever lost after the "`ss`" to "`tf`" conversion.

For a system F in either the state space form or the transfer function form, commands

```
>> [A,B,C,D]=ssdata(F);
>> [num,den]=tfdata(F);
```

give back the parameter matrices of its state space description and the numerator and denomination coefficients of its transfer function respectively. Owing to the nonuniqueness of the state space model for a given transfer function, one may wonder which choice the first command takes if F is in transfer function form. It turns out that the same canonical form as that chosen by the command "`ss(F)`" is also chosen here.

The use of the LTI system variable brings much convenience. To find out the poles and zeros of system F, instead of computing the denominator and numerator polynomials of transfer function $F(s)$ and then using the command "roots", we can simply do

```
>> pole(F);
```

and

```
>> zero(F);
```

Here the actual form of F is immaterial. To compute the sum and product of systems, we can simply run

```
>> F+G;
```

and

```
>> F*G;
```

Here F and G may take different forms. The following commands have obvious meanings:

```
>> F-G;
```

and

```
>> F/G;
```

When doing the last operation, one may run into trouble when G is strictly proper and either F or G is in state space form since the inverse of a strictly proper system cannot be represented by a state space system. Nevertheless, no problem will arise if both F and G are in transfer function form unless $G(s) = 0$. The feedback connection of two systems are computed easily:

```
>> feedback(F,G)
```

computes

$$\frac{F}{1+FG}$$

and

```
>> feedback(F,G,1)
```

computes

$$\frac{F}{1-FG}.$$

LTI system variables give another way of representing and operating polynomials, simply by interpreting them as transfer functions with 1 as the denominator polynomials. In this case, the operations on polynomials, such as addition and multiplication, can be done in a way closer to natural language. Again, take

$$p(s) = s^3 + 2s^2 + 3s + 4, \quad q(s) = 5s + 6$$

as an example. The commands

```
>> p=tf([1 2 3 4],1);
>> q=tf([5 6],1);
```

assign p and q to be appropriate transfer functions with 1 as denominators, i.e., polynomials. Addition and multiplication can then be done in the following natural way without worrying about dimension compatibility and complications caused by vector multiplication:

```
>> p+q
>> p*q
```

2.7 SIMULATION AND IMPLEMENTATION OF SYSTEMS

We have seen how to model a physical system using differential equations and how to convert the model into a transfer function. In many situations, we also need to carry out the inverse process. For example, in system simulation, we often need to build a physical system with a given transfer function to observe the behavior of the system represented by the transfer function. In control implementation, we need to build a physical controller from the designed controller transfer function to connect it with the plant to form a feedback loop. The process of building a real physical system with a given transfer function is called **realization**. Unlike the modeling process of finding transfer functions from physical systems, the inverse realization process is highly nonunique, in terms of the kind of physical components used and the many possible configurations and structures. Although nowadays such a job is more and more accomplished by computer software, the traditional way of using hardware components is still of great theoretical and practical value. In the majority of such system simulation and implementation, op-amp circuits are used.

2.7.1 Hardware simulation and implementation

Let a system be given by a proper nth order transfer function

$$G(s) = \frac{b(s)}{a(s)} = \frac{b_0 s^n + b_1 s^{n-1} + \cdots + b_n}{a_0 s^n + a_1 s^{n-1} + \cdots + a_n}, \quad a_0 \neq 0.$$

The op-amp circuit shown in Figure 2.17 gives a realization of $G(s)$. To show this, notice that

$$a_0 x^{(n)}(t) = -a_1 x^{(n-1)}(t) - \cdots - a_n x(t) + u(t).$$

Taking the Laplace transforms with zero initial conditions, we get

$$a_0 s^n X(s) = -a_1 s^{n-1} X(s) - \cdots - a_n X(s) + U(s).$$

This gives us

$$a(s)X(s) = U(s).$$

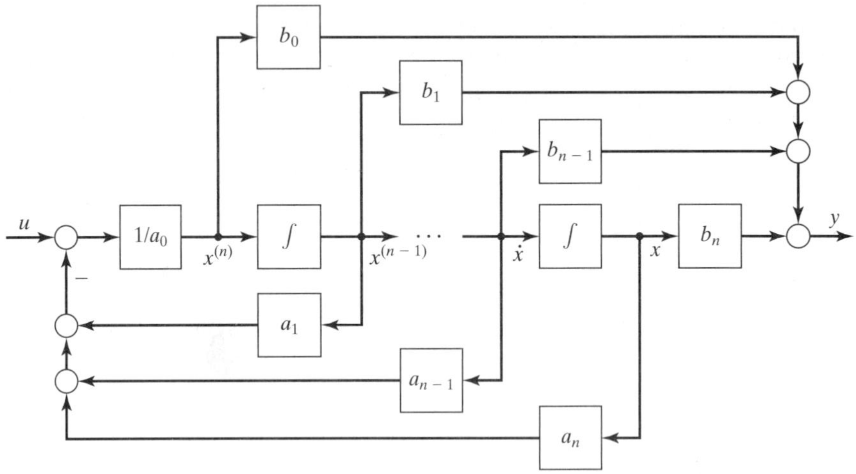

FIGURE 2.17: Controller form realization.

Also, notice that

$$y(t) = b_0 x^{(n)}(t) + b_1 x^{(n-1)}(t) + \cdots + b_n x(t).$$

Taking the Laplace transforms with zero initial conditions, we get

$$Y(s) = b_0 s^n X(s) + b_1 s^{n-1} X(s) + \cdots + b_n X(s) = b(s)X(s).$$

Hence

$$\frac{Y(s)}{U(s)} = \frac{b(s)}{a(s)}.$$

This realization is called a **controller form realization**.

The op-amp circuit in Figure 2.17 has a natural state space model. If we follow our tradition of assigning the voltages across capacitors, which are hidden in the integrators, then the state vector becomes

$$\boldsymbol{x}(t) = \begin{bmatrix} x^{(n-1)}(t) \\ \vdots \\ \dot{x}(t) \\ x(t) \end{bmatrix}.$$

The corresponding state space model is

$$\dot{\boldsymbol{x}}(t) = \begin{bmatrix} -\frac{a_1}{a_0} & \cdots & -\frac{a_{n-1}}{a_0} & -\frac{a_n}{a_0} \\ 1 & \cdots & 0 & 0 \\ \vdots & \ddots & \vdots & \vdots \\ 0 & \cdots & 1 & 0 \end{bmatrix} \boldsymbol{x}(t) + \begin{bmatrix} \frac{1}{a_0} \\ 0 \\ \vdots \\ 0 \end{bmatrix} u(t)$$

$$y(t) = \begin{bmatrix} b_1 - b_0\frac{a_1}{a_0} & \cdots & b_{n-1} - b_0\frac{a_{n-1}}{a_0} & b_n - b_0\frac{a_n}{a_0} \end{bmatrix} \boldsymbol{x}(t) + \frac{b_0}{a_0}u(t).$$

This state space model, of course, gives a transfer function exactly the same as the given one.

Another realization of the same system is given by the circuit shown in Figure 2.18. To show this, notice that

$$\begin{aligned} a_0 y(t) &= x_1(t) + b_0 u(t) \\ \dot{x}_1(t) &= x_2(t) - a_1 y(t) + b_1 u(t) \\ &\vdots \\ \dot{x}_{n-1}(t) &= x_n(t) - a_{n-1} y(t) + b_{n-1} u(t) \\ \dot{x}_n(t) &= -a_n y(t) + b_n u(t). \end{aligned}$$

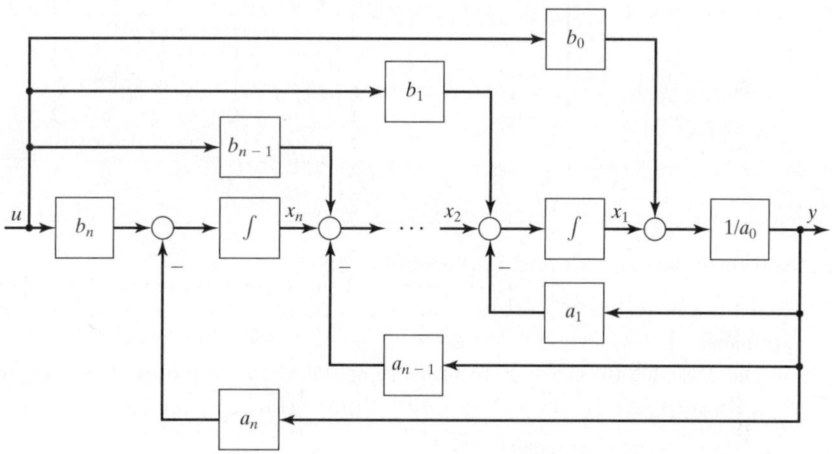

FIGURE 2.18: Observer form realization.

Taking the Laplace transforms with zero initial conditions, we get

$$\begin{aligned} a_0 Y(s) &= X_1(s) + b_0 U(s) \\ sX_1(s) &= X_2(s) - a_1 Y(s) + b_1 U(s) \\ &\vdots \\ sX_{n-1}(s) &= X_n(s) - a_{n-1} Y(s) + b_{n-1} U(s) \\ sX_n(s) &= -a_n Y(s) + b_n U(s). \end{aligned}$$

Multiplying the above equations by $s^n, s^{n-1}, \ldots, s, 1$, respectively, and adding them altogether, one can see that the variables $X_1(s), \ldots, X_n(s)$ are all cancelled and the resulting equation is

$$a(s)Y(s) = b(s)U(s).$$

So we also get

$$\frac{Y(s)}{U(s)} = \frac{b(s)}{a(s)}.$$

This realization is called the **observer form realization.** It also has a state space model. Let us again assign the voltage across capacitors as state variables. The state vector then becomes

$$\boldsymbol{x}(t) = \begin{bmatrix} x_1(t) \\ x_2(t) \\ \vdots \\ x_n(t) \end{bmatrix}.$$

The corresponding state space model is

$$\dot{\boldsymbol{x}}(t) = \begin{bmatrix} -\dfrac{a_1}{a_0} & 1 & \cdots & 0 \\ \vdots & \vdots & \ddots & \vdots \\ -\dfrac{a_{n-1}}{a_0} & 0 & \cdots & 1 \\ -\dfrac{a_n}{a_0} & 0 & \cdots & 0 \end{bmatrix} \boldsymbol{x}(t) + \begin{bmatrix} b_1 - b_0\dfrac{a_1}{a_0} \\ \vdots \\ b_{n-1} - b_0\dfrac{a_{n-1}}{a_0} \\ b_n - b_0\dfrac{a_n}{a_0} \end{bmatrix} u(t)$$

$$y(t) = \begin{bmatrix} \dfrac{1}{a_0} & 0 & \cdots & 0 \end{bmatrix} \boldsymbol{x}(t) + \frac{b_0}{a_0} u(t).$$

2.7.2 Software simulation and implementation

MATLAB provides certain tools for the numerical computation of system responses. For a system represented by a variable `G`, regardless of whether it is in the transfer function form or state space form, to compute its impulse response, i.e., its response to the unit impulse input $\delta(t)$, one can use

```
>> [y,t]=impulse(G);
```

To find its step response, i.e., its response to the unit step input $\sigma(t)$, one can use

```
>> [y,t]=step(G);
```

To calculate a system response with respect to a more general input signal than impulse and step, one can use a MATLAB command `lsim`. For example, the following sequence of commands gives the sinusoidal response of the system:

```
>> t=0:0.1:10;
>> u=sin(t);
>> y=lsim(G,u,t);
```

Another software product associated with MATLAB is SIMULINK. It can be used to simulate an interconnected system, represented by a block diagram. Let us demonstrate its use by a couple of examples.

EXAMPLE 2.16

Consider the unity feedback system shown in Figure 2.19 with a loop transfer function

$$L(s) = \frac{1}{s(s+1)}.$$

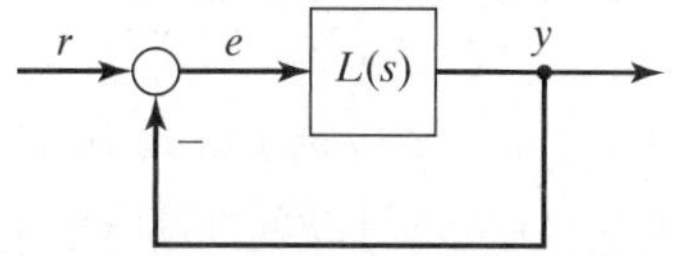

FIGURE 2.19: For Examples 2.16 and 2.17.

The time responses of this system with respect to various kinds of command signals are simulated using a SIMULINK diagram as shown in Figure 2.20 and are plotted in Figure 2.21.

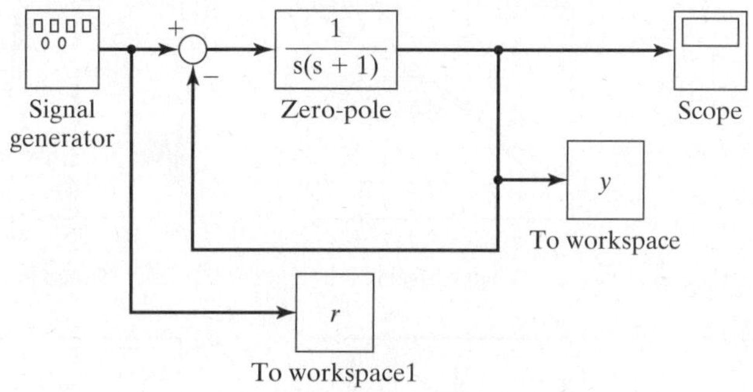

FIGURE 2.20: SIMULINK diagram for Example 2.16.

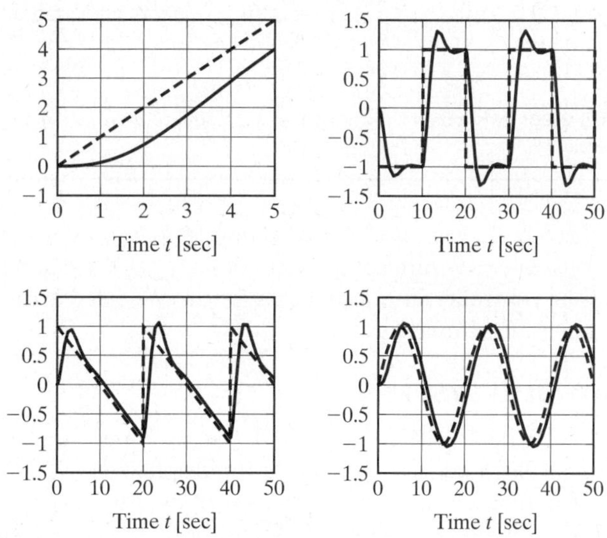

FIGURE 2.21: Responses to a ramp signal, a square wave, a sawtooth signal, and a sine wave for Example 2.16.

EXAMPLE 2.17

Consider the unity feedback system shown in Figure 2.19 with

$$L(s) = \frac{5(s+1)}{s^2(s+2)}.$$

Then a similar SIMULINK block can be built, and the time response of the system with respect to a ramp and a square wave of 0.05 Hz are shown in Figure 2.22. It is noted that there is no steady-state error in tracking a ramp for this system. However, this system is lightly damped, so the transient performance is rather poor.

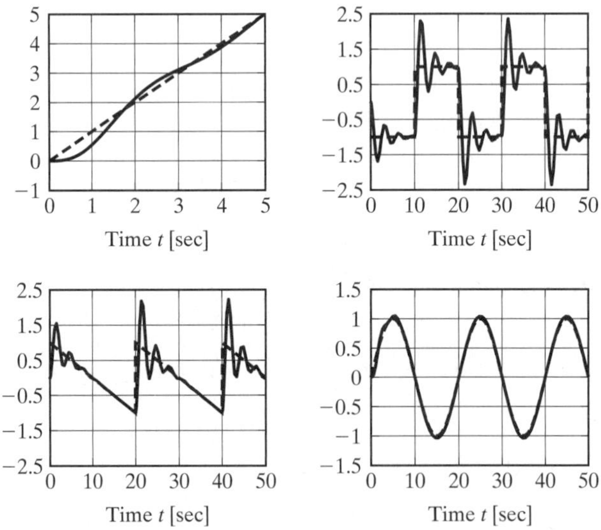

FIGURE 2.22: Responses to a ramp signal, a square wave, a sawtooth signal, and a sine wave for Example 2.17.

SIMULINK can also be used to implement a controller in a digital form, i.e., it can be used to generate computer codes so that the computer with the codes can serve as the controller in a feedback system. We refer to the SIMULINK manual on how this can be done.

2.8 MISO AND SIMO SYSTEMS

So far we have dealt with mostly SISO systems. A general system might be MIMO, i.e., it might have multiple inputs and multiple outputs. We will leave the theory of MIMO feedback control systems to a more advanced course. In this book, we will occasionally deal with MISO systems or SIMO systems for two reasons. One is that MISO systems and SIMO systems are frequently seen in practice and they can be dealt with using techniques not much beyond the theory for SISO systems. The other reason is that in the control of SISO systems, such as in the regulation control, we often need to use MISO systems, which lead to better performance. To simplify notation and understanding, we will study only double-input–single-output (DISO) systems and single-input–double-output (SIDO) systems. The general theory of MISO and SIMO systems extends in a rather trivial way from that of DISO and SIDO systems. The MISO and SIMO systems that we are going to deal with and use in this book are actually DISO and SIDO systems.

An LTI DISO system has two inputs and one output. Its block diagram takes the form in Figure 2.23 and its transfer function takes the form

$$\boldsymbol{G}(s) = \begin{bmatrix} G_1(s) & G_2(s) \end{bmatrix}$$

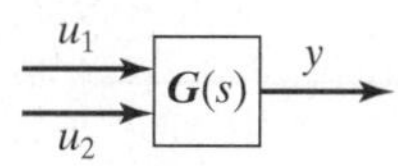

FIGURE 2.23: A DISO system.

which has the meaning

$$Y(s) = \begin{bmatrix} G_1(s) & G_2(s) \end{bmatrix} \begin{bmatrix} U_1(s) \\ U_2(s) \end{bmatrix} = G_1(s)U_1(s) + G_2(s)U_2(s).$$

Here, we follow our convention of writing vectors or matrices using bold letters.

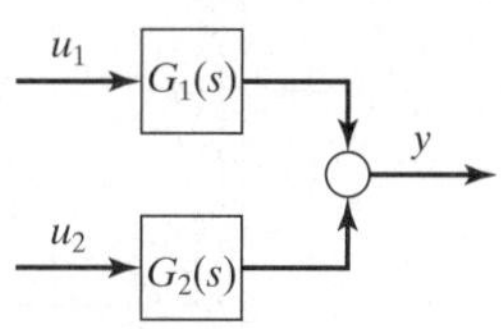

FIGURE 2.24: A misleading way of viewing a DISO system.

Although one may identify a DISO system by a connection of two SISO subsystems as shown in Figure 2.24, this view would do more harm than good. It is useful only in the algebraic manipulation of the input/output relations, but might lead to erroneous results in stability analysis, simulation, and implementation. Thus we say that Figure 2.24 is only algebraically equivalent to Figure 2.23. It is better to view a DISO system as an inseparable single system. Hence we prefer to write

$$\begin{aligned} \boldsymbol{G}(s) &= \frac{\boldsymbol{b}(s)}{a(s)} = \frac{1}{a(s)} \begin{bmatrix} b_1(s) & b_2(s) \end{bmatrix} \\ &= \frac{\begin{bmatrix} b_{10}s^n + \cdots + b_{1n} & b_{20}s^n + \cdots + b_{2n} \end{bmatrix}}{a_0 s^n + a_1 s^{n-1} + \cdots + a_n}, \qquad a_0 \neq 0. \end{aligned}$$

Here $a(s)$ is the common denominator of $G_1(s)$ and $G_2(s)$, i.e., the least common multiple of the denominators of $G_1(s)$ and $G_2(s)$. A typical question one might ask is what the order of this system is. Let us take the view that the order of a system is the minimum number of integrators needed in its op-amp circuit realization. If we realize the two subsystems $G_1(s)$ and $G_2(s)$ separately and then connect them as in Figure 2.24, then we may need $2n$ integrators in the worst case. However, a more economical realization is given in Figure 2.25 where only n integrators are used. Hence the order of the system is n, instead of $2n$. The significance of the realization in Figure 2.25 is not just to save integrators. More importantly, it has eliminated many redundant signals which might behave in unexpected and undesirable ways if a DISO system is realized as the sum of two separate systems.

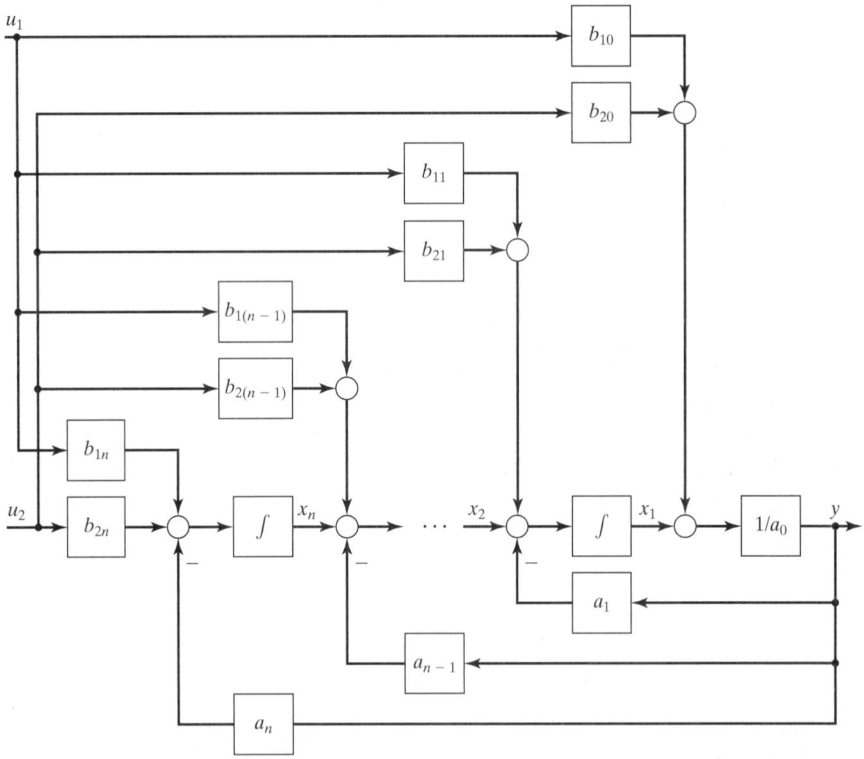

FIGURE 2.25: Realization of a DISO system.

When SIMULINK, or any other software, is used to simulate or implement the system, one should always keep in mind that a DISO system is a single integrated system, not the sum of two SISO systems.

Similarly, an LTI SIDO system has one input and two outputs. Its block diagram takes the form in Figure 2.26

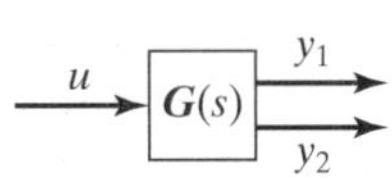

FIGURE 2.26: A SIDO system.

and its transfer function takes the form

$$\boldsymbol{G}(s) = \begin{bmatrix} G_1(s) \\ G_2(s) \end{bmatrix},$$

which has the meaning

$$\begin{bmatrix} Y_1(s) \\ Y_2(s) \end{bmatrix} = \begin{bmatrix} G_1(s) \\ G_2(s) \end{bmatrix} U(s)$$

or

$$Y_1(s) = G_1(s)U(s), \quad Y_2(s) = G_2(s)U(s).$$

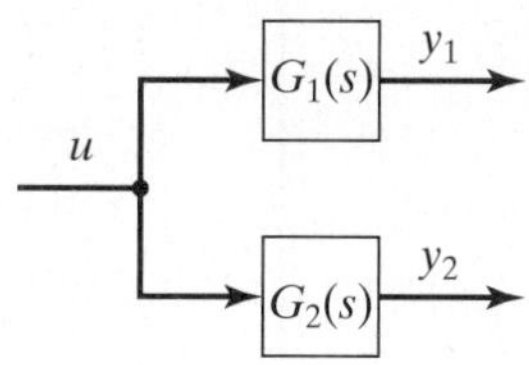

FIGURE 2.27: A misleading way to view a SIDO system.

Again, we should be careful in viewing a SIDO system as a connection of two SISO subsystems shown in Figure 2.27. This view is useful only in the algebraic manipulation of input/output relations, but might lead to erroneous results in stability analysis, simulation, and implementation. So we say that Figure 2.27 is only algebraically equivalent to Figure 2.26. Hence we prefer to write

$$\boldsymbol{G}(s) = \frac{\boldsymbol{b}(s)}{a(s)} = \frac{1}{a(s)} \begin{bmatrix} b_1(s) \\ b_2(s) \end{bmatrix} = \frac{\begin{bmatrix} b_{10}s^n + \cdots + b_{1n} \\ b_{20}s^n + \cdots + b_{2n} \end{bmatrix}}{a_0 s^n + a_1 s^{n-1} + \cdots + a_n}, \quad a_0 \neq 0.$$

Let us again take the view that the order of a system is the minimum number of integrators needed in its op-amp circuit realization. If we realize the two subsystems $G_1(s)$ and $G_2(s)$ separately and then connect them as in Figure 2.27, then we may need $2n$ integrators in maximum. However, a more economical realization is given in Figure 2.28 where only n integrators are used. Hence the order of the system is n, instead of $2n$. Again, the realization in Figure 2.28 has eliminated many redundant signals which might behave in unexpected and undesirable ways if a SIDO system is realized as the interconnection of two separate systems.

2.9 MODELING OF CLOSED-LOOP SYSTEMS

Two types of closed-loop systems will be used in this book. One is the **feedback system for stabilization** as shown in Figure 2.29. The other is the **feedback system for regulation** as shown in Figure 2.30.

Let us first analyze the feedback system for stabilization. The transfer function from w_1 to y_1 is $\dfrac{P(s)C(s)}{1+P(s)C(s)}$. Here $1+P(s)C(s)$ has to be nonzero in order for the transfer function to be meaningful. If we try to find the transfer functions from any external signal to any internal signal, we can see that we always need $1+P(s)C(s)$ to be nonzero.

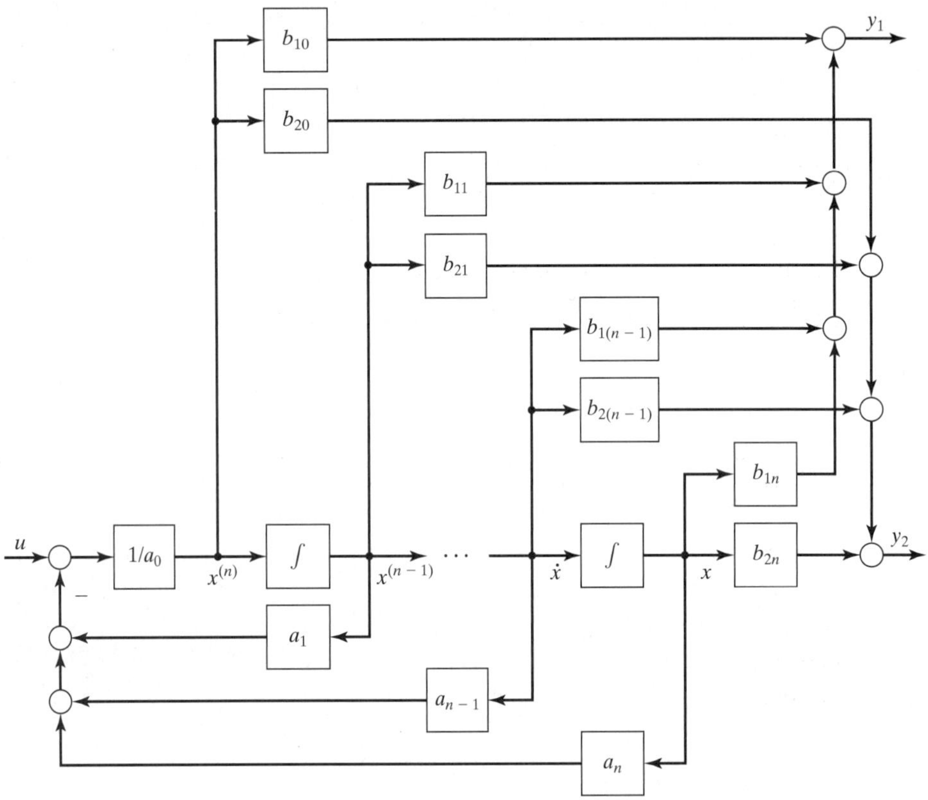

FIGURE 2.28: Realization of a SIDO system.

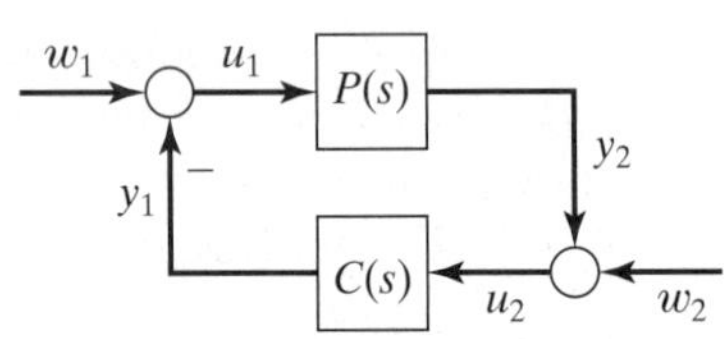

FIGURE 2.29: Feedback system for stabilization.

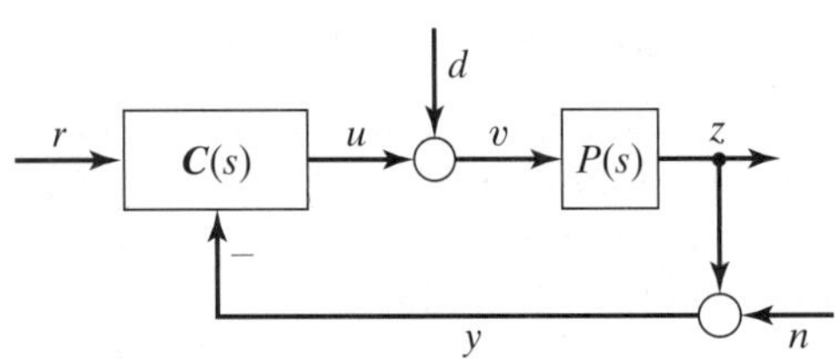

FIGURE 2.30: Feedback system for regulation.

Definition 2.18. The closed-loop system shown in Figure 2.29 is said to be **well posed** if $1 + P(s)C(s) \not\equiv 0$. Otherwise, it is said to be **ill posed**.

An ill-posed system is not meaningful, does not work properly, and should be avoided.

EXAMPLE 2.19

If $P(s) = 1$ and $C(s) = -1$, then the system shown in Figure 2.29 is ill posed. In this case, the internal signals cannot be uniquely determined from the external signals.

Proposition 2.20. The closed-loop system shown in Figure 2.29 is well posed if $P(s)$ and $C(s)$ are proper and at least one of them is strictly proper.

Proof. If $P(s)$ and $C(s)$ are proper, then $|P(\infty)| < \infty$ and $|C(\infty)| < \infty$. Furthermore, if at least one of $P(s)$ and $C(s)$ is strictly proper, then $P(\infty) = 0$ or $C(\infty) = 0$. This shows $P(\infty)C(\infty) = 0$, i.e., $1 + P(\infty)C(\infty) = 1$. Therefore $1 + P(s)C(s) \not\equiv 0$. □

In most applications, the condition in Proposition 2.20 is satisfied. When the closed-loop system is well posed, we can find the transfer function from one of the external signals $w_1(t), w_2(t)$ to one of the internal signals $u_1(t)$, $u_2(t)$, $y_1(t)$, $y_2(t)$. For example, the transfer function from $w_1(t)$ to $y_2(t)$ is

$$\frac{P(s)}{1 + P(s)C(s)}.$$

It is more convenient to write down all the transfer functions in a compact matrix form as

$$\begin{bmatrix} U_1(s) \\ U_2(s) \\ Y_1(s) \\ Y_2(s) \end{bmatrix} = \begin{bmatrix} \dfrac{1}{1 + P(s)C(s)} & \dfrac{-C(s)}{1 + P(s)C(s)} \\ \dfrac{P(s)}{1 + P(s)C(s)} & \dfrac{1}{1 + P(s)C(s)} \\ \dfrac{P(s)C(s)}{1 + P(s)C(s)} & \dfrac{C(s)}{1 + P(s)C(s)} \\ \dfrac{P(s)}{1 + P(s)C(s)} & \dfrac{-P(s)C(s)}{1 + P(s)C(s)} \end{bmatrix} \begin{bmatrix} W_1(s) \\ W_2(s) \end{bmatrix} \tag{2.28}$$

where an element of the big 4×2 matrix is simply the transfer function from the corresponding external signal to the corresponding internal signal. Although there are eight different possible input/output pairs, the transfer functions between these input/output pairs contain some repetitions. Essentially there are only four different transfer functions:

$$\frac{1}{1 + P(s)C(s)}, \quad \frac{P(s)}{1 + P(s)C(s)}, \quad \frac{C(s)}{1 + P(s)C(s)}, \quad \frac{P(s)C(s)}{1 + P(s)C(s)}. \tag{2.29}$$

These four different transfer functions are called the **gang of four**. Among the transfer functions in the gang of four, we also denote

$$S(s) = \frac{1}{1+P(s)C(s)} \quad \text{and} \quad T(s) = \frac{P(s)C(s)}{1+P(s)C(s)},$$

and call them sensitivity function and complementary sensitivity function, respectively. These two closed-loop transfer functions are related by

$$S(s) + T(s) = 1.$$

Let us now consider the feedback system for regulation. In this case $\boldsymbol{C}(s)$ is a DISO system. Hence it can be denoted by

$$U(s) = \begin{bmatrix} C_1(s) & -C_2(s) \end{bmatrix} \begin{bmatrix} R(s) \\ Y(s) \end{bmatrix}.$$

Here we put a minus sign in front of $C_2(s)$ because this suggests a negative feedback. This controller $\boldsymbol{C}(s)$ is also called a **two-degree-of-freedom**, or simply a **2DOF**, controller since it actually involves two transfer functions $C_1(s)$ and $C_2(s)$. From a pure algebraic point of view, the feedback system in Figure 2.30 is equivalent to the system shown in Figure 2.31. However, we will later see that analytically they are not equivalent and Figure 2.31 should not be used to build the actual feedback system.

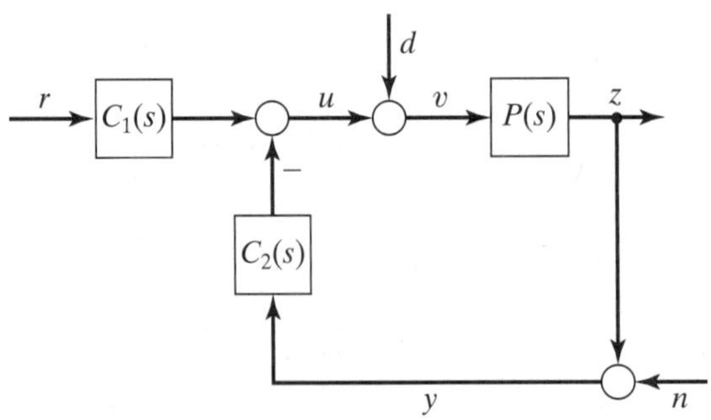

FIGURE 2.31: Algebraic equivalence of Figure 2.30.

From Figure 2.31, it can be seen that a stabilization system formed by $P(s)$ and $C_2(s)$ is a subsystem of the regulation system. Therefore, in order for the system in Figure 2.30 to work properly, the stabilization system formed by $P(s)$ and $C_2(s)$ has to work properly.

Definition 2.21. The closed-loop system shown in Figure 2.30 is said to be well posed if the feedback system for stabilization formed by $P(s)$ and $C_2(s)$ is well posed.

If the closed-loop system is well posed, we can find each of the transfer functions from external signals $d(t), n(t), r(t)$ to internal signals $u(t), v(t), y(t), z(t)$.

Again we can put these transfer functions into a compact matrix form

$$\begin{bmatrix} U(s) \\ V(s) \\ Y(s) \\ Z(s) \end{bmatrix} = \begin{bmatrix} \frac{-P(s)C_2(s)}{1+P(s)C_2(s)} & \frac{-C_2(s)}{1+P(s)C_2(s)} & \frac{C_1(s)}{1+P(s)C_2(s)} \\ \frac{1}{1+P(s)C_2(s)} & \frac{-C_2(s)}{1+P(s)C_2(s)} & \frac{C_1(s)}{1+P(s)C_2(s)} \\ \frac{P(s)}{1+P(s)C_2(s)} & \frac{1}{1+P(s)C_2(s)} & \frac{P(s)C_1(s)}{1+P(s)C_2(s)} \\ \frac{P(s)}{1+P(s)C_2(s)} & \frac{-P(s)C_2(s)}{1+P(s)C_2(s)} & \frac{P(s)C_1(s)}{1+P(s)C_2(s)} \end{bmatrix} \begin{bmatrix} D(s) \\ N(s) \\ R(s) \end{bmatrix}. \tag{2.30}$$

We see that there are 12 elements in the matrix corresponding to the 12 possible input/output pairs, but again there are repetitions among the 12 elements. Essentially, there are only six different transfer functions, the gang of four formed by $P(s)$ and $C_2(s)$ as well as two additional ones involving $C_1(s)$:

$$\frac{C_1(s)}{1+P(s)C_2(s)} \quad \text{and} \quad \frac{P(s)C_1(s)}{1+P(s)C_2(s)}.$$

The 2DOF controller also takes some other equivalent forms. Figure 2.32 shows a two-loop feedback system. Here $\boldsymbol{H}(s)$ is a DISO system with transfer function

$$\boldsymbol{H}(s) = \begin{bmatrix} H_1(s) & H_2(s) \end{bmatrix}.$$

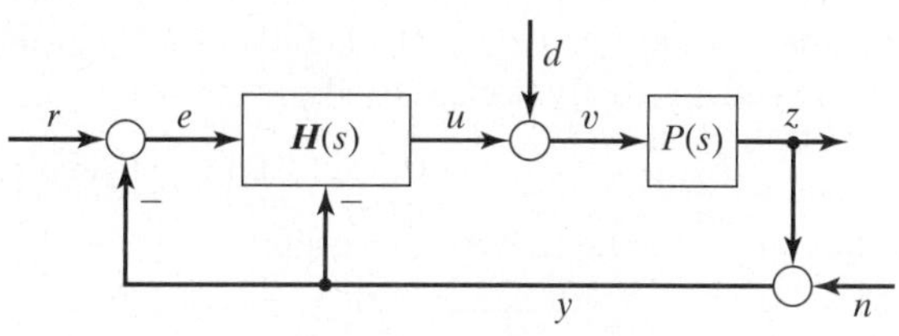

FIGURE 2.32: Two-loop feedback system.

Hence

$$U(s) = \begin{bmatrix} H_1(s) & -H_2(s) \end{bmatrix} \begin{bmatrix} E(s) \\ Y(s) \end{bmatrix}.$$

The system has an inner loop consisting of $H_2(s)$ and $P(s)$ and an outer loop consisting of $H_1(s)$ and the inner loop. This might be more easily seen from its algebraic equivalence shown in Figure 2.33. It can be easily shown that if

$$H_1(s) = C_1(s) \quad \text{and} \quad H_2(s) = C_2(s) - C_1(s),$$

then it is equivalent to the 2DOF controller.

Another equivalent form of the DISO controller is the feedback plus feedforward configuration shown in Figure 2.34. Here $\boldsymbol{F}(s)$ is a DISO system with transfer function

$$\boldsymbol{F}(s) = \begin{bmatrix} F_1(s) & F_2(s) \end{bmatrix}.$$

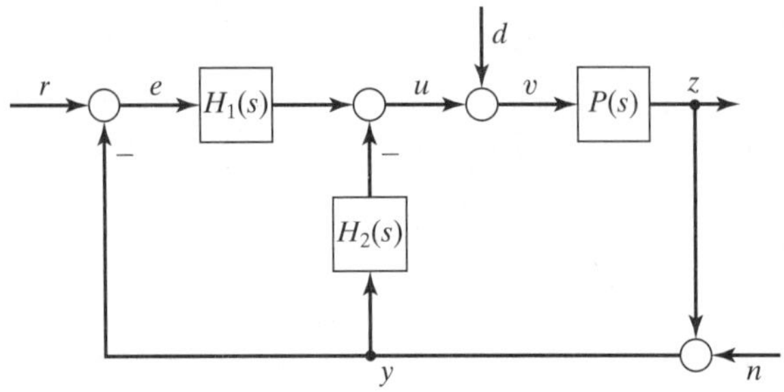

FIGURE 2.33: Algebraic equivalence of Figure 2.32.

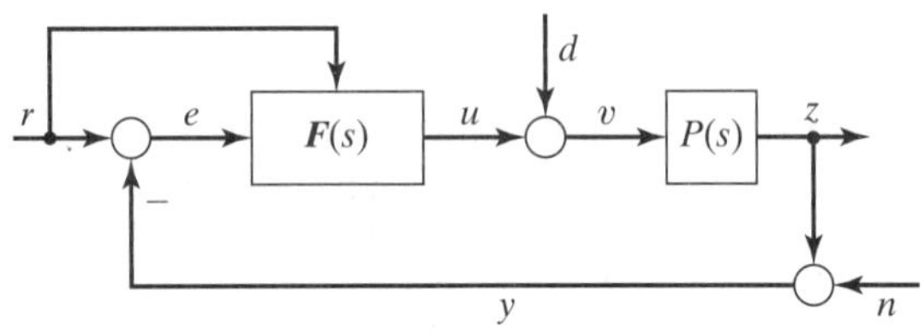

FIGURE 2.34: Feedback plus feedforward system.

Hence

$$U(s) = \begin{bmatrix} F_1(s) & F_2(s) \end{bmatrix} \begin{bmatrix} R(s) \\ E(s) \end{bmatrix}.$$

This system has a feedback loop consisting of $F_2(s)$ and $P(s)$ and a feedforward path with transfer function $F_1(s)$. It also has an algebraic equivalence as shown in Figure 2.35. It can be easily seen that if

$$F_1(s) = C_1(s) - C_2(s) \quad \text{and} \quad F_2(s) = C_2(s),$$

then it is equivalent to the 2DOF controller.

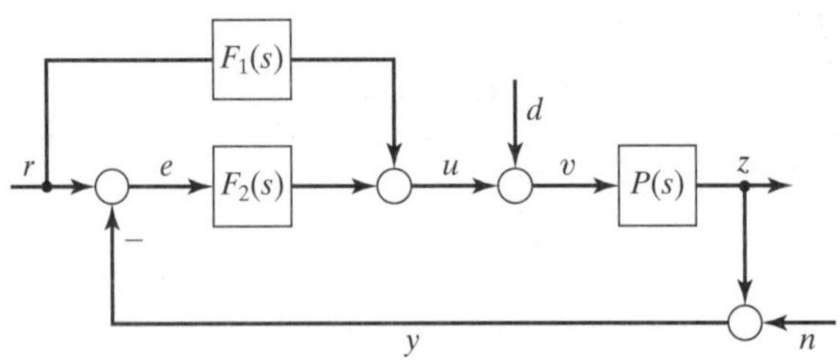

FIGURE 2.35: Algebraic equivalence of Figure 2.34.

Yet another equivalent form of the 2DOF controller is the observer-based feedback system shown in Figure 2.36. It is so called since the controller $\boldsymbol{O}(s)$ takes both the input and the output of the plant and has the capability to observe the internal behavior of the plant from external signals. It can be seen that if

$$\boldsymbol{O}(s) = \begin{bmatrix} O_1(s) & O_2(s) \end{bmatrix} = \begin{bmatrix} \dfrac{1 - C_1(s)}{C_1(s)} & \dfrac{C_2(s)}{C_1(s)} \end{bmatrix},$$

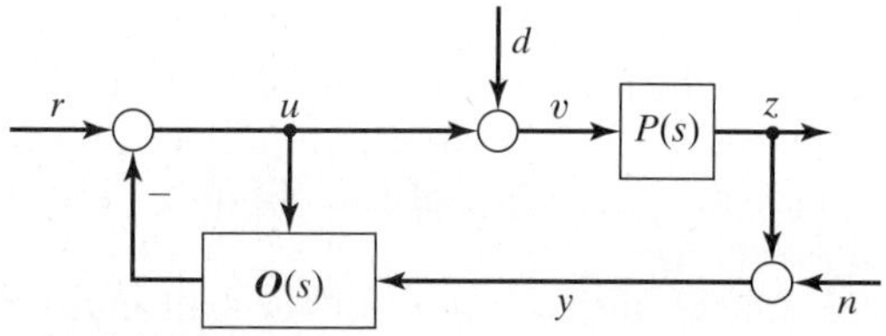

FIGURE 2.36: Observer based feedback system.

then it is equivalent to the 2DOF controller.

A widely used special case of the regulation system shown in Figure 2.30 is the unity feedback system shown in Figure 2.37, which is obtained from Figure 2.30 by setting $C_1(s) = C_2(s) = C(s)$.

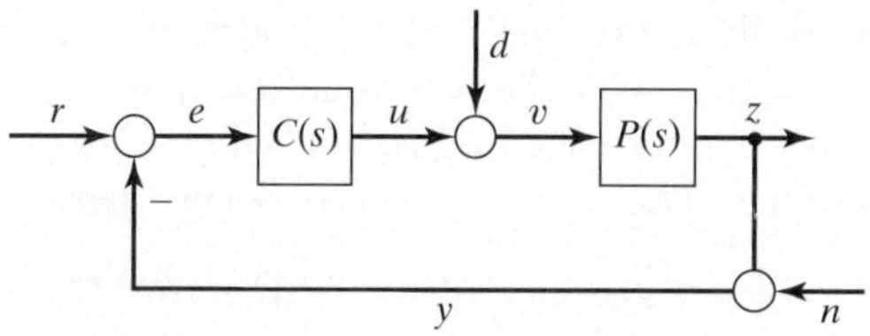

FIGURE 2.37: Unity feedback system.

It is easy to see that this system is well posed if and only if the system for stabilization in Figure 2.29 is well posed.

2.10 CASE STUDIES

In this section we take a close look at two mechanical systems that we often see in laboratories. The first is a ball and beam system. The second is an inverted pendulum.

2.10.1 Ball and beam system

The ball and beam system consists of a ball rolling along a tilting rail, as shown in Figure 2.38. The rail (also called the beam) is made up of two parts: a steel rod and a resistance bar. When a steel ball rolls on these two components, it acts as a wiper similar to that of a potentiometer. Voltage is applied on both ends of the resistance bar, and thus the distance of the ball is found by measuring the voltage

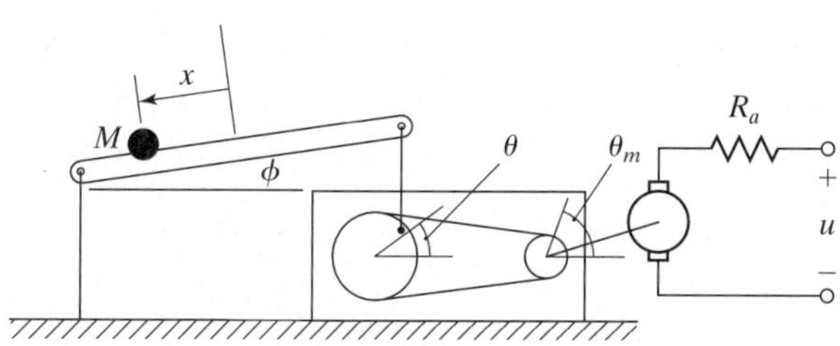

FIGURE 2.38: A ball and beam system.

on the steel bar. A terminal block is fixed on one end of the beam as support. A mechanical arm which translates the rotatory motion of a big disk into vertical motion is on the other end of the beam. Thus, the beam is pivoted around the support by controlling the angle of the big disk. The big disk is then driven by a geared motor assembly.

A precise mathematical model of the ball and beam system is complicated. Fortunately, approximate models are sufficient for feedback control. Let us derive a simple approximate model of the ball and beam system. Roughly, the ball and beam model can be decomposed into four parts: the ball and beam model, the angle transfer mechanism model, the gear model, and the motor model. The ball and beam model relates the position $x(t)$ of the ball with the tilt angle $\phi(t)$ of the beam, whereas the angle transfer mechanism model relates the tilt angle $\phi(t)$ with the rotation angle $\theta(t)$ of the big disc. Finally, the motor in the ball and beam system is internally controlled and its model relates the voltage input $u(t)$ to the motor control system with the rotational angle $\theta_m(t)$.

Let us approximate the ball rolling on the beam by a point mass sliding in a frictionless surface. Then Newton's second law gives

$$M\ddot{x}(t) = Mg \sin \phi(t)$$

where M is the mass of the ball and g is the acceleration of gravity. Assume that the tilt angle $\phi(t)$ is small. Then we get the linearized equation

$$\ddot{x}(t) = g\phi(t).$$

Taking the Laplace transform we get

$$\frac{X(s)}{\Phi(s)} = \frac{g}{s^2}. \tag{2.31}$$

The relationship between the tilt angle $\phi(t)$ and the rotational angle $\theta(t)$ of the big disk is nonlinear but static. It can be approximated by

$$\frac{\Phi(s)}{\Theta(s)} = \frac{\phi(t)}{\theta(t)} = \frac{R}{L}, \tag{2.32}$$

where R is the radius of the motor disk and L is the length of the beam.

The angular displacement $\theta(t)$ of the big disk and that of the rotor of the motor, denoted by $\theta_m(t)$, is related by the gear ratio K_g as follows

$$\frac{\Theta(s)}{\Theta_m(s)} = K_g.$$

Finally, the motor transfer function is taken as an approximation of the one given in Example 2.12 by assuming $L_a \approx 0$ and $K_b \approx 0$:

$$\frac{\Theta_m(s)}{U(s)} = \frac{K_t}{R_a s(Js + K_f)}.$$

Notice that the motor dynamics and the ball dynamics are actually coupled since the load inertia also depends on the ball location and the ball motion depends on the motor velocity. However, the above approximation is good enough for the control purpose.

As a result, the whole ball and beam system can be approximated by the cascade connection of four systems: the ball and beam part, the angle transfer part, the gear part, and the motor part, as shown in Figure 2.39.

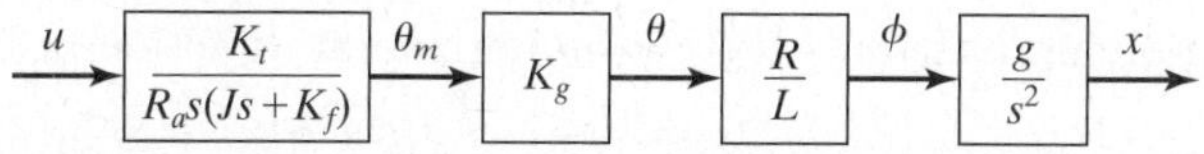

FIGURE 2.39: Block diagram of the ball and beam system.

The ball and beam system has two measurements: the disc angle $\theta(t)$ and the ball position $x(t)$. Therefore, the whole system is SIDO and is given by the input/output relation

$$\begin{bmatrix} \Theta(s) \\ X(s) \end{bmatrix} = \frac{1}{R_a s^3 (Js + K_f)} \begin{bmatrix} K_t K_g s^2 \\ K_t K_g R g / L \end{bmatrix} U(s).$$

In a particular case, we have $R_a = 10\,[\Omega]$, $J = 0.75 \times 10^{-3}$ [Nm sec^2/rad], $L = 0.4$ [m], $R = 0.04$ [m], $K_t = 7.5 \times 10^{-3}$[Nm/A], $K_f = 37.5 \times 10^{-3}$ [Nm sec/rad], and $K_g = 1/75$. Plugging in all these parameters, we get the input/output relation of a typical ball and beam system

$$\begin{bmatrix} \Theta(s) \\ X(s) \end{bmatrix} = \frac{1}{s^3(s+50)} \begin{bmatrix} s^2/75 \\ g/750 \end{bmatrix} U(s)$$

where g is equal to 9.8 [m/sec^2].

2.10.2 Inverted pendulum system

The inverted pendulum mimics a game we often played in our childhood: balancing a long stick upward on our finger tip. As shown in Figure 2.40, an inverted pendulum system consists of a moving cart with a rod mounted on the top. Unlike the childhood game where our fingers move in a horizontal plane and the stick can fall in all directions, the cart moves linearly along a straight rail and the rod can only fall either to the front or to the back of the cart.

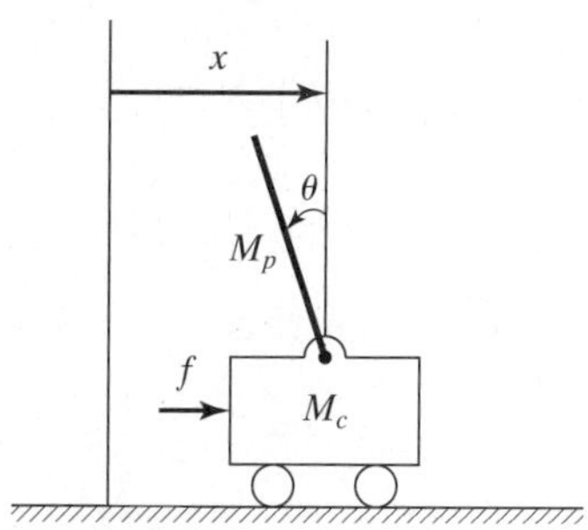

FIGURE 2.40: Inverted pendulum system.

The system consists of a cart and a rod. The cart, with a mass M_c, slides on a stainless steel shaft and is equipped with a motor. A rod, with an evenly distributed mass M_p and a length L, is mounted on the cart whose axis of rotation is perpendicular to the direction of the motion of the cart. The cart position $x(t)$ and the pendulum angle $\theta(t)$ can be measured. The input is the force $f(t)$ applied on the cart.

Applying Newton's law to the horizontal direction of Figure 2.40, we get

$$f(t) = M_c \frac{d^2 x(t)}{dt^2} + M_p \frac{d^2[x(t) - L/2 \sin\theta(t)]}{dt^2}.$$

This gives

$$(M_p + M_c)\ddot{x}(t) - \frac{M_p L}{2}\ddot{\theta}(t)\cos\theta(t) + \frac{M_p L}{2}\dot{\theta}^2(t)\sin\theta(t) = f(t). \tag{2.33}$$

Note that the moment of inertia of the rod with respect to the pivot point can be computed as

$$J = \int_0^L \ell^2 \left(\frac{M_p}{L} d\ell\right) = \frac{1}{3} M_p L^2.$$

Writing the rotational version of Newton's second law for the pendulum around the pivot point, we obtain

$$J\ddot{\theta}(t) = M_p \ddot{x}(t)\cos\theta(t)\frac{L}{2} + M_p g \sin\theta(t)\frac{L}{2}.$$

This gives

$$\frac{2}{3}L\ddot{\theta}(t) - \ddot{x}(t)\cos\theta(t) - g\sin\theta(t) = 0. \tag{2.34}$$

Therefore the differential equation model of the system is given by

$$(M_p + M_c)\ddot{x}(t) - \frac{1}{2}M_p L\ddot{\theta}(t)\cos\theta(t) + \frac{1}{2}M_p L\dot{\theta}^2(t)\sin\theta(t) = f(t)$$

$$-\ddot{x}(t)\cos\theta(t) + \frac{2}{3}L\ddot{\theta}(t) - g\sin\theta(t) = 0.$$

This is a highly nonlinear system. A straightforward way to linearize it around $x(t) = 0, \dot{x}(t) = 0, \theta(t) = 0, \dot{\theta}(t) = 0$ is to remove the second- and higher-order terms. This gives

$$(M_p + M_c)\ddot{x}(t) - \frac{1}{2}M_p L\ddot{\theta}(t) = f(t)$$

$$-\ddot{x}(t) + \frac{2}{3}L\ddot{\theta}(t) - g\theta(t) = 0.$$

Taking Laplace transform on both sides of the equations, we get

$$(M_p + M_c)s^2 X(s) - \frac{1}{2}M_p L s^2 \Theta(s) = F(s)$$

$$-s^2 X(s) + \frac{2}{3}L s^2 \Theta(s) - g\Theta(s) = 0.$$

Taking $X(s)$ and $\Theta(s)$ as the outputs, and $F(s)$ as the input, we get

$$\begin{bmatrix} X(s) \\ \Theta(s) \end{bmatrix} = \frac{1}{(M_p + M_c)s^2(2Ls^2/3 - g) - M_pLs^4/2} \begin{bmatrix} 2Ls^2/3 - g \\ s^2 \end{bmatrix} F(s)$$

$$= \frac{1}{s^2[(M_p + 4M_c)Ls^2 - 6(M_p + M_c)g]} \begin{bmatrix} 4Ls^2 - 6g \\ 6s^2 \end{bmatrix} F(s).$$

Therefore, the vector transfer function of the system is

$$\boldsymbol{P}(s) = \begin{bmatrix} P_x(s) \\ P_\theta(s) \end{bmatrix} = \frac{1}{s^2[(M_p + 4M_c)Ls^2 - 6(M_p + M_c)g]} \begin{bmatrix} 4Ls^2 - 6g \\ 6s^2 \end{bmatrix}.$$

As an example, let us have a virtual inverted pendulum with $M_c = 1\,[\text{kg}], M_p = 2\,[\text{kg}], L = 1\,[\text{m}]$. This gives

$$\boldsymbol{P}(s) = \frac{1}{s^2(s^2 - 3g)} \begin{bmatrix} 2/3s^2 - g \\ s^2 \end{bmatrix} \qquad (2.35)$$

where $g = 9.8\,[\text{m/sec}^2]$.

PROBLEMS

2.1. Consider the circuit shown in Figure 2.41. Let the input be $v_i(t)$ and output be $v_o(t)$. Obtain a state space model and the transfer function of this system.

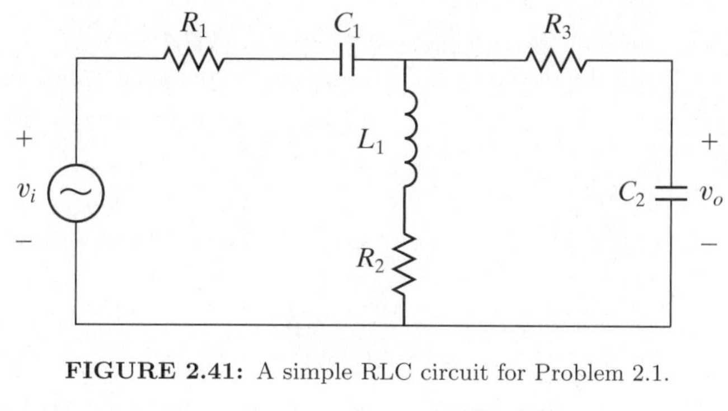

FIGURE 2.41: A simple RLC circuit for Problem 2.1.

2.2. Consider the mechanical system shown in Figure 2.42. Here $u(t)$ is an external force applied to the mass M, $y(t)$ is the displacement of the mass with respect

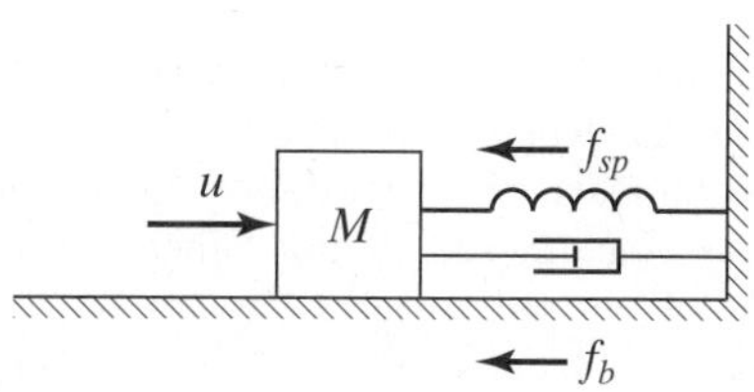

FIGURE 2.42: A mass and spring system for Problem 2.2.

to the position when the spring is relaxed. The spring force and friction force are given respectively by

$$f_{sp}(t) = k(1 + ay^2(t))y(t), \qquad f_b(t) = b\dot{y}(t).$$

1. Write the differential equation model of this system.
2. Write a state space description of the system.
3. Is the system linear? If it is not linear, linearize it around the operating point with $u_0 = 0$.
4. Find the transfer function of the linearized system.

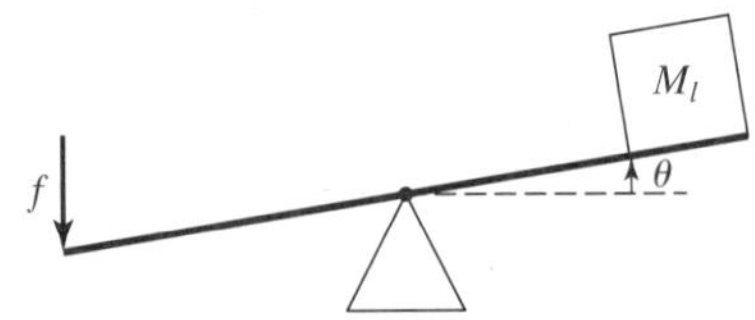

FIGURE 2.43: A see-saw system for Problem 2.3.

2.3. Consider the see-saw system shown in Figure 2.43. The beam has length L with an evenly distributed mass M_b. A mass M_l sits on one end. A vertically downward force $f(t)$ is applied on the other end. We are concerned with the angular displacement $\theta(t)$ of the beam with the horizontal line.

1. Write the differential equation model of this system with input $f(t)$ and output $\theta(t)$.
2. Is this system linear? If not, linearize it about the operating point with $\theta_0 = 0$.
3. Obtain the transfer function from $\Delta f(t)$ to $\Delta\theta(t)$, where $\Delta f(t)$ and $\Delta\theta(t)$ are the deviations of $f(t)$ and $\theta(t)$ from their values at the operating point.

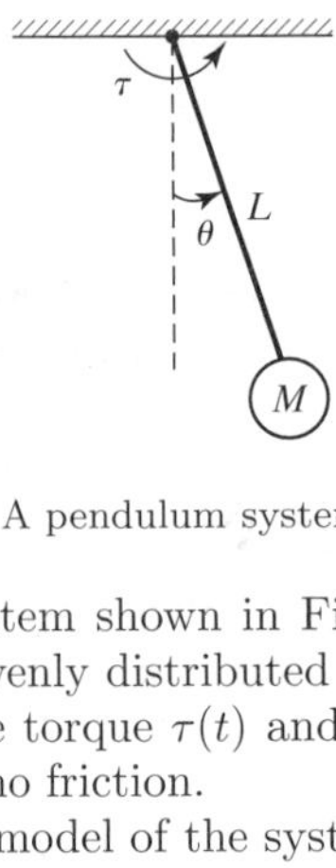

FIGURE 2.44: A pendulum system for Problem 2.4.

2.4. Consider the pendulum system shown in Figure 2.44. The pendulum consists of a rod of length L with evenly distributed mass m and a point mass M at its lower end. The input is the torque $\tau(t)$ and the output is the angular position $\theta(t)$. Assume that there is no friction.

a. Derive the state space model of the system.
b. Is the system linear? If not, linearize it around the operating point with $\theta_0 = 0$.
c. Compute the transfer function of the linearized model.

2.5. Consider the system shown in Figure 2.45. ϕ is a continuously differentiable nonlinear function satisfying $\phi(0) = 0$. Write a state space equation of the

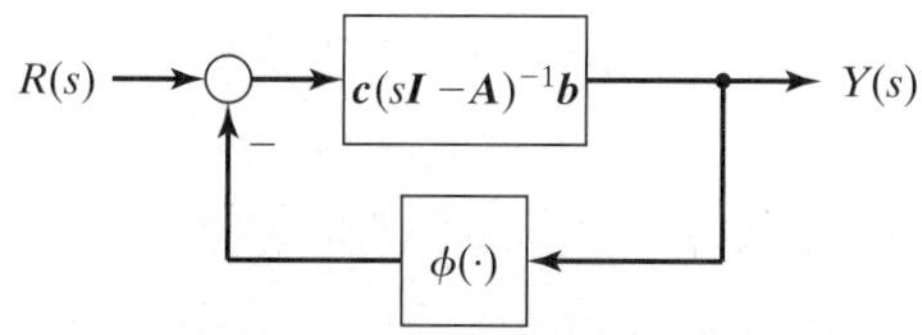

FIGURE 2.45: Linear system with nonlinear memoryless feedback for Problem 2.5.

closed-loop system. Linearize it around an equilibrium point with zero output. Write the transfer function of the linearized closed-loop system.

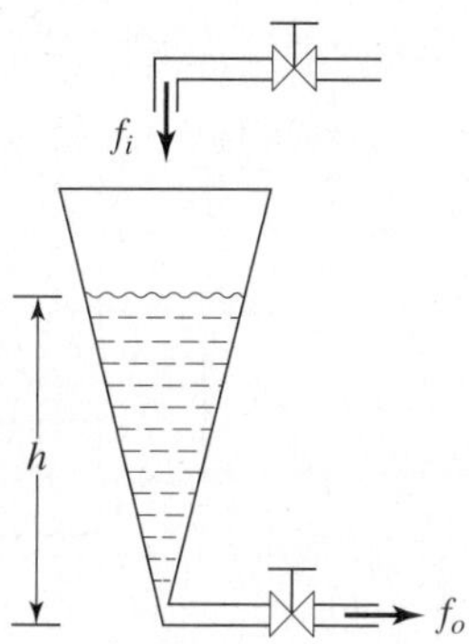

FIGURE 2.46: A conical water tank for Problem 2.6.

2.6. Consider a water tank in the shape of an ice cream cone shown in Figure 2.46. The height is 4 [m] and the top diameter is 2 [m]. Assume that the input is the inflow $f_i(t)$ and the output is the water level $h(t)$. It is also known that the outflow is $f_o(t) = \sqrt{3h(t)}$.

1. Choose a state variable of the system and write a state space model of the system.
2. Is the system linear? If not, linearize it around the equilibrium point with $h_0 = 3$.
3. Find the transfer function of the linearized system.

2.7. Consider a ball shaped water tank with radius R shown in Figure 2.47. Assume that the input is the inflow $f_i(t)$ and the output is the water level $h(t)$. It is also known that the outflow is $f_o(t) = \alpha$.

1. Write a state space model of the system.
2. Is the system linear? If not, linearize it when the tank is half full.
3. Find the transfer function of the linearized system.
4. What are the poles and zeros of the system?

2.8. Consider Figure 2.48. Assume that a state space model of $P(s)$ is

$$\dot{\boldsymbol{x}}(t) = \boldsymbol{A}\boldsymbol{x}(t) + \boldsymbol{b}u(t)$$

$$y(t) = \boldsymbol{c}\boldsymbol{x}(t)$$

and K is a pure gain. Obtain a state space model of the closed-loop system with input $r(t)$ and output $y(t)$.

2.9. Find the closed-loop transfer function of the system shown in Figure 2.49. Choose $F(s)$ such that the closed-loop transfer function is 1.

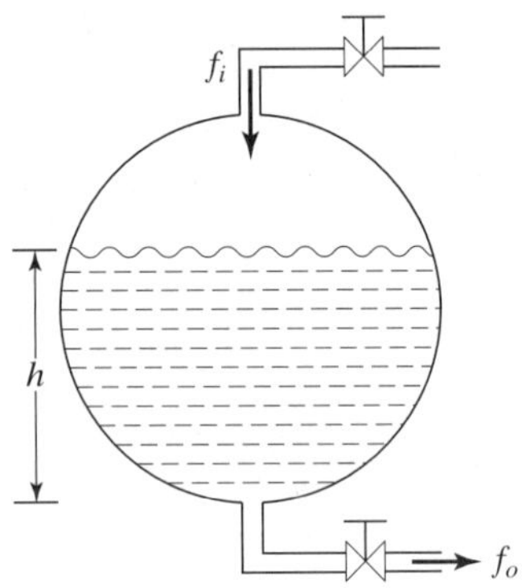

FIGURE 2.47: A ball shaped water tank for Problem 2.7.

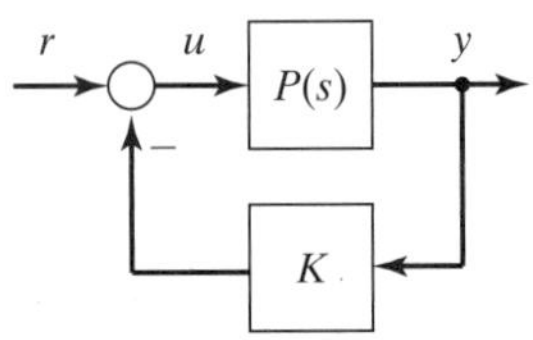

FIGURE 2.48: Unity feedback system with a proportional feedback for Problem 2.8.

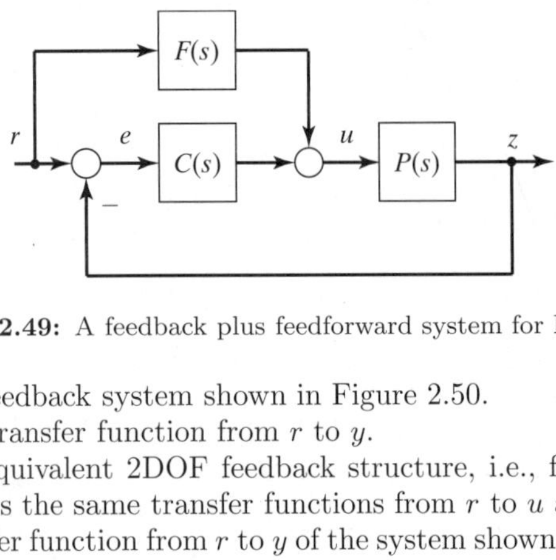

FIGURE 2.49: A feedback plus feedforward system for Problem 2.9.

2.10. Consider the feedback system shown in Figure 2.50.

1. Find the transfer function from r to y.
2. Find its equivalent 2DOF feedback structure, i.e., find a 2DOF controller which gives the same transfer functions from r to u and from y to u.

2.11. Find the transfer function from r to y of the system shown in Figure 2.51. Choose $H(s)$ so that the transfer function is equal to 1.

2.12. Find the transfer function from r to y of the system shown in Figure 2.52.

MATLAB PROBLEMS

2.13. In MATLAB, there is a third form to describe a SISO LTI system in addition to the state space form and the transfer function form. This third form is called the zero-pole-gain form. The command to create or convert to a system described by the zero-pole-gain form is `zpk`. Learn the use of the command `zpk` and the related command `zpkdata` from the online help by using the `help` command.

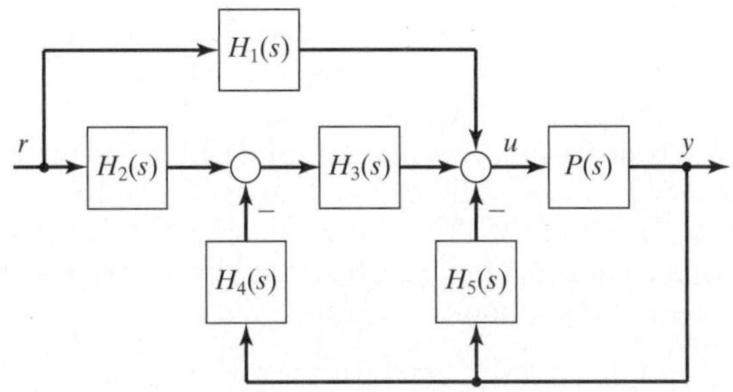

FIGURE 2.50: A complicated feedback system for Problem 2.10.

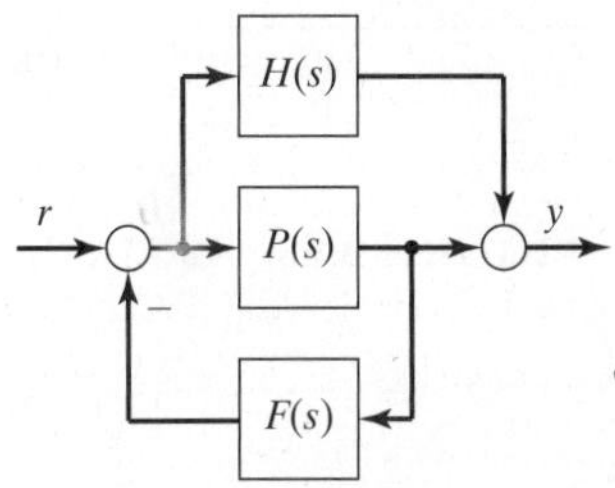

FIGURE 2.51: A partial feedback system for Problem 2.11.

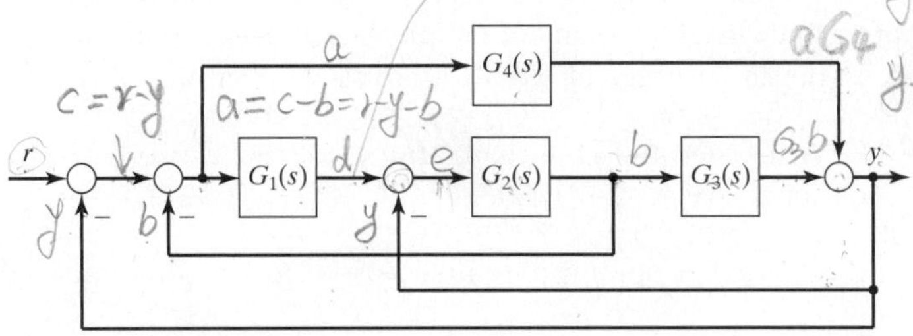

FIGURE 2.52: A complicated feedback system for Problem 2.12.

2.14. Use MATLAB to solve this problem. A flexible beam system is described by the following state space equation

$$\dot{\boldsymbol{x}}(t) = \begin{bmatrix} -1.7860 & 15.9674 & -1.6930 & 13.1487 \\ -1.1773 & -15.2644 & -1.3548 & -14.5145 \\ 3.3427 & -6.8798 & 3.4513 & -3.5889 \\ 1.5017 & 16.7058 & 1.1717 & 13.4250 \end{bmatrix} \boldsymbol{x}(t) + \begin{bmatrix} -0.9618 \\ 1.3011 \\ 0.1081 \\ -1.1984 \end{bmatrix} u(t)$$

$$y(t) = \begin{bmatrix} 29.8608 & 35.9450 & 29.4689 & 17.7162 \end{bmatrix} \boldsymbol{x}(t).$$

1. Form a system variable representing this system. (Use ss.)
2. Find the transfer function of the system. (Use tf.)
3. Find the poles, zeros, and the DC gain of the system. (Use zpk.)
4. Write a state space equation of the system from its transfer function using the formulas in Section 2.7.

5. Obtain a state space equation of the system from its transfer function by using ss. Compare the result with the original state space equation and the result of part 4. Are they the same? Explain why.

2.15. Build a transfer function block in SIMULINK to implement a MISO system. Use your new block to build the system shown in Figure 2.53(a) and also build the system in Figure 2.53(b) using the existing blocks in SIMULINK. Simulate the step responses of two systems and compare. Pay attention to the signals before the summing point in Figure 2.53(b).

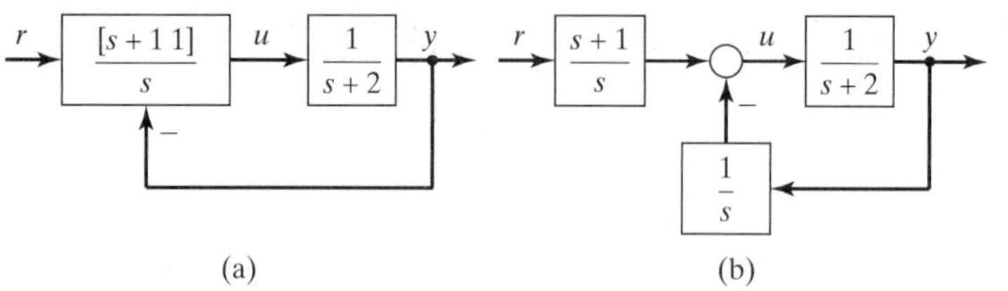

FIGURE 2.53: For MATLAB Problem 2.15.

NOTES AND REFERENCES

Modeling of a physical system usually involves specialized knowledge in the areas where the system falls in. For simple systems, such as the ones discussed in this chapter, knowledge on elementary physics is usually sufficient.
More in-depth coverage of circuit analysis is given in

C. A. Desoer and E. S. Kuh, *Basic Circuit Theory*, McGraw-Hill International, Auckland, 1969.

For the analysis of op-amp circuits, as well as detailed knowledge on DC motors, see

R. J. Smith and R. C. Dorf, *Circuits, Devices, and Systems*, 5th Edition, John Wiley & Sons, New York, 1992.

A good reference on mechanical system modeling is

A. Bedford and W. Fowler, *Engineering Mechanics*, 3rd Edition, Prentice Hall, Upper Saddle River, NewJersey, 2002.

The use of the gang of four for the four transfer functions in (2.29) first appeared in

K. J. Åström and R. Murray, *Feedback Systems: An Introduction for Scientists and Engineers*, Princeton University Press, Princeton and Oxford, 2008.

CHAPTER

3

Stability and Stabilization

A fundamental requirement of a feedback system is to ensure that all signals in the system stay well behaved when all input signals (including external disturbances, references, noises, etc.) are well behaved. In this chapter, we shall study how to address this requirement by introducing some stability concepts. In particular, we shall define the concept of bounded input and bounded output (BIBO) stability and give several well-known stability criteria such as the Routh stability criterion and the Routh–Hurwitz stability criterion. We shall also consider the robust stability of a special type of system characterized by interval polynomials using the Kharitonov theorem. The internal stability concept for feedback closed-loop systems, both for stabilization and for regulation, is then introduced and characterized. In the second half of the chapter, the issue of designing stabilizing controllers is considered. We will see how to design a stabilizing controller using the pole placement technique, which is to design a controller $C(s)$ for the given plant $P(s)$ so that the closed-loop system has the desired poles. This can be easily done by solving a polynomial Diophantine equation, which is essentially a matrix linear equation. In this chapter, we will also address a theoretically extremely important problem: the characterization of all stabilizing controllers. This problem seeks a way to express the set of all stabilizing controllers for a given plant in a simple way so that all the design freedom is explicitly exhibited. This greatly facilitates the determination of a particular stabilizing controller in the practical design. The solution to this problem is given by a formula parameterizing the set of all stabilizing controllers for a given plant, which was developed in the 1970s and is now considered as one of the most important milestones of the postmodern control theory.

3.1 CONCEPT OF STABILITY

The idea of system stability is to make sure that when its input is a well-behaved signal its output is also a well-behaved signal. There are several possible ways to define the stability of a system. We will use one of the weakest and the simplest definitions. A signal $x(t)$ is said to be bounded if there is a positive real number M such that

$$|x(t)| \leq M$$

for all $t \in [0, \infty)$. In this case, M is said to be a bound of signal $x(t)$. If a signal is bounded, then its bounds are not unique. Very often we are interested in the smallest such bound, called the **peak magnitude** or **amplitude of the signal**, which will be denoted by[1]

$$\|x(t)\|_\infty = \sup_{t \in [0,\infty)} |x(t)|.$$

Now consider the control system shown in Figure 3.1, where $u(t)$ is the input, $y(t)$ is the output, and $G(s)$ is the transfer function. We assume that the transfer function $G(s)$ is a rational function.

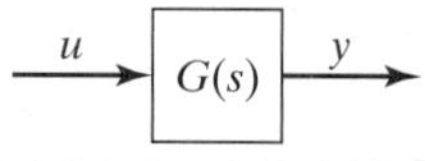

FIGURE 3.1: A control system.

Definition 3.1. A system is said to be **bounded-input–bounded-output (BIBO) stable** if for each bounded input the corresponding output is bounded.

A BIBO stable system guarantees that the output generated by a bounded input is always bounded. Since we will only deal with BIBO stability in most parts of the book, we will omit the word BIBO when talking about stability. Directly from the definition we see that to show that a system is unstable we only need to find a bounded input that produces an unbounded output. On the other hand, it is much harder to show that a system is stable since it is not enough to show that the outputs are bounded for a few bounded test inputs. Instead, we have to establish, analytically and logically, that the output is bounded for every arbitrary bounded input.

[1]There is a reason for us to use sup instead of max here. For example, 1 is the smallest upper bound for $x(t) = 1 - e^{-t}, t \geq 0$ but this bound is never achieved with any finite $t \geq 0$. Thus $\sup_{t\in[0,\infty)} |x(t)| = 1$ but $\max_{t\in[0,\infty)} |x(t)|$ does not exist.

EXAMPLE 3.2

We wish to know if a system with transfer function $G(s) = \dfrac{1}{s+\gamma}$ is stable.

If $\gamma < 0$, let $u(t) = \sigma(t)$, the unit step function, which is obviously bounded. Then,

$$y(t) = \mathcal{L}^{-1}\left[\frac{1}{s\,(s+\gamma)}\right] = \mathcal{L}^{-1}\left[\frac{1}{\gamma}\left(\frac{1}{s} - \frac{1}{s+\gamma}\right)\right] = \frac{1}{\gamma}(1 - e^{-\gamma t})\sigma(t),$$

which goes to ∞ when t goes to ∞, and hence is unbounded. This shows that the system is unstable.

If $\gamma = 0$, the system is called an **integrator**. Again let $u(t) = \sigma(t)$. Then,

$$y(t) = \mathcal{L}^{-1}\left[\frac{1}{s^2}\right] = t\sigma(t),$$

which goes to ∞ as t goes to ∞. This shows that the system is again unstable.

If $\gamma > 0$, the impulse response of $G(s)$ is $g(t) = e^{-\gamma t}\sigma(t)$. Then,

$$|y(t)| = \left|\int_0^t g(\tau)u(t-\tau)d\tau\right| \le \int_0^t |g(\tau)||u(t-\tau)|d\tau.$$

If $|u(t)| \le M$ for all $t \in [0, \infty)$, then

$$|y(t)| \le M\int_0^\infty |g(\tau)|d\tau = M\int_0^\infty e^{-\gamma\tau}d\tau = \frac{M}{\gamma}.$$

Hence, as long as $u(t)$ is bounded by M, then $y(t)$ is bounded by M/γ. This shows that the system is stable.

In summary, a system with transfer function $G(s) = \dfrac{1}{s+\gamma}$ is stable if $\gamma > 0$ and is unstable if $\gamma \le 0$.

Using the procedure in the example to test the stability of a system is often tedious. We desire to have a simple stability test.

Theorem 3.3. An LTI system with impulse response function $g(t)$ is stable if and only if

$$\int_0^\infty |g(t)|dt < \infty. \tag{3.1}$$

Proof. Assume that the input $u(t)$ is bounded by M; then

$$|y(t)| = \left|\int_0^t g(\tau)u(t-\tau)d\tau\right| \le \int_0^t |g(\tau)||u(t-\tau)|d\tau \le M\int_0^\infty |g(\tau)|d\tau.$$

If (3.1) is satisfied, choose $N = M\int_0^\infty |g(\tau)|d\tau$, and then the input bounded by M generates the output bounded by N; i.e., the system is stable. This shows the sufficiency.

If (3.1) is not satisfied, $\int_0^\infty |g(t)|dt = \infty$. For any $M > 0$, choose t_0 so that

$$\int_0^{t_0} |g(\tau)|d\tau > M$$

and choose

$$u(t) = \begin{cases} 1, & \text{if } g(t_0 - t) \geq 0; \\ -1, & \text{if } g(t_0 - t) < 0 \end{cases}$$

for $0 \leq t \leq t_0$. Then,

$$|y(t_0)| = \left| \int_0^{t_0} g(t_0 - \tau)u(\tau)d\tau \right| = \int_0^{t_0} |g(t_0 - \tau)|d\tau = \int_0^{t_0} |g(\tau)|d\tau > M.$$

This shows that for an input $u(t)$ bounded by 1, the output can be arbitrarily large. Hence, the system is not stable. This proves the necessity. □

A signal is said to be absolutely integrable over an interval if the integral of the absolute value of the signal over the interval is finite. Hence, a linear system is stable if its impulse response is absolutely integrable over $[0, \infty)$. This theorem makes it a bit easier to check the stability of a system.

EXAMPLE 3.4

Let us again consider $G(s) = \dfrac{1}{s + \gamma}$. The impulse response of the system is

$$g(t) = e^{-\gamma t}\sigma(t).$$

This gives

$$\int_0^\infty |g(t)|dt = \int_0^\infty e^{-\gamma t}dt = \begin{cases} \dfrac{1}{\gamma} & \text{if } \gamma > 0 \\ \infty & \text{if } \gamma \leq 0. \end{cases}$$

Therefore, the system is stable if $\gamma > 0$ and unstable if $\gamma \leq 0$. This conclusion agrees with that in Example 3.2.

However, it is still not a very easy task to compute the integral of the absolute value of a function. The following theorem further simplifies the task of checking the stability of an LTI system.

Theorem 3.5. A system with transfer function $G(s)$ is stable if and only if $G(s)$ is proper and all poles of $G(s)$ have negative real parts.

Proof. If the system is stable, then

$$\int_0^\infty |g(t)|dt < \infty.$$

This implies that for all s with $\text{Re}(s) \geq 0$,

$$|G(s)| = \left| \int_0^\infty g(t)e^{-st}dt \right| \leq \int_0^\infty |g(t)|dt < \infty,$$

i.e., $G(s)$ is bounded in the right half of the complex plane. Therefore $G(s)$ has to be proper and free of poles in the right half of the complex plane.

If $G(s)$ is proper and all poles of $G(s)$ have negative real parts, then $G(s)$ has the following partial fractional expansion:

$$G(s) = \sum_i G_i(s)$$

where $G_i(s)$ takes three possible forms:

$$G_i(s) = a \quad \text{or} \quad \frac{b}{(s+\gamma)^m} \quad \text{or} \quad \frac{q(s)}{[(s+\rho)^2+\omega^2]^l}$$

where $\gamma > 0$, $\rho > 0$, and $q(s)$ is an lth order polynomial of s. Hence,

$$g_i(t) = \mathcal{L}^{-1}[G_i(s)] = a\delta(t) \text{ or } e^{-\gamma t}p(t) \text{ or } e^{-\rho t}[p_1(t)\cos\omega t + p_2(t)\sin\omega t]$$

where $p(t), p_1(t), p_2(t)$ are polynomials of t. Since for each of these possibilities

$$\int_0^\infty |g_i(t)|dt < \infty$$

following the fact that the exponential functions $e^{-\gamma t}$ and $e^{-\rho t}$ converge to zero much faster than the speed in which $p(t), p_1(t), p_2(t)$ go to infinity, it follows that

$$\int_0^\infty |g(t)|dt \leq \sum_i \int_0^\infty |g_i(t)|dt < \infty.$$

This shows that the system is stable. □

The requirement on the properness of the transfer function for the stability of a system is important. Because of this, an ideal differentiator, whose transfer function is s, is unstable. A system with transfer function

$$G(s) = K_P + K_I\frac{1}{s} + K_D s,$$

which is known as a **proportional-integral-derivative (PID)** device, is also unstable. Any other system with a nonproper transfer function is unstable. However, a nonproper system is only a mathematical abstraction. Practically all physical systems, no matter whether they are natural or manmade, are causal and hence cannot have nonproper transfer functions. This is because systems with transfer functions like s or s^2 (**ideal differentiator** or **ideal double differentiator**) are noncausal operations. Nonproper transfer functions practically represent systems with very fast dynamics. For example, a practical differentiator we often use has a transfer function $\dfrac{s}{Ts+1}$ with a very small positive T. Therefore, a practical differentiator is stable.

Because of Theorems 3.3 and 3.5, we can talk about the stability of an impulse response and that of a transfer function. Naturally, an impulse response is said to be stable if it is the impulse response of a stable system, i.e., it is integrable; a transfer function is said to be stable if it is the transfer function of a stable system, i.e., it is proper and all its poles have negative real parts. In other words, the stability of an LTI system, the stability of its impulse response $g(t)$, and the stability of its transfer function $G(s)$ are all equivalent concepts. We will refer to these concepts interchangeably since no confusion is likely to arise.

Since the poles of a transfer function $G(s)$ are the roots of its denominator polynomial, with Theorem 3.5 the determination of the stability of an LTI system is essentially equivalent to checking the root positions of a polynomial. Consequently, it is also meaningful to talk about the stability of a polynomial.

Definition 3.6. A polynomial is said to be stable if all of its roots have negative real parts.

3.2 ROUTH CRITERION

To determine the stability of a polynomial, one can simply compute the roots of the polynomial using, for example, MATLAB. However, when some coefficients are functions of parameters, computing the roots for all possible parameter variations becomes difficult or impossible. Nevertheless, it is possible to determine the stability of a polynomial without computing the roots.

Consider a polynomial

$$a(s) = a_0 s^n + a_1 s^{n-1} + \cdots + a_n, \qquad a_0 > 0. \tag{3.2}$$

It can always be factored as

$$a(s) = a_0(s+\gamma_1)(s+\gamma_2)\cdots(s+\gamma_m)[(s+\rho_1)^2+\omega_1^2]\cdots[(s+\rho_l)^2+\omega_l^2]$$

where γ_i, $i = 1, \ldots, m$, and ρ_i, ω_i, $i = 1, \ldots, l$, are all real. Notice that $-\gamma_i$ are the real roots of $a(s)$ and $-\rho_i \pm j\omega_i$ are the complex pairs of roots of $a(s)$. If $a(s)$ is stable, then γ_i, $i = 1, \ldots, m$, and ρ_i, $i = 1, \ldots, l$, are positive. Since a_i, $i = 1, \ldots, n$, are sums of products of $\gamma_i, \rho_i, \omega_i^2$, they must be positive. This leads to the following theorem.

Theorem 3.7. If $a(s)$ is stable, then $a_i > 0$, $i = 1, \ldots, n$.

Theorem 3.7 gives a necessary condition for stability, which can be used for the initial stability screening. If one of the coefficients is zero or negative, then the polynomial is unstable. However, the condition is by no means sufficient, except for first-order and second-order polynomials. If all coefficients are positive, no immediate conclusion can be reached, in general.

EXAMPLE 3.8

Consider a polynomial

$$a(s) = s^3 + s^2 + 4s + 30 = (s+3)(s^2 - 2s + 10).$$

This polynomial has all positive coefficients but it is not stable. Its roots are -3 and $1 \pm 3j$.

In this section, we present a method to determine the stability of a polynomial without explicitly solving for its roots. First, for polynomial (3.2) let us construct the so-called **Routh table** as shown in Table 3.1.

s^n	$r_{00} = a_0$	$r_{01} = a_2$	$r_{02} = a_4$	$r_{03} = a_6$	$\cdots$
s^{n-1}	$r_{10} = a_1$	$r_{11} = a_3$	$r_{12} = a_5$	$r_{13} = a_7$	$\cdots$
s^{n-2}	r_{20}	r_{21}	r_{22}	r_{23}	$\cdots$
s^{n-3}	r_{30}	r_{31}	r_{32}	r_{33}	$\cdots$
$\vdots$	$\vdots$	$\vdots$	$\vdots$	$\vdots$	
s^2	$r_{(n-2)0}$	$r_{(n-2)1}$			
s^1	$r_{(n-1)0}$				
s^0	r_{n0}				

TABLE 3.1: Routh table.

The first two rows come directly from the coefficients of $a(s)$. Each of the other rows is computed from its two preceding rows as

$$r_{ij} = -\frac{1}{r_{(i-1)0}} \det \begin{bmatrix} r_{(i-2)0} & r_{(i-2)(j+1)} \\ r_{(i-1)0} & r_{(i-1)(j+1)} \end{bmatrix} = \frac{r_{(i-1)0}r_{(i-2)(j+1)} - r_{(i-2)0}r_{(i-1)(j+1)}}{r_{(i-1)0}}.$$

Here, i goes from 2 to n and j goes from 0 to $\lfloor \frac{n-i}{2} \rfloor$. For example,

$$\begin{aligned} r_{20} &= -\frac{1}{r_{10}} \det \begin{bmatrix} r_{00} & r_{01} \\ r_{10} & r_{11} \end{bmatrix} = \frac{r_{10}r_{01} - r_{00}r_{11}}{r_{10}} = \frac{a_1a_2 - a_0a_3}{a_1}, \\ r_{21} &= -\frac{1}{r_{10}} \det \begin{bmatrix} r_{00} & r_{02} \\ r_{10} & r_{12} \end{bmatrix} = \frac{r_{10}r_{02} - r_{00}r_{12}}{r_{10}} = \frac{a_1a_4 - a_0a_5}{a_1}, \\ r_{22} &= -\frac{1}{r_{10}} \det \begin{bmatrix} r_{00} & r_{03} \\ r_{10} & r_{13} \end{bmatrix} = \frac{r_{10}r_{03} - r_{00}r_{13}}{r_{10}} = \frac{a_1a_6 - a_0a_7}{a_1}. \end{aligned}$$

When computing the last element of a certain row of the Routh table, one may find that the preceding row is one element short of what we need. For example, when we compute r_{n0}, we need $r_{(n-1)1}$ but $r_{(n-1)1}$ is not an element of the Routh table. In this case, we can simply augment the preceding row by a 0 in the end and keep the computation going. Keep in mind that this augmented 0 is not considered as part of the Routh table. Equivalently, whenever $r_{(i-1)(j+1)}$ is missing, simply let $r_{ij} = r_{(i-2)(j+1)}$. For example, r_{n0} can be computed as

$$r_{n0} = -\frac{1}{r_{(n-1)0}} \det \begin{bmatrix} r_{(n-2)0} & r_{(n-2)1} \\ r_{(n-1)0} & \mathbf{0} \end{bmatrix} = r_{(n-2)1}.$$

Another way to generate the Routh table is illustrated in the following example.

EXAMPLE 3.9

Consider

$$a(s) = a_0 s^5 + a_1 s^4 + a_2 s^3 + a_3 s^2 + a_4 s + a_5.$$

The Routh table can be constructed as follows.

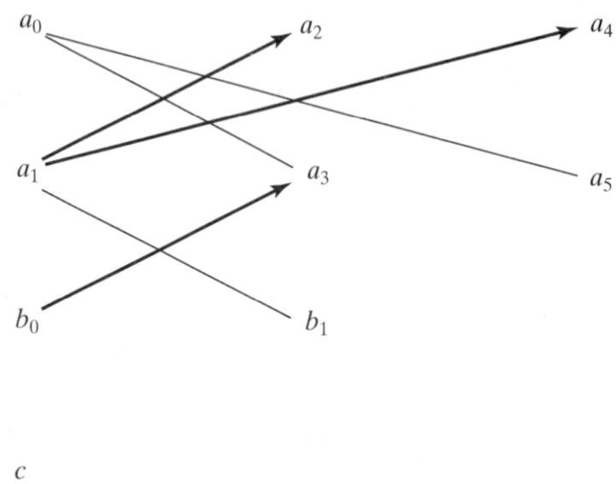

The b_0 in the table is obtained by first multiplying along the arrow $(a_1 \times a_2)$ and then subtracting $a_0 \times a_3$ and finally dividing by a_1:

$$b_0 = \frac{a_1 a_2 - a_0 a_3}{a_1}.$$

b_1 and c can be obtained similarly as

$$b_1 = \frac{a_1 a_4 - a_0 a_5}{a_1}$$

and

$$c = \frac{b_0 a_3 - a_1 b_1}{b_0}.$$

Edward John Routh (1831–1907) was born in Quebec (Canada) and moved to England at the age of three. He attended University College School and then entered University College, London, in 1847 with a scholarship. There he studied under Augustus De Morgan, whose influence led to him to decide on a career in mathematics. Routh obtained his BA (1849) and MA (1853) degrees in London. Then in 1854, he obtained a BA (Cantab.) degree (Senior Wrangler) and Smith's prize, followed by an MA degree in 1857. Routh became the most famous of the Cambridge "coaches" of students preparing for the Mathematical Tripos examination of the University of Cambridge in its heydays in the middle of the nineteenth century. Over a period of 22 years from 1862, he coached the Senior Wrangler for every year, and in 1854 was himself a Senior Wrangler, beating James Clerk Maxwell. He published famous advanced treatises that became standard applied mathematics texts, including *A Treatise on Stability of a Given State of Motion* (1877). His work on dynamic stability won him the Adams Prize in 1877. He was elected Fellow of the Royal Society on June 6, 1872.

Theorem 3.10 (Routh Stability Criterion). The following three statements are equivalent:

1. $a(s)$ is stable.
2. All elements of the Routh table are positive, i.e., $r_{ij} > 0$, $i = 0, 1, \ldots, n$, $j = 0, 1, \ldots, \lfloor \frac{n-i}{2} \rfloor$.
3. All elements in the first column of the Routh table are positive, i.e., $r_{i0} > 0$, $i = 0, 1, \ldots, n$.

In general, the Routh table cannot be completely constructed when some elements in the first column are zero. In this case, there is no need to complete the rest of the table since we already know from the Routh criterion that the polynomial is unstable.

At the end of this section, we will give a simple proof of the Routh stability criterion. Let us first see several examples on the use of this criterion.

EXAMPLE 3.11

Consider

$$a(s) = s^4 + 10s^3 + 35s^2 + 50s + 24.$$

We can construct the corresponding Routh table as

s^4	1	35	24
s^3	10	50	
s^2	$r_{20} = 30$	$r_{21} = 24$	
s^1	$r_{30} = 42$		
s^0	$r_{40} = 24$		

where r_{20}, r_{21}, r_{30}, and r_{40} are computed as

$$r_{20} = \frac{-1}{10} \det \begin{bmatrix} 1 & 35 \\ 10 & 50 \end{bmatrix} = 30, \qquad r_{21} = \frac{-1}{10} \det \begin{bmatrix} 1 & 24 \\ 10 & \mathbf{0} \end{bmatrix} = 24,$$

$$r_{30} = \frac{-1}{30} \det \begin{bmatrix} 10 & 50 \\ 30 & 24 \end{bmatrix} = 42, \qquad r_{40} = \frac{-1}{42} \det \begin{bmatrix} 30 & 24 \\ 42 & \mathbf{0} \end{bmatrix} = 24.$$

Note that the missing terms in the table are replaced by 0.

Since every element in the first column of the table is positive, we conclude that the system is stable. Indeed,

$$a(s) = (s+1)(s+2)(s+3)(s+4).$$

EXAMPLE 3.12

Consider

$$a_1(s) = s^3 + 2s^2 + 3s + 10.$$

We can construct the corresponding Routh table as

s^3	1	3
s^2	2	10
s^1	−2	
s^0	10	

Since the first column contains a negative entry, it follows that $a_1(s)$ is unstable. Actually, this conclusion can be made after we see the −2 entry in the Routh table; there is no need to complete the whole Routh table.

Sometimes, a negative entry first appears in a position not in the first column. In this situation, the equivalence of statements 1 and 2 in the Routh criterion becomes useful. Consider

$$a_2(s) = s^5 + 10s^4 + 35s^3 + 50s^2 + 24s + 300.$$

The corresponding (partial) Routh table is

s^5	1	35	24
s^4	10	50	300
s^3	$r_{20} = 30$	$r_{21} = -6$	

As soon as we encounter the negative entry in the Routh table, we can conclude that $a_2(s)$ is unstable.

Now, consider another polynomial

$$a_3(s) = s^4 + 3s^3 + 3s^2 + 3s + 2.$$

We can construct the corresponding Routh table as

s^4	1	3	2
s^3	3	3	
s^2	2	2	
s^1	$r_{30} = 0$		
s^0	r_{40}		

Note that since $r_{30} = 0$, the next element r_{40} cannot be computed. Of course, we can stop at this point to conclude that $a_3(s)$ is unstable since the elements in the first column are not all positive. In fact, it has two roots on the imaginary axis: $a_3(s) = (s+1)(s+2)(s^2+1)$.

As we have pointed out earlier, the most useful application of the Routh criterion is to determine the stability of a polynomial when some parameters are involved.

EXAMPLE 3.13

Consider a polynomial

$$a(s) = s^4 + 2s^3 + 4s^2 + 4s + K.$$

Determine the value of K such that $a(s)$ is stable.

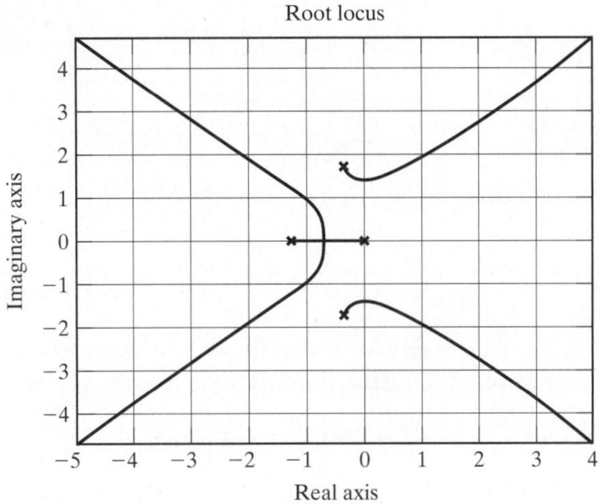

FIGURE 3.2: Root locus for the polynomial in Example 3.13 when K varies from 0 to ∞.

The Routh table corresponding to this polynomial is

s^4	1	4	K
s^3	2	4	
s^2	2	K	
s^1	$4-K$		
s^0	K		

Therefore, $a(s)$ is stable if $K > 0$ and $4 - K > 0$. This implies that the necessary and sufficient condition for the stability of $a(s)$ is $0 < K < 4$.

Using MATLAB, one can compute all possible root locations when K varies from 0 to ∞. These root locations are plotted in Figure 3.2. This plot is called the **root locus**, which will be discussed in detail in Chapter 5.

The rest of this section is dedicated to the proof of the Routh stability criterion. It is for curious minds and is not necessary for understanding the materials to follow. Also, the proof shares some ideas with the root-locus analysis to be introduced in Chapter 5.

A set of polynomials, each corresponding to a row in the Routh table, can be defined in the following way:

$$\begin{aligned}
r_0(s) &= r_{00}s^n + r_{01}s^{n-2} + r_{02}s^{n-4} + \cdots \\
r_1(s) &= r_{10}s^{n-1} + r_{11}s^{n-3} + r_{12}s^{n-5} + \cdots \\
&\vdots \\
r_{n-1}(s) &= r_{(n-1)0}s \\
r_n(s) &= r_{n0}.
\end{aligned}$$

The construction of the Routh table means that

$$r_{i+1}(s) = \frac{r_{i0}r_{i-1}(s) - r_{(i-1)0}sr_i(s)}{r_{i0}} = r_{i-1}(s) - \frac{r_{(i-1)0}}{r_{i0}}sr_i(s) \tag{3.3}$$

for $i = 1, 2, \ldots, n-1$. Another set of polynomials can be defined by taking the sum of two consecutive polynomials above:

$$\begin{aligned}
a_1(s) &= r_0(s) + r_1(s) = r_{00}s^n + r_{10}s^{n-1} + \cdots \\
a_2(s) &= r_1(s) + r_2(s) = r_{10}s^{n-1} + r_{20}s^{n-2} + \cdots \\
&\vdots \\
a_n(s) &= r_{n-1}(s) + r_n(s) = r_{(n-1)0}s + r_{n0}.
\end{aligned}$$

Clearly $a_1(s) = a(s)$. Each $a_i(s)$ has order $n - i + 1$ and the Routh table of $a_i(s)$ is precisely the remaining part of that of $a(s)$ after the first few rows are removed.

Fact 3.14. Assume that $a_i(s)$ has a positive leading coefficient, i.e., $r_{(i-1)0} > 0$. Then $a_i(s)$ is stable if and only if $r_{i0} > 0$ and $a_{i+1}(s)$ is stable.

Proof. Consider the polynomial $p_K(s) = a_i(s) + Ka_{i+1}(s)$. Since the roots of a polynomial are continuous functions of its coefficients, as K sweeps from 0 to ∞ the roots of $p_K(s)$ move continuously from the roots of $a_i(s)$ to those of $a_{i+1}(s)$ and ∞. More details on how the roots move can be found in Section B.2. We first show that if either $a_i(s)$ or $a_{i+1}(s)$ is stable, then as K sweeps from 0 to ∞, the continuous locus formed by the roots of $p_K(s)$ does not cross the imaginary axis, i.e., $p_K(j\omega) \neq 0$ for all $K \in [0, \infty)$ and $\omega \in \mathbb{R}$. Let us assume the opposite, i.e., for some $K \in [0, \infty)$ and $\omega \in \mathbb{R}$, the equality $p_K(j\omega) = 0$ holds, since, with equality (3.3),

$$\begin{aligned}
p_K(j\omega) &= a_i(j\omega) + Ka_{i+1}(j\omega) \\
&= r_{i-1}(j\omega) + r_i(j\omega) + Kr_i(j\omega) + Kr_{i+1}(j\omega) \\
&= \left[(1+K)r_{i-1}(j\omega) - K\frac{r_{(i-1)0}}{r_{i0}}j\omega r_i(j\omega)\right] + (1+K)r_i(j\omega).
\end{aligned}$$

If $n-i$ is odd, then $r_i(j\omega)$ is real and $r_{i-1}(j\omega)$ is imaginary. If $n-i$ is even, then $r_i(j\omega)$ is imaginary and $r_{i-1}(j\omega)$ is real. In either case, the two terms above are real and imaginary, respectively. Hence $p_K(j\omega) = 0$ implies $r_i(j\omega) = 0$ and $r_{i-1}(j\omega) = 0$, which further implies $r_{i+1}(j\omega) = 0$ by virtue of (3.3). The definitions of $a_i(s)$ and $a_{i+1}(s)$ then give $a_i(j\omega) = a_{i+1}(j\omega) = 0$, which is impossible if either $a_i(s)$ or $a_{i+1}(s)$ is assumed to be stable.

Now assume that $a_i(s)$ is stable. It follows from Theorem 3.7 that $r_{i0} > 0$. Also, since the roots of $p_K(s)$ do not cross the imaginary axis as K sweeps from 0 to ∞, the roots of $a_{i+1}(s)$ must be on the same side of the imaginary axis as those of $a_i(s)$, i.e., $a_{i+1}(s)$ is stable.

Conversely, assume $r_{i0} > 0$ and $a_{i+1}(s)$ is stable. Notice that $a_{i+1}(s)$ has $n-i$ roots and $a_i(s)$ has $n-i+1$ roots. As K sweeps from 0 to ∞, the $n-i$ roots of $p_K(s)$ move continuously from $n-i$ roots of $a_i(s)$ toward those of $a_{i+1}(s)$, and the remaining root of $p_K(s)$ moves continuously from the remaining root of $a_i(s)$ to ∞. Denote the root moving toward the ∞ by s_K. Then

$$0 = \frac{1}{s_K^{n-i}} p_K(s_K) = r_{(i-1)0}s_K + r_{i0} + \cdots + K(r_{i0} + r_{(i-1)0}s_K^{-1} + \cdots)$$

where the omitted terms are all negligible. We see that s_K has to be negative to counteract the other positive terms. If the roots do not cross the imaginary axis during the sweep, then the $n-i$ roots of $a_i(s)$ leading to those of $a_{i+1}(s)$ must be on the same side of the imaginary axis as those of $a_{i+1}(s)$, which are on the left-hand side, and furthermore the root of $a_i(s)$ going to infinity must also be on the same side of a large negative number. This shows that $a_i(s)$ is stable. □

With Fact 3.14, the equivalence of statements 1 and 3 in the Routh stability criterion follows from a mathematical induction. Since $a(s)$ is assumed to have a positive leading coefficient, the following chain of equivalence gives what we want:

$$\begin{aligned}
a(s) = a_1(s) \text{ is stable} &\iff r_{10} > 0 \text{ and } a_2(s) \text{ is stable}\\
&\iff r_{10} > 0,\ r_{20} > 0 \text{ and } a_3(s) \text{ is stable}\\
&\ \ \vdots\\
&\iff r_{10} > 0, r_{20} > 0, \ldots r_{(n-1)0} > 0 \text{ and } a_n(s) \text{ is stable}\\
&\iff r_{10} > 0, r_{20} > 0, \ldots, r_{n0} > 0.
\end{aligned}$$

The last equivalence is due to the simple fact that $a_n(s)$ is a first-order polynomial with coefficients $r_{(n-1)0}$ and r_{n0}. The above induction also shows that statement 1 implies statement 2 because of Theorem 3.7, whereas it is obvious that statement 2 implies statement 3. Therefore, all three statements in the Routh stability criterion are equivalent.

3.3 OTHER STABILITY CRITERIA

The **Hurwitz matrix** corresponding to polynomial (3.2) is an $n \times n$ matrix defined as

$$H = \begin{bmatrix} a_1 & a_3 & a_5 & a_7 & \cdots & 0 \\ a_0 & a_2 & a_4 & a_6 & \ddots & \vdots \\ 0 & a_1 & a_3 & a_5 & \ddots & \vdots \\ 0 & a_0 & a_2 & a_4 & \ddots & \vdots \\ \vdots & \ddots & \ddots & \ddots & \ddots & \vdots \\ 0 & \cdots & \cdots & \cdots & \cdots & a_n \end{bmatrix}.$$

The first two rows are obtained directly from the coefficients of $a(s)$, possibly augmented with zeros. The subsequent two rows are obtained by shifting the preceding two rows toward the right and filling the emptied spaces with zeros. Let Δ_i be the ith principal minor of H, i.e., the determinant of the $i \times i$ submatrix of H in the left-upper corner of H:

$$\Delta_1 = a_1,\ \Delta_2 = \det \begin{bmatrix} a_1 & a_3 \\ a_0 & a_2 \end{bmatrix}, \ldots,\ \Delta_i = \det \begin{bmatrix} a_1 & a_3 & \cdots & \cdots \\ a_0 & a_2 & \cdots & \cdots \\ \vdots & \vdots & \ddots & \vdots \\ \vdots & \vdots & \cdots & a_i \end{bmatrix}.$$

The determinants Δ_i, $i = 1, 2, \ldots, n$, are called the **Hurwitz determinants**. They are closely related to the Routh table. First, notice that H can be rewritten as

$$H = \begin{bmatrix} r_{10} & r_{11} & r_{12} & r_{13} & \cdots \\ r_{00} & r_{01} & r_{02} & r_{03} & \ddots \\ 0 & r_{10} & r_{11} & r_{12} & \ddots \\ 0 & r_{00} & r_{01} & r_{02} & \ddots \\ \vdots & \ddots & \ddots & \ddots & \ddots \end{bmatrix}$$

where r_{ij} are elements of the Routh table. If we subtract $\frac{r_{00}}{r_{10}}$ times the $(2i-1)$th row of H from its $2i$th row, then H is transformed to the following matrix:

$$\begin{bmatrix} r_{10} & r_{11} & r_{12} & r_{13} & \cdots \\ 0 & r_{20} & r_{21} & r_{22} & \ddots \\ 0 & r_{10} & r_{11} & r_{12} & \ddots \\ 0 & 0 & r_{20} & r_{21} & \ddots \\ \vdots & \ddots & \ddots & \ddots & \ddots \end{bmatrix},$$

and this transformation leaves the leading principal minors invariant. Now we subtract $\frac{r_{10}}{r_{20}}$ times the $2i$th row of this new matrix from its $(2i+1)$th row, and then the above matrix is converted into the following matrix:

$$\begin{bmatrix} r_{10} & r_{11} & r_{12} & r_{13} & \cdots \\ 0 & r_{20} & r_{21} & r_{22} & \ddots \\ 0 & 0 & r_{30} & r_{31} & \ddots \\ 0 & 0 & r_{20} & r_{21} & \ddots \\ \vdots & \ddots & \ddots & \ddots & \ddots \end{bmatrix}.$$

Continuing this process, we will ultimately reach the following upper triangular matrix:

$$\begin{bmatrix} r_{10} & r_{11} & r_{12} & r_{13} & \cdots & 0 \\ 0 & r_{20} & r_{21} & r_{22} & \ddots & \vdots \\ 0 & 0 & r_{30} & r_{31} & \ddots & \vdots \\ 0 & 0 & 0 & r_{40} & \ddots & \vdots \\ \vdots & \ddots & \ddots & \ddots & \ddots & \vdots \\ 0 & \cdots & \cdots & \cdots & \cdots & r_{n0} \end{bmatrix}.$$

All the transformations keep the leading principal minors invariant. Therefore, we see that Δ_i are related to the elements of the Routh table in the following way:

$$\Delta_1 = r_{10}, \quad \Delta_2 = r_{10}r_{20}, \quad \Delta_3 = r_{10}r_{20}r_{30}, \ldots.$$

It can be easily seen that the elements $r_{10}, r_{20}, r_{30}, \ldots$ in the first column of the Routh table are all positive if and only if the Hurwitz determinants $\Delta_1, \Delta_2, \Delta_3, \ldots$ are all positive. This immediately gives a new stability criterion.

Theorem 3.15 (Routh–Hurwitz Stability Criterion). The polynomial $a(s)$ is stable if and only if $\Delta_i > 0$, $i = 1, 2, \ldots, n$.

Though obtained by Hurwitz independently, this theorem is called the Routh–Hurwitz stability criterion because it is closely related to the Routh criterion as described above. Hurwitz's contribution was also fairly recognized in view of a common practice to call a stable polynomial a **Hurwitz polynomial.**

Adolf Hurwitz (1859–1919) was born in Hildesheim (now Lower Saxony), Germany. Hurwitz entered the University of Munich in 1877. He spent one year there attending lectures by Felix Klein, before spending the academic year 1877–1878 at the University of Berlin, where he attended classes by Kummer, Weierstrass, and Kronecker, after which he returned to Munich. In October 1880, Felix Klein moved to the University of Leipzig. Hurwitz followed him there and became a doctoral student under Klein's direction, finishing a dissertation on elliptic modular functions in

1881. Following two years at the University of Gottingen, in 1884 he was invited to become an Extraordinary Professor at the Albertus Universitat in Konigsberg (today the Kant Russian State University); there he encountered the young David Hilbert and Hermann Minkowski, on whom he had a major influence. Following the departure of Frobenius, Hurwitz took up a chair at the Eidgenossische Polytechnikum Zurich (today the ETH Zurich) in 1892, and remained there for the rest of his life. He was one of the early masters of the Riemann surface theory, and used it to prove many of the foundational results on algebraic curves: for instance, Hurwitz's automorphisms theorem. This work anticipates a number of later theories, such as the general theory of algebraic correspondences, Hecke operators, and the Lefschetz fixed-point theorem. Soon after he went to Zurich he was asked a question by Aurel Stodola, one of his colleagues, concerning when an nth-degree polynomial with real coefficients

$$f(x) = a_0 x^n + a_1 x^{n-1} + \cdots + a_n$$

with a positive leading coefficient $a_0 > 0$ has only roots with negative real parts. Hurwitz solved this problem completely, showing that the condition held if and only if a certain sequence of determinants is all positive. He published this in the paper *Uber die Bedingungen, unter welchen eine Gleichung nur Wurzeln mit negativen reellen Theilen besitzt*, which appeared in *Mathematische Annalen* in 1895.

Computing all leading principal minors of H might not be easy. Some savings in computation are possible using the following Liénard and Chipart criterion.

Theorem 3.16 (Liénard and Chipart Stability Criterion). If $a_i > 0$, $i = 0, 1, \ldots, n$, then the following statements are equivalent

1. $a(s)$ is stable.
2. $\Delta_1 > 0,\ \Delta_3 > 0,\ \Delta_5 > 0,\ \ldots.$
3. $\Delta_2 > 0,\ \Delta_4 > 0,\ \Delta_6 > 0,\ \ldots.$

The Liénard and Chipart criterion means that for a polynomial whose coefficients are all positive, if the Hurwitz determinants of odd order are positive, then those of even order are also positive, and vice versa. This reduces the number of Hurwitz determinants that need to be evaluated to roughly half of n.

EXAMPLE 3.17

Let us consider the same polynomial as in Example 3.13:

$$a(s) = s^4 + 2s^3 + 4s^2 + 4s + K.$$

We now wish to use the Routh–Hurwitz and the Liénard and Chipart stability criteria to determine the value of K such that $a(s)$ is stable. Here

$$H = \begin{bmatrix} 2 & 4 & 0 & 0 \\ 1 & 4 & K & 0 \\ 0 & 2 & 4 & 0 \\ 0 & 1 & 4 & K \end{bmatrix}.$$

Hence

$$\Delta_1 = 2, \Delta_2 = 4, \Delta_3 = 16 - 4K, \Delta_4 = K(16 - 4K).$$

By the Routh–Hurwitz stability criterion, we require $16 - 4K > 0$ and $K(16 - 4K) > 0$. This gives $0 < K < 4$.

By the Liénard and Chipart stability criterion, we only need to check Δ_1 and Δ_3, as well as all coefficients of the polynomial $a(s)$. This gives the condition $16 - 4K > 0$ and $K > 0$, which is equivalent to $0 < K < 4$.

The conditions given by all stability criteria are the same.

There is, nevertheless, another stability criterion, which is closely related to the Routh stability criterion and which will be called an alternative version of the Routh criterion. Again, consider the polynomial (3.2), out of which we can define two polynomials

$$\begin{aligned} r_0(s) &= a_0 s^n + a_2 s^{n-2} + a_4 s^{n-4} + \cdots \\ r_1(s) &= a_1 s^{n-1} + a_3 s^{n-3} + a_5 s^{n-5} + \cdots . \end{aligned}$$

One of them contains only the terms with even powers of s, and the other contains only the terms with odd powers of s. Let us decompose the ratio of $r_0(s)$ over $r_1(s)$ as

$$\frac{r_0(s)}{r_1(s)} = \alpha_1 s + \frac{r_2(s)}{r_1(s)}. \tag{3.4}$$

Here, the first term is the polynomial part of the ratio, and the second term is the strictly proper part of the ratio. Put in another way, $\alpha_1 s$ is the quotient and $r_2(s)$ is the remainder of $r_0(s)$ divided by $r_1(s)$. The reason for the polynomial part, or the quotient, to contain only one term linear in s is because of the special structure of $r_0(s)$ and $r_1(s)$. It is easy to see $\alpha_1 = a_0/a_1$. It can also be seen that $r_2(s)$ is of the form

$$r_2(s) = r_{20} s^{n-2} + r_{21} s^{n-4} + \cdots ,$$

i.e., it contains only the terms with even powers of s or only odd powers of s. Now $r_1(s)$ and $r_2(s)$ can be treated in the same way as $r_0(s)$ and $r_1(s)$. This gives

$$\frac{r_1(s)}{r_2(s)} = \alpha_2 s + \frac{r_3(s)}{r_2(s)} \tag{3.5}$$

where

$$\alpha_2 = \frac{a_1}{r_{20}} \text{ and } r_3(s) = r_{30} s^{n-3} + r_{31} s^{n-5} + \cdots .$$

Combining (3.4) and (3.5), we get

$$\frac{r_0(s)}{r_1(s)} = \alpha_1 s + \cfrac{1}{\alpha_2 s + \cfrac{r_3(s)}{r_2(s)}}.$$

Continuing this process, we obtain a recursive formula

$$\frac{r_{i-1}(s)}{r_i(s)} = \alpha_i s + \frac{r_{i+1}(s)}{r_i(s)}, i = 1, 2, \ldots, n$$

where $r_{n+1}(s) = 1$. Combining all the recursive formulas, we get

$$\frac{r_0(s)}{r_1(s)} = \alpha_1 s + \cfrac{1}{\alpha_2 s + \cfrac{1}{\alpha_3 s + \ddots + \cfrac{1}{\alpha_{n-1} s + \cfrac{1}{\alpha_n s}}}}. \tag{3.6}$$

Here $\alpha_1, \alpha_2, \cdots, \alpha_n$ are called the **atoms** of $a(s)$. And the expression in (3.6) is called a **continued fractional expansion** of $r_0(s)/r_1(s)$.

One may have noticed that $r_0(s)$ and $r_1(s)$ are polynomials formed from the first and the second rows of the Routh table corresponding to polynomial $a(s)$. To obtain α_1 and $r_2(s)$, we can do the following:

$$\begin{aligned}\frac{r_0(s)}{r_1(s)} &= \frac{a_0 s^n + \frac{a_0}{a_1} a_3 s^{n-2} + \frac{a_0}{a_1} a_5 s^{n-4} + \cdots}{a_1 s^{n-1} + a_3 s^{n-3} + a_5 s^{n-5} + \cdots} \\ &\quad + \frac{(a_2 - \frac{a_0}{a_1} a_3) s^{n-2} + (a_4 - \frac{a_0}{a_1} a_5) s^{n-4} + \cdots}{a_1 s^{n-1} + a_3 s^{n-3} + a_5 s^{n-5} + \cdots} \\ &= \frac{a_0}{a_1} s + \frac{(a_2 - \frac{a_0}{a_1} a_3) s^{n-2} + (a_4 - \frac{a_0}{a_1} a_5) s^{n-4} + \cdots}{a_1 s^{n-1} + a_3 s^{n-3} + a_5 s^{n-5} + \cdots}.\end{aligned}$$

This shows that the coefficients of $r_2(s)$ exactly match the elements of the third row of the Routh table. Further examination reveals that the polynomials $r_i(s)$ are exactly the same as the ones formed by the rows of the Routh table of $a(s)$:

$$r_i(s) = r_{i0} s^{n-i} + r_{i1} s^{n-i-2} + \cdots$$

and hence α_i are given by

$$\alpha_i = \frac{r_{(i-1)0}}{r_{i0}}.$$

It is easily seen that $r_{10}, r_{20}, \ldots, r_{n0}$ are positive if and only if $\alpha_1, \alpha_2, \ldots, \alpha_n$ are positive. Immediately, we obtain yet another stability criterion.

Theorem 3.18 (Alternative version of the Routh stability criterion). The polynomial $a(s)$ is stable if and only if all of its atoms are positive, i.e., $\alpha_i > 0$, $i = 1, 2, \ldots, n$.

EXAMPLE 3.19

Let us consider the polynomial as in Examples 3.13 and 3.17:

$$a(s) = s^4 + 2s^3 + 4s^2 + 4s + K$$

using the alternative version of the Routh stability criterion. For this polynomial

$$r_0(s) = s^4 + 4s^2 + K$$

$$r_1(s) = 2s^3 + 4s.$$

First

$$\frac{r_0(s)}{r_1(s)} = \frac{1}{2}s + \frac{2s^2 + K}{2s^3 + 4s}.$$

This gives $\alpha_1 = 1/2$ and $r_2(s) = 2s^2 + K$. In the next step,

$$\frac{r_1(s)}{r_2(s)} = s + \frac{(4-K)s}{2s^2 + K}.$$

This gives $\alpha_2 = 1$ and $r_3(s) = (4-K)s$. Finally,

$$\frac{r_2(s)}{r_3(s)} = \frac{2}{4-K}s + \frac{K}{(4-K)s}.$$

This gives $\alpha_3 = \frac{2}{4-K}$ and $\alpha_4 = \frac{4-K}{K}$. Combining the above procedures, we have the continued fraction

$$\frac{r_0(s)}{r_1(s)} = \frac{1}{2}s + \cfrac{1}{s + \cfrac{1}{\frac{2}{4-K}s + \cfrac{1}{\frac{4-K}{K}s}}}.$$

By the alternative version of the Routh stability criterion, we require

$$\frac{2}{(4-K)} > 0 \text{ and } \frac{4-K}{K} > 0.$$

This again gives $0 < K < 4$.

For low-order polynomials, explicit stability conditions can be obtained by applying any of the above stability criteria directly to the polynomials.

Corollary 3.20.

1. A second-order polynomial $a(s) = s^2 + a_1 s + a_2$ is stable if and only if $a_1 > 0$ and $a_2 > 0$.
2. A third-order polynomial $a(s) = s^3 + a_1 s^2 + a_2 s + a_3$ is stable if and only if $a_1 > 0$, $a_3 > 0$, and $a_1 a_2 - a_3 > 0$.
3. A fourth-order polynomial $a(s) = s^4 + a_1 s^3 + a_2 s^2 + a_3 s + a_4$ is stable if and only if $a_2 > 0$, $a_3 > 0$, $a_4 > 0$, and $a_1 a_2 a_3 - a_4 a_1^2 - a_3^2 > 0$.

3.4 ROBUST STABILITY

In many applications, the parameters of a system are not exactly given. We may only know that the parameters are in certain ranges. It is often required to determine whether a system is stable for all possible values of its parameters.

Consider a set of polynomials:

$$\mathcal{P} = \{a(s) = a_0 s^n + a_1 s^{n-1} + \cdots + a_{n-1}s + a_n : a_i \in [\underline{a}_i, \overline{a}_i]\}.$$

This set of polynomials is often called an **interval polynomial** since the coefficients are taken from intervals. Our problem is to determine whether or not all members of $\mathcal{P}$ are stable.

Theorem 3.21 (Kharitonov Theorem on an Interval Polynomial). All members of $\mathcal{P}$ are stable if and only if the following four polynomials are stable:

$$\begin{aligned}
a_1(s) &= \underline{a}_0 s^n + \underline{a}_1 s^{n-1} + \overline{a}_2 s^{n-2} + \overline{a}_3 s^{n-3} + \underline{a}_4 s^{n-4} + \cdots \\
a_2(s) &= \underline{a}_0 s^n + \overline{a}_1 s^{n-1} + \overline{a}_2 s^{n-2} + \underline{a}_3 s^{n-3} + \underline{a}_4 s^{n-4} + \cdots \\
a_3(s) &= \overline{a}_0 s^n + \underline{a}_1 s^{n-1} + \underline{a}_2 s^{n-2} + \overline{a}_3 s^{n-3} + \overline{a}_4 s^{n-4} + \cdots \\
a_4(s) &= \overline{a}_0 s^n + \overline{a}_1 s^{n-1} + \underline{a}_2 s^{n-2} + \underline{a}_3 s^{n-3} + \overline{a}_4 s^{n-4} + \cdots .
\end{aligned}$$

We will postpone the proof of this theorem to the end of this section. The coefficients of the four special polynomials in the Khoritonov theorem are formed by the extreme values, i.e., the upper or lower bounds, of the coefficient intervals of the interval polynomial. The first two coefficients start from the four possible combinations of extreme values and the subsequent coefficients follow the rule of switching from two upper (or lower) bounds to two lower (or upper) bounds consecutively.

EXAMPLE 3.22

Consider a set of polynomials

$$\mathcal{P} = \{s^4 + a_1 s^3 + a_2 s^2 + a_3 s + a_4 : a_1 \in [3,4], a_2 \in [3,4], a_3 \in [2,3], a_4 \in [0.5,1]\}.$$

Is every polynomial in this set stable? The four polynomials required by the Kharitonov theorem are

$$\begin{aligned}
a_1(s) &= s^4 + 3s^3 + 4s^2 + 3s + 0.5 \\
a_2(s) &= s^4 + 4s^3 + 4s^2 + 2s + 0.5 \\
a_3(s) &= s^4 + 3s^3 + 3s^2 + 3s + 1 \\
a_4(s) &= s^4 + 4s^3 + 3s^2 + 2s + 1.
\end{aligned}$$

By using any of the stability criteria in the previous sections or numerical computation, it is easy to check that all four polynomials are stable in this case. Therefore, the original interval polynomial is stable.

The rest of this section is dedicated to the proof of the Kharitonov theorem. Again, it is for curious minds only. Some preparations are needed first.

Let us consider the image of $a(j\omega)$ as ω goes from 0 to ∞. As ω goes from 0 to ∞, the image of $a(j\omega)$ forms a continuous path in the complex plane. If $a(j\omega) \neq 0$, then $\angle a(j\omega)$, the argument or phase of $a(j\omega)$, is well defined as a continuous function.

Fact 3.23. If $a(s)$ is stable, then $\angle a(j\omega)$ is a strictly increasing function of $\omega \in [0, \infty)$.

Proof. If $a(s)$ is stable, then

$$a(s) = a_0 \prod_{i=1}^{n} (s - z_i)$$

where $\operatorname{Re} z_i < 0$. This implies that

$$\angle a(j\omega) = \sum_{i=1}^{n} \angle(j\omega - z_i).$$

Since $\angle(j\omega - z_i)$ is strictly increasing, as illustrated by z_2, z_3, z_4 in Figure 3.3, so is $\angle a(j\omega)$. □

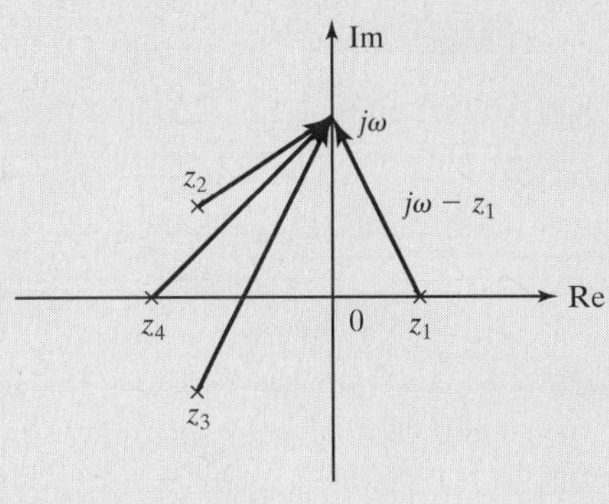

FIGURE 3.3: $\angle(j\omega - z)$ for various z's.

Now denote the total increment of $\angle a(j\omega)$ as ω runs from 0 to ∞ by $\Delta\angle a(j\omega)$, i.e.,

$$\Delta\angle a(j\omega) = \lim_{\omega\to\infty} \angle a(j\omega) - \angle a(0).$$

For example, it is easy to see from Figure 3.3 that $\Delta\angle(j\omega - z_1) = -\pi/2$, $\Delta\angle(j\omega - z_4) = \pi/2$, and $\Delta\angle(j\omega - z_2)(j\omega - z_3) = \pi$.

Fact 3.24. $a(s)$ is stable if and only if $a(j\omega) \neq 0$ for all $\omega \in [0, \infty)$ and

$$\Delta\angle a(j\omega) = n\frac{\pi}{2}.$$

Proof. Since $a(s)$ can be factored as

$$a(s) = a_0 \prod_{i=1}^{n} (s - z_i),$$

it follows that

$$\Delta\angle a(j\omega) = \sum_{i=1}^{n} \Delta\angle(j\omega - z_i).$$

If $a(s)$ is stable, then $\operatorname{Re} z_i < 0$. Hence, if z_i is real, then

$$\Delta\angle(j\omega - z_i) = \frac{\pi}{2},$$

as illustrated by z_4 in Figure 3.3. If z_i is not real, then $z_j = z_i^*$ for some $j \neq i$ and

$$\Delta\angle(j\omega - z_i) + \Delta\angle(j\omega - z_j) = \pi,$$

as illustrated by z_2 and z_3 in Figure 3.3. This shows $\Delta\angle a(j\omega) = n\pi/2$. If $a(s)$ is unstable, then either $a(j\omega) = 0$ for some $\omega \in [0, \infty)$ or at least one of z_i has a positive real part. In the latter case, if this z_i is real, then

$$\Delta\angle(j\omega - z_i) = -\frac{\pi}{2},$$

as illustrated by z_1 in Figure 3.3, and if this z_i is not real, then there is another $z_j = z_i^*$ satisfying

$$\Delta\angle(j\omega - z_i) + \Delta\angle(j\omega - z_j) = -\pi.$$

This implies that

$$\Delta\angle a(j\omega) < n\frac{\pi}{2}.$$

□

From this fact, we see that we can test the stability of a polynomial by plotting the frequency plot of $a(s)$ and checking whether it circulates the origin $n/4$ times in the counterclockwise direction. There might not be any computational advantage in using this stability test, but conceptually it is the one used in the following proof of the Kharitonov theorem.

The necessity of the theorem is obvious. We need only to prove the sufficiency. The idea in the proof is that if the four special polynomials circulate the origin $n/4$ times in the counterclockwise direction, so does any polynomial in the set $\mathcal{P}$. Define

$$f_1(s) = \underline{a}_n + \overline{a}_{n-2}s^2 + \underline{a}_{n-4}s^4 + \cdots$$

$$f_2(s) = \overline{a}_n + \underline{a}_{n-2}s^2 + \overline{a}_{n-4}s^4 + \cdots$$

$$g_1(s) = \underline{a}_{n-1}s + \overline{a}_{n-3}s^3 + \underline{a}_{n-5}s^5 + \cdots$$

$$g_2(s) = \overline{a}_{n-1}s + \underline{a}_{n-3}s^3 + \overline{a}_{n-5}s^5 + \cdots$$

and

$$h_{kl}(s) = f_k(s) + g_l(s) \qquad k, l = 1, 2.$$

Notice that the four polynomials $a_i(s)$, $i = 1, 2, 3, 4$, are exactly $h_{kl}(s)$, k, $l = 1, 2$. Then, notice that $f_k(j\omega)$ is real and $g_l(j\omega)$ is imaginary. Therefore,

$$\mathrm{Re}\,[h_{k1}(j\omega)] = \mathrm{Re}\,[h_{k2}(j\omega)] = f_k(j\omega)$$
$$j\mathrm{Im}\,[h_{1l}(j\omega)] = j\mathrm{Im}\,[h_{2l}(j\omega)] = g_l(j\omega).$$

This implies that

$$\{h_{11}(j\omega), h_{12}(j\omega), h_{21}(j\omega), h_{22}(j\omega)\}$$

form the corners of a rectangle with sides parallel to the axes in the complex plane, as shown in Figure 3.4.

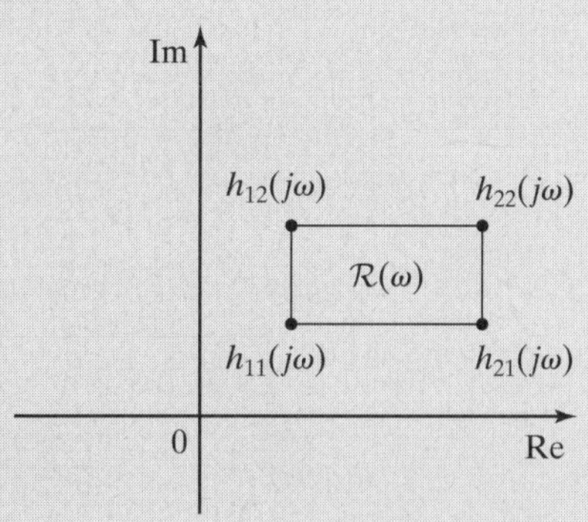

FIGURE 3.4: Possible values of $a(j\omega)$ for $a(s) \in \mathcal{P}$.

Let us call the set of points bounded by this rectangle as $\mathcal{R}(\omega)$. Let $a(s) \in \mathcal{P}$.

$$\begin{aligned} a(j\omega) &= [a_n + a_{n-2}(j\omega)^2 + a_{n-4}(j\omega)^4 + \cdots] \\ &\quad + [a_{n-1}(j\omega) + a_{n-3}(j\omega)^3 + a_{n-5}(j\omega)^5 + \cdots] \\ &= (a_n - a_{n-2}\omega^2 + a_{n-4}\omega^4 - \cdots) + j(a_{n-1}\omega - a_{n-3}\omega^3 + a_{n-5}\omega^5 + \cdots). \end{aligned}$$

Then, for $\omega \in [0, \infty)$,

$$\begin{aligned} f_1(j\omega) = \underline{a}_n - \overline{a}_{n-2}\omega^2 + \underline{a}_{n-4}\omega^4 - \cdots \le \mathrm{Re}\,[a(j\omega)] \\ \le \overline{a}_n - \underline{a}_{n-2}\omega^2 + \overline{a}_{n-4}\omega^4 - \cdots = f_2(j\omega) \end{aligned}$$

and

$$\begin{aligned} g_1(j\omega)/j = \underline{a}_{n-1}\omega - \overline{a}_{n-3}\omega^3 + \underline{a}_{n-5}\omega^5 - \cdots \le \mathrm{Im}\,[a(j\omega)] \\ \le \overline{a}_{n-1}\omega - \underline{a}_{n-3}\omega^3 + \overline{a}_{n-5}\omega^5 - \cdots = g_2(j\omega)/j. \end{aligned}$$

This shows that $a(j\omega) \in \mathcal{R}(\omega)$. Now assume that h_{kl}, $k, l = 1, 2$, are stable. Then, as ω goes from 0 to ∞, each corner of $\mathcal{R}(\omega)$ encircles the origin continuously. Assume that the origin breaks into $\mathcal{R}(\omega)$ as $\mathcal{R}(\omega)$ moves. It must break through a point on the boundary of $\mathcal{R}(\omega)$. This point cannot be one of the four corners of $\mathcal{R}(\omega)$ since the four polynomials associated with the corners are stable. Suppose that this point is in the middle of one of the four bounding segments, as shown in Figure 3.5. Then according to Fact 3.23, the end points of this segment will move in the counterclockwise direction. This will certainly contradict the fact that the rectangles $\mathcal{R}(\omega)$ have sides parallel to the axes of the complex plane for all ω. This shows that as ω goes from 0 to ∞, the rectangles $\mathcal{R}(\omega)$ will not intersect with the origin and will encircle the origin in a similar way as their corners. Since the four corners will encircle the origin $n/4$ times, so will any point inside the rectangle. This shows $a(j\omega)$ will not intersect with the origin and will encircle the origin $n/4$ times. Hence $a(s)$ is stable.

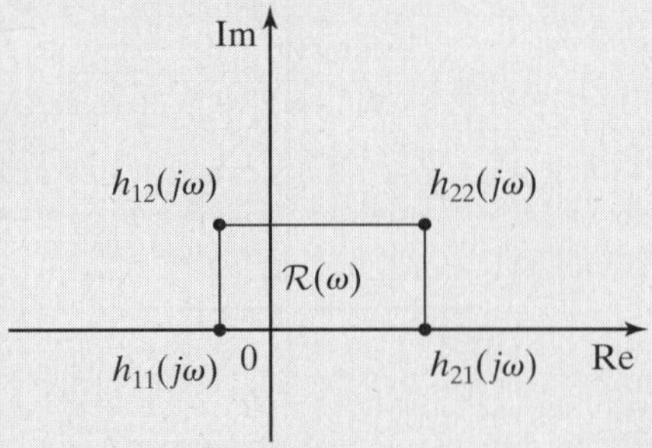

FIGURE 3.5: Impossible situation for the values of $a(j\omega)$ for $a(s) \in \mathcal{P}$.

3.5 STABILITY OF CLOSED-LOOP SYSTEMS

We have studied the stability of a system with an input and an output, without considering where the system comes from. One fundamental purpose of control is to design a controller so that a closed-loop system is stable and possibly satisfies other performance specifications. What is the stability of a closed-loop system then? Consider the closed-loop system for stabilization shown in Figure 3.6.

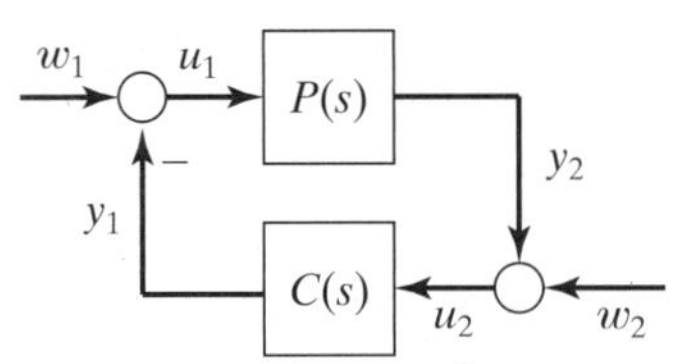

FIGURE 3.6: Feedback system for stabilization.

Definition 3.25. The closed-loop system shown in Figure 3.6 is said to be **internally stable** if the eight transfer functions from w_i, $i = 1, 2$, to u_j and y_j, $j = 1, 2$, are stable[2].

The motivation for this definition is to ensure that all internal signals $u_1(t)$, $u_2(t)$, $y_1(t)$, $y_2(t)$ are all bounded as long as the external signals $w_1(t), w_2(t)$, which usually represent disturbances, noises, or excitations, are bounded.

The eight transfer functions are given in (2.28). There are repetitions among the eight transfer functions. Essentially, there are only four different transfer functions, which are the gang of four:

$$\frac{1}{1+P(s)C(s)}, \quad \frac{P(s)}{1+P(s)C(s)}, \quad \frac{C(s)}{1+P(s)C(s)}, \quad \frac{P(s)C(s)}{1+P(s)C(s)}. \tag{3.7}$$

Hence we only need to check the stability of these four transfer functions when testing the closed-loop stability. Furthermore, if we have

$$P(s) = \frac{b(s)}{a(s)} \quad \text{and} \quad C(s) = \frac{q(s)}{p(s)}$$

where $a(s)$ and $b(s)$ are coprime polynomials and so are $p(s)$ and $q(s)$, then it is easy to see that the gang of four can be expressed as

$$\begin{aligned}
\frac{1}{1+P(s)C(s)} &= \frac{a(s)p(s)}{a(s)p(s)+b(s)q(s)} \\
\frac{P(s)}{1+P(s)C(s)} &= \frac{b(s)p(s)}{a(s)p(s)+b(s)q(s)} \\
\frac{C(s)}{1+P(s)C(s)} &= \frac{a(s)q(s)}{a(s)p(s)+b(s)q(s)} \\
\frac{P(s)C(s)}{1+P(s)C(s)} &= \frac{b(s)q(s)}{a(s)p(s)+b(s)q(s)}.
\end{aligned}$$

We see that all the four transfer functions seem to have the same denominator $a(s)p(s) + b(s)q(s)$. This may lead us to expect that the closed-loop system would be internally stable if and only if one of the four different transfer functions is stable. One may even expect that the closed-loop system would be internally stable if and only if the polynomial $a(s)p(s) + b(s)q(s)$ is stable. These expectations are indeed correct in most cases.

Theorem 3.26. Let $P(s) = \dfrac{b(s)}{a(s)}$ and $C(s) = \dfrac{q(s)}{p(s)}$ be proper transfer functions and assume that $1 + P(\infty)C(\infty) \neq 0$. Then the following statements are equivalent:

1. The closed-loop system is internally stable.
2. The polynomial $a(s)p(s) + b(s)q(s)$ is stable.

[2]Recall that a transfer function is stable if and only if it is proper and the denominator polynomial is stable. Hence $\frac{s^2}{s+1}$ is not stable since it is not proper.

3. There is no unstable pole/zero cancellation in forming the product $P(s)$ $C(s)$ and any one of the four transfer functions in (3.7) is stable.

Because of the importance of polynomial $a(s)p(s) + b(s)q(s)$ in determining closed-loop stability and other closed-loop properties, we make the following definition.

Definition 3.27. For a feedback system shown in Figure 3.6 with $P(s) = \frac{b(s)}{a(s)}$ and $C(s) = \frac{q(s)}{p(s)}$, the polynomial $a(s)p(s) + b(s)q(s)$ is called its **characteristic polynomial**.

The characteristic polynomial of a feedback system is very important in feedback system analysis.

When the assumptions in Theorem 3.26 are not satisfied, one has to exercise caution when asserting the internal stability of a closed-loop system. The following example shows several pathological cases when the assumptions in Theorem 3.26 are not satisfied.

EXAMPLE 3.28

1. Suppose $P(s) = \frac{1}{s-1}$ and $C(s) = \frac{s-1}{s+1}$. Then

$$\frac{1}{1+P(s)C(s)} = \frac{s+1}{s+2}, \qquad \frac{P(s)}{1+P(s)C(s)} = \frac{s+1}{(s-1)(s+2)},$$
$$\frac{C(s)}{1+P(s)C(s)} = \frac{s-1}{s+2}, \qquad \frac{P(s)C(s)}{1+P(s)C(s)} = \frac{1}{s+2}.$$

This shows that the closed-loop system is unstable even though three out of four different closed-loop transfer functions, except

$$\frac{P(s)}{1+P(s)C(s)},$$

are stable.

2. Suppose $P(s) = \frac{s-2}{s+1}$ and $C(s) = \frac{1}{s-2}$. Then

$$\frac{1}{1+P(s)C(s)} = \frac{s+1}{s+2}, \qquad \frac{P(s)}{1+P(s)C(s)} = \frac{s-2}{s+2},$$
$$\frac{C(s)}{1+P(s)C(s)} = \frac{(s+1)}{(s+2)(s-2)}, \qquad \frac{P(s)C(s)}{1+P(s)C(s)} = \frac{1}{s+2}.$$

This shows that the closed-loop system is unstable even though three out of the four different closed-loop transfer functions, except

$$\frac{C(s)}{1+P(s)C(s)},$$

are stable.

3. Suppose $P(s) = \dfrac{\omega_n^2}{s^2 + 2\zeta\omega_n s + \omega_n^2}$ and $C(s) = K_D s + K_P$. Such a controller is called a **proportional-derivative (PD)** controller. We have

$$\frac{P(s)C(s)}{1+P(s)C(s)} = \frac{(K_D s + K_P)\omega_n^2}{s^2 + (2\zeta\omega_n + K_D\omega_n^2)s + (1+K_P)\omega_n^2}$$

which can be made stable by designing K_D and K_P, but

$$\frac{C(s)}{1+P(s)C(s)} = \frac{(s^2 + 2\zeta\omega_n s + \omega_n^2)(K_D s + K_P)}{s^2 + (2\zeta\omega_n + K_D\omega_n^2)s + (1+K_P)\omega_n^2}$$

which is unstable as long as $K_D \neq 0$ since the transfer function is not bounded at $s \to \infty$. This conclusion may puzzle advocates of PD and PID control. However, if the differentiation term in the controller is replaced by a practical differentiator, i.e., if $C(s)$ is replaced by $K_D \dfrac{s}{Ts+1} + K_P$ with a very small $T > 0$, then both $P(s)$ and $C(s)$ are proper, $1 + P(\infty)C(\infty) = 1$, and the closed-loop characteristic polynomial becomes

$$Ts[s^2 + 2\zeta\omega_n s + (1+K_D)\omega_n^2] + s^2 + (2\zeta\omega_n + K_D\omega_n^2)s + (1+K_P)\omega_n^2.$$

By the material in Section B.2 in Appendix B, we know that for small $T > 0$, the roots of this polynomial are those close to the roots of

$$s^2 + (2\zeta\omega_n + K_D\omega_n^2)s + (1+K_P)\omega_n^2$$

together with an additional one on the negative real axis. By Theorem 3.26, we can conclude that when the practical PD controller is used, we can always design K_P and K_D so that the closed-loop system is internally stable.

4. Suppose $P(s) = \dfrac{s+2}{s+1}$ and $C(s) = -1$. Then

$$\frac{P(s)}{1+P(s)C(s)} = s+2, \quad \frac{C(s)}{1+P(s)C(s)} = s+1.$$

This shows that the closed-loop system is unstable even though $a(s)p(s) + b(s)q(s) = -1$ is stable and both the plant and the controller are proper.

This example shows that several things can go wrong if we follow our intuition to check only one of the possible transfer functions or check only the polynomial $a(s)p(s) + b(s)q(s)$ to decide closed-loop stability. First, there may be unstable pole/zero cancellations in forming $P(s)C(s)$, i.e., one of the unstable pole (or zero) of $P(s)$ is exactly the same as a zero (or pole) of $C(s)$. These poles and zeros are cancelled when $P(s)C(s)$ is formed. The cancellation can occur for poles and zeros at the "infinity," as in case 3 of Example 3.28, when a strictly proper plant is connected to a nonproper controller. This sort of cancellation will definitely lead to an unstable closed loop, though some of the closed-loop transfer functions might still be stable. Second, the order of $a(s)p(s) + b(s)q(s)$ may drop below that of $a(s)p(s)$ or $b(s)q(s)$ because of cancellation of the leading coefficients of the two terms. This causes one of the four transfer functions to become nonproper. Therefore, in general, in order to know the stability of the closed-loop system, it is not sufficient to check carelessly the stability of only one transfer function among the four different transfer functions. Nor is it sufficient to check carelessly the stability of polynomial $a(s)p(s) + b(s)q(s)$. One has to observe the assumptions in Theorem 3.26.

The following theorem is sometimes useful.

Theorem 3.29. The following statements are equivalent:

1. The closed-loop system is internally stable.
2. The transfer functions
$$\frac{P(s)}{1+P(s)C(s)} \quad \text{and} \quad \frac{C(s)}{1+P(s)C(s)}$$
are stable.
3. The inequality
$$\deg[a(s)p(s) + b(s)q(s)] \geq \max\{\deg a(s)q(s), \deg b(s)p(s)\} \tag{3.8}$$
holds and the polynomial $a(s)p(s) + b(s)q(s)$ is stable.

The proof of Theorem 3.29 is left as an extra credit problem.

Theorem 3.29 indicates that among all different closed-loop transfer functions in the feedback system shown in Figure 3.6, only two, namely,

$$\frac{P(s)}{1+P(s)C(s)} \quad \text{and} \quad \frac{C(s)}{1+P(s)C(s)},$$

are essential as far as closed-loop stability is concerned. We only need to check the stability of these two transfer functions to verify the internal stability of the closed-loop system. On the other hand, Example 3.28 shows that it is, in general, impossible to further reduce the number of transfer functions that are required to check for stability of the feedback system.

Theorem 3.29 also indicates that if one is to use the stability of $a(s)p(s) + b(s)q(s)$ to infer closed-loop stability, an additional minor condition on polynomial degrees has to be checked.

Nevertheless, further simplification is possible if additional conditions are imposed on the plant or controller transfer functions to rule out the pathological cases when there are unstable pole and zero cancellations in forming $P(s)C(s)$ or when there is an algebraic loop in the feedback system with gain -1.

Corollary 3.30. Let $P(s)$ and $C(s)$ be proper transfer functions and assume that $1 + P(\infty)C(\infty) \neq 0$.

1. Assume that $P(s)$ is stable. Then the closed-loop system is internally stable if and only if $\dfrac{C(s)}{1+P(s)C(s)}$ is stable.

2. Assume that $C(s)$ is stable. Then the closed-loop system is internally stable if and only if $\dfrac{P(s)}{1+P(s)C(s)}$ is stable.

Proof. We shall prove only part 1. The proof for the part 2 is similar.

Assume that $P(s)$ and $\dfrac{C(s)}{1+P(s)C(s)}$ are both stable. Then it follows that

$$\frac{1}{1+P(s)C(s)} = 1 - \frac{P(s)C(s)}{1+P(s)C(s)} = 1 - P(s) \times \frac{C(s)}{1+P(s)C(s)}$$

is stable. Hence

$$\frac{P(s)}{1+P(s)C(s)} = P(s) \times \frac{1}{1+P(s)C(s)}$$

is stable. Thus by Theorem 3.29, the closed-loop system is internally stable. □

The closed-loop system structure for regulation, also called **2DOF controller structure**, is shown in Figure 3.7.

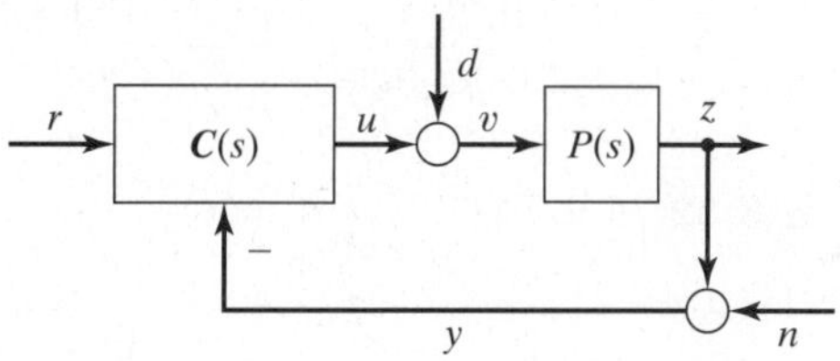

FIGURE 3.7: 2DOF feedback control system.

Definition 3.31. The closed-loop system shown in Figure 3.7 is said to be internally stable if the 12 transfer functions from $d(t), n(t), r(t)$ to $u(t)$, $v(t)$, $y(t)$, $z(t)$ are all stable.

The 12 transfer functions are given in (2.30). Again, there are quite a few repetitions among the 12 transfer functions. Six of them are different

$$\frac{1}{1+P(s)C_2(s)}, \quad \frac{P(s)}{1+P(s)C_2(s)}, \quad \frac{C_2(s)}{1+P(s)C_2(s)}, \quad \frac{P(s)C_2(s)}{1+P(s)C_2(s)} \tag{3.9}$$

and

$$\frac{C_1(s)}{1+P(s)C_2(s)}, \quad \frac{P(s)C_1(s)}{1+P(s)C_2(s)}. \tag{3.10}$$

Let

$$P(s) = \frac{b(s)}{a(s)}$$

and

$$\boldsymbol{C}(s) = \begin{bmatrix} C_1(s) & C_2(s) \end{bmatrix} = \frac{1}{p(s)} \begin{bmatrix} q_1(s) & q_2(s) \end{bmatrix}$$

where $a(s)$ and $b(s)$ are coprime polynomials and $p(s)$ and $\begin{bmatrix} q_1(s) & q_2(s) \end{bmatrix}$ are coprime polynomials. Note that $p(s)$ and $q_1(s)$ (or $q_2(s)$) are not necessarily coprime. For example, $p(s) = (s+1)(s+2)$, $q_1(s) = s+1$, and $q_2(s) = s+2$. Then $p(s)$, $q_1(s)$ and $q_2(s)$ are coprime polynomials but $p(s)$ and $q_1(s)$ are not coprime. Nor are $p(s)$ and $q_2(s)$.

Hence

$$
\begin{aligned}
\frac{1}{1+P(s)C_2(s)} &= \frac{a(s)p(s)}{a(s)p(s)+b(s)q_2(s)} \\
\frac{P(s)}{1+P(s)C_2(s)} &= \frac{b(s)p(s)}{a(s)p(s)+b(s)q_2(s)} \\
\frac{C_2(s)}{1+P(s)C_2(s)} &= \frac{a(s)q_2(s)}{a(s)p(s)+b(s)q_2(s)} \\
\frac{P(s)C_2(s)}{1+P(s)C_2(s)} &= \frac{b(s)q_2(s)}{a(s)p(s)+b(s)q_2(s)} \\
\frac{C_1(s)}{1+P(s)C_2(s)} &= \frac{a(s)q_1(s)}{a(s)p(s)+b(s)q_2(s)} \\
\frac{P(s)C_1(s)}{1+P(s)C_2(s)} &= \frac{b(s)q_1(s)}{a(s)p(s)+b(s)q_2(s)}.
\end{aligned}
$$

Theorem 3.32. Let $P(s) = \dfrac{b(s)}{a(s)}$ and

$$\boldsymbol{C}(s) = \begin{bmatrix} C_1(s) & C_2(s) \end{bmatrix} = \frac{1}{p(s)} \begin{bmatrix} q_1(s) & q_2(s) \end{bmatrix}$$

be proper transfer functions and assume that $1 + P(\infty)C_2(\infty) \neq 0$. Then the following statements are equivalent:

1. The 2DOF closed-loop system is internally stable.
2. The polynomial $a(s)p(s) + b(s)q_2(s)$ is stable.
3. There is no unstable pole/zero cancellation in forming the product $P(s)$ $C_2(s)$ and any one of the six transfer functions in (3.9) and (3.10) is stable.

Since we have shown that the stability of $\dfrac{P(s)}{1+P(s)C_2(s)}$ and $\dfrac{C_2(s)}{1+P(s)C_2(s)}$ implies that of $\dfrac{P(s)C_2(s)}{1+P(s)C_2(s)}$ and $\dfrac{1}{1+P(s)C_2(s)}$, it follows that we only need to check the stability of

$$\frac{P(s)}{1+P(s)C_2(s)}, \quad \frac{C_2(s)}{1+P(s)C_2(s)}, \quad \frac{C_1(s)}{1+P(s)C_2(s)}, \quad \frac{P(s)C_1(s)}{1+P(s)C_2(s)}.$$

In summary, we obtain the following closed-loop stability test.

Theorem 3.33. The following statements are equivalent:

1. The closed-loop system shown in Figure 3.7 is internally stable.
2. The four transfer functions
$$\frac{P(s)}{1+P(s)C_2(s)}, \quad \frac{C_2(s)}{1+P(s)C_2(s)}, \quad \frac{C_1(s)}{1+P(s)C_2(s)}, \quad \frac{P(s)C_1(s)}{1+P(s)C_2(s)}$$
are stable.
3. The inequality
$$\begin{aligned}&\deg[a(s)p(s)+b(s)q_2(s)]\\ &\geq \max\{\deg a(s)q_2(s), \deg b(s)p(s), \deg a(s)q_1(s), \deg b(s)q_1(s)\} \quad (3.11)\end{aligned}$$
holds and the polynomial $a(s)p(s)+b(s)q_2(s)$ is stable.

In most applications, transfer functions $C_1(s)$, $C_2(s)$, and $P(s)$ are all proper and at least one of $C_2(s)$ and $P(s)$ is strictly proper. In this case, the condition (3.11) is always satisfied; hence we only need to check the stability of the characteristic polynomial $a(s)p(s)+b(s)q_2(s)$.

We have shown that the system in Figure 3.7 is algebraically equivalent to that shown in Figure 3.8. However, they are not analytically equivalent since the system in Figure 3.8 has an extra internal signal $w(t)$. The internal stability of the system in Figure 3.7 does not imply that of the system in Figure 3.8.

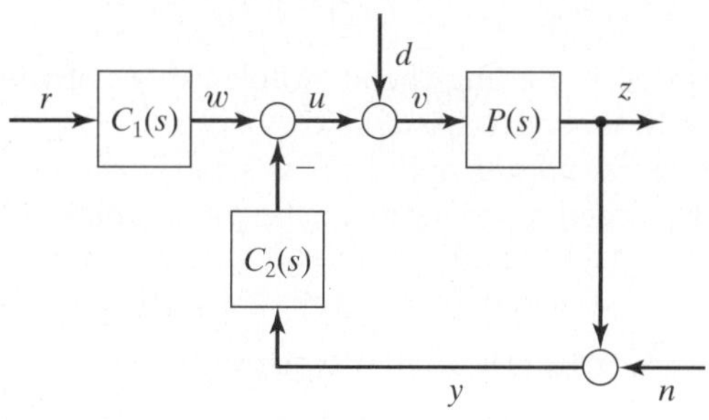

FIGURE 3.8: Algebraic equivalence of Figure 3.7.

EXAMPLE 3.34

Let $\boldsymbol{C}(s) = \frac{1}{s}\begin{bmatrix} s+1 & 1 \end{bmatrix}$ and $P(s) = \frac{1}{s+2}$. Then the system in Figure 3.7 is internally stable, but the transfer function from r to w in Figure 3.8 is $\frac{s+1}{s}$, which is not stable. This means that if we build this feedback system using the structure in Figure 3.8, then some internal signals will become unbounded for bounded external signals. Therefore, Figures 3.7 and 3.8 are not analytically equivalent (in terms of stability) and we should not use the structure in Figure 3.8 to realize the system in Figure 3.7.

For the widely used unity feedback system shown in Figure 3.9, its internal stability is equivalent to that of the feedback system for stabilization shown in Figure 3.6.

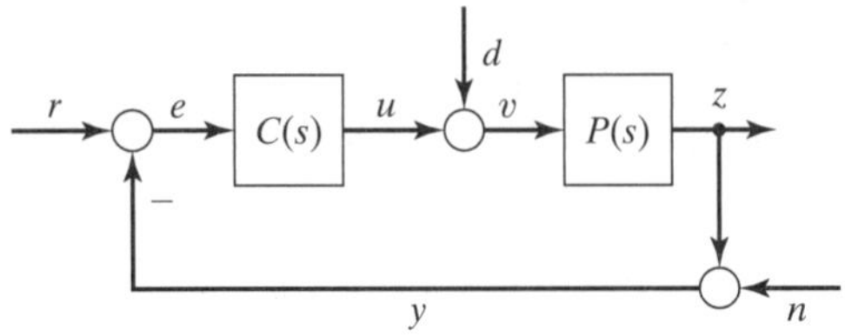

FIGURE 3.9: Unity feedback system.

Finally, let us see how the Routh stability criterion can be used to determine the ranges of parameters to ensure closed-loop stability for a feedback system.

EXAMPLE 3.35

Consider the unity feedback control system shown in Figure 3.9 with

$$P(s) = \frac{10}{(s+1)(s+2)}.$$

Let $C(s)$ be a **proportional-integral (PI)** controller, i.e.,

$$C(s) = K_P + \frac{K_I}{s}.$$

We are interested in finding the allowable ranges of the parameters K_P and K_I so that the closed-loop system is stable.

Since $C(s)$ is proper and $P(s)$ is strictly proper, we only need to check the stability of the closed-loop characteristic polynomial. The characteristic polynomial is

$$d(s) = s^3 + 3s^2 + (2 + 10K_P)s + 10K_I.$$

The corresponding Routh table is constructed as

s^3	1	$(2+10K_P)$
s^2	3	$10K_I$
s^1	$\frac{3(2+10K_P)-10K_I}{3}$	
s^0	$10K_I$	

By Routh criterion, we need

$$\frac{3(2+10K_P)-10K_I}{3} > 0, \quad 10K_I > 0$$

to ensure the stability of the system. That is, we need

$$3 + 15K_P - 5K_I > 0, \quad K_I > 0.$$

The stability region of K_I and K_P is plotted in Figure 3.10.

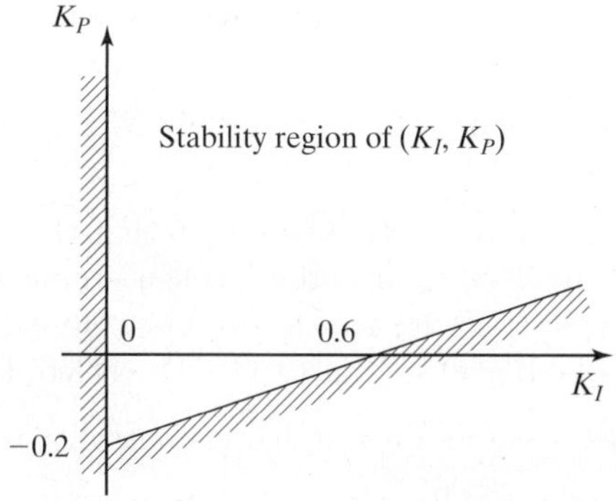

FIGURE 3.10: Stability region for Example 3.35.

3.6 POLE PLACEMENT DESIGN

The next few sections concern stabilizing controller design. We start with the stabilization problem shown in Figure 3.11, in which the plant $P(s)$ is given and the controller $C(s)$ is to be designed so that the closed-loop system is internally stable.

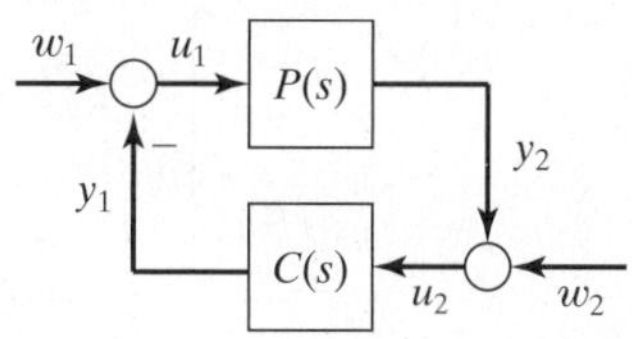

FIGURE 3.11: Feedback system for stabilization.

The idea of pole placement is to design the controller so that the poles of the closed-loop system, i.e., the roots of the closed-loop characteristic polynomial are placed in the prespecified desirable location. In the following we will assume that the plant is proper. We will consider the pole placement problems using proper controllers and strictly proper controllers.

Let a plant be given by

$$P(s) = \frac{b(s)}{a(s)} = \frac{b_0 s^n + b_1 s^{n-1} + \cdots + b_n}{a_0 s^n + a_1 s^{n-1} + \cdots + a_n} \tag{3.12}$$

where $a(s)$ and $b(s)$ are coprime and $a_0 \neq 0$. We first consider proper controllers of the form

$$C(s) = \frac{q(s)}{p(s)} = \frac{q_0 s^m + q_1 s^{m-1} + \cdots + q_m}{p_0 s^m + p_1 s^{m-1} + \cdots + p_m} \tag{3.13}$$

where $p(s)$ and $q(s)$ are coprime and $p_0 \neq 0$. Then the closed-loop characteristic polynomial is

$$a(s)p(s) + b(s)q(s) = c(s) \tag{3.14}$$

where $c(s)$ is a polynomial of degree not exceeding $n+m$:

$$c(s) = c_0 s^{n+m} + c_1 s^{n+m-1} + \cdots + c_{n+m}.$$

It follows from the materials in Section 3.5 that the closed-loop system is internally stable if and only if $c_0 \neq 0$ and $c(s)$ is a stable polynomial. Therefore, if we can arbitrarily specify an $(n+m)$th order stable polynomial $c(s)$, and then choose polynomials $p(s)$ and $q(s)$ so that (3.14) is satisfied, then we will be able to stabilize the closed-loop system and a stabilizing controller is given by $C(s) = \dfrac{q(s)}{p(s)}$.

Equation (3.14), with given $a(s), b(s), c(s)$ and unknown $p(s), q(s)$, is called a polynomial **Diophantine equation**. Under what condition does a Diophantine equation have a solution? How can one find a solution if it exists? We are set to answer these questions.

Following Appendix B.2, define two $(n+m) \times m$ matrices

$$\boldsymbol{T}(a(s),m) = \overbrace{\begin{bmatrix} a_0 & 0 & \cdots & 0 \\ a_1 & a_0 & \ddots & \vdots \\ \vdots & a_1 & \ddots & 0 \\ a_n & \vdots & \ddots & a_0 \\ 0 & a_n & \ddots & a_1 \\ \vdots & \vdots & \ddots & \vdots \\ 0 & 0 & \cdots & a_n \end{bmatrix}}^{m \text{ columns}}, \quad \boldsymbol{T}(b(s),m) = \overbrace{\begin{bmatrix} b_0 & 0 & \cdots & 0 \\ b_1 & b_0 & \ddots & \vdots \\ \vdots & b_1 & \ddots & 0 \\ b_n & \vdots & \ddots & b_0 \\ 0 & b_n & \ddots & b_1 \\ \vdots & \vdots & \ddots & \vdots \\ 0 & 0 & \cdots & b_n \end{bmatrix}}^{m \text{ columns}}.$$

These matrices have the property that elements in each of their diagonals are equal. Such matrices are called **Toeplitz matrices**. Further, define

$$\boldsymbol{S}(a(s),b(s),m) = \begin{bmatrix} \boldsymbol{T}(a(s),m) & \boldsymbol{T}(b(s),m) \end{bmatrix},$$

which is an $(n+m)\times 2m$ matrix. When $m < n$, it is a tall matrix. When $m = n$, it is a square matrix. When $m > n$, it is a wide matrix. It follows from Theorem B.3 that the Sylvester's resultant matrix $\boldsymbol{S}(a(s),b(s),n)$ is nonsingular if and only if polynomials $a(s)$ and $b(s)$ are coprime.

What about $\boldsymbol{S}(a(s),b(s),m)$ when $m \neq n$? Well, if $a(s)$ and $b(s)$ are coprime, then $\boldsymbol{S}(a(s),b(s),m)$ has full column rank when $m < n$ and has full row rank when $m > n$. The reader might wish to test his/her linear algebra skill by proving this statement based on Theorem B.3.

Now we are ready to carry out the pole placement. Notice that by comparing the coefficients of both sides of the Diophantine equation (3.14), we can write it as a set of linear equations in the matrix form

$$
\boldsymbol{S}(a(s), b(s), m+1) \begin{bmatrix} p_0 \\ \vdots \\ p_m \\ q_0 \\ \vdots \\ q_m \end{bmatrix} = \begin{bmatrix} c_0 \\ c_1 \\ \vdots \\ c_{n+m-1} \\ c_{n+m} \end{bmatrix}. \tag{3.15}
$$

If $m+1 < n$, then the set of linear equations (3.15) has more equations than unknowns and the Sylvester's resultant matrix $\boldsymbol{S}(a(s), b(s), m+1)$ has full column rank. Hence, if $c(s)$ is arbitrarily chosen, then the set of linear equations (3.15) is unlikely to have a solution. If $m+1 = n$, then the set of equations has the same number of equations as unknowns and the coefficient matrix $\boldsymbol{S}(a(s), b(s), m+1)$ is nonsingular following Theorem B.3. Hence, no matter how $c(s)$ is chosen, there exists a unique solution to (3.15). If $m+1 > n$, then the set of equations has more unknowns than equations; i.e., $\left[\ \boldsymbol{T}(a(s), m+1) \quad \boldsymbol{T}(b(s), m+1)\ \right] = \boldsymbol{S}(a(s), b(s), m+1)$ is a wide matrix. This matrix has full row rank. Therefore, in this case, (3.15) has infinitely many solutions for an arbitrary $c(s)$.

In summary, there exists an mth order controller so that the closed-loop poles are arbitrarily assigned to a set of $n+m$ complex numbers if and only if $m \geq n-1$. If this condition is satisfied and the desired closed-loop characteristic polynomial $c(s)$ is given, then the controller is obtained by solving the set of linear equations.

In the case when a pole placement controller exists, it can be found by setting up and solving the set of linear equations (3.15), especially in cases when the plant is of high order and when computer-aided design is used. For a low-order plant and for hand computation, it is often more convenient to find the pole placement controllers by directly comparing the coefficients of both sides of (3.14).

EXAMPLE 3.36

Let

$$
P(s) = \frac{s-1}{s(s-2)}.
$$

Design a first-order controller so that the closed-loop poles are

$$
-1, \quad -2, \quad -3,
$$

or, equivalently, so that the closed-loop characteristic polynomial is

$$
c(s) = (s+1)(s+2)(s+3) = s^3 + 6s^2 + 11s + 6.
$$

Denote

$$
C(s) = \frac{q_0 s + q_1}{p_0 s + p_1}.
$$

Then the set of equations (3.15) is

$$\begin{bmatrix} 1 & 0 & 0 & 0 \\ -2 & 1 & 1 & 0 \\ 0 & -2 & -1 & 1 \\ 0 & 0 & 0 & -1 \end{bmatrix} \begin{bmatrix} p_0 \\ p_1 \\ q_0 \\ q_1 \end{bmatrix} = \begin{bmatrix} 1 \\ 6 \\ 11 \\ 6 \end{bmatrix}.$$

The solution is

$$p_0 = 1, \quad p_1 = -25, \quad q_0 = 33, \quad q_1 = -6.$$

Therefore the required controller is

$$C(s) = \frac{33s - 6}{s - 25}.$$

We can also set the equations by comparing the coefficients of

$$s(s-2)(p_0 s + p_1) + (s-1)(q_0 s + q_1) = s^3 + 6s^2 + 11s + 6.$$

This gives

$$p_0 = 1, \quad -2p_0 + p_1 + q_0 = 6, \quad -2p_1 - q_0 + q_1 = 11, \quad -q_1 = 6.$$

Solving this set of linear equations, we get the same controller coefficients. It appears that the latter method is simpler for hand computation.

We have seen that to achieve arbitrary pole placement, the controller order m has to be at least $n-1$, where n is the plant order. However, we may be able to place the closed-loop poles to certain particular positions by using controllers with an order less than $n-1$, as shown in the following example.

EXAMPLE 3.37

Let

$$P(s) = \frac{1}{s(s+2)}.$$

Suppose that we wish to use a zeroth order controller $C(s) = q_0/p_0$ to place the closed-loop poles to -1 and -1, i.e., to make the closed-loop characteristic polynomial

$$c(s) = s(s+2)p_0 + q_0 = (s+1)^2.$$

The solution exists with $p_0 = q_0 = 1$, i.e., $C(s) = 1$. What happened to the above Diophantine equation is that although it results in a set of linear equations with more equations than unknowns, the equations are not independent because of the particular $c(s)$.

It is of interest to notice what happens if we insist on using a first-order controller

$$C(s) = \frac{q_0 s + q_1}{p_0 s + p_1}$$

to place the closed-loop poles to $-1, -1, -2$. In this case, the Diophantine equation is

$$s(s+2)(p_0 s + p_1) + (q_0 s + q_1) = (s+1)^2(s+2).$$

The solution is

$$p_0 = 1, \quad p_1 = 2, \quad q_0 = 1, \quad p_1 = 2,$$

which gives

$$C(s) = \frac{s+2}{s+2} = 1.$$

It is then debatable whether $C(s)$ is of first order or whether the closed-loop system is of third order. It is probably unimportant to have an absolutely clear interpretation here. Readers can exercise their own judgement.

This example also enforces the fact that the lowest order of a stabilizing controller for an nth order plant might be lower than $n-1$, although the lowest order of controllers for arbitrary pole placement is $n-1$. Since the order of the controller is directly related to controller complexity and is an important specification in controller design, it is desirable to use a low-order controller to stabilize a plant. A systematic way to find the lowest order of the stabilizing controllers for a given plant is not yet available. However, there are useful *ad hoc* methods to design low-order stabilizing controllers, which work for many plants encountered in practice. These methods will be introduced in later chapters.

The pole placement controller is unique if and only if $m = n-1$. If $m > n-1$, then pole placement controllers become nonunique for a given set of desired closed-loop poles. The reason for this is because when m increases, the number of controller parameters increases twice as fast as the number of coefficients of the closed-loop characteristic polynomial, leading to a set of equations with more unknowns than equations.

Next, we consider pole placement using strictly proper controllers. Strictly proper controllers are more advantageous in some applications, as will be seen in later sections. In this case, a controller has the form

$$C(s) = \frac{q(s)}{p(s)} = \frac{q_1 s^{m-1} + \cdots + q_m}{p_0 s^m + p_1 s^{m-1} + \cdots + p_m} \tag{3.16}$$

where $p(s)$ and $q(s)$ are coprime and $p_0 \neq 0$. The closed-loop characteristic polynomial is still

$$a(s)p(s) + b(s)q(s) := c(s) = c_0 s^{n+m} + c_1 s^{n+m-1} + \cdots + c_{n+m}.$$

By comparing coefficients, the Diophantine equation becomes a set of equations of thc form

$$\begin{cases} p_0 = c_0/a_0 \\ \boldsymbol{S}(a(s), b(s), m) \begin{bmatrix} p_1 \\ \vdots \\ p_m \\ q_1 \\ \vdots \\ q_m \end{bmatrix} = \begin{bmatrix} c_1 - a_1 c_0/a_0 \\ \vdots \\ c_n - a_n c_0/a_0 \\ c_{n+1} \\ \vdots \\ c_{n+m} \end{bmatrix} \end{cases}. \tag{3.17}$$

In summary, there exists an mth order, strictly proper controller so that the closed-loop poles are arbitrarily assigned to a set of $n+m$ complex numbers if and only if $m \geq n$. If this condition is satisfied and the desired closed-loop characteristic polynomial $c(s)$ is given, then the controller is obtained by solving the set of linear equations.

EXAMPLE 3.38

Again, let

$$P(s) = \frac{s-1}{s(s-2)}.$$

Design a second-order, strictly proper controller so that the closed-loop poles are

$$-1 \pm j1, \quad -2 \pm j2,$$

or, equivalently, so that the closed-loop characteristic polynomial is

$$c(s) = (s^2 + 2s + 2)(s^2 + 4s + 8) = s^4 + 6s^3 + 18s^2 + 24s + 16.$$

Denote

$$C(s) = \frac{q_1 s + q_2}{p_0 s^2 + p_1 s + p_2}.$$

Then the set of equations (3.17) is

$$\begin{cases} p_0 = 1 \\ \begin{bmatrix} 1 & 0 & 0 & 0 \\ -2 & 1 & 1 & 0 \\ 0 & -2 & -1 & 1 \\ 0 & 0 & 0 & -1 \end{bmatrix} \begin{bmatrix} p_1 \\ p_2 \\ q_1 \\ q_2 \end{bmatrix} = \begin{bmatrix} 6-(-2) \\ 18-0 \\ 24 \\ 16 \end{bmatrix} \end{cases}.$$

Its solution is

$$p_0 = 1, \quad p_1 = 8, \quad p_2 = -74, \quad q_1 = 108, \quad q_2 = -16.$$

Therefore the required controller is

$$C(s) = \frac{108s - 16}{s^2 + 8s - 74}.$$

One can also set the equations from

$$s(s-2)(p_0 s^2 + p_1 s + p_2) + (s-1)(q_1 s + q_2) = s^4 + 6s^3 + 18s^2 + 24s + 16.$$

This gives

$$p_0 = 1, \quad -2p_0 + p_1 = 6, \quad -2p_1 + p_2 + q_1 = 18, \quad -2p_2 - q_1 + q_2 = 24, \quad -q_2 = 16.$$

The same solution

$$p_0 = 1, \quad p_1 = 8, \quad p_2 = -74, \quad q_1 = 108, \quad q_2 = -16$$

is obtained. Again, for hand computation, the latter method appears more convenient. It is noted that the controller itself is unstable even though the closed-loop system is stable. It will also be shown in later chapters that this system cannot be stabilized by any stable controller.

Again, notice that the strictly proper pole placement controller is unique if $m = n$ and is nonunique if $m > n$.

3.7 ALL STABILIZING CONTROLLERS*

Again, let

$$P(s) = \frac{b(s)}{a(s)} = \frac{b_0 s^n + b_1 s^{n-1} + \cdots + b_n}{a_0 s^n + a_1 s^{n-1} + \cdots + a_n}$$

where $a(s)$ and $b(s)$ are coprime and $a_0 \neq 0$. Since any useful controller has to at least stabilize $P(s)$, we are interested in studying the set of all stabilizing controllers. We will denote this set by $\mathcal{S}(P)$. Let

$$C_0(s) = \frac{q(s)}{p(s)} = \frac{q_0 s^m + q_1 s^{m-1} + \cdots + q_m}{p_0 s^m + p_1 s^{m-1} + \cdots + p_m},$$

where $p(s)$ and $q(s)$ are coprime and $p_0 \neq 0$, be any stabilizing controller, i.e.,

$$c(s) = a(s)p(s) + b(s)q(s)$$

is stable. Factorize $c(s)$ as $c(s) = f(s)h(s)$ such that $\deg f(s) = n$ and $\deg h(s) = m$. Let

$$M(s) = \frac{a(s)}{f(s)}, \quad N(s) = \frac{b(s)}{f(s)}, \quad X(s) = \frac{p(s)}{h(s)}, \quad Y(s) = \frac{q(s)}{h(s)}.$$

Then $M(s), N(s), X(s), Y(s)$ are all stable transfer functions satisfying

$$P(s) = \frac{N(s)}{M(s)}, \quad C_0(s) = \frac{Y(s)}{X(s)}$$

and

$$M(s)X(s) + N(s)Y(s) = 1.$$

Theorem 3.39 (Parametrization of all stabilizing controllers).

$$\mathcal{S}(P) = \left\{ C(s) = \frac{Y(s) + M(s)Q(s)}{X(s) - N(s)Q(s)} : Q(s) \text{ is an arbitrary stable system} \right\}.$$

This theorem says that every stabilizing controller has the form shown in Figure 3.12 or 3.13, where $Q(s)$ is an arbitrary stable system. In the actual implementation of the controller, if a satisfactory $Q(s)$ has already been chosen, we can either explicitly compute the transfer function of $C(s)$ using the formula given in Theorem 3.39 and then connect the controller to the plant as in Figure 3.11, or build up and connect together the different parts of the controller as shown in Figure 3.12 or 3.13.

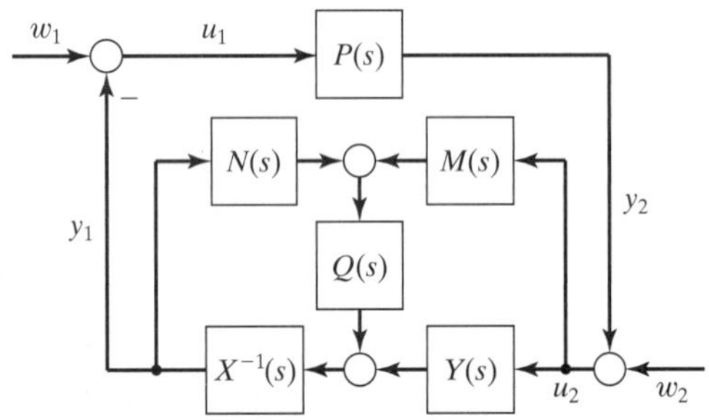

FIGURE 3.12: All stabilizing controller (1).

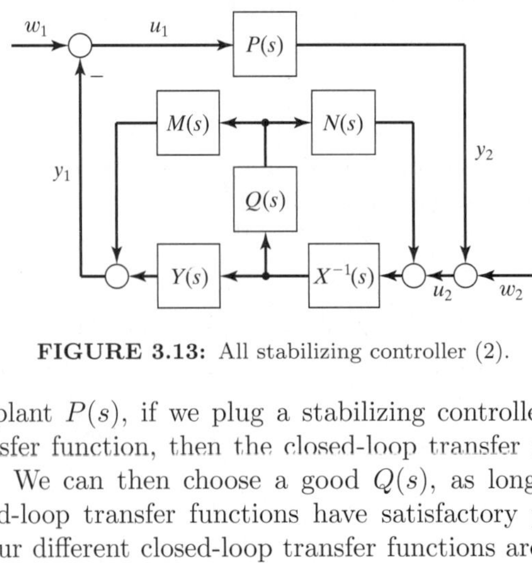

FIGURE 3.13: All stabilizing controller (2).

For a fixed plant $P(s)$, if we plug a stabilizing controller of this form into a closed-loop transfer function, then the closed-loop transfer function becomes a function of $Q(s)$. We can then choose a good $Q(s)$, as long as it is stable, so that certain closed-loop transfer functions have satisfactory properties. Rather surprisingly, all four different closed-loop transfer functions are affine functions of $Q(s)$, as given in the following formula:

$$\begin{bmatrix} \dfrac{P(s)}{1+P(s)C(s)} & \dfrac{1}{1+P(s)C(s)} \\ \dfrac{P(s)C(s)}{1+P(s)C(s)} & \dfrac{C(s)}{1+P(s)C(s)} \end{bmatrix}$$
$$= \begin{bmatrix} N(s)X(s) & M(s)X(s) \\ N(s)Y(s) & M(s)Y(s) \end{bmatrix} - \begin{bmatrix} N(s) \\ -M(s) \end{bmatrix} Q(s) \begin{bmatrix} N(s) & M(s) \end{bmatrix}. \quad (3.18)$$

This makes choosing $Q(s)$ rather easy in many applications. Examples of such applications are given in Section 9.4.

EXAMPLE 3.40

Let us consider the double integrator plant $P(s) = 1/s^2$. It is easy to check that

$$C_0(s) = \frac{2\sqrt{2}s + 1}{s^2 + 2\sqrt{2}s + 4}$$

is a stabilizing controller. Under this controller, the closed-loop characteristic polynomial is

$$s^2(s^2 + 2\sqrt{2}s + 4) + 2\sqrt{2}s + 1 = (s^2 + \sqrt{2}s + 1)^2.$$

Let $f(s) = h(s) = s^2 + \sqrt{2}s + 1$ and

$$M(s) = \frac{a(s)}{f(s)} = \frac{s^2}{s^2 + \sqrt{2}s + 1}, \qquad N(s) = \frac{b(s)}{f(s)} = \frac{1}{s^2 + \sqrt{2}s + 1},$$

$$X(s) = \frac{p(s)}{h(s)} = \frac{s^2 + 2\sqrt{2}s + 4}{s^2 + \sqrt{2}s + 1}, \qquad Y(s) = \frac{q(s)}{h(s)} = \frac{2\sqrt{2}s + 1}{s^2 + \sqrt{2}s + 1}.$$

Then, all stabilizing controllers for the double integrator plant are given by

$$C(s) = \frac{\dfrac{2\sqrt{2}s + 1}{s^2 + \sqrt{2}s + 1} + \dfrac{s^2}{s^2 + \sqrt{2}s + 1}Q(s)}{\dfrac{s^2 + 2\sqrt{2}s + 4}{s^2 + \sqrt{2}s + 1} - \dfrac{1}{s^2 + \sqrt{2}s + 1}Q(s)} \tag{3.19}$$

where $Q(s)$ can be any stable system. If we choose $Q(s) = 0$, then we get

$$C(s) = \frac{2\sqrt{2}s + 1}{s^2 + 2\sqrt{2}s + 4}, \tag{3.20}$$

which is the original stabilizing controller $C_0(s)$. We can also choose $Q(s) = K$, which is an arbitrary static system. In this case, a family of stabilizing controllers is given by

$$C(s) = \frac{Ks^2 + 2\sqrt{2}s + 1}{s^2 + 2\sqrt{2}s + 4 - K}$$

for arbitrary $K \in \mathbb{R}$. We can even choose $Q(s) = 1 + \sqrt{2} + \dfrac{2(1 - \sqrt{2})}{s + 1}$. Then,

$$C(s) = \frac{(1 + \sqrt{2})s + 1}{s + 1 + \sqrt{2}}. \tag{3.21}$$

Although the above process gives all stabilizing controllers for a given plant, nothing is said about whether any particular one is better than another. At this point, the only intuition that we can exercise is that a lower order controller is probably easier to implement than a higher order controller. In this sense, stabilizing controller (3.21) is better than stabilizing controller (3.20).

Let us now parameterize the set of stabilizing controllers starting from a different initial stabilizing controller. Let us choose

$$C_0(s) = \frac{(1+\sqrt{2})s+1}{s+1+\sqrt{2}}.$$

Then the characteristic polynomial is

$$s^2(s+1+\sqrt{2}) + (1+\sqrt{2})s + 1 = (s^2+\sqrt{2}s+1)(s+1).$$

Let $f(s) = s^2+\sqrt{2}s+1$ and $h(s) = s+1$, and let

$$M(s) = \frac{a(s)}{f(s)} = \frac{s^2}{s^2+\sqrt{2}s+1}, \qquad N(s) = \frac{b(s)}{f(s)} = \frac{1}{s^2+\sqrt{2}s+1},$$

$$X(s) = \frac{p(s)}{h(s)} = \frac{s+1+\sqrt{2}}{s+1}, \qquad Y(s) = \frac{q(s)}{h(s)} = \frac{(1+\sqrt{2})s+1}{s+1}.$$

Then, all stabilizing controllers are also given by

$$C(s) = \frac{\dfrac{(1+\sqrt{2})s+1}{s+1} + \dfrac{s^2}{s^2+\sqrt{2}s+1}Q(s)}{\dfrac{s+1+\sqrt{2}}{s+1} - \dfrac{1}{s^2+\sqrt{2}s+1}Q(s)} \tag{3.22}$$

where $Q(s)$ can be any stable system. Parametrizations (3.22) and (3.19) have different appearances but they actually give the same set of controllers. For example, taking $Q(s) = -1-\sqrt{2}-\dfrac{2(1-\sqrt{2})}{s+1}$ in (3.22) gives controller (3.20), and taking $Q(s) = 0$ gives controller (3.21).

EXAMPLE 3.41

Let us consider a bi-proper plant $P(s) = \dfrac{s+1}{s+2}$. Something special arises in such cases. Since $P(s)$ is stable, a particular stabilizing controller is the zero controller $C_0(s) = 0$. For this particular initial stabilizing controller, we have

$$p(s) = 1, \quad q(s) = 0, \quad f(s) = a(s), \quad h(s) = 1.$$

Therefore,

$$M(s) = \frac{a(s)}{f(s)} = 1, \qquad N(s) = \frac{b(s)}{f(s)} = P(s) = \frac{s+1}{s+2},$$

$$X(s) = \frac{p(s)}{h(s)} = 1, \qquad Y(s) = \frac{q(s)}{h(s)} = 0$$

and all stabilizing controllers are given by

$$C(s) = \frac{Q(s)}{1-P(s)Q(s)}$$

where $Q(s)$ can be any stable system. If we choose $Q(s) = 1$, then another particular stabilizing controller is given by $C(s) = s + 2$, which is a PD controller and is nonproper. It sounds a bit awkward from a practical point of view, but theoretically there is nothing wrong if only closed-loop stability is concerned.

Let us briefly consider the reason why Theorem 3.39 is true. If a controller is given by

$$C(s) = \frac{Y(s) + M(s)Q(s)}{X(s) - N(s)Q(s)}$$

for some stable $Q(s)$, then all the closed-loop transfer functions are given by (3.18), which are all stable since they involve only the products and sums of stable transfer functions. Hence the closed-loop system is stable. On the other hand, if $C(s)$ is a stabilizing controller, then $C(s)$ can be written as

$$C(s) = \frac{Y(s) + M(s)Q(s)}{X(s) - N(s)Q(s)}$$

with

$$Q(s) = \frac{X(s)C(s) - Y(s)}{N(s)C(s) + M(s)}.$$

In this case, the following closed-loop transfer functions

$$\begin{bmatrix} \dfrac{1}{1+P(s)C(s)} & \dfrac{-C(s)}{1+P(s)C(s)} \\ \dfrac{P(s)}{1+P(s)C(s)} & \dfrac{1}{1+P(s)C(s)} \end{bmatrix}$$

$$= \begin{bmatrix} M(s)X(s) & -M(s)Y(s) \\ N(s)X(s) & M(s)X(s) \end{bmatrix} - \begin{bmatrix} M(s) \\ N(s) \end{bmatrix} Q(s) \begin{bmatrix} N(s) & M(s) \end{bmatrix} \quad (3.23)$$

have to be stable. Multiply (3.23) from the left by $\begin{bmatrix} X(s) & Y(s) \end{bmatrix}$ and from the right by $\begin{bmatrix} Y(s) \\ X(s) \end{bmatrix}$. Remember that $M(s)X(s) + N(s)Y(s) = 1$. We then get

$$Q(s) = \begin{bmatrix} X(s) & Y(s) \end{bmatrix} \begin{bmatrix} M(s)X(s) & -M(s)Y(s) \\ N(s)X(s) & M(s)X(s) \end{bmatrix} \begin{bmatrix} Y(s) \\ X(s) \end{bmatrix}$$

$$- \begin{bmatrix} X(s) & Y(s) \end{bmatrix} \begin{bmatrix} \dfrac{1}{1+P(s)C(s)} & \dfrac{-C(s)}{1+P(s)C(s)} \\ \dfrac{P(s)}{1+P(s)C(s)} & \dfrac{1}{1+P(s)C(s)} \end{bmatrix} \begin{bmatrix} Y(s) \\ X(s) \end{bmatrix}$$

which has to be stable since it only involves the products and sums of stable transfer functions. This implies that $C(s)$ has to belong to the set given in Theorem 3.39.

If $P(s)$ is stable, then $C_0(s) = 0$ is one of its stabilizing controllers. In this case, $c(s) = a(s)$. We choose $f(s) = a(s)$ and $h(s) = 1$. Hence

$$M(s) = 1, \quad N(s) = P(s), \quad X(s) = 1, \quad Y(s) = 0.$$

Then the following corollary follows.

Corollary 3.42. If $P(s)$ is stable, then

$$\mathcal{S}(P) = \left\{ C(s) = \frac{Q(s)}{1 - P(s)Q(s)} : Q(s) \text{ is an arbitrary stable system} \right\}.$$

In this case, Figures 3.12 and 3.13 become Figure 3.14. If the controller is implemented using the structure shown in Figure 3.14, then the controller contains a model of the plant. Hence this controller structure is often called the **internal model control**.

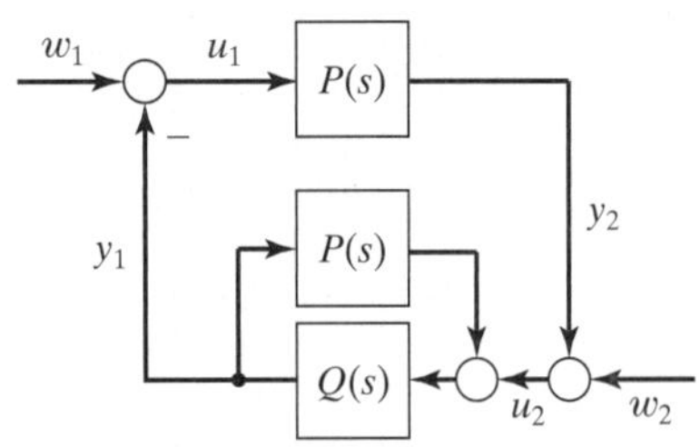

FIGURE 3.14: Internal model stabilization.

In this case, the four different closed-loop transfer functions, expressed in terms of $Q(s)$, are given by the following formula:

$$\begin{bmatrix} \dfrac{P(s)}{1+P(s)C(s)} & \dfrac{1}{1+P(s)C(s)} \\ \dfrac{P(s)C(s)}{1+P(s)C(s)} & \dfrac{C(s)}{1+P(s)C(s)} \end{bmatrix}$$

$$= \begin{bmatrix} P(s) & 1 \\ 0 & 0 \end{bmatrix} - \begin{bmatrix} P(s) \\ -1 \end{bmatrix} Q(s) \begin{bmatrix} P(s) & 1 \end{bmatrix}.$$

3.8 ALL STABILIZING 2DOF CONTROLLERS*

In this section, we extend the parametrization of all stabilizing controllers in the last section to 2DOF feedback systems as shown in Figure 3.15.

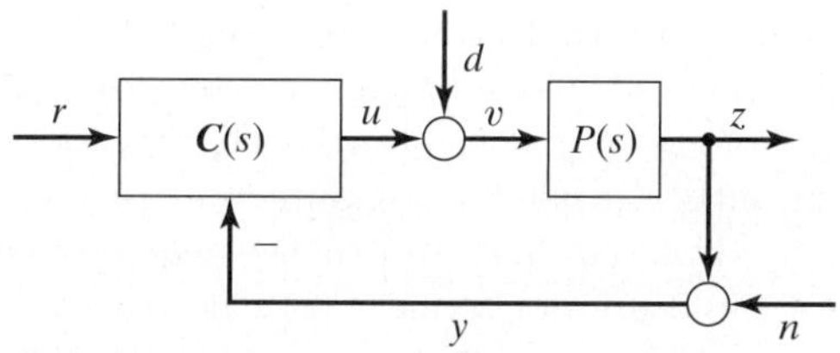

FIGURE 3.15: 2DOF feedback control system.

Let

$$P(s) = \frac{b(s)}{a(s)} = \frac{b_0 s^n + b_1 s^{n-1} + \cdots + b_n}{a_0 s^n + a_1 s^{n-1} + \cdots + a_n}$$

where $a(s)$ and $b(s)$ are coprime and $a_0 \neq 0$. Since any useful 2DOF controller has to at least stabilize $P(s)$, i.e., give an internally stable closed-loop system, we are interested in studying the set of all 2DOF controllers that give stable closed-loop systems. We will denote this set by $\mathcal{T}(P)$; i.e.,

$$\mathcal{T}(P) = \{\boldsymbol{C}(s) = [\ C_1(s) \quad C_2(s)\ ] : \text{the system in Figure 3.15 is internally stable.}\}.$$

Let

$$C_0(s) = \frac{q(s)}{p(s)} = \frac{q_0 s^m + q_1 s^{m-1} + \cdots + q_m}{p_0 s^m + p_1 s^{m-1} + \cdots + p_m},$$

where $p(s)$ and $q(s)$ are coprime and $p_0 \neq 0$, be any stabilizing controller, i.e.,

$$c(s) = a(s)p(s) + b(s)q(s)$$

is stable. Factorize $c(s)$ as $c(s) = f(s)h(s)$ such that $\deg f(s) = n$ and $\deg h(s) = m$. Let

$$M(s) = \frac{a(s)}{f(s)}, \quad N(s) = \frac{b(s)}{f(s)}, \quad X(s) = \frac{p(s)}{h(s)}, \quad Y(s) = \frac{q(s)}{h(s)}.$$

Then $M(s), N(s), X(s), Y(s)$ are all stable transfer functions satisfying

$$P(s) = \frac{N(s)}{M(s)}, \quad C_0(s) = \frac{Y(s)}{X(s)}$$

and

$$M(s)X(s) + N(s)Y(s) = 1.$$

Theorem 3.43 (Parametrization of all stabilizing 2DOF controllers).

$$\mathcal{T}(P) = \left\{ \boldsymbol{C}(s) = \left[\frac{Q_1(s)}{X(s) - N(s)Q_2(s)} \quad \frac{Y(s) + M(s)Q_2(s)}{X(s) - N(s)Q_2(s)} \right] : Q_1(s) \text{ and } Q_2(s) \text{ are two arbitrary stable systems} \right\}. \quad (3.24)$$

This theorem says that every stabilizing 2DOF controller has the form shown in Figure 3.16, where $Q_1(s)$ and $Q_2(s)$ are arbitrary, stable systems. In the actual implementation of the controller, if satisfactory $Q_1(s)$ and $Q_2(s)$ have already been chosen, we can either explicitly compute the transfer function of $\boldsymbol{C}(s)$ using the formula given in Theorem 3.43 and then connect the controller to the plant as in Figure 3.15, or build up and connect together the different parts of the controller as shown in Figure 3.16.

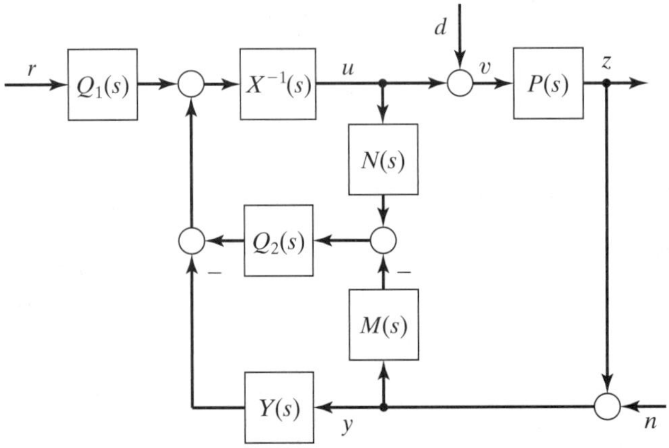

FIGURE 3.16: All stabilizing 2DOF controllers.

For a fixed plant $P(s)$, if we plug a stabilizing 2DOF controller of this form into a closed-loop transfer function, then the closed-loop transfer function becomes a function of $Q_1(s)$ and $Q_2(s)$. We can then choose good $Q_1(s)$ and $Q_2(s)$, as long as they are stable, to meet the design specification. It turns out that all different closed-loop transfer functions are affine functions of either $Q_1(s)$ or $Q_2(s)$, but not both. The four transfer functions from r to the internal variables u, v, y, z depend only on $Q_1(s)$:

$$\begin{bmatrix} \dfrac{C_1(s)}{1+P(s)C_2(s)} \\[2ex] \dfrac{C_1(s)}{1+P(s)C_2(s)} \\[2ex] \dfrac{P(s)C_1(s)}{1+P(s)C_2(s)} \\[2ex] \dfrac{P(s)C_1(s)}{1+P(s)C_2(s)} \end{bmatrix} = \begin{bmatrix} M(s) \\[2ex] M(s) \\[2ex] N(s) \\[2ex] N(s) \end{bmatrix} Q_1(s). \tag{3.25}$$

Only two of them are distinct. The eight transfer functions from d, n to u, v, y, z depend only on $Q_2(s)$:

$$\begin{bmatrix} \dfrac{-P(s)C_2(s)}{1+P(s)C_2(s)} & \dfrac{-C_2(s)}{1+P(s)C_2(s)} \\ \dfrac{1}{1+P(s)C_2(s)} & \dfrac{-C_2(s)}{1+P(s)C_2(s)} \\ \dfrac{P(s)}{1+P(s)C_2(s)} & \dfrac{1}{1+P(s)C_2(s)} \\ \dfrac{P(s)}{1+P(s)C_2(s)} & \dfrac{-P(s)C_2(s)}{1+P(s)C_2(s)} \end{bmatrix}$$

$$= \begin{bmatrix} -N(s)Y(s) & -M(s)Y(s) \\ M(s)X(s) & -M(s)Y(s) \\ N(s)X(s) & M(s)Y(s) \\ N(s)X(s) & -N(s)X(s) \end{bmatrix} - \begin{bmatrix} M(s) \\ M(s) \\ N(s) \\ N(s) \end{bmatrix} Q_2(s) \begin{bmatrix} N(s) & M(s) \end{bmatrix} \tag{3.26}$$

Only four of them are distinct. This makes choosing $Q_1(s)$ and choosing $Q_2(s)$ decoupled and rather convenient.

EXAMPLE 3.44

Again, consider the double integrator plant $P(s) = 1/s^2$. Let us choose

$$C_0(s) = \frac{(1+\sqrt{2})s+1}{s+1+\sqrt{2}}.$$

Then the characteristic polynomial is

$$s^2(s+1+\sqrt{2}) + (1+\sqrt{2})s + 1 = (s^2+\sqrt{2}s+1)(s+1)$$

Let $f(s) = s^2 + \sqrt{2}s + 1$ and $h(s) = s + 1$, and let

$$M(s) = \frac{a(s)}{f(s)} = \frac{s^2}{s^2+\sqrt{2}s+1}, \qquad N(s) = \frac{b(s)}{f(s)} = \frac{1}{s^2+\sqrt{2}s+1},$$

$$X(s) = \frac{p(s)}{h(s)} = \frac{s+1+\sqrt{2}}{s+1}, \qquad Y(s) = \frac{q(s)}{h(s)} = \frac{(1+\sqrt{2})s+1}{s+1}.$$

Then, all stabilizing 2DOF controllers are given by

$$\boldsymbol{C}(s) = \frac{\begin{bmatrix} Q_1(s) & \dfrac{(1+\sqrt{2})s+1}{s+1} + \dfrac{s^2}{s^2+\sqrt{2}s+1}Q_2(s) \end{bmatrix}}{\dfrac{s+1+\sqrt{2}}{s+1} - \dfrac{1}{s^2+\sqrt{2}s+1}Q_2(s)} \tag{3.27}$$

where $Q_1(s)$ and $Q_2(s)$ are any stable systems. For example, taking

$$Q_1(s) = \frac{s+1+\sqrt{2}}{s+1} \quad \text{and} \quad Q_2(s) = 0$$

in (3.27) gives

$$\boldsymbol{C}(s) = \begin{bmatrix} 1 & \dfrac{(1+\sqrt{2})s+1}{s+1+\sqrt{2}} \end{bmatrix}$$

and taking

$$Q_1(s) = \frac{(1+\sqrt{2})s+1}{s+1} \quad \text{and} \quad Q_2(s) = 0$$

gives the unity feedback controller

$$\boldsymbol{C}(s) = \frac{(1+\sqrt{2})s+1}{s+1+\sqrt{2}} \begin{bmatrix} 1 & 1 \end{bmatrix}.$$

The reason why Theorem 3.43 is true is again quite simple. If a 2DOF controller is given by

$$\boldsymbol{C}(s) = \begin{bmatrix} \dfrac{Q_1(s)}{X(s) - N(s)Q_2(s)} & \dfrac{Y(s) + M(s)Q_2(s)}{X(s) - N(s)Q_2(s)}, \end{bmatrix} \tag{3.28}$$

where $Q_1(s)$ and $Q_2(s)$ are any stable systems, then all the closed-loop transfer functions are given by (3.25) and (3.26), which are all stable since they involve only the products and sums of stable transfer functions. Hence the closed-loop system is stable. On the other hand, if $\boldsymbol{C}(s)$ is a stabilizing 2DOF controller, a similar argument as in Section 3.7 shows that $\boldsymbol{C}(s)$ has to be in the form (3.28) with stable $Q_1(s)$ and $Q_2(s)$.

If $P(s)$ is stable, then $C_0(s) = 0$ is one of its stabilizing controllers. In this case, $c(s) = a(s)$. We choose $f(s) = a(s)$ and $h(s) = 1$. Hence

$$M(s) = 1, \quad N(s) = P(s), \quad X(s) = 1, \quad Y(s) = 0.$$

Then the following corollary follows:

Corollary 3.45. If $P(s)$ is stable, then

$$\mathcal{T}(P) = \left\{ \boldsymbol{C}(s) = \begin{bmatrix} \dfrac{Q_1(s)}{1 - P(s)Q_2(s)} & \dfrac{Q_2(s)}{1 - P(s)Q_2(s)} \end{bmatrix} : \right.$$
$$\left. Q_1(s) \text{ and } Q_2(s) \text{ are arbitrary stable systems} \right\}.$$

In this case, Figure 3.16 becomes Figure 3.17. If the controller is implemented using the structure shown in Figure 3.17, then the controller contains a model of the plant. Hence this controller structure can be called the **2DOF internal model control**.

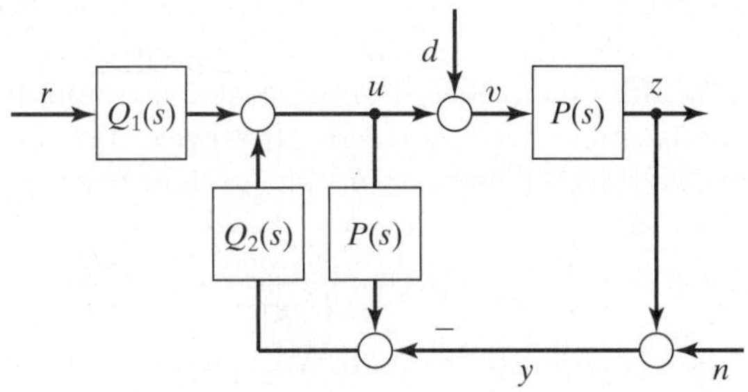

FIGURE 3.17: 2DOF internal model control.

In this case, the transfer functions from r to u, z are $Q_1(s)$ and $P(s)Q_1(s)$, respectively. The other four transfer functions are given by the following formula:

$$\begin{bmatrix} \frac{P(s)}{1+P(s)C_2(s)} & \frac{1}{1+P(s)C_2(s)} \\ \frac{P(s)C_2(s)}{1+P(s)C_2(s)} & \frac{C_2(s)}{1+P(s)C_2(s)} \end{bmatrix}$$

$$= \begin{bmatrix} P(s) & 1 \\ 0 & 0 \end{bmatrix} - \begin{bmatrix} P(s) \\ -1 \end{bmatrix} Q_2(s) \begin{bmatrix} P(s) & 1 \end{bmatrix}. \quad (3.29)$$

3.9 CASE STUDIES

3.9.1 Ball and beam system

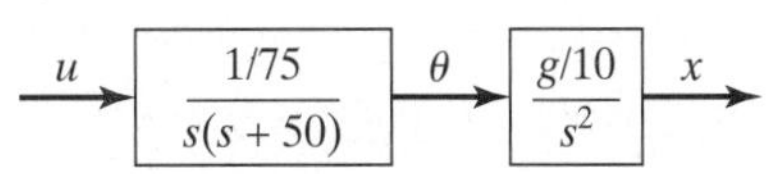

FIGURE 3.18: Block diagram of the ball and beam system.

In Chapter 2, we obtained the mathematical model of a ball and beam system. The block diagram is redrawn in Figure 3.18. The transfer function is given by

$$\begin{bmatrix} \Theta(s) \\ X(s) \end{bmatrix} = \begin{bmatrix} P_\theta(s) \\ P_x(s) \end{bmatrix} U(s) = \frac{1}{s^3(s+50)} \begin{bmatrix} s^2/75 \\ g/750 \end{bmatrix} U(s).$$

The system is a SIDO system with input $u(t)$ and outputs $\theta(t)$ and $x(t)$. Since we have only SISO system stability and stabilization theory available, we attempt to see whether we can turn the stabilization problem of the ball and beam system to stabilization problems of SISO systems. The system has three unstable poles, all at the origin. It will be interesting to ask whether we can stabilize the whole system by using only one output. If we can, then the problem becomes a SISO stabilization problem, the available theory can be used, and one sensor can be saved. It is easily seen that we cannot stabilize the system using only the output $\theta(t)$ since the system from $u(t)$ to $\theta(t)$ contains only one unstable pole and the whole system has three unstable poles. No matter how we design such a SISO controller, the other two unstable poles are outside the feedback loop and cannot be stabilized. However, it is possible to use output $x(t)$ alone to stabilize the whole system. The system from $u(t)$ to $x(t)$ has the transfer function

$$P_x(s) = \frac{X(s)}{U(s)} = \frac{g/750}{s^3(s+50)}.$$

This system is a fourth-order system. We can use a third-order controller to place all closed-loop poles to the left-hand side of the complex plane. However, deciding the desired closed-loop pole locations is not a simple matter. Some understanding of the relationship between pole locations and the closed-loop system performance is needed, which will be covered in later chapters. It usually involves tuning and trial and error. The contents of the subsequent chapters will give certain guidelines in selecting the closed-loop poles. In particular, when optimal control is studied, the best closed-loop poles will be decided by the design algorithms. At this moment, for the purpose of illustration, let us simply choose the following pole locations:

$$-1, -1, -1, -1, -1, -1, -50.$$

Notice that one of the closed-loop poles is chosen to be the same as one of the open-loop poles. The reason is that if this is a good pole, there is no reason to consume control effort to move it. The other poles are chosen all at -1 for lack of better choices at this moment. With a third-order controller

$$C_x(s) = \frac{q_0s^3 + q_1s^2 + q_2s + q_3}{p_0s^3 + p_1s^2 + p_2s + p_3},$$

the Diophantine equation is

$$s^3(s+50)(p_0s^3 + p_1s^2 + p_2s + p_3) + \frac{g}{750}(q_0s^3 + q_1s^2 + q_2s + q_3) = (s+50)(s+1)^6$$

which has the solution

$$p_0 = 1, \qquad p_1 = 6, \qquad p_2 = 15, \qquad p_3 = 20,$$

$$q_0 = 1,148, \qquad q_1 = 5.786 \times 10^4, \qquad q_2 = 2.304 \times 10^4, \qquad q_3 = 3,827.$$

Therefore the designed controller $C_x(s)$ is

$$C_x(s) = \frac{1148s^3 + 5.786 \times 10^4 s^2 + 2.304 \times 10^4 s + 3827}{s^3 + 6s^2 + 15s + 20}.$$

Although it is possible to stabilize the whole system using only one sensor output, it may not be desirable to do so from the viewpoint of achieving good performance when there are indeed two sensors. We are not ready to talk about system performance rigorously at this stage, but some intuition may apply. If there is any disturbance occurring in the DC motor part $P_\theta(s)$ so that some of its variables deviate from their desirable values, it will take a significant amount of time for the information to propagate to the $x(t)$ sensor. The controller will take action only after this sensor realizes that things are going wrong. After all this time, the abnormality in $P_\theta(s)$ might have worsened. This partially explains why big control effort is needed to stabilize the system from the $x(t)$ sensor alone.

Now let us use both sensors in the stabilization. One possible scheme is that we first use a $\theta(t)$ feedback controller $C_\theta(s)$ to stabilize $P_\theta(s)$ and then use an $x(t)$ feedback controller $C_x(s)$ to stabilize the cascaded system of $\dfrac{P_\theta(s)}{1+P_\theta(s)C_\theta(s)}$ and $\dfrac{g/10}{s^2}$. The schematic block diagram of the closed-loop system is shown in Figure 3.19.

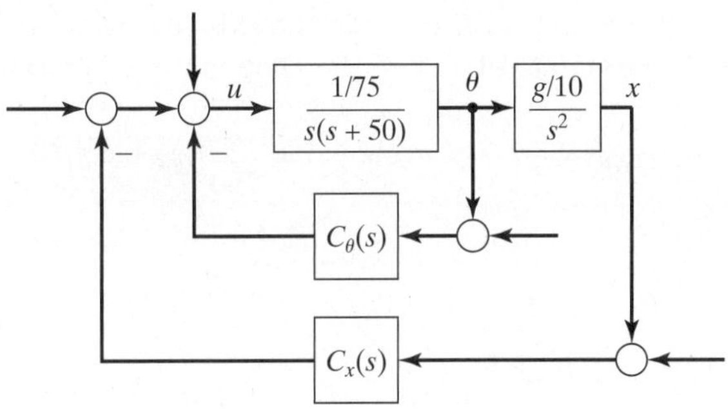

FIGURE 3.19: Block diagram of the ball and beam control system.

It is easy to stabilize $P_\theta(s)$. It is a second-order system. Let us choose a first-order pole placement controller

$$C_\theta(s) = \frac{q_0 s + q_1}{p_0 s + p_1}$$

to place the poles to $-25, -25, -50$. The Diophantine equation becomes

$$s(s+50)(p_0 s + p_1) + 1/75(q_0 s + q_1) = (s+25)^2(s+50).$$

Its solution is

$$p_0 = 1, \quad p_1 = 50, \quad q_0 = 46875, \quad q_1 = 2343750 = 46875 \times 50.$$

Hence the designed controller is

$$C_\theta(s) = \frac{46875(s+50)}{s+50} = 46875.$$

It is rather surprising that the controller order drops to zero because of cancellation. This is a positive by-product of the selection of the inner closed-loop poles. With this design of $C_\theta(s)$, the order of the inner closed-loop system

$$\frac{P_\theta(s)}{1+P_\theta(s)C_\theta(s)} = \frac{1/75}{(s+25)^2}$$

is 2, instead of 3 if other stable poles are selected. This drop of order simplifies the design of controller $C_x(s)$, in addition to making the implementation of the controllers easier. Controller $C_x(s)$ is to be designed to stabilize

$$\frac{g/10}{s^2}\frac{P_\theta(s)}{1+P_\theta(s)C_\theta(s)} = \frac{g/750}{s^2(s+25)^2}.$$

A pole placement controller for arbitrary pole placement would have order 3. Let

$$C_x(s) = \frac{q_0 s^3 + q_1 s^2 + q_2 s + q_3}{p_0 s^3 + p_1 s^2 + p_2 s + p_3},$$

and let us choose the closed-loop poles as

$$-25, \quad -25, \quad -1, \quad -1, \quad -1 \quad -1 \quad -1.$$

Again, here we do not wish to move the existing two poles at -25. The rest of the poles are all chosen at -1. Then the Diophantine equation is

$$s^2(s+25)^2(p_0 s^3 + p_1 s^2 + p_2 s + p_3) + \frac{g}{750}(q_0 s^3 + q_1 s^2 + q_2 s + q_3) = (s+25)^2(s+1)^5.$$

The solution is

$$p_0 = 1, \quad p_1 = 5, \quad p_2 = 10, \quad p_3 = 10,$$

$$q_0 = 382.7, \quad q_1 = 1.921 \times 10^4, \quad q_2 = 2.430 \times 10^5, \quad q_3 = 4.783 \times 10^4.$$

Therefore the designed controller $C_x(s)$ is

$$\begin{aligned} C_x(s) &= \frac{382.7s^3 + 1.921 \times 10^4 s^2 + 2.430 \times 10^5 s + 4.783 \times 10^4}{s^3 + 5s^2 + 10s + 10} \\ &= \frac{382.7(s+25)^2(s+0.2679)}{(s+2.1213)(s^2+1.6108s+1.7456)}. \end{aligned}$$

We are not as lucky as in the design of $C_\theta(s)$; the order drop in $C_x(s)$ does not happen here. The designed controller $C_x(s)$ is of third order. The overall controller is also of third order since $C_\theta(s)$ is of zeroth order. It turns out that the high order of the controller is not necessary in controlling this system. In Chapter 5, we will see how a lower order controller can be designed.

3.9.2 Inverted pendulum system

The linearized model of an inverted pendulum system, as obtained in Section 2.10, has transfer function

$$\boldsymbol{P}(s) = \begin{bmatrix} P_x(s) \\ P_\theta(s) \end{bmatrix} = \frac{1}{s^2(s^2-3g)} \begin{bmatrix} 2/3s^2 - g \\ s^2 \end{bmatrix},$$

which is again a fourth-order SIDO system with poles at $0, 0, \sqrt{3g}, -\sqrt{3g}$. It is highly unstable. It is easily seen that the system cannot be stabilized by using only the $\theta(t)$ output since

$$P_\theta(s) = \frac{1}{s^2 - 3g}$$

which contains only one out of the three unstable poles. Is it possible to stabilize the system using $x(t)$ output alone? This corresponds to balancing a long stick on our finger tip by only looking at the finger tip. This does not seem to be possible. Rather surprisingly, from a theoretical point of view, it is possible to stabilize the inverted pendulum system using the $x(t)$ output alone since the transfer function from the input $f(t)$ to $x(t)$

$$P_x(s) = \frac{2/3s^2 - g}{s^2(s^2 - 3g)}$$

contains all unstable poles of the system. Let us try this scheme for now, i.e., design a feedback controller from $x(t)$ to $f(t)$ to stabilize the inverted pendulum system. Figure 3.20 gives the block diagram of the closed-loop system.

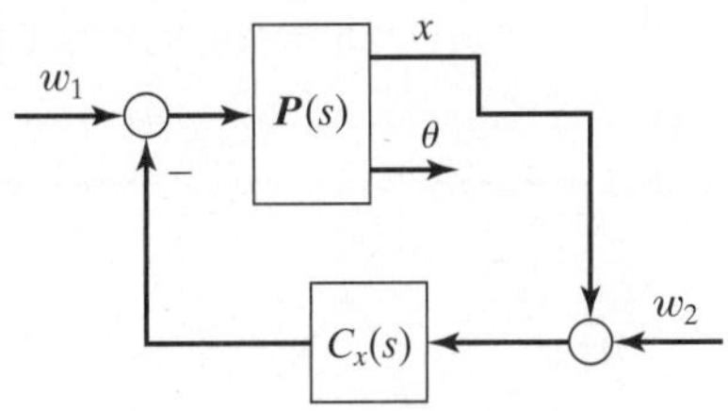

FIGURE 3.20: Stabilization of inverted pendulum using only the output $x(t)$.

Let us carry out a pole placement design using a third-order controller

$$C_x(s) = \frac{q_0 s^3 + q_1 s^2 + q_2 s + q_3}{p_0 s^3 + p_1 s^2 + p_2 s + p_3}$$

to place all closed-loop poles to the following set of values:

$$-6, -0.5 \pm j4, -2 \pm j2.5, -8.5 \pm j4.5.$$

This leads to the Diophantine equation

$$s^2(s^2 - 3g)(p_0 s^3 + p_1 s^2 + p_2 s + p_3) + (0.67s^2 - g)(q_0 s^3 + q_1 s^2 + q_2 s + q_3)$$
$$= (s+6)(s^2 + s + 16.25)(s^2 + 4s + 10.25)(s^2 + 17s + 92.5).$$

The solution is

$$p_0 = 1, \qquad p_1 = 28, \qquad p_2 = -1434, \qquad p_3 = -5963,$$
$$q_0 = 2705, \qquad q_1 = 13640, \qquad q_2 = -7567, \qquad q_3 = -9433.$$

Then the designed controller is

$$C_x(s) = \frac{2705s^3 + 13640s^2 - 7567s - 9433}{s^3 + 28s^2 - 1434s \quad 5963}.$$

Is this a good design? Let us take a look at the simulation. In order for the simulation to make sense physically, we consider the impulse responses of the systems from $w_1(t)$ to $x(t)$ and from $w_1(t)$ to $\theta(t)$. Physically, what we wish to know from the simulation is how the cart position and pendulum angle respond when one gives the cart a kick. The impulse responses are shown in Figure 3.21, which are quite oscillatory and take a long time to settle. One can observe that the controller parameters take extreme values. More critically, the controller is unstable. Though internal stability of a feedback system does not necessarily require the controller to be stable, stable controllers are often desirable. In a later chapter, we will show that $P_x(s)$ does not have a stable stabilizing controller. This implies that if the signal $x(t)$ alone is used for feedback control, though it is possible to design a "theoretical" stabilizing controller, the design becomes impossible if we further require the controller to be stable. Later on, we will justify in various ways that $P_x(s)$ is a bad plant to control simply because it has an unstable pole at $\sqrt{3g}$ and an unstable zero at $\sqrt{3g/2}$, which are too close to each other. It is impossible to achieve good performance when controlling such a system. This partially explains why we do not only look at the base of the stick when trying to balance a long stick using our hand.

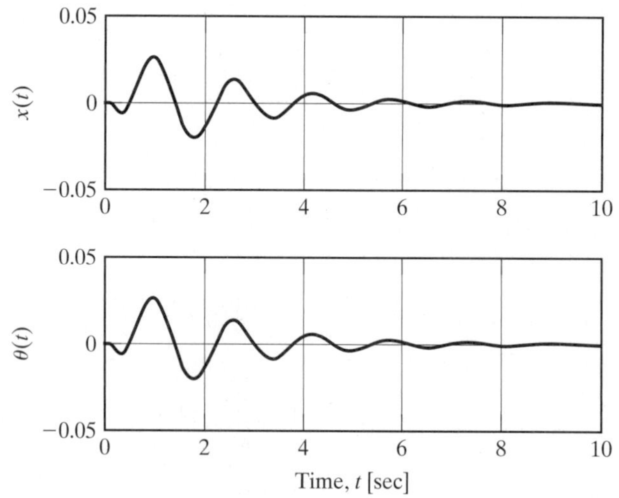

FIGURE 3.21: Impulse responses of an inverted pendulum system when controlled by $C_x(s)$.

What do we really look at when we try to balance a long stick using our hand? We look at the upper tip of the stick. If we mimic this human behavior, we should neither use $\theta(t)$ feedback alone nor $x(t)$ feedback alone. Rather, we should feed back the (horizontal) position of the tip, which is

$$z(t) = x(t) - L\sin\theta(t)$$

After linearization, we get

$$z(t) = x(t) - L\theta(t)$$

where for the particular system we consider $L = 1$. If we take $z(t)$ as the output, then the system transfer function from $f(t)$ to $z(t)$ becomes

$$P_z(s) = P_x(s) - P_\theta(s) = -\frac{1/3s^2 + g}{s^2(s^2 - 3g)}.$$

The zero $\sqrt{3g/2}$ of $P_x(s)$ on the positive real axis near the positive real pole $\sqrt{3g}$ is no longer a zero of $P_z(s)$. The plant $P_z(s)$ is much easier to control. Let us now design a controller to stabilize the system $P_z(s)$. The block diagram of the feedback system using this scheme is shown in Figure 3.22.

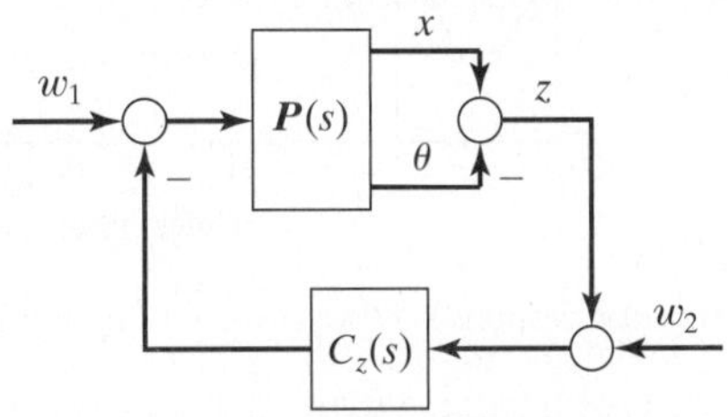

FIGURE 3.22: Stabilization of inverted pendulum using output $x(t) - L\theta(t)$.

Again we shall use pole placement design with a third-order controller

$$C_z(s) = \frac{q_0 s^3 + q_1 s^2 + q_2 s + q_3}{p_0 s^3 + p_1 s^2 + p_2 s + p_3}$$

to place the closed-loop poles to the same places as above. The Diophantine equation is

$$\begin{aligned} s^2(s^2 - 3g)(p_0 s^3 + p_1 s^2 + p_2 s + p_3) - (1/3s^2 + g)(q_0 s^3 + q_1 s^2 + q_2 s + q_3) \\ = (s + 6)(s^2 + s + 16.25)(s^2 + 4s + 10.25)(s^2 + 17s + 92.5). \end{aligned}$$

Its solution is

$$\begin{aligned} &p_0 = 1, && p_1 = 28, && p_2 = 47.25, && p_3 = 1015, \\ &q_0 = -966.5, && q_1 = -6337, && q_2 = -7567, && q_3 = -9433. \end{aligned}$$

The designed controller is

$$C_z(s) = -\frac{966.5s^3 + 6337s^2 + 7567s + 9433}{s^3 + 28s^2 + 47.25s + 1015}.$$

We can immediately observe that the controller parameters are less extreme. The simulation of the impulse responses of $x(t)$ and $\theta(t)$ when the impulse occurs at $w_1(t)$, shown in Figure 3.23, demonstrates that this is indeed a better design. We can also observe that the controller now is stable with poles at -27.62, $-0.1900 \pm j6.060$.

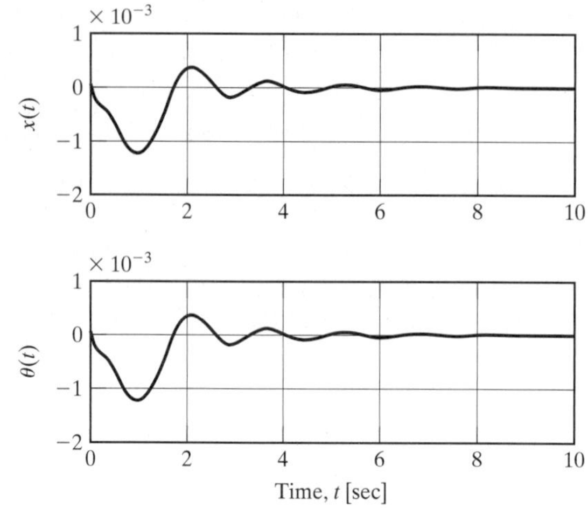

FIGURE 3.23: Impulse responses of an inverted pendulum system when controlled by $C_z(s)$.

The previous two designs are based on SISO pole placement theory. What if we carry out a genuine SIDO design? Though we do not have a general SIDO theory covered in the book, let us extend the idea in the SISO theory to this particular SIDO system. The feedback system structure for the stabilization of a SIDO system is shown in Figure 3.24. Since the plant is SIDO, the controller is DISO. Let us write the controller as

$$\boldsymbol{C}(s) = \begin{bmatrix} C_x(s) & C_\theta(s) \end{bmatrix} = \frac{\begin{bmatrix} q_x(s) & q_\theta(s) \end{bmatrix}}{p(s)}.$$

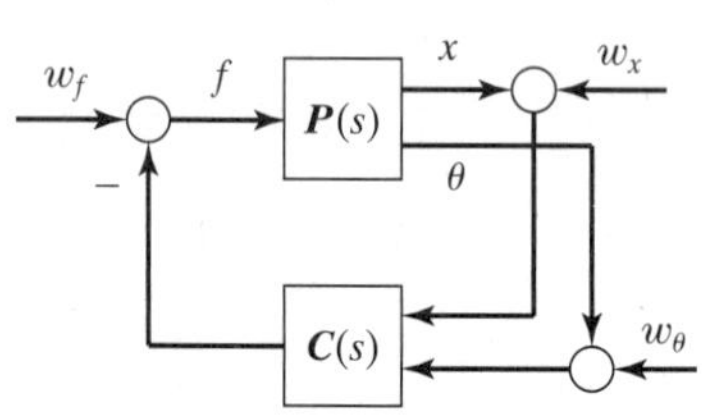

FIGURE 3.24: The feedback system structure for stabilizing an inverted pendulum.

Then the closed-loop characteristic polynomial is

$$c(s) = a(s)p(s) + b_x(s)q_x(s) + b_\theta(s)q_\theta(s).$$

What is the smallest order of the controller to carry out an arbitrary pole placement? The SISO result does not apply here. Let us see if the zeroth order controller works. In this case, the controller is of the form

$$\boldsymbol{C}(s) = \frac{\begin{bmatrix} q_{x0} & q_{\theta 0} \end{bmatrix}}{p_0}$$

which has three parameters, whereas the closed-loop characteristic polynomial $c(s)$ has degree 4 with five coefficients. Hence, there is not enough design freedom in arbitrarily assigning $c(s)$ so as to arbitrarily place the closed-loop poles. Therefore, zeroth order controllers do not work.

Next let us try first-order controllers. In this case, the controller is of the form

$$\boldsymbol{C}(s) = \frac{\left[\begin{array}{cc} q_{x0}s + q_{x1} & q_{\theta 0}s + q_{\theta 1} \end{array}\right]}{p_0 s + p_1}$$

which has six parameters. Because $c(s)$ is of degree 5, it has six free coefficients. What a perfect match! This shows that we can indeed use a first-order controller to place the closed-loop poles arbitrarily. Now assume that the desired closed-loop poles are $-6, -2 \pm j2.5, -8.5 \pm j4.5$. Then the polynomial equation needed for the pole placement is

$$s^2(s^2 - 3g)(p_0 s + p_1) + (2/3s^2 - g)(q_{x0}s + q_{x1}) + s^2(q_{\theta 0}s + q_{\theta 1}) \\ = (s+6)(s^2 + 4s + 10.25)(s^2 + 17s + 92.5).$$

This equation is usually not called a Diophantine equation since it has three unknown polynomials. However, it can be solved in the same way as the Diophantine equation by comparing the coefficients. The solution is

$$p_0 = 1, \quad p_1 = 27, \quad q_{x0} = -430.0, \quad q_{x1} = -580.5, \quad q_{\theta 0} = 612.9, \quad q_{\theta 1} = 2750.$$

Hence the designed controller is

$$\boldsymbol{C}(s) = \frac{\left[\begin{array}{cc} -430.0s - 580.5 & 612.9s + 2750 \end{array}\right]}{s + 27}.$$

The genuine SIDO design yields a controller with a lower order. The simulation of the closed-loop impulse responses is given in Figure 3.25. Here again, the unit

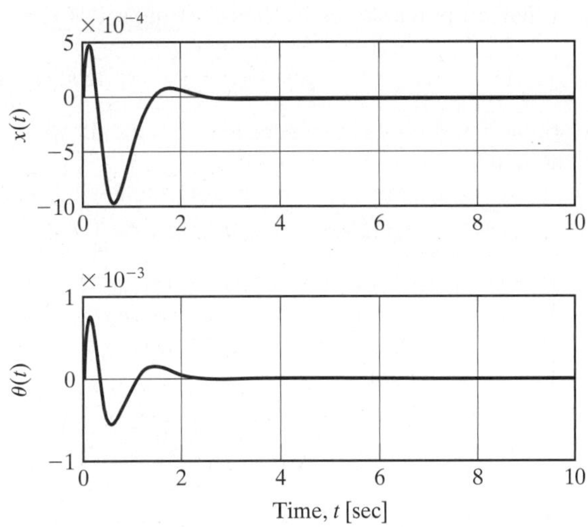

FIGURE 3.25: Impulse responses of an inverted pendulum system when controlled by the DISO controller $\boldsymbol{C}(s)$.

impulse comes in at $w_f(t)$. Under this controller, the reactions of $x(t)$ and $\theta(t)$ to a unit impulse at $w_f(t)$ are much smaller than those under the previous controllers. This gives a strong motivation for the exploration of MIMO system control theory. Though it serves as a good start for studying feedback control, the SISO system control theory has severe insufficiencies in applications.

PROBLEMS

3.1. The system with transfer function $\frac{\omega_n^2}{s^2+\omega_n^2}$ is known to be unstable. Find a bounded input such that the output is unbounded.

3.2. An ideal differentiator is a system whose output is the derivative of its input. The transfer function of an ideal differentiator is s. Show that a differentiator is unstable by constructing a bounded input so that the output is unbounded.

3.3. Check the stability of the following polynomials:

1. $p_1(s) = s^4 + 2s^3 + 9s^2 + 2s + 5$.
2. $p_2(s) = s^5 + 5s^4 + 12s^3 + 24s^2 + 32s + 16$.
3. $p_3(s) = s^3 + (2K+1)s^2 + 3Ks + 5$.

3.4. Find the range of K such that the following polynomial is stable:

$$s^4 + s^3 + Ks^2 + s + 1.$$

3.5. Validate Corollary 3.20 using the Liénard and Chipart stability criterion.

3.6. The complexity of an algorithm is often measured by the number of multiplications needed to complete the computation as a function of the problem size. If we determine the stability of a polynomial by computing its roots, give an estimate on the complexity as a function of the polynomial order n. (You may need to consult a book on numerical computation in order to do this.) If we determine the stability by using the Routh criterion, estimate the complexity as a function of n. Which method has the lower complexity?

3.7. Show that all polynomials in the following set of third-order polynomials

$$\{s^3 + a_1s^2 + a_2s + a_3 : a_1 \in [\underline{a}_1, \overline{a}_1], a_2 \in [\underline{a}_2, \overline{a}_2], a_3 \in [\underline{a}_3, \overline{a}_3]\}.$$

are stable if and only if two special polynomials are stable. Which are these two polynomials?

3.8. Determine whether every member of the following sets of polynomials is stable:

1. $\mathcal{P} = \{a_0s^4 + a_1s^3 + a_2s^2 + a_3s + a_4 : a_0, a_1 - 1, a_2 - 7, a_3 - 1, a_4/2 \in [1,2]\}$.
2. $\mathcal{P} = \{s^4 + a_1s^3 + a_2s^2 + a_3s + a_4 : a_1 \in [2,3], a_2 \in [10,20], a_3, a_4 \in [1,2]\}$.
3. $\mathcal{P} = \{s^4 + a_1s^3 + a_2s^2 + a_3s + a_4 : a_1, a_2/2, a_3 \in [2,3], a_4 \in [2,4]\}$.
4. $\mathcal{P} = \{s^4 + a_1s^3 + a_2s^2 + a_3s + a_4 : a_1, a_2/2, a_3, a_4 + 1 \in [2,3]\}$.
5. $\mathcal{P} = \{s^4 + a_1s^3 + a_2s^2 + a_3s + a_4 : a_1, a_2 \in [3,4],\ a_3 \in [2,3], a_4 \in [0.5,1]\}$.

3.9. Consider a polynomial

$$a(s) = a_0s^3 + a_1s^2 + a_2s + a_3.$$

1. If $a_0 \in [1,4], a_1 \in [1,3], a_2 \in [1,2], a_3 \in [0.1,0.5]$, is $a(s)$ stable for all possible coefficients?
2. If $a_0 = 1+3K, a_1 = 1+2K, K \in [0,1], a_2 \in [1,2], a_3 \in [0.1,0.5]$, is $a(s)$ stable for all possible coefficients?

3.10. One disadvantage in using Fact 3.24 to test stability is that $a(j\omega)$ has a large magnitude when ω is large, and consequently one needs a huge piece of paper to plot $a(j\omega)$. To eliminate this disadvantage, we can plot $a(-j\omega)/a(j\omega)$ instead. This plot will not grow too big unless $a(s)$ has roots on the imaginary axis, which implies that $a(s)$ is unstable. Show that $a(s)$ is stable if and only if

$$\Delta\angle\left[\frac{a(-j\omega)}{a(j\omega)}\right] = -n\pi$$

where $\deg a(s) = n$.

3.11. Is a feedback system for stabilization consisting of the following plant and controller internally stable?

1. $P(s) = \dfrac{s-3}{(s+1)(s+2)}$, $C(s) = \dfrac{1}{s-3}$.
2. $P(s) = \dfrac{1}{(s+1)(s+2)}$, $C(s) = 3 + \dfrac{2}{s} + s$.
3. $P(s) = \dfrac{s+1}{s}$, $C(s) = -\dfrac{s+2}{s+1}$.

3.12. Consider a feedback system for stabilization.

1. Is it internally stable if $P(s) = \dfrac{s+1}{s+2}$ and $C(s) = -1$?
2. Is it internally stable if $P(s) = \dfrac{s-2}{s+3}$ and $C(s) = \dfrac{1}{s-2}$?
3. Is it internally stable if $P(s) = \dfrac{1}{s+1}$ and $C(s) = \dfrac{s+1}{s}$?

3.13. A feedback system for stabilization consists of an uncertain plant $P(s) = \dfrac{b_1 s + b_2}{a_0 s^2 + a_1 s + a_2}$ where $b_1 \in [2,3]$, $b_2 \in [1,2]$, $a_0 = 1$, $a_1 \in [2,3]$, $a_2 = [4,6]$, and a double integrator controller $C(s) = \dfrac{1}{s^2}$. Is this system stable for all possible plant parameters?

3.14. Prove Theorem 3.26.

3.15. Consider the feedback system for stabilization. Let $P(s) = \dfrac{s^2+15}{s^3+s^2+10s+5}$ and $C(s) = K$. Find the range of K such that the feedback system is internally stable.

3.16. A unity negative feedback system has the forward path transfer function given by

$$L(s) = \frac{K(s+2)}{s(1+\tau s)(1+2s)}.$$

Determine and plot the region of stability in the $K-\tau$ plane for this system.

3.17. Consider a 2DOF feedback system. Let

$$P(s) = \frac{b}{s^3 + a_1 s^2 + a_2 s + a_3}$$

where $a_1 \in [1,2]$, $a_2 = 6$, $a_3 \in [1,2]$, $b \in [1,2]$ and

$$\mathbf{C}(s) = \frac{1}{s}\begin{bmatrix} s+2 & 2 \end{bmatrix}.$$

Is the closed-loop system internally stable for all possible values of the parameters?

3.18. Let $P(s) = \frac{1}{s^2}$. Design a stabilizing controller so that the closed-loop system has poles at $-1 \pm j$ and -2.

3.19. Let $P(s) = \frac{s-1}{s^2}$.

1. Is it possible to stabilize this system by pure proportional feedback? Why?
2. Design a feedback controller so that the closed-loop poles are -5 and $-1 \pm j$.

3.20. Let $P(s) = \frac{1}{s(s+1)}$. Design a second-order, strictly proper controller to place the closed-loop poles to $-1, -9, -9, -7$.

3.21. Let $P(s) = \frac{1/2}{s(s+1)}$. Design a stabilizing controller so that the closed-loop poles are $-1, -2, -3$.

3.22. Given a plant $P(s) = \frac{1}{(s+1)(s-1)}$, design a stabilizing controller so that the closed-loop system has poles at $-5, -2, -9$.

3.23. Characterize all stabilizing controllers for plant $P(s) = \frac{s+1}{s(s-1)}$ in terms of a free transfer function $Q(s)$. Set $Q(s) = K$ where K is a pure gain. Draw in the complex plane the trajectory of the poles of the closed-loop system for K varying from $-\infty$ to ∞. Observe that the trajectory is completely contained in the left-hand side of the complex plane. Explain why.

MATLAB PROBLEMS

3.24. Write a MATLAB program to implement the Routh stability criterion. It should be a function named "routh" whose input is a vector representing the coefficients of a polynomial and whose output should have two variables: the first is a string with possible values "stable" and "unstable"; and the second is a matrix containing the Routh table. Check the stability of the following polynomials using your program.

1. $s^8 + 7s^7 + 4s^6 + s^5 + 7s^4 + 2s^2 + s + 3$.
2. $s^5 + 10s^4 + 30s^3 + 80s^2 + 344s + 480$.
3. $s^6 + 22s^5 + 169s^4 + 547s^3 + 760s^2 + 331s + 42$.
4. $s^6 + 14s^5 + 62s^4 + 185s^3 + 446s^2 + 584s + 112$.

3.25. Write a MATLAB program to implement the Hurwitz stability criterion. It should be a function named "hurwitz" whose input is a vector representing the coefficients of a polynomial and whose output should have two variables: the first is a string with possible values "stable" and "unstable"; and the second is a vector containing all Hurwitz determinants. Check the stability of the polynomials in Problem 3.24 using your program.

3.26. Write a MATLAB program that determines the stability of an interval polynomial

$$\mathcal{P} = \{a_0 s^n + a_1 s^{n-1} + \cdots + a_{n-1}s + a_n : a_i \in [\underline{a}_i, \overline{a}_i]\}$$

using the program "routh" you wrote in Problem 3.24. We again require this program to be a function with two input variables: one is a vector representing the lower bound coefficients, and the other is a vector representing the upper bound coefficients. Its output should again be a string "stable" or "unstable." Check the stability of the interval polynomials in Problem 3.8 using your program.

3.27. Write a MATLAB program "poleplace" for pole placement. This should be a program whose inputs are $P(s)$ and the desired closed-loop poles and whose output is $C(s)$. The input should also contain an option that decides whether a proper or strictly proper controller should be used.

EXTRA CREDIT PROBLEMS

3.28. Prove the Liénard and Chipart stability criterion.
3.29. Prove Theorem 3.29.
3.30. Let

$$p(s) = p_0 s^n + p_1 s^{n-1} + \cdots + p_n$$

$$q(s) = q_0 s^n + q_1 s^{n-1} + \cdots + q_n.$$

The Hadamard product of $p(s)$ and $q(s)$ is defined as

$$p(s) \odot q(s) = p_0 q_0 s^n + p_1 q_1 s^{n-1} + \cdots + p_n q_n.$$

Show that if $p(s)$ and $q(s)$ are stable, then $p(s) \odot q(s)$ is stable.

NOTES AND REFERENCES

The British physicist J. C. Maxwell initiated the effort in establishing conditions under which all roots of a given polynomial lie in the left half plane:

J. C. Maxwell, "On governors," *Proceedings of the Royal Society of London*, vol. 16, pp. 270–283, 1868.

The Routh criterion was developed by the British applied mathematician E. J. Routh in

E. J. Routh, *A Treatise on the Stability of a Given State of Motion*, Macmillan, London, 1877.

E. J. Routh, *The Advanced Part of a Treatise on the Dynamics of a Rigid Body*, 6th Edition, Macmillan, London, 1905; reprint, Dover, New York, 1959.

The proof given by Routh is quite involved and is usually omitted in feedback control textbooks. There have been continuous efforts in finding simpler proofs, such as in

P. C. Parks, "A new proof of the Routh–Hurwitz stability criterion using the second method of Lyapunov," *Proceedings of the Cambridge Philosophical Society*, vol. 58, pp. 694–702, 1962.

H. Chapellat, M. Mansour, and S. P. Bhattacharyya, "Elementary proofs of some classical stability criteria," *IEEE Transactions on Education*, vol. 33, pp. 232–239, 1990.

M. Margaliot and G. Langholz, "The Routh–Hurwitz array and realization of characteristic polynomials," *IEEE Transactions on Automatic Control*, vol. 45, pp. 2424–2426, 2000.

The simple proof presented in Section 3.2 first appeared in

K. J. Åström, *Introduction to Stochastic Control Theory*, Academic Press, New York, 1970.

It was rediscovered at least a couple of times by

G. Meinsma, "Elementary proof of the Routh–Hurwitz test," *Systems & Control Letters*, vol. 25, pp. 227–242, 1995.

A. Ferrante, A. Lepschy, and U. Viaro, "A simple proof of the Routh test," *IEEE Transactions on Automatic Control*, vol. 44, pp. 1306–1309, 1999.

The Routh–Hurwitz stability criterion was given by the German mathematician A. Hurwitz independently in

A. Hurwitz, "Über die Bedingungen unter welchen eine Gleichung nur Wurzeln mit negativen reellen Eeilen besitzt," *Mathematische Annalen*, vol. 46, pp. 273–284, 1895.

It is commonly referred to as the Routh–Hurwitz criterion because of its close connection with the Routh criterion.
The Liénard and Chipart criterion was named after the two French mathematicians Liénard and Chipart.

A. Liénard and M. H. Chipart, "Sur la signe de la partie réelle des racines d'une équations algébrique," *Journal de Mathématiques Pures et Appliquées*, vol. 10, pp. 291–346, 1914.

The Routh stability criterion and other stability criteria can be extended in a few directions. The first such direction is to determine algebraically the number of roots of a given polynomial in the left-hand side of the complex plane or more ambitiously the inertia, the triple of integers $\{n_-, n_0, n_+\}$ representing the numbers of roots with negative real parts, zero real parts, and positive real parts, respectively, of a polynomial. The second such direction is to consider polynomials with complex coefficients. The most authoritative reference on stability criteria and their relationship with other mathematical problems is

F. R. Gantmacher, *Theory of Matrices*, Vol. 2, Clelsea, New York, 1960.

This book also contains a rather involved proof of the Liénard and Chipart criterion. The intention of Problem 3.28 is to find a simple proof that only uses tools within the scope of this book.
The stability problem of interval polynomial was proposed and its surprisingly simple solution was given by Russian mathematician V. Kharitonov.

V. Kharitonov, "Asymptotic stability an equilibrium position of a family of systems of differential equations," *Differential Uravnen*, vol. 14, pp. 2086–2088, 1978.

The original proof was complicated. Efforts were made to simplify the proof, such as in

N. K. Bose, "A system theoretic approach to stability of sets of polynomials," *Contemporary Mathematics*, vol. 47, pp. 25–34, 1985.

K. S. Yeung and S. S. Wang, "A simple proof of Kharitonov's theorem," *IEEE Transactions on Automatic Control*, vol. 32, pp. 822–823, 1987.

H. Chapellat and S. P. Bhattacharyya, "An alternative proof of Kharitonov's theorem," *IEEE Transactions on Automatic Control*, vol. 34, pp. 448–450, 1989.

The proof presented in Section 3.4 is due to

R. J. Minnichelli, J. J. Anagnost, and C. Desoer, "An elementary proof of Kharitonov's stability theorem with extensions," *IEEE Transactions on Automatic Control*, vol. 34, pp. 995–998, 1989.

The Kharitonov theorem stimulated a surge of activities in research in the 1980s on the stability of polynomials or systems whose parameters are only known to sit in certain sets. Extensive and in-depth coverage of these activities is contained in

B. R. Barmish, *New Tools for Robustness of Linear Systems*, Macmillan, Yew York, 1994.

S. P. Bhattacharyya, H. Chapellat, and L. H. Keel, *Robust Control, the Parametric Approach*, Prentice Hall, Upper Saddle River, 1995.

The parametrization of all stabilizing controllers was first called the Youla parametrization in recognition of its "discovery" in

D. C. Youla, H. A. Jabr, and J. J. Bongiorno Jr, "Modern Wiener–Hopf design of optimal controllers –Part II: the multivariable case," *IEEE Transactions on Automatic Control*, vol. AC-21, pp. 319–318, 1976.

It was later called the Youla–Kuěcera parametrization in recognition of its earlier "discovery" in

V. Kucěra, "Stability of discrete linear systems," *Preprints of the 6th IFAC World Congress*, Boston, MA, 1975.

The earliest known discovery occurred much sooner in

V. B. Larin, K. I. Naumenko, and V. N. Suntsev, *Spectral Methods for Synthesis of Linear Systems with Feedback* (in Russian), Naukova Dumka, Kiev, USSR, 1971.

Should we call it Youla–Kučera–Larin parametrization, or Larin–Kucěra–Youla parametrization, or simply Larin parametrization? What about LNS parametrization, to credit the first discoverers?
The parametrization of all stabilizing 2DOF controllers, presented in Section 3.8, was introduced in

> C. A. Desoer and C. L. Gustafson, "Algebraic theory of linear multivariable feedback systems," *IEEE Transactions on Automatic Control*, vol. AC-29, pp. 909–917, 1984.
>
> D. C. Youla, H. A. Jabr, and J. J. Bongiorno Jr, "A feedback theory of two-degree-of freedom optimal Wiener-Hopf design," *IEEE Transactions on Automatic Control*, vol. AC-30, pp. 652–665, 1985.

A good exposition of the materials in both Sections 3.7 and 3.8 is contained in

> M. Vidyasagar, *Control System Synthesis: a Factorization Approach*, MIT Press, Cambridge, MA, 1985.

The result in Problem 3.30 is due to

> J. Garloff and D. G. Wagner, "Hadamard products of stable polynomials are stable," *Journal of Mathematical Analysis and Applications*, vol. 202, pp. 797–809, 1996.

CHAPTER 4

Time-Domain Analysis

The purpose of system analysis is to find out how a system behaves and what quality it has. The result of analysis tells us whether the system satisfies our specifications and, if not, gives us insights on how to design a controller to improve its performance. This chapter provides certain tools to assess the performance of a system from a time-domain point of view. First, we look at the system response under some typical input signals, such as an impulse, a step, sinusoids, etc. We then give a detailed study of the step response of a system. The transient and steady-state responses of a system to a unit step input tell us a great deal about the system performance and will be studied in several sections. We will also study the effect of system parameter or even structure changes on the step response, which gives us insights into feedback controller design or system approximation. Finally, we will examine how the strength of a signal or a system can be measured and computed. Such quantitative measures of signals and systems are of great importance in system optimization.

4.1 RESPONSES TO TYPICAL INPUT SIGNALS

In Chapter 2, we saw how the response of a system can be obtained from either hardware or software simulation. For some typical input signals, the response of a system can also be obtained analytically. We know that the Laplace transform of the system response is the product of its transfer function and the Laplace transform of its input:

$$Y(s) = G(s)U(s).$$

Therefore, the time function of the response can be obtained by taking the inverse Laplace transform if we know the transfer function and the input. For example, the (unit) **impulse response** of system $G(s)$ is

$$y(t) = \mathcal{L}^{-1}[G(s)],$$

its (unit) **step response** is

$$y(t) = \mathcal{L}^{-1}\left[G(s)\frac{1}{s}\right],$$

and its response to a sinusoidal input $u(t) = (\sin \omega t)\sigma(t)$ is

$$y(t) = \mathcal{L}^{-1}\left[G(s)\frac{\omega}{s^2 + \omega^2}\right].$$

EXAMPLE 4.1

Consider a first-order system

$$G(s) = \frac{T}{Ts + 1}.$$

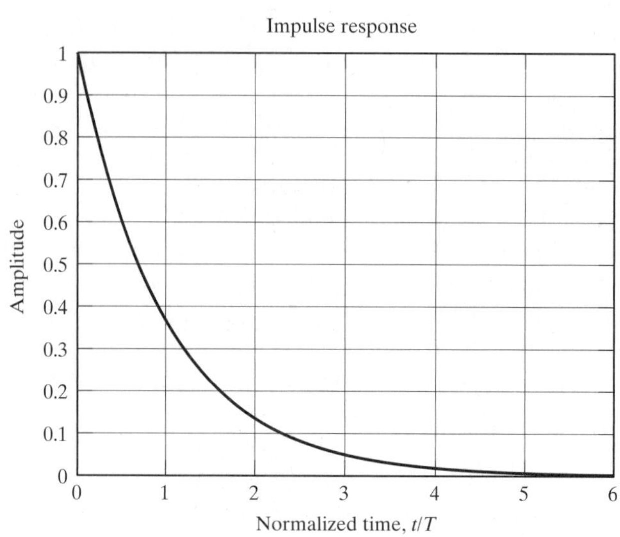

FIGURE 4.1: Impulse response of a standard first-order system.

Here T is called the **time constant** of the system. The system has a pole at $-1/T$. Its impulse response is

$$y(t) = \mathcal{L}^{-1}\left[\frac{T}{Ts+1}\right] = e^{-t/T}\sigma(t).$$

The impulse response versus the normalized time t/T is plotted in Figure 4.1. The impulse response always decays from 1 to 0 exponentially in a monotonic manner. We see that the speed of the decay of the impulse response is determined by the time constant T: the smaller T is, the faster the impulse response decays to zero.

EXAMPLE 4.2

Consider a standard second-order system

$$G(s) = \frac{\omega_n^2}{s^2 + 2\zeta\omega_n s + \omega_n^2}$$

where $\omega_n > 0$ is called the **natural frequency** and $\zeta \geq 0$ is called the **damping ratio**.

Depending on the value of the damping ratio ζ, the system is classified as **undamped** ($\zeta = 0$), **underdamped** ($0 < \zeta < 1$), **critically damped** ($\zeta = 1$), or **overdamped** ($\zeta > 1$) cases. Notice that in the undamped case the system is actually unstable; we only include this case here for the purpose of comparison. The pole configuration of the system is shown in Figure 4.2 and is given as follows:

1. $\zeta = 0$: $G(s)$ has two imaginary axis poles at $\pm j\omega_n$;
2. $0 < \zeta < 1$: $G(s)$ has two complex poles at $-\zeta\omega_n \pm j\omega_d$, where $\omega_d = \omega_n\sqrt{1-\zeta^2}$ is called the **damped frequency**;
3. $\zeta = 1$: $G(s)$ has two repeated poles at $-\zeta\omega_n$;
4. $\zeta > 1$: $G(s)$ has two real poles at $p_1 = -\zeta\omega_n - \omega_n\sqrt{\zeta^2-1}$ and $p_2 = -\zeta\omega_n + \omega_n\sqrt{\zeta^2-1}$.

The impulse response of the system is

$$y(t) = \mathcal{L}^{-1}\left[\frac{\omega_n^2}{s^2 + 2\zeta\omega_n s + \omega_n^2}\right]$$

$$= \begin{cases} \dfrac{\omega_n}{\sqrt{1-\zeta^2}} e^{-\zeta\omega_n t} \sin\sqrt{1-\zeta^2}\,\omega_n t & \text{if } 0 \leq \zeta < 1 \\ \omega_n^2 t e^{-\omega_n t} & \text{if } \zeta = 1 \\ \dfrac{\omega_n}{2\sqrt{\zeta^2-1}}\left[e^{-(\zeta-\sqrt{\zeta^2-1})\omega_n t} - e^{-(\zeta+\sqrt{\zeta^2-1})\omega_n t}\right] & \text{if } \zeta > 1 \end{cases}$$

for $t \geq 0$. The impulse responses for $\zeta = 0, 0.1, 0.3, 0.6, 1.0, 1.5,$ and 2.1 are plotted in Figure 4.3.

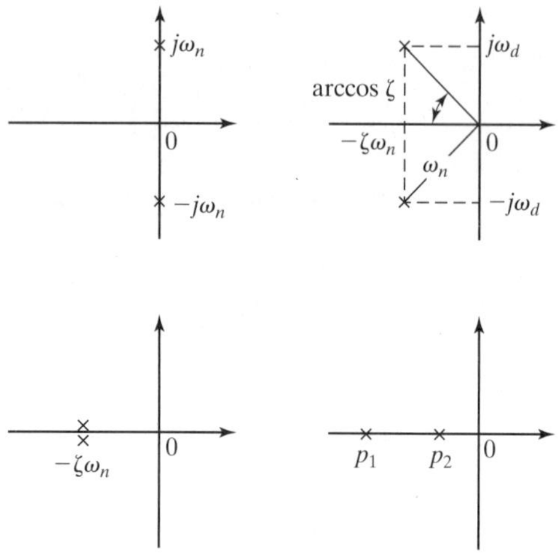

FIGURE 4.2: Pole configuration of a second-order system.

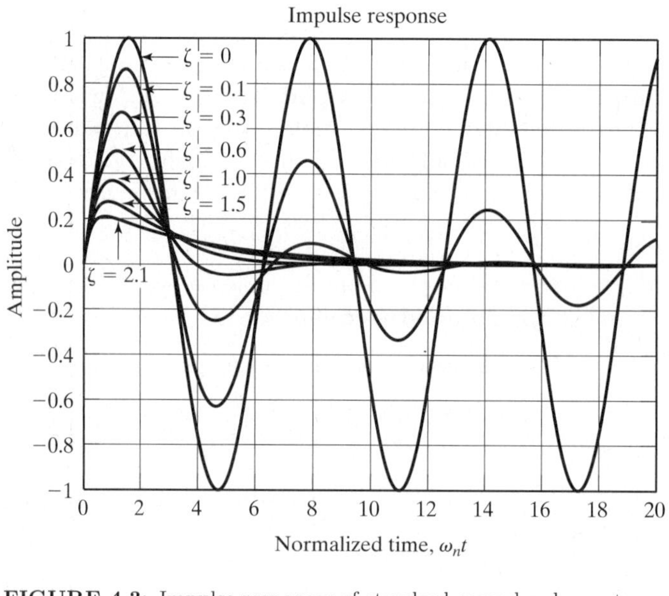

FIGURE 4.3: Impulse responses of standard second-order systems.

The impulse response of a dynamic system has great theoretic value. It is directly related to the system transfer function. It plays an important role in stability analysis. Theorem 3.3 can be rephrased as follows: a system is stable if its impulse response is absolutely integrable. Later on, we will see that the speed by which the impulse response decays to zero is often taken as a measure of relative stability of the system. We will also use the impulse response to define the strength of a system. However, impulse responses are often overlooked in classical control theory since they cannot be reproduced in real hardware simulations. In the computer era, generating impulse responses in computer simulations is not a problem at all.

The step response of a dynamic system will be studied in detail in the next sections. The response of a system to sinusoidal signals is an important subject and will be covered in Chapter 6.

Consider the dynamic system shown in Figure 4.4 with transfer function $G(s)$.

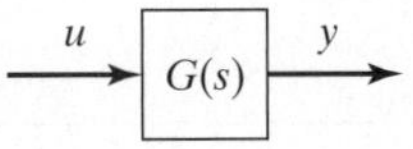

FIGURE 4.4: A dynamic system.

The time response of a stable system with respect to any input signals can be separated into two parts: the **transient response**, denoted by $y_t(t)$, which goes to zero as $t \to \infty$, and the **steady-state response**, denoted by $y_{ss}(t)$, which does not go to zero as $t \to \infty$. For example, consider the system shown in Figure 4.4 with input $u(t)$ and output $y(t)$. Let $G(s) = \dfrac{1}{s+1}$ and $u(t) = t\sigma(t)$. Then

$$Y(s) = \frac{1}{s^2(s+1)}, \quad \Rightarrow \quad y(t) = (e^{-t} - 1 + t)\sigma(t).$$

Thus the transient response and the steady-state response of the system are given respectively as

$$y_t(t) = e^{-t}, \quad y_{ss}(t) = -1 + t.$$

Let $G(s)$ be a stable transfer function and $u(t)$ be an input signal of the system. We shall now consider the steady-state responses of the system with respect to some typical classes of input signals. Note that

$$Y(s) = G(s)U(s).$$

Step input: In this case,

$$u(t) = \sigma(t), \quad U(s) = \frac{1}{s}.$$

Then we have

$$Y(s) = G(s)\frac{1}{s} = \frac{G(0)}{s} + \text{the strictly proper and stable part}$$

and for $t \geq 0$,

$$y(t) = G(0) + \text{the exponentially decaying part.}$$

Thus

$$\boxed{y_{ss}(t) = G(0)\sigma(t).}$$

Ramp input: In this case,

$$u(t) = t\sigma(t), \quad U(s) = \frac{1}{s^2}.$$

Then we have

$$Y(s) = G(s)\frac{1}{s^2} = \frac{G'(0)}{s} + \frac{G(0)}{s^2} + \text{the strictly proper and stable part}$$

where $G'(s)$ denotes the first derivative of $G(s)$ and for $t \geq 0$,

$$y(t) = G'(0) + G(0)t + \text{the exponentially decaying part.}$$

Thus

$$\boxed{y_{ss}(t) = [G'(0) + G(0)t]\sigma(t).}$$

Acceleration input: In this case,

$$u(t) = \frac{1}{2}t^2\sigma(t), \quad U(s) = \frac{1}{s^3}.$$

Then we have

$$Y(s) = G(s)\frac{1}{s^3} = \frac{G''(0)}{2s} + \frac{G'(0)}{s^2} + \frac{G(0)}{s^3} + \text{the strictly proper and stable part}$$

where $G''(s)$ denotes the second derivative of $G(s)$, and for $t \geq 0$

$$y(t) = \frac{1}{2}G''(0) + G'(0)t + G(0)\frac{t^2}{2} + \text{the exponentially decay part.}$$

Thus

$$\boxed{y_{ss}(t) = \left[\frac{1}{2}G''(0) + G'(0)t + G(0)\frac{t^2}{2}\right]\sigma(t).}$$

Sinusoidal input: In this case, $u(t) = (\sin \omega t)\ \sigma(t)$, i.e.,

$$U(s) = \frac{\omega}{s^2 + \omega^2}.$$

Then

$$\begin{aligned} Y(s) &= G(s)U(s) = G(s)\frac{\omega}{s^2 + \omega^2} \\ &= \frac{-G(-j\omega)}{2}\frac{1}{s + j\omega} + \frac{G(j\omega)}{2}\frac{1}{s - j\omega} + \text{the strictly proper and stable part} \\ &= \frac{s\ \operatorname{Im} G(j\omega) + \omega \operatorname{Re} G(j\omega)}{s^2 + \omega^2} + \text{the strictly proper and stable part} \end{aligned}$$

and for $t \geq 0$,

$$y(t) = \operatorname{Im} G(j\omega)\cos \omega t + \operatorname{Re} G(j\omega)\sin \omega t + \text{the exponentially decaying part.}$$

Thus

$$y_{ss}(t) = \operatorname{Im} G(j\omega) \cos \omega t + \operatorname{Re} G(j\omega) \sin \omega t = |G(j\omega)| \sin(\omega t + \angle G(j\omega))$$

where $\angle G(j\omega) = \tan^{-1} \dfrac{\operatorname{Im} G(j\omega)}{\operatorname{Re} G(j\omega)}$ is the phase of $G(j\omega)$ and $|G(j\omega)|$ is the magnitude of $G(j\omega)$.

4.2 STEP RESPONSE ANALYSIS

We are particularly interested in how well a system responds to a step input (in particular, the unit step input $\sigma(t)$). The output of the system in this case is called the **(unit) step response**.

We always assume that the system is stable. We also assume that the DC gain of the system is nonzero. Because of this, the step response has a nonzero steady state $y(\infty)$. The case when the DC gain is zero is not very interesting. Our statements will be valid for both positive and negative DC gain cases, but we only need to visualize the case when the DC gain is 1 since, by proper scaling, all other cases can be converted to the case of the unit DC gain.

Figure 4.5 shows several typical step responses. These step response samples are obtained from the following systems:

$$G_1(s) = \frac{1}{s+1}, \qquad G_2(s) = \frac{1}{s^2+s+1},$$
$$G_3(s) = \frac{1-s}{s^2+2s+1}, \qquad G_4(s) = \frac{1-s}{s^2+s+1},$$
$$G_5(s) = \frac{(1-s)^2}{(s+1)^3}, \qquad G_6(s) = \frac{(1-s)^2}{(s^2+s+1)(s+1)}.$$

The position of the step response in the array of plots in Figure 4.5 exactly matches the position of the system in the above equation array. In the later part of this chapter, we shall discuss in greater detail why these responses are of such shapes.

The performance of the system response to a step input is usually measured in terms of rise time, settling time, percentage overshoot (PO), and so on. These performance criteria, illustrated in Figure 4.6, are defined as follows:

Rise time t_r: In the case when the step response is monotonic, the rise time t_r is the time taken by the step response to rise from 10 to 90% of the final value. In the case when the step response is not monotonic, a precise definition of the rise time is difficult since the step response can hit 10 and/or 90% of the final value at several time instances. In this case, the rise time is usually defined as the time between the last time instance when the step response is equal to the 10% of its final value and the first time instance when the step response is equal to the 90% of its final value.

Settling time t_s: This is the time taken by the system to settle within a desired error tolerance Δ of the final value. For example, Figure 4.6 shows the time

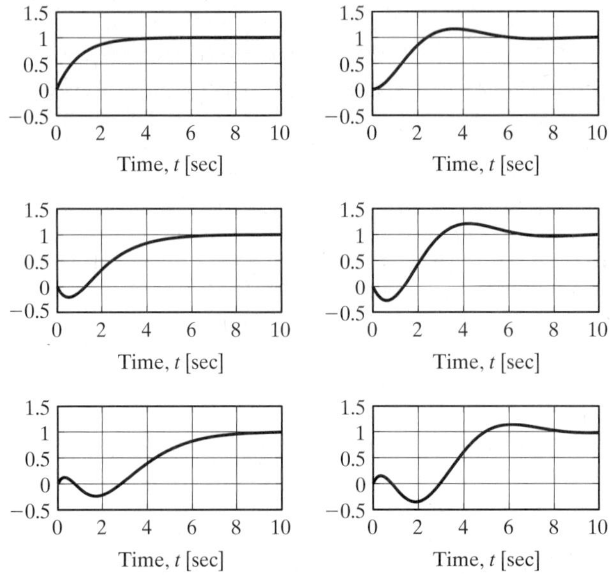

FIGURE 4.5: Typical step responses.

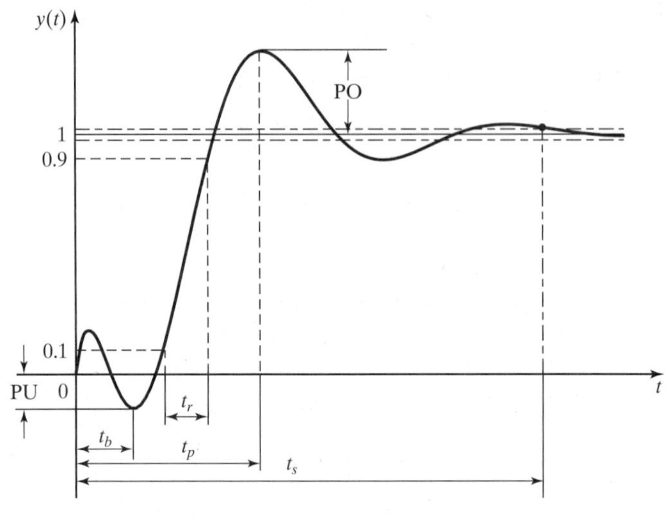

FIGURE 4.6: Definition of performance criteria.

taken by the system to settle within 2% of the final value. $\Delta = 2\%$ and $\Delta = 5\%$ are the most widely used error tolerances in practice.

Peak time t_p and percentage overshoot (PO): The step response is said to have an overshoot if there exists $t > 0$ such that $y(t) > y(\infty)$ in the case when $y(\infty) > 0$ or $y(t) < y(\infty)$ in the case when $y(\infty) < 0$. The time taken by $y(t)/y(\infty)$ to reach its maximum value is called peak time, denoted by t_p. When the step response does not have overshoot, i.e., when $y(t)$ never goes beyond $y(\infty)$, we also say $t_p = \infty$. The PO is defined as

$$\text{PO} = \frac{y(t_p) - y(\infty)}{y(\infty)} \times 100\%.$$

It is usually used as a smoothness measure of the system response.

Bottom time t_b and percentage undershoot (PU): The step response is said to have an undershoot if there exists $t > 0$ such that $y(t)/y(\infty) < 0$. The time when $y(t)/y(\infty)$ reaches its minimum value is called bottom time, denoted by t_b. In the case when the bottom time does not exist, i.e., when $y(t)$ never goes into the opposite direction from $y(\infty)$, we also say that $t_b = 0$. The PU is defined as

$$\text{PU} = \frac{-y(t_b)}{y(\infty)} \times 100\%.$$

In Figure 4.5, step response 1 has neither overshoot nor undershoot, step response 2 has overshoot but no undershoot, step responses 3 and 5 have undershoot but no overshoot, and step responses 4 and 6 have both overshoot and undershoot.

For general systems of arbitrary order, we can obtain the values of these performance criteria by simulation. For example, we can plot the step response as in Figure 4.6 and read off the values from the plot. However, for simple first-order and second-order systems, we can obtain these values analytically.

Let us first consider a first-order system with transfer function

$$G(s) = \frac{1}{Ts + 1}$$

where $T > 0$ is called the time constant of the system. Let $r(t)$ be a unit step input $r(t) = \sigma(t)$. The Laplace transform of the step response is given by

$$Y(s) = G(s)\Sigma(s) = \frac{1}{Ts + 1}\,\frac{1}{s}.$$

Take partial fractional expansion of $Y(s)$, and then take the inverse Laplace transform. Then we get

$$y(t) = 1 - e^{-t/T}.$$

The physical meanings of time constant T are $1/T = \dot{y}(0^+)$ and $y(T) = 1 - e^{-1} = 0.632$.

The time response of $y(t)$ is shown in Figure 4.7. In particular, we have

$$y(3T) = 0.9502, \quad y(4T) = 0.9817.$$

Thus, we have

$$t_s \approx \begin{cases} 3T, & \text{if } \Delta = 5\% \\ 4T, & \text{if } \Delta = 2\%. \end{cases} \tag{4.1}$$

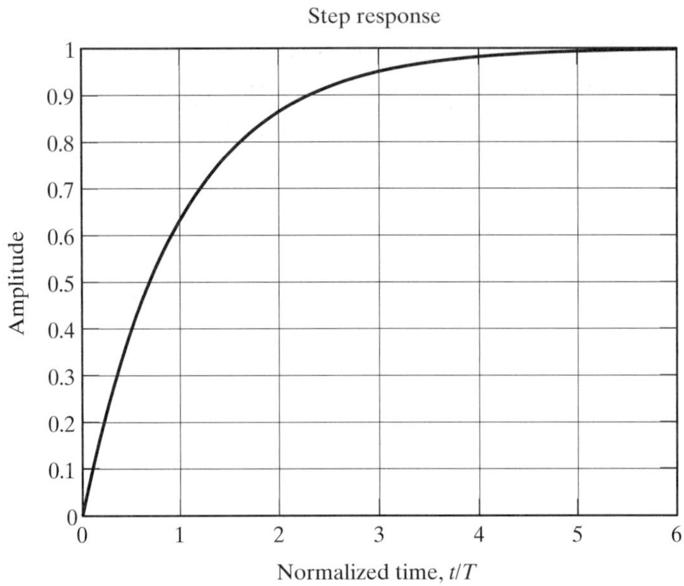

FIGURE 4.7: Step response of a standard first-order system.

Let t_1 and t_2 be the time such that $y(t_1) = 0.1$ and $y(t_2) = 0.9$, respectively, i.e.,

$$1 - e^{-t_1/T} = 0.1, \quad 1 - e^{-t_2/T} = 0.9.$$

Then $t_r = t_2 - t_1$. Thus we get the rise time

$$\boxed{t_r = (\ln 9)T \approx 2.2T.}$$

It is clear that the step response has no overshoot and no undershoot.

Next, let us consider a standard second-order system with transfer function

$$G(s) = \frac{\omega_n^2}{s^2 + 2\zeta\omega_n s + \omega_n^2}.$$

As we have seen in Example 4.2, here $\omega_n > 0$ is called the **natural frequency** and $\zeta \geq 0$ is called the **damping ratio**.

We shall now look at the step response in the following four cases depending on the value of the damping ratio ζ.

1. Undamped case ($\zeta = 0$):

$$Y(s) = \frac{\omega_n^2}{s(s^2 + \omega_n^2)} = \frac{1}{s} - \frac{s}{s^2 + \omega_n^2}.$$

Thus we have

$$y(t) = 1 - \cos\omega_n t.$$

2. Underdamped case ($0 < \zeta < 1$):

$$Y(s) = \frac{\omega_n^2}{s(s^2 + 2\zeta\omega_n s + \omega_n^2)}$$
$$= \frac{1}{s} - \frac{s + \zeta\omega_n}{(s + \zeta\omega_n)^2 + \omega_d^2} - \frac{\zeta}{\sqrt{1-\zeta^2}} \frac{\omega_d}{(s + \zeta\omega_n)^2 + \omega_d^2}.$$

Taking the inverse Laplace transform, we get

$$y(t) = 1 - \frac{e^{-\zeta\omega_n t}}{\sqrt{1-\zeta^2}} \left(\sqrt{1-\zeta^2} \cos \omega_d t + \zeta \sin \omega_d t \right)$$
$$= 1 - \frac{e^{-\zeta\omega_n t}}{\sqrt{1-\zeta^2}} \sin\left(\omega_d t + \arccos \zeta\right).$$

3. Critically damped case ($\zeta = 1$):

$$Y(s) = \frac{\omega_n^2}{s(s + \omega_n)^2} = \frac{1}{s} - \frac{1}{s + \omega_n} - \frac{\omega_n}{(s + \omega_n)^2}.$$

Taking the inverse Laplace transform, we get

$$y(t) = 1 - e^{-\omega_n t}(1 + \omega_n t).$$

4. Overdamped case ($\zeta > 1$):

$$Y(s) = \frac{\omega_n^2}{s(s - p_1)(s - p_2)} = \frac{1}{s} + \frac{\omega_n}{2\sqrt{\zeta^2 - 1}} \left(\frac{1}{p_2} \frac{1}{s - p_2} - \frac{1}{p_1} \frac{1}{s - p_1} \right).$$

Taking the inverse Laplace transform, we get

$$y(t) = 1 - \frac{1}{2\sqrt{\zeta^2-1}} \frac{e^{-(\zeta - \sqrt{\zeta^2-1})\omega_n t}}{\zeta - \sqrt{\zeta^2-1}} + \frac{1}{2\sqrt{\zeta^2-1}} \frac{e^{-(\zeta + \sqrt{\zeta^2-1})\omega_n t}}{\zeta + \sqrt{\zeta^2-1}}.$$

Figure 4.8 illustrates some typical step responses of the system for the above cases. Among all the cases, the underdamped case is the most interesting. In this case, we can derive the performance criteria explicitly.

First let us consider the peak time t_p and PO. A necessary condition for $y(t)$ to be the maximum is $\dot{y}(t) = 0$, i.e.,

$$\dot{y}(t) = \mathcal{L}^{-1}\left[sY(s)\right] = \mathcal{L}^{-1}\left[\frac{\omega_n^2}{s^2 + 2\zeta\omega_n s + \omega_n^2}\right]$$
$$= \frac{\omega_n}{\sqrt{1-\zeta^2}} e^{-\zeta\omega_n t} \sin \omega_d t = 0$$

which gives $\omega_d t = 0, \pi, 2\pi, \ldots$. Thus, the peak time is given by

$$t_p = \frac{\pi}{\omega_d}.$$

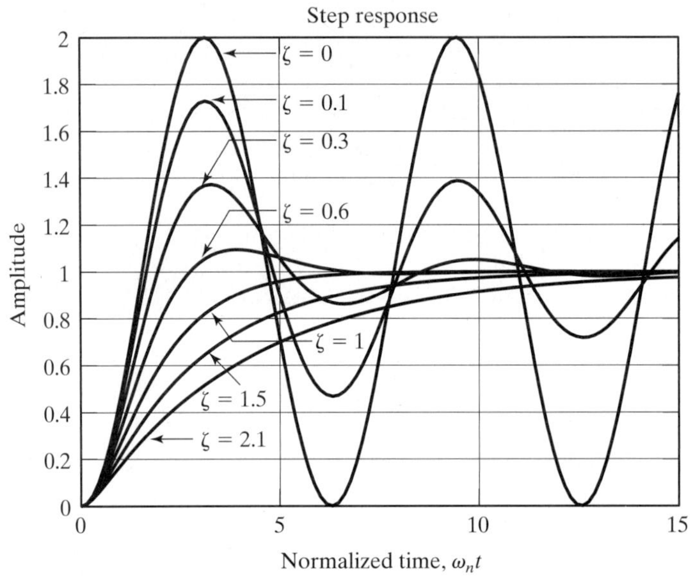

FIGURE 4.8: Step responses of second-order systems.

The step response at time t_p is

$$y(t_p) = 1 + e^{-\zeta\omega_n t_p} = 1 + e^{-\zeta\pi/\sqrt{1-\zeta^2}}.$$

Thus the PO is

$$\text{PO} = \frac{y(t_p) - y(\infty)}{y(\infty)} \times 100\% = e^{-\zeta\pi/\sqrt{1-\zeta^2}} \times 100\%.$$

The relationship between ζ and PO is shown in Figure 4.9.

The settling time t_s can be estimated by noting that $y(t)$ is bounded within the envelope curves $1 \pm e^{-\zeta\omega_n t}/\sqrt{1-\zeta^2}$, i.e.,

$$1 - \frac{e^{-\zeta\omega_n t}}{\sqrt{1-\zeta^2}} \leq y(t) \leq 1 + \frac{e^{-\zeta\omega_n t}}{\sqrt{1-\zeta^2}},$$

as shown in Figure 4.10. Thus the system is guaranteed to be settled if

$$\frac{e^{-\zeta\omega_n t}}{\sqrt{1-\zeta^2}} \leq \Delta$$

where Δ is the tolerance. Thus we can find an upper bound of the settling time as

$$t_s \leq \frac{-\ln(\Delta\sqrt{1-\zeta^2})}{\zeta\omega_n}.$$

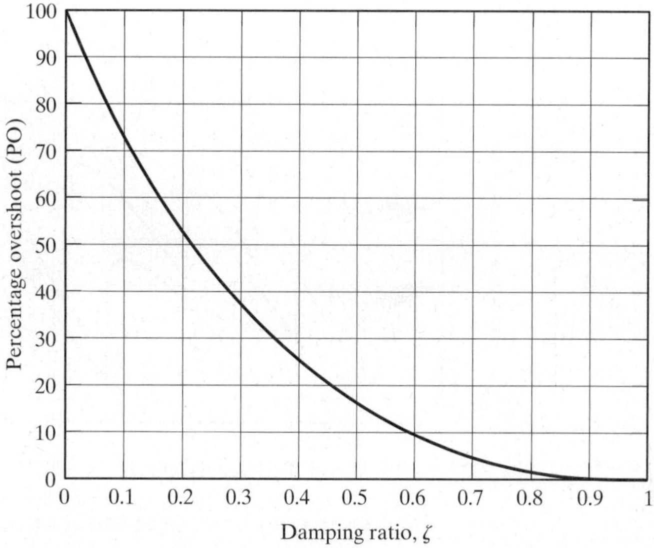

FIGURE 4.9: Percentage overshoot of a second-order system.

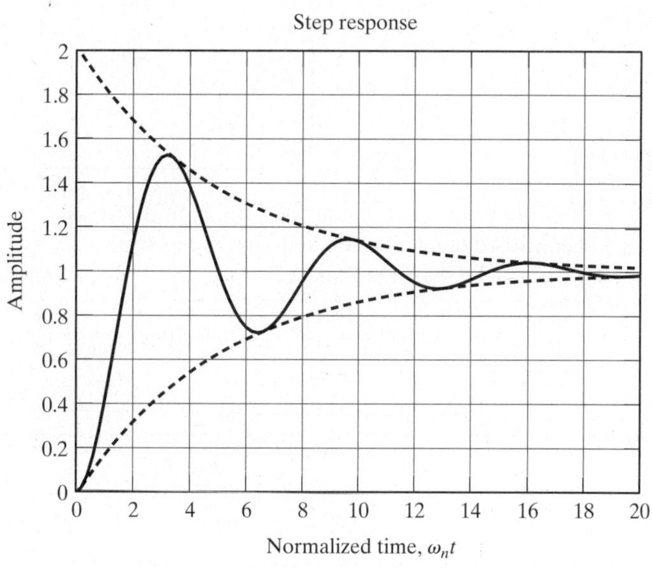

FIGURE 4.10: Envelopes of the step response of a second-order system.

Set $\Delta = 0.02$, i.e., within 2% of the final value, and we get

$$t_s \leq \frac{3.912 - 0.5\ln(1-\zeta^2)}{\zeta\omega_n} \leq \begin{cases} \dfrac{4.06}{\zeta\omega_n}, & \zeta \leq 0.5 \\[2ex] \dfrac{4.74}{\zeta\omega_n}, & \zeta \leq 0.9. \end{cases}$$

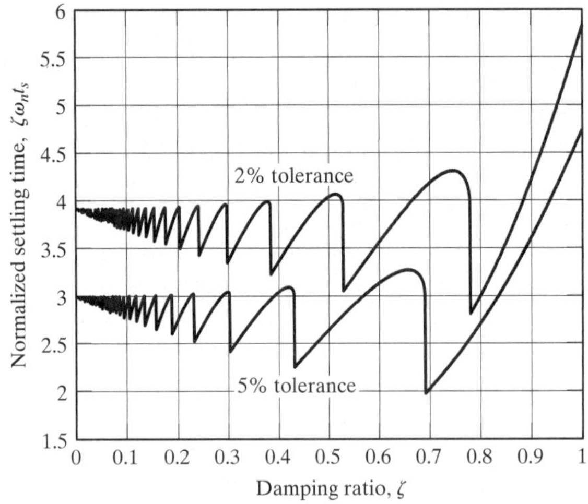

FIGURE 4.11: Settling time of a second-order system.

The exact settling time for each damping ratio is shown in Figure 4.11.

It is easy to see that a good estimate of the settling time is

$$t_s = \begin{cases} \dfrac{4}{\zeta\omega_n} & \text{if } \Delta = 2\% \\ \dfrac{3}{\zeta\omega_n} & \text{if } \Delta = 5\%. \end{cases}$$

It is possible to find the exact rise time t_r numerically, but for an underdamped standard second-order system we usually approximate t_r by the time of the first instance at which the step response reaches 100% of its final value. This gives

$$1 \approx 1 - \frac{1}{\sqrt{1-\zeta^2}} e^{-\zeta\omega_n t_r} \cos(\omega_d t_r - \arcsin\zeta),$$

which gives

$$t_r \approx \frac{\pi/2 + \arcsin\zeta}{\omega_d}.$$

EXAMPLE 4.3

Let

$$G(s) = \frac{10^2}{s^2 + 4s + 10^2}.$$

Then $\omega_n = 10$, $\zeta = 0.2$, and $\omega_d = \omega_n\sqrt{1-\zeta^2} = 9.798$. Hence $t_p = \pi/\omega_d = 0.32$, $\text{PO} = e^{-\zeta\pi/\sqrt{1-\zeta^2}} \times 100\% = 0.5266 \times 100\% = 52.66\%$, $t_s \approx 4/(\zeta\omega_n) = 4/2 = 2$ (with 2% tolerance). The rise time is approximately $t_r \approx 0.16$.

There are very few meaningful systems in the real world that are of first or second order in the standard form (although a lot of systems can be approximately modeled using first- and second-order systems). Most systems can be better modeled using high-order transfer functions (i.e., having more than two poles) and have finite zeros. It is therefore important to understand how these additional poles and zeros affect the system response. We shall analyze two special cases in the rest of this section where an additional pole or zero is respectively added to a standard second-order underdamped system.

We first consider the case when a pole is added to a standard second-order underdamped system. Consider the system shown in Figure 4.12.

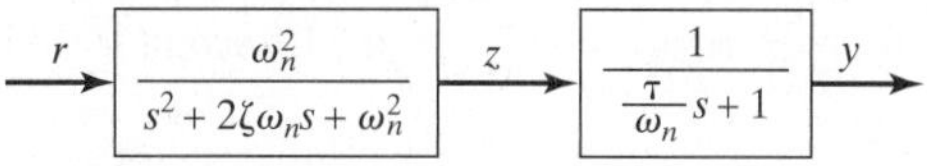

FIGURE 4.12: Adding a pole.

In order to make the results independent of ω_n, we have scaled the additional pole by ω_n. The system can be regarded as a cascade of a second-order system and a first-order system with the output of the second-order system as the input of the first-order one. Hence it is expected that the response of the system will be slower and smoother than the response of the standard second-order system because of the additional low-pass filtering. The simulation results are shown in Figure 4.13. These simulations show that an additional pole slows down the system response and reduces the overshoot.

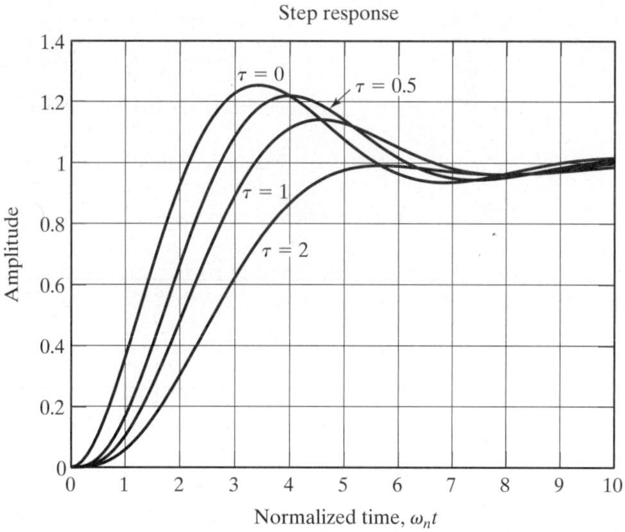

FIGURE 4.13: Effects of a pole on step response with $\zeta = 0.4$.

Next, we consider the case when a zero is added to a standard second-order, underdamped system. First we assume that the zero added is a minimum phase zero, as shown in Figure 4.14.

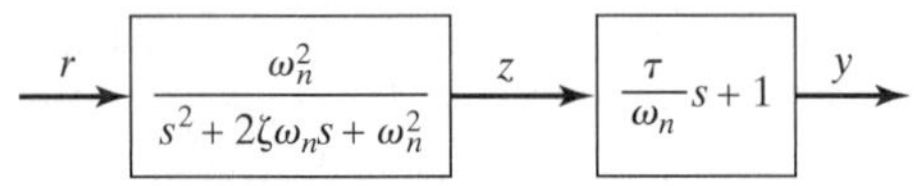

FIGURE 4.14: Adding a minimum phase zero.

It is easy to see that

$$y(t) = z(t) + \frac{\tau}{\omega_n}\dot{z}(t).$$

Since $\dot{z}(t) > 0$ for $t < t_p$, an immediate conclusion is that the system will have a shorter rise time, a shorter peak time, and a larger overshoot than the system without the zero.

The simulation results for zero in different locations are shown in Figure 4.15.

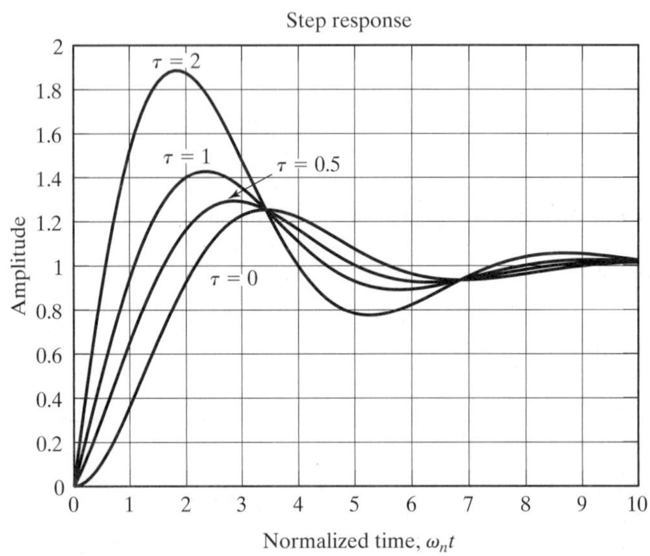

FIGURE 4.15: Effects of a zero on step response with $\zeta = 0.4$.

Now suppose that the additional zero is a nonminimum phase zero as shown in Figure 4.16.

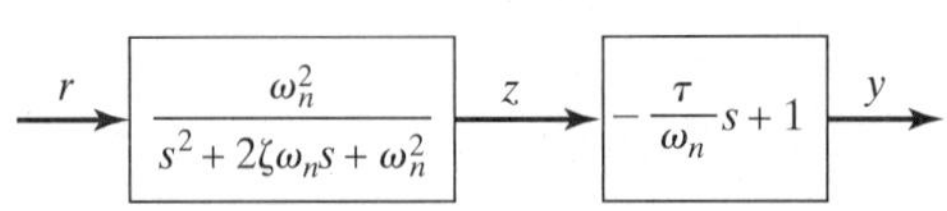

FIGURE 4.16: Adding a nonminimum phase zero.

Then we have

$$y(t) = z(t) - \frac{\tau}{\omega_n}\dot{z}(t).$$

Since $\dot{z}(t) > 0$ for $t < t_p$, the step response shows the undershoot behavior. This is very undesirable in many control systems. The simulation results for a system with a typical nonminimum phase zero are shown in Figure 4.17.

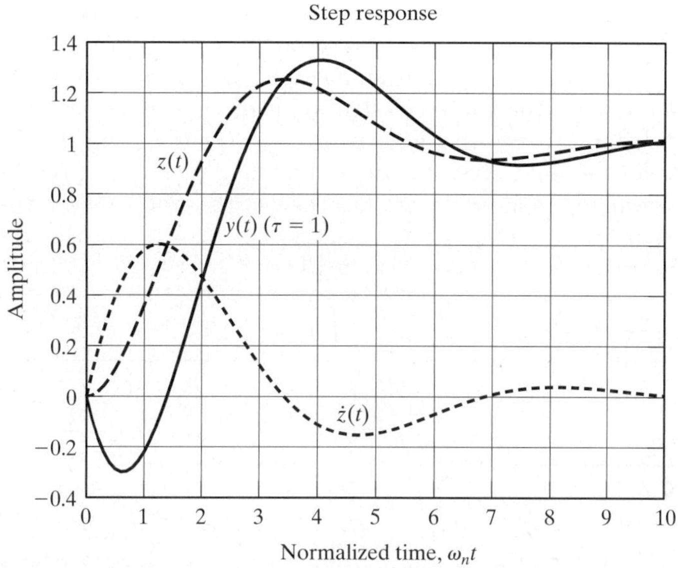

FIGURE 4.17: Effects of a nonminimum phase zero on step response with $\zeta = 0.4$.

4.3 DOMINANT POLES AND ZEROS

In general, it is very difficult to find analytic formulas to evaluate the performance of a high-order system. However, in many cases, there are natural separations among system poles and zeros. For example, some poles and zeros are much closer to the imaginary axis than others, as shown in Figure 4.18. Recall that, for a first-order or a second-order system the real parts of the poles essentially determine the settling time of the system. Thus those poles and zeros far from the imaginary axis will have little impact on the settling time and other performance specifications. Hence good low-order approximations can be found by appropriately neglecting the less significant poles and zeros so that the analysis can be simplified and the performance formulas for lower-order systems can be applied. Typically, if $D/d > 5 \sim 10$, we shall call those poles (zeros) that are close to the imaginary axis and with no zeros (poles) close by as **dominant poles (zeros)**, and those poles (zeros) far from the imaginary axis as **nondominant poles (zeros)**.

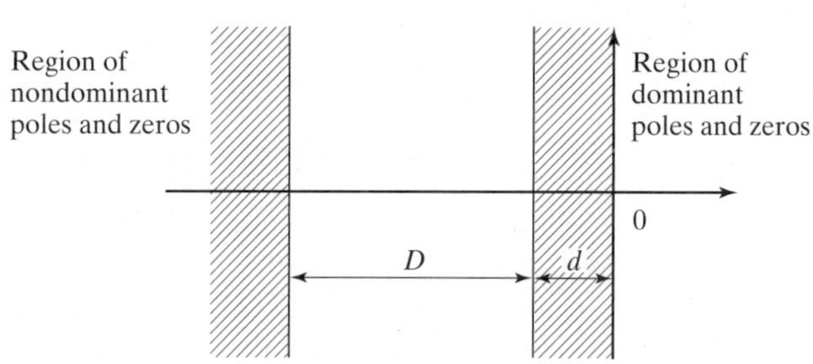

FIGURE 4.18: Regions of dominant and nondominant poles and zeros.

Typically, if the poles are dominated by a pair of complex poles, then the system can be approximated by a second-order system obtained from the dominant poles. Of course, computer simulation can be used to find the exact response. However, from the control design point of view, it is very useful to establish the relationship between the performance specifications and the pole locations.

For example, given the desired settling time t_s and the PO requirements, we can estimate the approximate locations of the dominant closed-loop poles from

$$t_s = \frac{4}{\zeta\omega_n}, \qquad \text{PO} = e^{-\zeta\pi/\sqrt{1-\zeta^2}} \times 100\%$$

as

$$\zeta\omega_n \geq \frac{4}{t_s}, \qquad \zeta \geq \frac{-\ln(\text{PO})}{\sqrt{\pi^2 + (\ln \text{PO})^2}}.$$

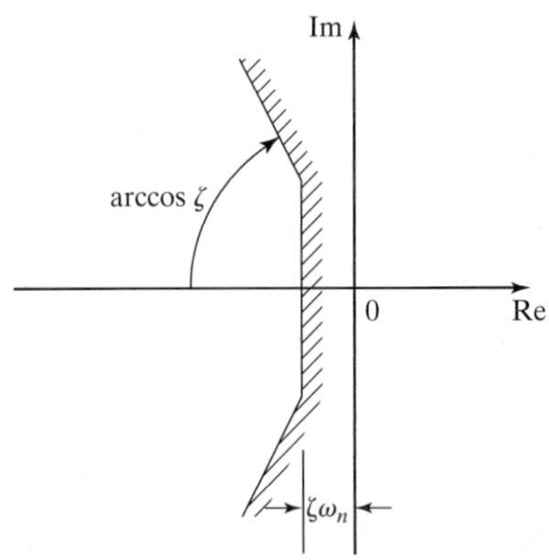

FIGURE 4.19: Desired dominant pole region.

The desired dominant pole locations are shown in Figure 4.19. The desired dominant pole locations can be used to guide our design so that suitable controllers are designed to place the more dominant poles in the desired regions and less dominant poles far away.

EXAMPLE 4.4

A unity feedback system is shown in Figure 4.20 with loop transfer function

$$L(s) = \frac{10(s+10)}{s(s+3)(s+5)}.$$

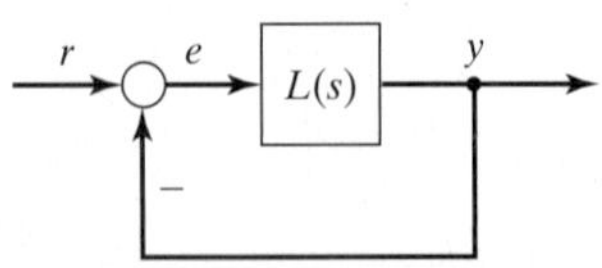

FIGURE 4.20: Example 4.4.

Then the closed-loop transfer function from $r(t)$ to $y(t)$ is given by

$$T(s) = \frac{L(s)}{1+L(s)} = \frac{10(s+10)}{(s+6.5182)\left[(s+0.7409)^2+3.8461^2\right]}.$$

Clearly, the pair of poles at $-0.7409 \pm j3.8461$ are the dominant poles of the closed-loop system. Thus system performance can be estimated on the basis of this pair of poles since the zero is also far away compared to this pair of poles.

$$T(s) \approx \frac{10 \times 10}{6.5182\left[(s+0.7409)^2+3.8461^2\right]} = \frac{3.9168^2}{s^2+2\times 0.189 \times 3.9168s + 3.9168^2}.$$

This pair of poles has a natural frequency $\omega_n = 3.92$ and a damping ratio $\zeta = 0.189$. Thus it is expected that the step response will have a settling time

$$t_s = \frac{4}{\zeta\omega_n} = 5.399$$

with 2% tolerance, and a PO

$$\mathrm{PO} = e^{-\zeta\pi/\sqrt{1-\zeta^2}}100\% = 54.63\%.$$

These estimates are confirmed from numerical simulation:

```
>> L=tf(10*[1,10],[1,8,15,0]);    T=feedback(L,1);

>> ltiview('step',T)
```

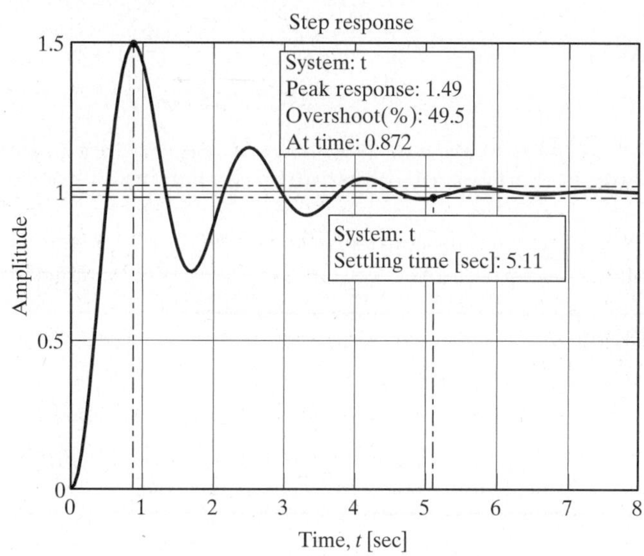

FIGURE 4.21: Step response of Example 4.4.

The step response of the system is shown in Figure 4.21.

4.4 STEADY-STATE RESPONSE AND SYSTEM TYPE

In this section, we will first study the steady-state response of an internally stable unity feedback system, as shown in Figure 4.22, subject to several typical input signals in the reference, disturbance, and noise.

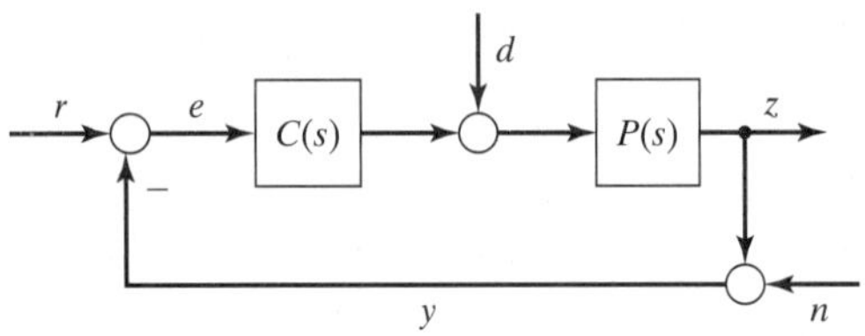

FIGURE 4.22: Unity feedback system.

We start with the steady-state tracking response. In this case, we assume that $d(t) = 0$ and $n(t) = 0$. Hence the feedback system can be simplified to the one shown in Figure 4.23. Here, we lumped $P(s)$ and $C(s)$ to a system $L(s) = P(s)C(s)$, which is called the **loop gain** or **loop transfer function**, and ignored the disturbance $d(t)$ and noise $n(t)$.

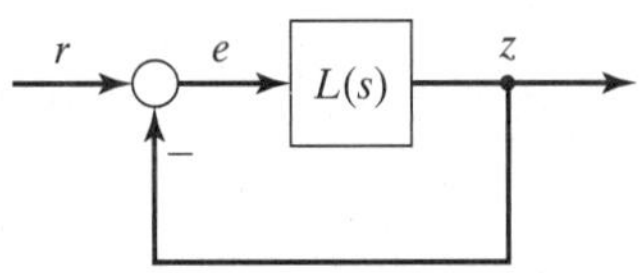

FIGURE 4.23: A simple unity feedback system.

Write

$$L(s) = \frac{1}{s^k} L_0(s)$$

where $L_0(s)$ has no zeros or poles at 0. Then k is said to be the type of $L(s)$. In other words, the **type of a system** is the number of cascaded integrators in the loop.

Normally, we only consider the case when $k \geq 0$, since if $k < 0$, then the DC gain of $L(s)$ is zero and the system cannot be used to track even a step signal.

EXAMPLE 4.5

Systems

$$\frac{1}{s+1}, \quad \frac{1}{s}, \quad \frac{s+1}{s^2(s-1)}$$

are type 0, type 1, and type 2 systems, respectively.

We often wish to know the steady-state error of the unity feedback system in Figure 4.23 for inputs of the form $r(t) = (t^l/l!)\sigma(t)$. Such an input is the unit step, the unit ramp, or the unit parabolic signal when l is equal to 0, 1, or 2, respectively. It turns out that the type of the loop transfer function $L(s)$, assumed to be k, qualitatively determines the steady-state error.

We will use the final value theorem to analyze the steady-state error. Since the feedback system is assumed to be internally stable, the transfer function from the reference $r(t)$ to the error $e(t)$

$$\frac{E(s)}{R(s)} = \frac{1}{1+L(s)} = \frac{s^k}{s^k + L_0(s)}$$

has all poles on the left-hand side of the complex plane and has k repeated zeros at the origin. The Laplace transform of the error is

$$E(s) = \frac{1}{1+L(s)}R(s) = \frac{s^k}{s^k + L_0(s)}\frac{1}{s^{l+1}}.$$

If $k \geq l$,

$$E(s) = \frac{s^{k-l-1}}{s^k + L_0(s)}$$

which has only stable poles except possibly a simple pole at the origin. By using the final value theorem (Theorem A.20)

$$e(\infty) = \lim_{s\to 0} sE(s) = \lim_{s\to 0}\frac{s^{k-l}}{s^k + L_0(s)} = \begin{cases} 0 & \text{if } k > l \\ \dfrac{1}{1+L_0(0)} & \text{if } k = l = 0 \\ \dfrac{1}{L_0(0)} & \text{if } k = l > 0. \end{cases}$$

If $k < l$, then $l+1-k \geq 2$ and it follows from the partial fractional expansion that

$$E(s) = \frac{c_1}{s^{l+1-k}} + \frac{c_2}{s^{l-k}} + \cdots + \frac{c_{l-k}}{s} + E_0(s)$$

where $c_1 = L_0(0) \neq 0$ and the poles of $E_0(s)$ are all stable. This means

$$\begin{aligned} e(t) &= \mathcal{L}^{-1}[E(s)] \\ &= \frac{c_1}{(l-k)!}t^{l-k} + \frac{c_2}{(l-k-1)!}t^{l-k-1} + \cdots + c_{l-k} + \mathcal{L}^{-1}[E_0(s)] \to \infty \end{aligned}$$

as $t \to \infty$.

In summary,

$$\lim_{t\to\infty} e(t) = \begin{cases} 0 & \text{if } k > l \\ \dfrac{1}{1+L_0(0)} & \text{if } k = l = 0 \\ \dfrac{1}{L_0(0)} & \text{if } k = l > 0 \\ \infty & \text{if } k < l. \end{cases} \tag{4.2}$$

This shows that in order to achieve asymptotic tracking for signal $r(t) = (t^l/l!)\sigma(t)$, the type k of the loop gain has to be greater than l. In particular, in order to achieve asymptotic tracking for a step signal, the type of the loop gain has to be greater than 0, i.e., it has to contain at least one cascaded integrator.

Another concept closely related to system type is the error constants.

Definition 4.6. Given transfer function $L(s)$, define

1. the **position error constant** $K_p = \lim_{s\to 0} L(s)$;
2. the **velocity error constant** $K_v = \lim_{s\to 0} sL(s)$;
3. the **acceleration error constant** $K_a = \lim_{s\to 0} s^2L(s)$.

Let $L(s)$ be of type k and $L(s) = \dfrac{1}{s^k}L_0(s)$. Table 4.1 gives the error constants for different values of k.

	$k=0$	$k=1$	$k=2$
K_p	$L_0(0)$	∞	∞
K_v	0	$L_0(0)$	∞
K_a	0	0	$L_0(0)$

TABLE 4.1: Error constants for loop gains of different types.

According to (4.2), the steady-state errors of unity feedback systems for various reference inputs when the loop transfer function is of various types are expressed in terms of the error constants of the loop transfer function in Table 4.2.

	$k=0$	$k=1$	$k=2$
$r(t)=\sigma(t)$	$\dfrac{1}{1+K_p}$	0	0
$r(t)=t\sigma(t)$	∞	$\dfrac{1}{K_v}$	0
$r(t)=\dfrac{t^2}{2}\sigma(t)$	∞	∞	$\dfrac{1}{K_a}$

TABLE 4.2: Steady-state error for different reference inputs and for systems with different types.

EXAMPLE 4.7

Consider the DC motor system with transfer function

$$P(s) = \frac{K_t}{L_aJs^3 + (R_aJ + K_fL_a)s^2 + (R_aK_f + K_tK_b)s}.$$

Let the controller be a proportional control $C(s) = K$. Then $L(s) = KP(s)$ is a type 1 system. Hence

$$K_p = \infty, \quad K_v = \frac{KK_t}{R_aK_f + K_tK_b}, \quad K_a = 0.$$

It then follows that

$$e(\infty) = \begin{cases} 0 & \text{if } r(t) = \sigma(t) \\ \dfrac{R_a K_f + K_t K_b}{K K_t} & \text{if } r(t) = t\sigma(t) \\ \infty & \text{if } r(t) = \dfrac{t^2}{2}\sigma(t). \end{cases}$$

Note that the error is always that of position, though the error constants are termed as those of position, velocity, and acceleration.

For this example, since the plant is always of type 1 even after small parameter perturbations, a simple proportional control can make the system track a step input.

Let us now consider the steady-state effect of the disturbance $d(t)$ on the unity feedback system as shown in Figure 4.22. When we consider the steady-state effect on the output due to the disturbance of the form $d(t) = t^l/l!$, we will see that it is the type of the controller $C(s)$, not the type of the loop transfer function $L(s) = P(s)C(s)$, that is important. For this purpose, we set $r(t) = 0$ and $n(t) = 0$, and redraw Figure 4.22 as Figure 4.24.

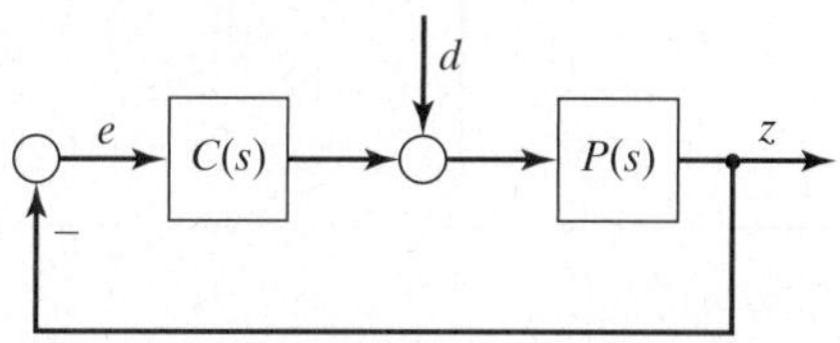

FIGURE 4.24: Feedback system subject to disturbance.

Write

$$C(s) = \frac{1}{s^k} C_0(s)$$

where $C_0(s)$ has no poles or zeros at the origin, i.e., $C(s)$ is of type k.

Again we will use the final value theorem to analyze the steady-state effect of the disturbance. The Laplace transform of the output is

$$Z(s) = \frac{P(s)}{1 + P(s)C(s)} D(s) = \frac{s^k P(s)}{s^k + P(s)C_0(s)} \frac{1}{s^{l+1}}.$$

If $k \geq l$,

$$Z(s) = \frac{s^{k-l-1} P(s)}{s^k + P(s)C_0(s)}$$

which has all stable poles except possibly a simple pole at the origin. By using the final value theorem

$$z(\infty) = \lim_{s \to 0} sZ(s) = \lim_{s \to 0} \frac{s^{k-l} P(s)}{s^k + P(s)C_0(s)} = \begin{cases} 0 & \text{if } k > l \\ \dfrac{P(0)}{1 + P(0)C_0(0)} & \text{if } k = l = 0 \\ \dfrac{1}{C_0(0)} & \text{if } k = l > 0. \end{cases}$$

If $k < l$, then $l + 1 - k \geq 2$ and

$$Z(s) = \frac{c_1}{s^{l+1-k}} + \frac{c_2}{s^{l-k}} + \cdots + \frac{c_{l-k}}{s} + Z_0(s)$$

where $c_1 \neq 0$ and $Z_0(s)$ has all poles on the left-hand side of the complex plane. This means that

$$\begin{aligned} z(t) &= \mathcal{L}^{-1}[Z(s)] \\ &= \frac{c_1}{(l-k)!}t^{l-k} + \frac{c_2}{(l-k-1)!}t^{l-k-1} + \cdots + c_{l-k} + \mathcal{L}^{-1}[Z_0(s)] \to \infty \end{aligned}$$

as $t \to \infty$.

In summary,

$$\lim_{t\to\infty} z(t) = \begin{cases} 0 & \text{if } k > l \\ \dfrac{P(0)}{1 + P(0)C_0(0)} & \text{if } k = l = 0 \\ \dfrac{1}{C_0(0)} & \text{if } k = l > 0 \\ \infty & \text{if } k < l. \end{cases}$$

This says that in order to reject disturbance of the form $t^l/l!$ asymptotically, the type of the controller, not that of the loop transfer function, has to be greater than l. In particular, in order to reject a step disturbance, the type of the controller has to be greater than 0, i.e., it has to contain at least one cascaded integrator.

EXAMPLE 4.8

Again consider the DC motor system with transfer function

$$P(s) = \frac{K_t}{L_a J s^3 + (R_a J + K_f L_a)s^2 + (R_a K_f + K_t K_b)s}.$$

Now assume that the system is subject to a load torque $d(t) = \sigma(t)$. In this case, in order to reject the disturbance completely, the controller has to be of at least type 1. The simplest such controller is an I (integral) controller of the form $C(s) = K_I/s$. However, it is impossible to find a K_I to stabilize the plant. So we should consider the next simplest type 1 controller, a PI controller of the form $C(s) = K_P + K_I/s$. It turns out that we are able to design K_P and K_I so that the closed-loop system is stable.

What happens if the feedback system is subject to measurement error, i.e., $n(t) = (t^l/l!)\sigma(t)$? In this case, we set $r(t) = 0, d(t) = 0$ in Figure 4.22 and redraw it as Figure 4.25. Here again $L(s) = P(s)C(s)$.

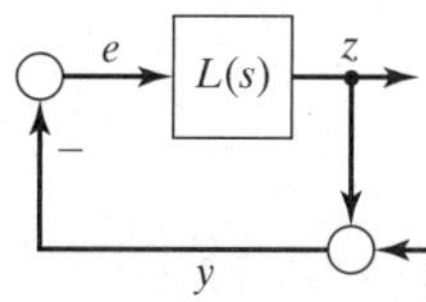

FIGURE 4.25: Feedback system subject to measurement noise.

We see that

$$Z(s) = \frac{-L(s)}{1+L(s)} N(s).$$

If $n(t) = \sigma(t)$, then

$$\lim_{t\to\infty} z(t) = \frac{-L(0)}{1+L(0)}.$$

This means that the steady-state error due to the measurement error is zero only if the loop transfer function $L(s)$ has a zero at the origin. However, if $L(0) = 0$, then $z(t)$ can never track a reference of the form $r(t) = t^l/l!$. This shows that the feedback system is not capable of rejecting measurement error of the form $n(t) = t^l/l!$ if it is required to track a reference of the same form. In order to achieve zero steady-state tracking error for a reference in the form of a step, ramp, etc., the measurement cannot contain an error in the form of a step, ramp, etc. This makes a lot of sense intuitively. If we wish to track a step, for example, and there is a step measurement error, then the controller can never know the true value of the output and hence cannot take appropriate actions.

Summarizing the above discussions, we come up with the following internal model principle for a unity feedback system.

Theorem 4.9.

1. It is impossible to design a unity feedback controller so that the feedback system asymptotically tracks a reference of the form $r(t) = t^{l_r}\sigma(t)$ and asymptotically rejects a measurement noise of the form $n(t) = t^{l_n}\sigma(t)$.
2. It is possible to design a unity feedback controller so that the feedback controller asymptotically tracks a reference of the form $r(t) = t^{l_r}\sigma(t)$ and asymptotically rejects a disturbance of the form $t^{l_d}\sigma(t)$ if and only if $b(0) \neq 0$. The solution requires that

   (a) $L(s) = P(s)C(s)$ is at least of type $l_r + 1$.

   (b) $C(s)$ is at least of type $l_d + 1$.

Now, we shall briefly consider the steady-state error when the reference is a sinusoid; i.e., $r(t) = (\sin \omega t)\sigma(t)$ for some frequency ω. Note that in this case $R(s) = \dfrac{\omega}{s^2+\omega^2}$ and

$$E(s) = \frac{1}{1+L(s)} \frac{\omega}{s^2+\omega^2}.$$

It is important to note that, in general, the final value theorem cannot be applied to compute the steady-state error since $E(s)$ in general has poles at $j\omega$ and $-j\omega$

unless $L(s)$ has poles at $j\omega$ and $-j\omega$. Nevertheless, since the closed-loop system is assumed to be stable, the sensitivity function

$$S(s) := \frac{1}{1 + L(s)}$$

is stable. Thus we can compute the steady-state response of the error as

$$e_{ss}(t) = |S(j\omega)| \sin(\omega t + \angle S(j\omega)).$$

Hence the steady-state error with respect to a sinusoidal input is no greater than $|S(j\omega)|$. It is also easy to see that the steady-state error is zero if and only if $S(j\omega) = 0$, i.e., $L(s)$ has a pair of poles at $j\omega$ and $-j\omega$.

EXAMPLE 4.10

Consider the feedback system shown in Figure 4.26. We shall now find the feasible parameter K so that the system is stable and the steady-state error with respect to a unit ramp input $r(t) = t\sigma(t)$ is no greater than 0.1.

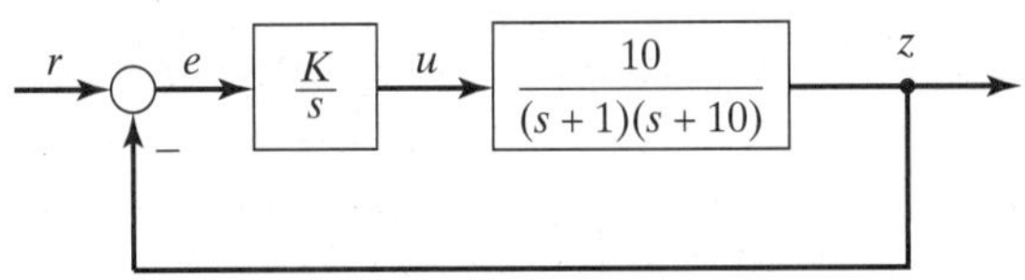

FIGURE 4.26: Example 4.10.

Since the closed-loop transfer function from r to e is

$$\frac{E(s)}{R(s)} = \frac{s(s+1)(s+10)}{s^3 + 11s^2 + 10s + 10K}$$

by Routh–Hurwitz stability criterion, the system is stable if and only if $K < 11$. Since

$$K_v = \lim_{s \to 0} sC(s)P(s) = \lim_{s \to 0} \frac{10K}{(s+1)(s+10)} = K,$$

the steady-state error of the system with respect to a ramp input is given by

$$e_{ss} = \frac{1}{K_v} = \frac{1}{K} \le 0.1$$

which gives $K \ge 10$. Thus we need $10 \le K < 11$.

Let $K = 10$. Then the steady-state error with respect to $r(t) = (\sin 2t)\sigma(t)$ is no greater than

$$\left|\frac{E(j2)}{R(j2)}\right| = \left|\frac{j2(j2+1)(j2+1)}{(j2)^3 + 11(j2)^2 + 10(j2) + 100}\right| = 0.7963.$$

4.5 INTERNAL MODEL PRINCIPLE

In the last section, we have seen that in order for a unity feedback system to track a step reference without steady-state error in the situation when the disturbance and measurement noise are zero, the forward path transfer function $L(s) = P(s)C(s)$ needs to include an integrator, i.e., to have a pole at the origin.

If $L(s)$ contains an integrator, the unity feedback system not only tracks the step function but also does so robustly, in the sense that the other parameters are irrelevant as far as steady-state tracking is concerned and as long as the closed-loop system is internally stable.

Since the Laplace transform of the unit step input is $1/s$, the requirement that the $L(s)$ is at least of type 1 can be rephrased as that the poles of the Laplace transform of the reference are contained in the poles of the forward path transfer function $L(s)$. Hence, the unity feedback system robustly tracks a step reference if and only if the poles of $L(s)$ contain the pole of the Laplace transform of reference. Similarly, the unity feedback system robustly tracks a ramp reference if and only if the poles of $L(s)$ contain the poles of the Laplace transform of the reference, which in this case are 0 repeated twice.

The robustness of the tracking critically depends on whether the type of the loop transfer function can be made robust, i.e., insensitive to parameter changes. The type of the loop transfer function is the sum of the type of the plant $P(s)$ and that of the controller $C(s)$. The type of the plant is usually determined by the structures of the physical system, not the parameters of the plant. For example, in the position control of the DC motor as in Example 2.12, the plant is always of type 1 no matter what the parameters R_a, L_a, K_t, K_b, K_f, and J are. The type of the controller can be made independent of the parameters as long as it is implemented appropriately. For example, if we use the op-amp implementation shown in Figures 2.17 and 2.18, just set $a_n = 0$ (disconnect the branch of the circuit), and then the controller will robustly have type 1.

In this section, we shall extend the above conclusion to a 2DOF controller structure as shown in Figure 4.27.

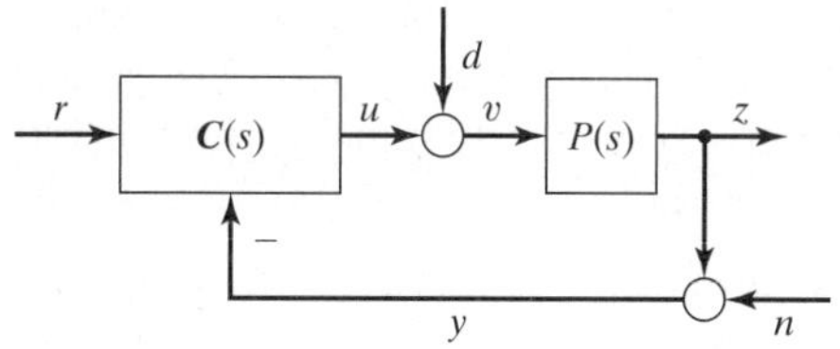

FIGURE 4.27: A 2DOF control system.

Assume

$$r(t) = \left(\frac{\alpha_0}{l!}t^l + \frac{\alpha_1}{(l-1)!}t^{l-1} + \cdots + \alpha_l\right)\sigma(t), \quad \alpha_0 \neq 0.$$

Such a signal is called a polynomial signal of order l. Let us also assume $d(t) = 0$ and $n(t) = 0$. Let

$$P(s) = \frac{b(s)}{a(s)}$$

and

$$C(s) = \begin{bmatrix} C_1(s) & C_2(s) \end{bmatrix} = \frac{1}{p(s)} \begin{bmatrix} q_1(s) & q_2(s) \end{bmatrix}.$$

Then the Laplace transform of the error $e(t) = r(t) - z(t)$ is

$$E(s) = \frac{a(s)p(s) + b(s)[q_2(s) - q_1(s)]}{a(s)p(s) + b(s)q_2(s)} \frac{\alpha_0 + \alpha_1 s + \cdots + \alpha_l s^l}{s^{l+1}}.$$

Hence if we require internal stability and steady-state tracking, then the closed-loop characteristic polynomial $a(s)p(s) + b(s)q_2(s)$ has to be stable and the numerator polynomial $a(s)p(s) + b(s)[q_2(s) - q_1(s)]$ has to contain s^{l+1} as a factor. There are many ways of designing $p(s)$, $q_1(s)$, and $q_2(s)$ to satisfy these requirements.

EXAMPLE 4.11

Assume that $P(s) = \dfrac{1}{s+1}$ and $C(s) = \begin{bmatrix} 2 & 1 \end{bmatrix}$. Also assume that $r(t) = \sigma(t)$. Then

$$E(s) = \frac{s}{s+2}\frac{1}{s} = \frac{1}{s+2}.$$

In this case, the 2DOF feedback system can track a step reference in the steady state when the disturbance and noise are not present. However, this tracking property is not robust if the plant is subject to a small perturbation to become $\dfrac{1}{s+1.1}$ or if the controller is subject to a small perturbation to become $\begin{bmatrix} 2.05 & 0.9 \end{bmatrix}$, though the internal stability of the closed-loop system is preserved under such small perturbations.

Example 4.11 shows a distinct feature of the 2DOF control structure different from the unity feedback structure: the requirements for steady-state tracking and robust steady-state tracking are different. In order to achieve robust tracking, we need to require s^{l+1} to be a factor of both $a(s)p(s)$ and $b(s)[q_2(s) - q_1(s)]$. Since $a(s)p(s)$ and $b(s)$ cannot share unstable roots, the requirement becomes that $a(s)p(s)$ and $q_2(s) - q_1(s)$ both have s^{l+1} as factors.

The analysis of the effects of the disturbance and the measurement noise for the 2DOF control case is the same as the unity feedback case. Hence, we arrive at the following internal model principle for the 2DOF control structure.

Theorem 4.12. The robust regulation problem for a polynomial reference of order l_r, a polynomial disturbance of degree l_d, a polynomial noise of degree l_n is solvable by using a 2DOF controller if and only if $P(0) \neq 0$ and $n(t) = 0$. The solution of the robust regulation problem requires that

1. $L(s) = P(s)C_2(s)$ is at least of type $l_r + 1$, and s^{l_r+1} is a factor of $q_1(s) - q_2(s)$.
2. $C_2(s)$ is at least of type $l_d + 1$.

One may now question how one can implement the controller so as to ensure robustly that s^{l_r+1} is a factor of $q_1(s) - q_2(s)$. Notice that this requirement is equivalent to

$$q_{1n} = q_{2n}, \quad q_{1(n-1)} = q_{2(n-1)}, \quad \ldots, \quad q_{1(n-l_r)} = q_{2(n-l_r)}. \tag{4.3}$$

If we use op-amp circuit implementation of the 2DOF controller, the circuit in Figure 2.25 can be modified to the one in Figure 4.28. We see that the requirement (4.3) is now guaranteed by the structure, not the parameter match, of the controller. The same trick can be used in software implementation of the 2DOF controller.

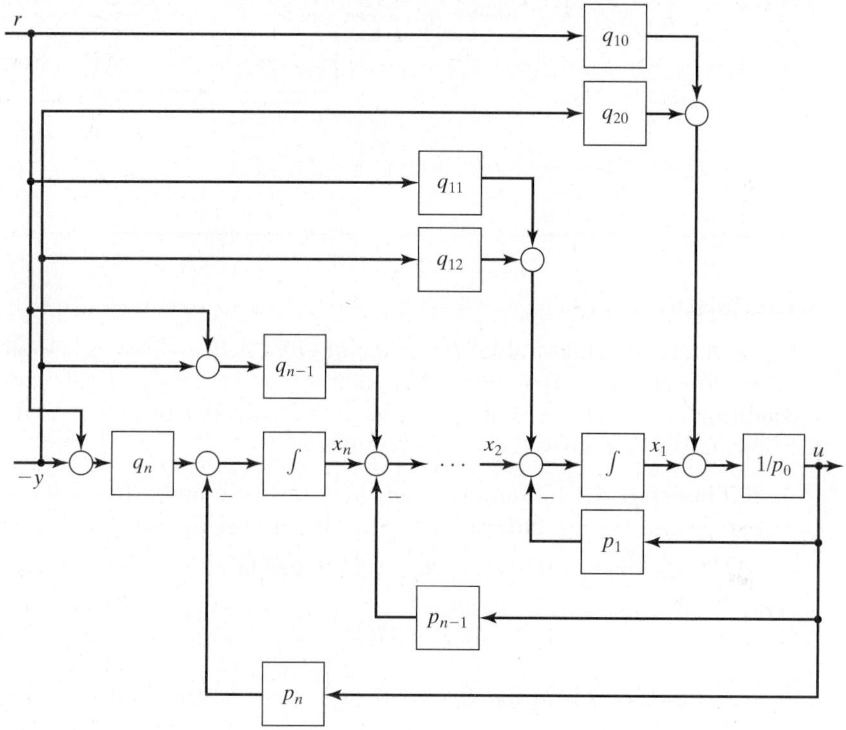

FIGURE 4.28: Realization of a 2DOF controller.

EXAMPLE 4.13

Let

$$P(s) = \frac{1}{s(s-1)}$$

and

$$\boldsymbol{C}(s) = \frac{1}{s+7}\begin{bmatrix} 2s+6 & 18s+6 \end{bmatrix}.$$

Then this controller achieves robust tracking of a step reference $R(s) = 1/s$ since $P(s)C_2(s)$ has a pole at 0 and $C_2(s) - C_1(s)$ has a zero at 0. However, this controller does not achieve robust disturbance rejection for a step disturbance $D(s) = 1/s$ since $C_2(s)$ does not have a pole at 0.

An op-amp implementation of this controller, which guarantees the robustness of the property that s divides $C_2(s) - C_1(s)$, is shown in Figure 4.29.

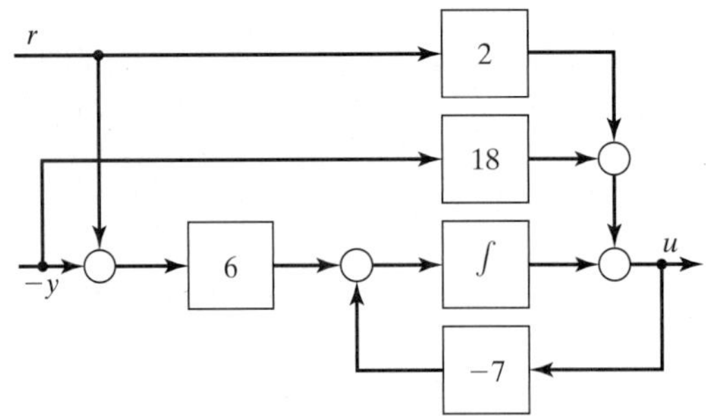

FIGURE 4.29: Op-amp implementation of the 2DOF controller in Example 4.13.

4.6 UNDERSHOOT

Undershoots are undesirable. It is then of great help to know when an undershoot occurs for a given system without simulating the step response. Giving precisely a condition on system parameters under which undershoot occurs is not easy. The following theorem gives a sufficient condition.

Theorem 4.14. Suppose that $G(s)$ is stable and $G(0) \neq 0$. Then its step response has an undershoot if $G(s)$ has a positive real zero.

Proof. Let $z > 0$ be a positive real zero of $G(s)$ and $y(t)$ be the step response. Then

$$Y(s) = G(s)\frac{1}{s} = \int_{0-}^{\infty} y(t)e^{-st}dt.$$

Notice that z is in the region of convergence of the Laplace transform of $y(t)$. Then

$$0 = G(z)\frac{1}{z} = \mathcal{L}[y(t)](z) = \int_{0-}^{\infty} y(t)e^{-zt}dt.$$

Assume $G(0) > 0$. Then $y(\infty) = G(0) > 0$. Since $y(t)e^{-zt} > 0$ for sufficiently large t, there must exist $t_0 > 0$ such that $y(t_0)e^{-zt_0} < 0$, i.e., $y(t_0) < 0$. A similar argument applies when $G(0) < 0$. □

EXAMPLE 4.15

Consider the following systems:

$$G_1(s) = \frac{(s-1)(s-2)}{(s+1)(s^2+2s+2)}, \quad G_2(s) = \frac{2-s}{s^2+2s+2}.$$

Both systems have positive real zeros, thus it is expected that their step responses, $y_1(t)$ and $y_2(t)$, will both have undershoots as shown in Figure 4.30. The undershoot of $y_2(t)$ is easy to see since $\dot{y}_2(0^+) = \lim_{s\to\infty} s^2 Y_2(s) = -1 < 0$ and $y_2(t)$ starts in the wrong direction from the beginning. On the other hand, the undershoot of $y_1(t)$ comes some time later since $\dot{y}_1(0^+) = 1 > 0$.

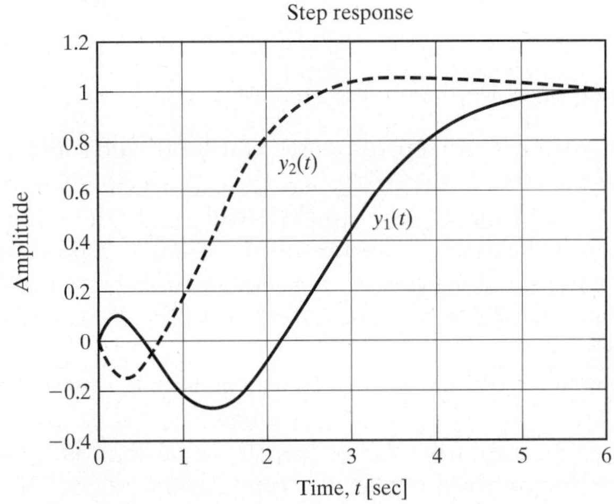

FIGURE 4.30: Step responses for systems $G_1(s)$ and $G_2(s)$ in Example 4.15.

Theorem 4.14 shows that positive real zeros are not desirable. One may conjecture that the complex zeros with positive real parts are not desirable either. This is indeed the case in many situations, though they do not necessarily lead to undershoot in the step response, see Problem 4.13. This problem also gives an example where the transfer function is free of positive real zeros but its step response has undershoot.

In the case when $G(s)$ has a positive real zero, not only does the undershoot occur but the PU also has a trade-off with the settling time t_s.

Theorem 4.16. Suppose that $G(s)$ is stable and $G(0) \neq 0$. If $G(s)$ has a positive real zero $z > 0$, then

$$\text{PU} \geq \frac{1-\Delta}{e^{zt_s}-1} \times 100\%$$

where Δ is the error tolerance used in defining the settling time t_s.

Proof. Let us prove for the case when $G(0) = 1$. The other cases follow by proper scaling. Let $y(t)$ be the step response. Then for $t \geq t_s$, it holds that $y(t) \geq 1 - \Delta$. We have

$$\int_0^\infty y(t)e^{-zt}dt = \int_0^{t_s} y(t)e^{-zt}dt + \int_{t_s}^\infty y(t)e^{-zt}dt = 0.$$

Since

$$\int_{t_s}^\infty y(t)e^{-zt}dt \geq (1-\Delta)\int_{t_s}^\infty e^{-zt}dt = (1-\Delta)\frac{e^{-zt_s}}{z}$$

and

$$\int_0^{t_s} y(t)e^{-zt}dt \geq y(t_b)\int_0^{t_s} e^{-zt}dt = y(t_b)\frac{1-e^{-zt_s}}{z},$$

it follows that

$$-y(t_b)\frac{1-e^{-zt_s}}{z} \geq (1-\Delta)\frac{e^{-zt_s}}{z}$$

which gives the inequality that we need. □

Undershoots in step responses can be divided into two types. The step response of a system is said to have a **type A undershoot** if it starts, at time 0, in the wrong direction opposite to the steady-state direction. An undershoot not of type A is called a **type B undershoot**. The undershoots in step responses 3 and 4 in Figure 4.5 are of type A, whereas those in step responses 5 and 6 in Figure 4.5 are of type B. In Example 4.15, $y_2(t)$ has a type A undershoot and $y_1(t)$ has a type B undershoot.

How can we capture, in a mathematically precise way, the phenomenon of starting in the wrong direction? Assume that the step response initially starts at value 0. If its initial derivative has the same sign as its steady-state value, then the step response does not have a type A undershoot. If its initial derivative has the opposite sign as its steady-state value, then it has a type A undershoot. What happens if the initial derivative is zero? In this case, the initial direction of the step response depends on its initial second derivative. In general, the initial direction of the step response depends on the sign of its first nonzero initial derivative, counted in the order of the first derivative, the second derivative, the third derivative, and so on. Mathematically, if a step response $y(t)$ satisfies $y^{(i)}(0^+) = 0$ for $i = 0, 1, 2, \ldots, k-1$ with $y^{(0)}(0^+) := y(0^+)$ but $y^{(k)}(0^+) \neq 0$, then it is said to have a type A undershoot if

$$y(\infty)y^{(k)}(0^+) < 0.$$

The order of the derivative at time 0^+ that one has to take in order to hit a nonzero value gives a measure of the flatness or smoothness of the step response at time 0. It happens that this order is exactly equal to the relative degree of $G(s)$; i.e., for a transfer function $G(s)$ with relative degree k, its step response $y(t)$ satisfies $y^{(i)}(0^+) = 0$ for $i = 0, 1, 2, \ldots, k-1$ but $y^{(k)}(0^+) \neq 0$. The reason for this is not difficult to see. Assume that $i \leq k$ and $y(0^+), \dot{y}(0^+), \ldots, y^{(i-1)}(0^+)$ are all equal to zero. Since

$$Y(s) = G(s)\frac{1}{s},$$

it follows that

$$\mathcal{L}[y^{(i)}(t)](s) = s^i Y(s) = s^{i-1}G(s).$$

By the initial value theorem,

$$y^{(i)}(0^+) = \lim_{s\to\infty} s^i G(s)$$

which is zero if $i < k$ and nonzero if $i = k$.

The following theorem characterizes precisely, in terms of simple system characteristics, the condition under which a type A undershoot occurs.

Theorem 4.17. Suppose that $G(s)$ is stable and $G(0) \neq 0$. Then its step response has a type A undershoot if and only if $G(s)$ has an odd number of positive real zeros.

Proof. Assume that the relative degree of $G(s)$ is k. Then $G(s)$ can always be written in the form

$$G(s) = K\frac{\prod_{i=1}^{l}(s-\alpha_i)\prod_{i=1}^{m}(s+\beta_i)\prod_{i=1}^{p}[(s+\sigma_i)^2+\omega_i^2]}{s^n + a_1 s^{n-1} + \cdots + a_n}$$

where $\alpha_i, \beta_i > 0$ and $l + m + 2p = n - k$. The number of positive real zeros is l. Now

$$y^{(k)}(0^+) = \lim_{s\to\infty} s^k G(s) = K$$

$$y(\infty) = G(0) = K\frac{\prod_{i=1}^{l}(-\alpha_i)\prod_{i=1}^{m}\beta_i\prod_{i=1}^{p}(\sigma_i^2+\omega_i^2)}{a_n}.$$

Since

$$a_n > 0, \quad \prod_{i=1}^{m}\beta_i > 0, \quad \prod_{i=1}^{p}(\sigma_i^2+\omega_i^2) > 0,$$

it follows that $y^{(k)}(0^+)y(\infty) < 0$ if and only if $\prod_{i=1}^{l}(-\alpha_i) < 0$, which is true if and only if l is odd. □

Usually a system operates in a closed-loop mode, such as shown in Figure 4.31. Thus we are interested in when an undershoot occurs if the plant $P(s)$ is given and $\boldsymbol{C}(s)$ is freely designable.

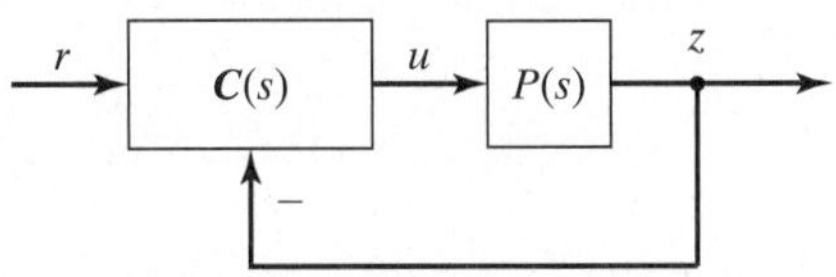

FIGURE 4.31: A 2DOF control system.

Let $P(s) = \dfrac{b(s)}{a(s)}$ and

$$\boldsymbol{C}(s) = \frac{1}{p(s)}\begin{bmatrix} q_1(s) & q_2(s) \end{bmatrix}$$

where $a(s)$ and $b(s)$ are coprime polynomials and $p(s)$, $q_1(s)$, and $q_2(s)$ are coprime polynomials. The transfer function from $r(t)$ to $z(t)$ is

$$\frac{b(s)q_1(s)}{a(s)p(s)+b(s)q_2(s)}.$$

If the closed-loop system is internally stable, then $a(s)p(s)+b(s)q_2(s)$ has no root with a positive real part. Any zero of $P(s)$, i.e., any root of $b(s)$, with a positive real part will show up as a zero of the closed-loop transfer function from $r(t)$ to $z(t)$.

This implies that if the plant $P(s)$ has a positive real zero, then no matter how you design the 2DOF controller $\boldsymbol{C}(s)$, undershoot will always occur. Furthermore, the PU would be worse if a fast settling time is required or, conversely, the settling time would be longer if a small PU is required. If undershoot is absolutely unacceptable, then plant redesign or modification should be carried out in order to remove all the positive real zeros.

4.7 OVERSHOOT

We shall now consider when a step response has overshoot. Let us consider the unity feedback system shown in Figure 4.32. When we study the step tracking property of this system, we usually assume that the type of the forward transfer function $L(s)$ is at least 1, since otherwise the system cannot track a step in the steady state.

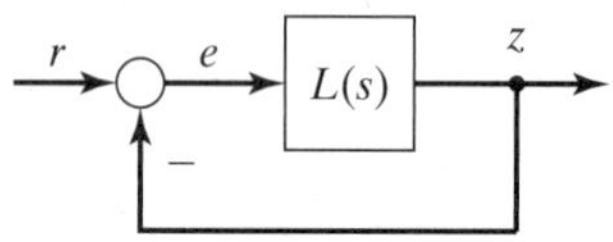

FIGURE 4.32: A unity feedback system with loop gain $L(s)$.

One possible cause of overshoot is the unstable open-loop poles.

Theorem 4.18. If $L(s)$ is at least of type 1 and has a positive real pole, then the step response of the system has an overshoot.

Proof. Without loss of generality, we assume that the step reference is the unit step. Then the Laplace transform of the error $e(t)$ is

$$E(s) = \frac{1}{1 + L(s)} \frac{1}{s}.$$

If $L(s)$ has a positive real pole at p, then

$$E(p) = \int_{0^-}^{\infty} e(t) e^{-pt} dt = 0.$$

Since $e(t)e^{-pt}$ is positive for sufficiently small t, there exists $t > 0$ such that $e(t)e^{-pt} < 0$, or, equivalently, $e(t) < 0$, i.e., $z(t)$ has an overshoot. □

In applications, the loop gain $L(s)$ consists of a plant $P(s)$ and a controller $C(s)$ as shown in Figure 4.33. Now assume that the plant $P(s)$ has a positive real pole. Since any controller $C(s)$ that makes the feedback system internally stable

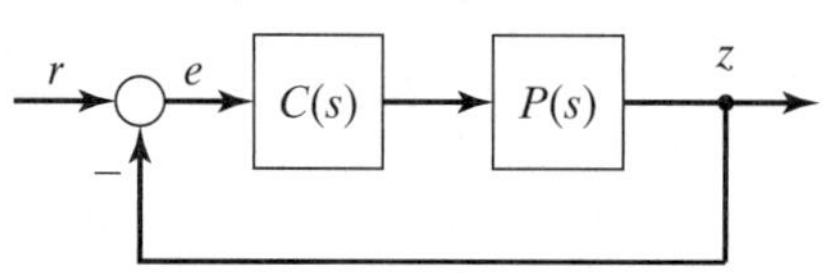

FIGURE 4.33: A unity feedback system with plant $P(s)$ and controller $C(s)$.

cannot cancel this pole, this pole will be a pole of $L(s) = P(s)C(s)$. This shows that no matter how we design a stabilizing controller $C(s)$, the overshoot always occurs whenever $L(s)$ is at least type 1.

Another possible cause of overshoot is the existence of too many integrators in a unity feedback loop.

Theorem 4.19. If the type of $L(s)$ is at least 2, then the step response of the system has an overshoot.

Proof. Again assume that the step reference is a unit step. The Laplace transform of the error $e(t)$ is

$$E(s) = \frac{1}{1+L(s)}\frac{1}{s}.$$

Hence the Laplace transform of the integral of error is

$$\mathcal{L}\left[\int_{0-}^{t} e(t)dt\right] = \frac{1}{1+L(s)}\frac{1}{s^2}.$$

By using the final value theorem, we see that

$$\int_{0-}^{\infty} e(t)dt = \lim_{t\to\infty}\int_{0-}^{t} e(t)dt = \lim_{s\to 0}\frac{1}{s}\frac{1}{1+L(s)} = 0.$$

Since $e(t) > 0$ for sufficiently small t, there must exist $t > 0$ such that $e(t) < 0$, i.e., $z(t)$ has an overshoot. □

We have seen in the previous sections that a higher system type of the loop transfer function means a better steady-state response. Now we see that this has the possible drawback of introducing overshoot, which is undesirable from the viewpoint of transient response.

Again, if the type of the plant $P(s)$ in Figure 4.33 is at least 2, then a stabilizing controller cannot reduce the type of the overall loop gain $L(s) = P(s)C(s)$. Hence no matter what stabilizing controller is used, the step response overshoot in the unity feedback of such a system always occurs.

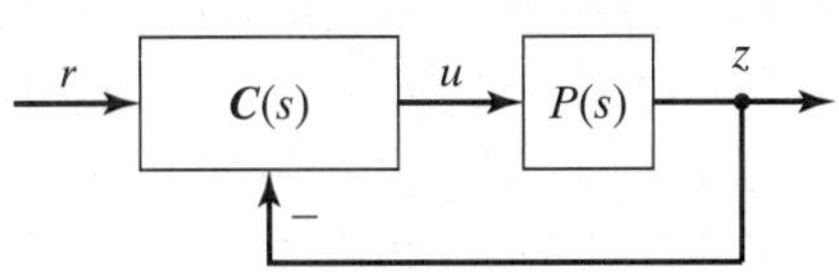

FIGURE 4.34: A 2DOF control system.

On the other hand, if we use the 2DOF tracking controller as shown in Figure 4.34 instead of the unity feedback, then it is possible to prevent overshoot from occurring even when $P(s)$ has a positive real pole and/or has a type higher than 2, as demonstrated in the following two examples. This gives an overwhelming advantage of the 2DOF controller over the unity feedback controller.

EXAMPLE 4.20

Let

$$P(s) = \frac{1}{s(s-1)}.$$

If we use the unity feedback as shown in Figure 4.33 with a controller

$$C(s) = \frac{18s+6}{s+7},$$

then the closed-loop system is stable, with transfer function from $r(t)$ to $z(t)$ being

$$\frac{18s+6}{s^3+6s^2+11s+6}.$$

Its step response is the solid curve in Figure 4.35, which has overshoot. This is expected since Theorem 4.18 has ruled out the possibility of overshoot-free step response no matter how the controller is designed. If we use a 2DOF controller as shown in Figure 4.34 with

$$\boldsymbol{C}(s) = \frac{1}{s+7}\left[\begin{array}{cc} 2s+6 & 18s+6 \end{array}\right],$$

then the transfer function from $r(t)$ to $z(t)$ of the closed-loop system is

$$\frac{2}{s^2+3s+2}.$$

Its step response is the dashed curve in Figure 4.35, which does not have an overshoot.

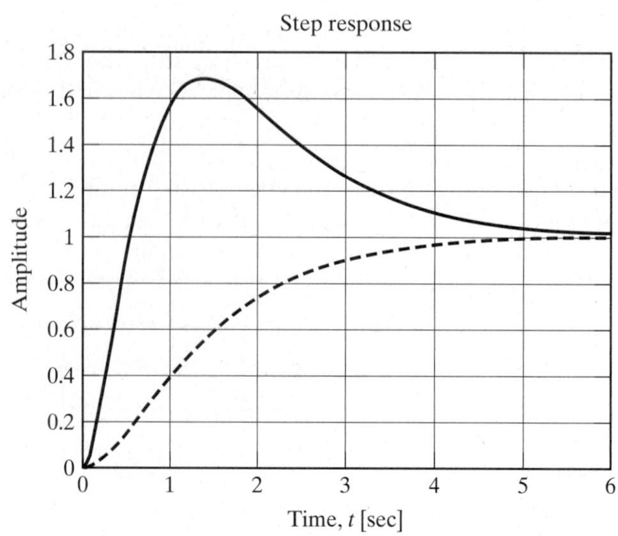

FIGURE 4.35: The step responses for two systems in Example 4.20.

EXAMPLE 4.21

Let

$$P(s) = \frac{2s+1}{s^2}.$$

Then by Theorem 4.19 the step response of a unity feedback system with this plant and any stabilizing controller has overshoot. If we use the 2DOF controller as in Figure 4.34 with

$$\boldsymbol{C}(s) = \frac{1}{2s+1}\begin{bmatrix} 1 & 2s+1 \end{bmatrix},$$

then the closed-loop transfer function from $r(t)$ to $z(t)$ is

$$\frac{Z(s)}{R(s)} = \frac{1}{(s+1)^2},$$

whose step response does not have overshoot because it is a critically damped, standard, second-order system.

The above two examples give the overwhelming advantage of the 2DOF controller over the unity feedback controller. It turns out that for any reasonable plant, it is always possible to avoid overshoot by designing a 2DOF controller.

Theorem 4.22. For any LTI plant $P(s)$ with $P(0) \neq 0$, there is a 2DOF controller $\boldsymbol{C}(s)$ such that the step response from the reference $r(t)$ to the output $z(t)$ does not have overshoot.

The rest of this section is dedicated to the proof of Theorem 4.22. The proof is constructive, so it also tells how such a 2DOF controller can be designed.

Here we use the parametrization of all stabilizing 2DOF controllers given in Section 3.8. Let the plant

$$P(s) = \frac{b(s)}{a(s)}$$

be proper with $\deg a(s) = n$ and

$$C_0(s) = \frac{q(s)}{p(s)}$$

where $\deg p(s) = m$ be an initial stabilizing feedback controller. The assumption that $P(0) \neq 0$ implies that $b(0) \neq 0$. Factorize $c(s) = a(s)p(s) + b(s)q(s)$ as $c(s) = f(s)h(s)$ such that $\deg f(s) = n$ and $\deg h(s) = m$. Let

$$M(s) = \frac{a(s)}{f(s)}, \quad N(s) = \frac{b(s)}{f(s)}, \quad X(s) = \frac{p(s)}{h(s)}, \quad Y(s) = \frac{q(s)}{h(s)}.$$

It follows from the theory in Section 3.8 that the set of all stabilizing 2DOF controllers is given by

$$\mathcal{T}(P) = \left\{ C(s) = \begin{bmatrix} \dfrac{Q_1(s)}{X(s) - N(s)Q_2(s)} & \dfrac{Y(s) + M(s)Q_2(s)}{X(s) - N(s)Q_2(s)} \end{bmatrix} : \right.$$

$$\left. Q_1(s) \text{ and } Q_2(s) \text{ are two arbitrary stable systems} \right\}$$

and the closed-loop transfer function from $r(t)$ to $z(t)$ is

$$G(s) = N(s)Q_1(s) = \frac{b(s)}{f(s)} Q_1(s)$$

which is independent of $Q_2(s)$. This shows that $Q_2(s)$ has nothing to do with the step response from $r(t)$ to $z(t)$ and hence can be designed independently.

Factorize $b(s)$ as $b(s) = b_s(s)b_u(s)$ where $b_s(s)$ is a stable polynomial and $b_u(s)$ is a polynomial whose roots are the unstable roots of $b(s)$. Construct

$$Q_1(s) = \frac{f(s)}{b_s(s)b_u(0)} \prod_{i=1}^{k} \frac{\alpha_i}{s + \alpha_i}$$

where $\alpha_i > 0$, $i = 1, 2, \ldots, k$, and the integer k is chosen so that $Q_1(s)$ is proper, i.e., $k \geq \deg f(s) - \deg b_s(s)$. With this $Q_1(s)$, we have

$$G(s) = \frac{b_u(s)}{b_u(0)} \prod_{i=1}^{k} \frac{\alpha_i}{s + \alpha_i}.$$

We have accomplished two objectives in designing this $Q_1(s)$. The first is to cancel the stable zeros of $N(s)$ using the stable poles of $Q_1(s)$. The second is to assign all poles of $G(s)$ to negative real numbers.

Let us first consider a number of special cases, which cover the majority of the practical cases.

Case 1: $b_u(s)$ is a constant, i.e., the plant is minimum phase.

In this case,

$$G(s) = \prod_{i=1}^{k} \frac{\alpha_i}{(s + \alpha_i)}.$$

The statement in Problem 4.18 implies that the step response of $G(s)$ is monotonic and hence is free of overshoot.

Case 2: $b_u(s) = s - z$ with $z > 0$, i.e., the plant only has one unstable zero z that is real and positive.

In this case,

$$G(s) = \frac{\alpha_1(s-z)}{-z(s+\alpha_1)} \prod_{i=2}^{k} \frac{\alpha_i}{s+\alpha_i}.$$

The conclusion in Problem 4.20.2 implies that $z(t)$ is free of overshoot.

Case 3: $b_u(s) = s^2 - 2\zeta\omega_n s + \omega_n^2$ with $\zeta \geq 0$, $\omega_n > 0$, i.e., the plant only has two complex conjugate or real unstable zeros.

In this case, set

$$\alpha_1 = \alpha_2 = \omega_n$$

Then

$$G(s) = \frac{s^2 - 2\zeta\omega_n s + \omega_n^2}{(s+\omega_n)^2} \prod_{i=3}^{k} \frac{\alpha_i}{s+\alpha_i}.$$

The conclusion in Problem 4.20.3 implies that $z(t)$ is free of overshoot.

In the general case, we cannot simply extend the design ideas above. We need to follow a different idea. Let us set $\alpha_i = \alpha$ for all i. Then the tracking error for a unit step reference is

$$E(s) = [1 - G(s)]\frac{1}{s} = \left[1 - \frac{b_u(s)\alpha^k}{b_u(0)(s+\alpha)^k}\right]\frac{1}{s} = \frac{b_u(0)(s+\alpha)^k - b_u(s)\alpha^k}{b_u(0)(s+\alpha)^k s}.$$

Notice that the numerator is

$$\begin{aligned}&b_u(0)[(s+\alpha)^k - \alpha^k] - [b_u(s) - b_u(0)]\alpha^k\\ &= b_u(0)s[(s+\alpha)^{k-1} + (s+\alpha)^{k-2}\alpha + \cdots + (s+\alpha)\alpha^{k-2} + \alpha^{k-1}] - [b_u(s) - b_u(0)]\alpha^k.\end{aligned}$$

Also notice that the second term above can be divided by s and hence, according to the partial fractional expansion formulas in Section A.3,

$$\frac{b_u(s) - b_u(0)}{b_u(0)(s+\alpha)^k s} = \frac{\beta_1(\alpha)}{s+\alpha} + \frac{\beta_2(\alpha)}{(s+\alpha)^2} + \cdots + \frac{\beta_k(\alpha)}{(s+\alpha)^k}$$

where $\beta_i(\alpha), i = 1, 2, \ldots, k$, are polynomial functions of α. This shows that

$$E(s) = \frac{\gamma_1(\alpha)}{s+\alpha} + \frac{\gamma_2(\alpha)}{(s+\alpha)^2} + \cdots + \frac{\gamma_k(\alpha)}{(s+\alpha)^k}$$

where for $i = 1, 2, \ldots, k$,

$$\gamma_i(\alpha) = \alpha^{i-1} - \beta_i(\alpha)\alpha^k = \alpha^{i-1}[1 - \beta_i(\alpha)\alpha^{k-i+1}].$$

As long as α is sufficiently small, $\gamma_i(\alpha), i = 1, 2, \ldots, k$, are nonnegative. Consequently,

$$e(t) = \mathcal{L}^{-1}[E(s)] = \left(\gamma_1(\alpha) + \gamma_2(\alpha)t + \cdots + \frac{\gamma_k(\alpha)}{(k-1)!}t^{k-1}\right) e^{-\alpha t}\sigma(t)$$

is always nonnegative, which means that $z(t)$ has no overshoot.

4.8 TIME-DOMAIN SIGNAL AND SYSTEM NORMS

In the analysis and design of control systems, it is often required to measure the strength of a signal or a system. Such a strength, in mathematical terms, is called a **norm**. We often use the norm of a certain signal, such as the tracking error or the plant input (when the reference signal is fixed), or a certain closed-loop system as the performance index of the control system. The purpose in the design is then to find the parameters of the controller so that the value of the performance index is small or minimized. There are several ways to define norms of signals and systems. In this section, we will first look at several time-domain norms. In Chapter 8, we will introduce the frequency-domain norms.

Definition 4.23. Let $x(t)$ be a signal.

1. The ∞-norm of $x(t)$ is defined as[1]

$$\|x(t)\|_\infty = \sup_{t \geq 0} |x(t)|.$$

2. The 2-norm of $x(t)$ is defined as

$$\|x(t)\|_2 = \left[\int_{0^-}^{\infty} x^2(t)dt\right]^{1/2}.$$

3. The 1-norm of $x(t)$ is defined as

$$\|x(t)\|_1 = \int_{0^-}^{\infty} |x(t)|dt.$$

All these norms have physical meanings. The ∞-norm is the peak magnitude or the amplitude of the signal. The square of the 2-norm is often called the **energy of the signal**. If $x(t)$ is the current or voltage of a resistive load, the square of the 2-norm of $x(t)$ is proportional to the total energy consumption of the load. If $x(t)$ is the velocity of a rigid body, then $\|x(t)\|_2^2$ is proportional to its kinetic energy. If $x(t)$ is the height of a point mass, then $\|x(t)\|_2^2$ is proportional to its potential energy. In the case when $x(t)$ is a certain error signal, then the square of the 2-norm

[1] Here "sup" stands for supremum. It means least upper bound.

of $x(t)$ is also called the **integral of the squared error** (ISE). The 1-norm is also called the action of the signal. If $x(t)$ represents the flow of a certain material or fuel, then the 1-norm of $x(t)$ is the total consumption of the material or the fuel. In the case that $x(t)$ is a certain error signal, then the 1-norm of $x(t)$ is also called the **integral of the absolute error (IAE)**.

We have seen the ∞-norm and the 1-norm in the study of system stability. A signal is said to be bounded if its ∞-norm is finite. Theorem 3.3 simply says that a system is stable if and only if the 1-norm of its impulse response is finite.

It can be easily seen that if $x(t)$ has unstable behavior, i.e., if it approaches infinity as time goes to infinity, or if it oscillates with exponentially increasing magnitude, its norms would be equal to infinity. For $\|x(t)\|_\infty$ to be finite, the signal $x(t)$ has to have a bounded behavior.

One might wonder what the norms of the impulse function $\delta(t)$ are. A momentary thought gives

$$\|\delta(t)\|_\infty = \infty, \quad \|\delta(t)\|_2 = \infty, \quad \|\delta(t)\|_1 = 1.$$

EXAMPLE 4.24

Consider the unity feedback system as shown in Figure 4.36.

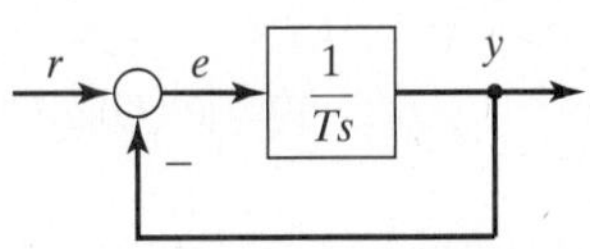

FIGURE 4.36: A first-order unity feedback system.

Assume $r(t)$ is a unit step. We are interested in finding various norms of $e(t)$. We know that

$$E(s) = \frac{T}{Ts+1}$$

and

$$e(t) = e^{-t/T}.$$

Direct computation gives

$$\|e(t)\|_\infty = 1, \ \|e(t)\|_2 = \left[\int_{0-}^{\infty} e^{-2t/T} dt\right]^{1/2} = \sqrt{\frac{T}{2}}, \ \|e(t)\|_1 = \int_{0-}^{\infty} e^{-t/T} dt = T.$$

In this case, the infinity norm is completely insensitive to the change of T, and is hence not suitable as a performance index, for it is unable to distinguish the good time constants from the bad ones. The other two norms agree with our common sense that the smaller the time constant, the smaller the error.

EXAMPLE 4.25

Consider the unity feedback system shown in Figure 4.37. We are interested in the different norms of the error signal when the reference is a unit step.

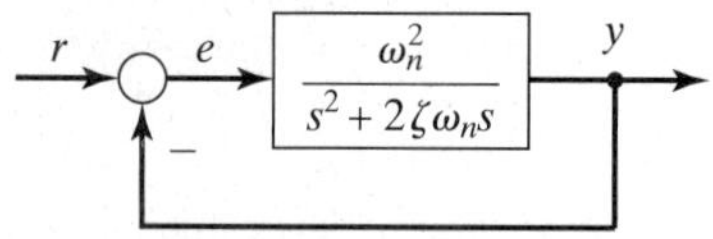

FIGURE 4.37: A second-order unity feedback system.

Analytically, we know

$$e(t) = \begin{cases} \dfrac{e^{-\zeta\omega_n t}}{\sqrt{1-\zeta^2}} \sin\left(\sqrt{1-\zeta^2}\omega_n t + \arccos\zeta\right) & 0<\zeta<1 \\ e^{-\omega_n t}(1+\omega_n t) & \zeta = 1 \\ \dfrac{1}{2\sqrt{\zeta^2-1}} \dfrac{e^{-\left(\zeta-\sqrt{\zeta^2-1}\right)\omega_n t}}{\zeta-\sqrt{\zeta^2-1}} - \dfrac{1}{2\sqrt{\zeta^2-1}} \dfrac{e^{-\left(\zeta+\sqrt{\zeta^2-1}\right)\omega_n t}}{\zeta+\sqrt{\zeta^2-1}} & \zeta > 1. \end{cases}$$

In this case, the norms $\|e(t)\|_\infty$ and $\|e(t)\|_2$ can be computed in a rather straightforward way:

$$\|e(t)\|_\infty = 1 \tag{4.4}$$

$$\|e(t)\|_2 = \sqrt{\left(\frac{1}{4\zeta}+\zeta\right)\frac{1}{\omega_n}}. \tag{4.5}$$

Cumbersome computation is needed for $\|e(t)\|_1$. The result is

$$\|e(t)\|_1 = \begin{cases} \dfrac{2\zeta}{\omega_n} + \dfrac{2e^{\zeta\arccos\zeta/\sqrt{1-\zeta^2}}}{\omega_n\left(e^{\zeta\pi/\sqrt{1-\zeta^2}}-1\right)} & \text{if } 0<\zeta<1 \\ \dfrac{2\zeta}{\omega_n} & \text{if } \zeta \ge 1. \end{cases} \tag{4.6}$$

The plots of all these norms with the normalized natural frequency ($\omega_n = 1$) versus the damping ratio are given in Figure 4.38. It is again seen that $\|e(t)\|_\infty$ is completely flat, and is therefore not an appropriate performance index in this case. For each of the other two norms, there is a best damping ratio. The best damping ratio for $\|e(t)\|_2$ is 0.5, and the minimum value of $\|e(t)\|_2$ is 1. The best damping ratio for $\|e(t)\|_1$ is 0.66 and the minimum value of $\|e(t)\|_1$ is 1.6.

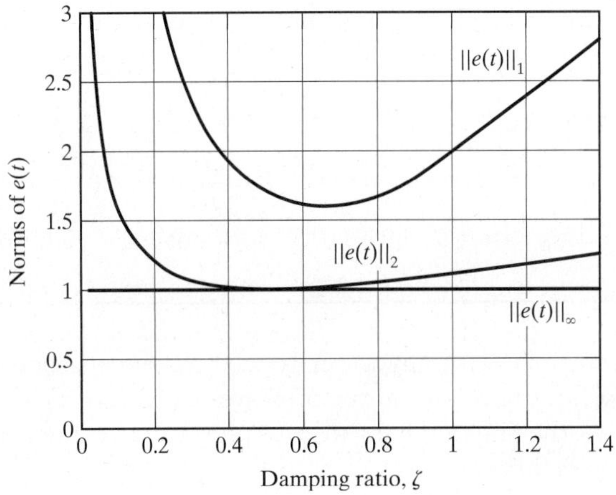

FIGURE 4.38: Various step response error norms for a standard second-order system.

Optimal control problems are often formulated as finding a controller or its parameters so that certain performance indices given by signal norms are minimized. An example is given as follows.

EXAMPLE 4.26

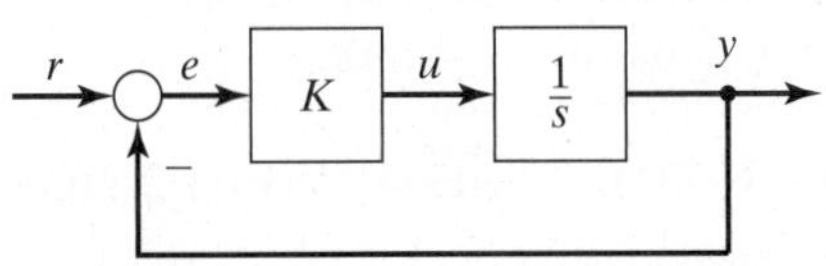

FIGURE 4.39: A first-order unity feedback system.

Consider the unity feedback system shown in Figure 4.39. Assume that $r(t)$ is a unit step. Design K to minimize

$$J = \|e(t)\|_2^2 + \|u(t)\|_2^2.$$

By using the result in Example 4.24,

$$\|e(t)\|_2 = \sqrt{\frac{1}{2K}}, \quad \|u(t)\|_2 = K\sqrt{\frac{1}{2K}}.$$

Then

$$J = \frac{1}{2K} + \frac{K}{2}.$$

To find the optimal K, we let

$$0 = \frac{dJ}{dK} = -\frac{1}{2K^2} + \frac{1}{2}.$$

This gives us $K = 1$. Therefore the optimal K is 1 and the optimal value of J is 1.

For this example, the optimal K depends on the performance index chosen. If we choose

$$J = \|e(t)\|_1 + \|u(t)\|_1,$$

then

$$J = \frac{1}{K} + 1.$$

Clearly, in this case the optimal K does not exist. The bigger the K, the smaller the J.

In principle, with any norm defined for signals there is a corresponding norm defined for systems: the norm of a system is the norm of its impulse response. However, in the time domain we will only use the 2-norm and 1-norm of systems. The formal definition is as follows.

Definition 4.27. Let a system have impulse response $g(t)$. Then

1. The time-domain 2-norm of the system is $\|g(t)\|_2$;
2. The time-domain 1-norm of the system is $\|g(t)\|_1$.

In general, the analytical computation of the norm of a signal or a system is difficult. Hence numerical simulation, integration, and computation are needed. Among all norms defined in this section, the 2-norm is probably the most commonly used, partly because it can be easily computed and handled. In the following section, we take a closer look at the 2-norm and give an algorithm for the computation of the 2-norm of a signal or a system.

4.9 COMPUTATION OF THE TIME-DOMAIN 2-NORM

In addition to the time-domain signal and system norms, another important concept is the time-domain inner product between signals or systems. The use of the inner product is crucial in the computation of the 2-norm.

Definition 4.28. Let $x(t)$ and $y(t)$ be signals. The inner product between $x(t)$ and $y(t)$ is defined as

$$\langle x(t), y(t)\rangle = \int_{0^-}^{\infty} x(t)y(t)dt.$$

Obviously,

$$\|x(t)\|_2^2 = \langle x(t), x(t)\rangle.$$

Let the Laplace transform of $x(t)$ be a rational function:

$$\mathcal{L}[x(t)] = X(s) = \frac{b(s)}{a(s)} = \frac{b_1 s^{n-1} + \cdots + b_n}{a_0 s^n + a_1 s^{n-1} + \cdots + a_n}, \quad a_0 > 0. \tag{4.7}$$

If a signal has a non-strictly proper Laplace transform or an unstable Laplace transform, its 2-norm is infinity. Hence we implicitly assume in (4.7) that $X(s)$ is strictly proper and we make an additional assumption that $a(s)$ is stable.

s^n	$r_{00} = a_0$	$r_{01} = a_2$	$r_{02} = a_4$	$r_{03} = a_6$	$\cdots$
s^{n-1}	$r_{10} = a_1$	$r_{11} = a_3$	$r_{12} = a_5$	$r_{13} = a_7$	$\cdots$
s^{n-2}	r_{20}	r_{21}	r_{22}	r_{23}	$\cdots$
s^{n-3}	r_{30}	r_{31}	r_{32}	r_{33}	$\cdots$
$\vdots$	$\vdots$	$\vdots$	$\vdots$	$\vdots$	
s^2	$r_{(n-2)0}$	$r_{(n-2)1}$			
s^1	$r_{(n-1)0}$				
s^0	r_{n0}				

TABLE 4.3: Routh table.

Construct the Routh table of stable polynomial $a(s)$ as in Table 4.3. Since $a(s)$ is stable, the Routh table can always be constructed until the end and all $r_{i0}, i = 0, 1, \ldots, n$, are positive. For each row (except the first one) of the Routh table, define a polynomial

$$\begin{aligned} r_1(s) &= r_{10}s^{n-1} + r_{11}s^{n-3} + \cdots \\ r_2(s) &= r_{20}s^{n-2} + r_{21}s^{n-4} + \cdots \\ &\vdots \\ r_{n-1}(s) &= r_{(n-1)0}s \\ r_n(s) &= r_{n0}. \end{aligned}$$

Also define

$$\alpha_i = \frac{r_{(i-1)0}}{r_{i0}}, \quad i = 1, 2, \ldots, n,$$

and functions

$$X_i(s) = \sqrt{2\alpha_i}\frac{r_i(s)}{a(s)}, \quad i = 1, 2, \ldots, n.$$

Finally, let $x_i(t)$ be the inverse Laplace transform of $X_i(s)$.

Theorem 4.29.

$$\langle x_i(t), x_j(t)\rangle = \begin{cases} 0 & i \neq j \\ 1 & i = j. \end{cases}$$

Another way to state Theorem 4.29 is that the functions $x_i(t), i = 1, 2, \ldots, n$, are orthonormal functions. We will postpone the proof of this theorem to Chapter 8.

EXAMPLE 4.30

Let us start with $a(s) = s^2 + 2\zeta\omega_n s + \omega_n^2$. The Routh table is as shown in Table 4.4.

s^2	1	ω_n^2
s^1	$2\zeta\omega_n$	
s^0	ω_n^2	

TABLE 4.4: The Routh table for Example 4.30.

This gives

$$r_1(s) = 2\zeta\omega_n s, \qquad \alpha_1 = \frac{1}{2\zeta\omega_n}$$

$$r_2(s) = \omega_n^2, \qquad \alpha_2 = \frac{2\zeta}{\omega_n}.$$

Hence

$$X_1(s) = \frac{2\sqrt{\zeta\omega_n}s}{s^2 + 2\zeta\omega_n s + \omega_n^2}, \quad X_2(s) = \frac{2\omega_n\sqrt{\zeta\omega_n}}{s^2 + 2\zeta\omega_n s + \omega_n^2}.$$

This gives

$$x_1(t) = \begin{cases} \dfrac{2\sqrt{\zeta\omega_n}}{\sqrt{1-\zeta^2}} e^{-\zeta\omega_n t} \cos\left(\sqrt{1-\zeta^2}\omega_n t + \arccos\zeta\right) & 0 < \zeta < 1 \\ 2\sqrt{\zeta\omega_n} e^{-\omega_n t}(1 - \omega_n t) & \zeta = 1 \\ \dfrac{\sqrt{\zeta\omega_n}}{\sqrt{\zeta^2 - 1}} \left[-\left(\zeta + \sqrt{\zeta^2-1}\right) e^{-\left(\zeta+\sqrt{\zeta^2-1}\right)\omega_n t} \right. & \\ \qquad \left. + \left(\zeta - \sqrt{\zeta^2-1}\right) e^{-\left(\zeta-\sqrt{\zeta^2-1}\right)\omega_n t} \right] & \zeta > 1 \end{cases}$$

and

$$x_2(t) = \begin{cases} \dfrac{2\sqrt{\zeta\omega_n}}{\sqrt{1-\zeta^2}} e^{-\zeta\omega_n t} \sin\sqrt{1-\zeta^2}\omega_n t & 0 < \zeta < 1 \\ 2\omega_n\sqrt{\zeta\omega_n} t e^{-\omega_n t} & \zeta = 1 \\ \dfrac{\sqrt{\zeta\omega_n}}{\sqrt{\zeta^2 - 1}} \left[e^{-\left(\zeta+\sqrt{\zeta^2-1}\right)\omega_n t} - e^{-\left(\zeta-\sqrt{\zeta^2-1}\right)\omega_n t} \right] & \zeta > 1. \end{cases}$$

Finally, it is a matter of (tedious) calculus to verify that

$$\langle x_1(t), x_2(t)\rangle = 0, \quad \|x_1(t)\|_2 = 1, \quad \|x_2(t)\|_2 = 1. \tag{4.8}$$

We leave this as an exercise.

If we expand $b(s)$ as

$$b(s) = \beta_1 r_1(s) + \beta_2 r_2(s) + \cdots + \beta_n r_n(s), \tag{4.9}$$

then

$$X(s) = \frac{\beta_1}{\sqrt{2\alpha_1}} X_1(s) + \frac{\beta_2}{\sqrt{2\alpha_2}} X_2(s) + \cdots + \frac{\beta_n}{\sqrt{2\alpha_n}} X_n(s).$$

Taking the inverse Laplace transforms, we get

$$x(t) = \frac{\beta_1}{\sqrt{2\alpha_1}}x_1(t) + \frac{\beta_2}{\sqrt{2\alpha_2}}x_2(t) + \cdots + \frac{\beta_n}{\sqrt{2\alpha_n}}x_n(t).$$

Consequently,

$$\begin{aligned}
\|x(t)\|_2^2 &= \left\langle \sum_{i=1}^{n} \frac{\beta_i}{\sqrt{2\alpha_i}}x_i(t), \sum_{i=1}^{n} \frac{\beta_i}{\sqrt{2\alpha_i}}x_i(t) \right\rangle \\
&= \sum_{i=1}^{n}\sum_{j=1}^{n} \frac{\beta_i}{\sqrt{2\alpha_i}}\frac{\beta_j}{\sqrt{2\alpha_j}}\langle x_i(t), x_j(t)\rangle \\
&= \sum_{i=1}^{n} \frac{\beta_i^2}{2\alpha_i}\langle x_i(t), x_i(t)\rangle \\
&= \frac{\beta_1^2}{2\alpha_1} + \frac{\beta_2^2}{2\alpha_2} + \cdots + \frac{\beta_n^2}{2\alpha_n}.
\end{aligned}$$

Therefore, $\|x(t)\|_2$ can be computed if the coefficients α_i, β_i, $i = 1, 2, \ldots, n$, can be obtained.

It seems that finding all β_i requires solving a set of linear equations obtained by comparing the coefficients in (4.9). These equations actually have special structures that lead to a tabular solution. Construct the **augmented Routh table** as in Table 4.5

r_{00}	r_{01}	$\cdots$	$q_{00} = b_1$	$q_{01} = b_3$	$\cdots$			
r_{10}	r_{11}	$\cdots$	r_{10}	r_{11}	$\cdots$	$q_{10} = b_2$	$q_{11} = b_4$	$\cdots$
r_{20}	r_{21}	$\cdots$	q_{20}	q_{21}	$\cdots$	r_{20}	r_{21}	$\cdots$
r_{30}	r_{31}	$\cdots$	r_{30}	r_{31}	$\cdots$	q_{30}	q_{31}	$\cdots$
$\vdots$	$\vdots$		$\vdots$	$\vdots$		$\vdots$	$\vdots$	
$r_{(n-1)0}$								
r_{n0}								

TABLE 4.5: Augmented Routh table.

The augmented Routh table is formed by adding two blocks to the right of the usual Routh table. The first added block (the middle block of the augmented Routh table) is constructed in the following way: the first row is directly from the coefficients $b_1, b_3, b_5, \ldots$, of $b(s)$; the second, fourth, sixth, ..., rows are copied from the corresponding rows of the Routh table; the third, fifth, seventh, ..., rows are obtained from their preceding two rows in exactly the same way as the rows of the Routh table:

$$q_{ij} = -\frac{1}{r_{(i-1)0}}\begin{vmatrix} q_{(i-2)0} & q_{(i-2)(j+1)} \\ r_{(i-1)0} & r_{(i-1)(j+1)} \end{vmatrix}. \tag{4.10}$$

The second added block (the right block of the augmented Routh table) is constructed in the following way: the first row is irrelevant; the second row is directly from the coefficients $b_2, b_4, b_6, \ldots$, of $b(s)$; the third, fifth, seventh, ..., rows are copied from the corresponding rows of the Routh table; the fourth, sixth, eighth, ..., rows are obtained from their preceding two rows in exactly the same way as the rows of the Routh table using formula (4.10).

In summary, the following algorithm gives the time-domain 2-norm of a signal $x(t)$ with a stable strictly proper Laplace transform $X(s)$ shown in (4.7).

Algorithm 4.31.

Step 1: Compute the augmented Routh table of $X(s) = \mathcal{L}[x(t)]$.

Step 2: Set $\alpha_i = \dfrac{r_{(i-1)0}}{r_{i0}}$ and $\beta_i = \dfrac{q_{(i-1)0}}{r_{i0}}$.

Step 3: $\|x(t)\|_2 = \sqrt{\dfrac{\beta_1^2}{2\alpha_1} + \dfrac{\beta_2^2}{2\alpha_2} + \cdots + \dfrac{\beta_n^2}{2\alpha_n}}$.

EXAMPLE 4.32

Consider a signal $x(t)$ with

$$\mathcal{L}[x(t)] = X(s) = \frac{b(s)}{a(s)} = \frac{s^3 + 2s^2 + 5s + 6}{s^4 + s^3 + 3s^2 + 2s + 1}.$$

The augmented Routh table of $X(s)$ is as in Table 4.6.

s^4	$a_0 = 1$	$a_2 = 3$	$a_4 = 1$	$b_1 = 1$	$b_3 = 5$		
s^3	$a_1 = 1$	$a_3 = 2$		1	2	$b_2 = 2$	$b_4 = 6$
s^2	1	1		3		1	1
s^1	1			1		4	
s^0	1					1	

TABLE 4.6: The augmented Routh table for Example 4.32.

Therefore,

$$\alpha_1 = \alpha_2 = \alpha_3 = \alpha_4 = 1$$

and

$$\beta_1 = 1, \quad \beta_2 = 2, \quad \beta_3 = 3, \quad \beta_4 = 4.$$

Hence

$$\|x(t)\|_2^2 = \frac{1^2}{2} + \frac{2^2}{2} + \frac{3^2}{2} + \frac{4^2}{2} = 15.$$

For signals whose Laplace transforms have low order, the following more explicit formulas are available and can be derived by applying Algorithm 4.31.

$$\left\| \mathcal{L}^{-1}\left[\frac{b_1}{a_0 s + a_1}\right] \right\|_2^2 = \frac{b_1^2}{2a_0 a_1} \tag{4.11}$$

$$\left\|\mathcal{L}^{-1}\left[\frac{b_1 s + b_2}{a_0 s^2 + a_1 s + a_2}\right]\right\|_2^2 = \frac{a_2 b_1^2 + a_0 b_2^2}{2a_0 a_1 a_2} \tag{4.12}$$

$$\left\|\mathcal{L}^{-1}\left[\frac{b_1 s^2 + b_2 s + b_3}{a_0 s^3 + a_1 s^2 + a_2 s + a_3}\right]\right\|_2^2 = \frac{a_2 a_3 b_1^2 + a_0 a_3 b_2^2 + a_0 a_1 b_3^2 - 2a_0 a_3 b_1 b_3}{2a_0 a_3 (a_1 a_2 - a_0 a_3)}. \tag{4.13}$$

EXAMPLE 4.33

Let us revisit Example 4.25, where $\|e(t)\|_2$ is computed using the definition. Let us now compute it using the algorithm in this section. Notice

$$\|e(t)\|_2 = \|\mathcal{L}^{-1}[E(s)]\|_2 = \left\|\mathcal{L}^{-1}\left[\frac{s + 2\zeta\omega_n}{s^2 + 2\zeta\omega_n s + \omega_n^2}\right]\right\|_2.$$

By (4.12),

$$\|e(t)\|_2 = \sqrt{\frac{1}{\omega_n}\left(\frac{1}{4\zeta} + \zeta\right)}.$$

The answer is the same but the computation is simpler.

EXAMPLE 4.34

Using (4.12), we immediately obtain

$$\left\|\mathcal{L}^{-1}\left[\frac{\omega_n^2}{s^2 + 2\zeta\omega_n s + \omega_n^2}\right]\right\|_2 = \frac{1}{2}\sqrt{\frac{\omega_n}{\zeta}}.$$

PROBLEMS

4.1. Let $G(s)$ be a stable transfer function with DC gain $G(0)$. What is the steady-state value of its response to input $u(t) = (5 + e^{-3t})\sigma(t)$?

4.2. Let $G(s)$ be a stable transfer function with DC gain $G(0)$. Let $g(t)$ be the unit impulse response of $G(s)$. Find

$$\int_{0^-}^{\infty} g(t)dt.$$

4.3. Obtain the closed-loop transfer function of the system as shown in Figure 4.40. Select the gains K_1 and K_2 so that the closed-loop system is critically damped with two equal poles at $s = -10$.

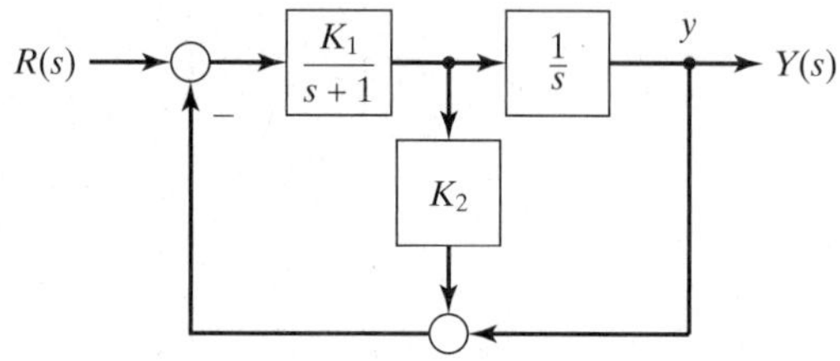

FIGURE 4.40: Output plus input feedback system for Problem 4.3.

4.4. Consider a unity feedback system as shown in Figure 4.41 with an open-loop transfer function

$$L(s) = \frac{2000}{s(s+5)(s+10)(s+15)}.$$

Estimate the rise time, settling time, and PO by using the dominant poles method and then compare the results with the results obtained by the simulation of the full-order system.

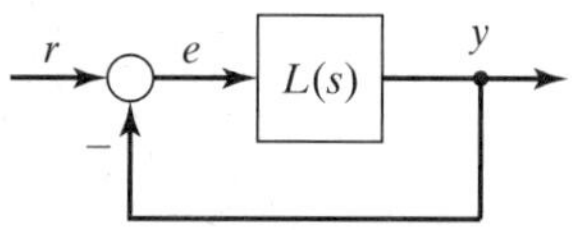

FIGURE 4.41: A simple unity feedback system for Problems 4.4, 4.5, 4.6, 4.9, and 4.15.

4.5. Repeat the last problem with

$$L(s) = \frac{4000(Ts+1)}{s(s+5)(s+10)(s+15)}$$

for $T = 0.25$ and $T = 0.025$.

4.6. Consider a unity feedback system as shown in Figure 4.41 with an open-loop transfer function

$$L(s) = \frac{K(s+10)}{s(s+2)(s+3)}.$$

Determine a K so that the system is stable and the steady-state error with respect to a unit ramp is no more than 0.5. With the chosen K, find the steady-state response of the system with respect to a sinusoidal input $r(t) = (\sin 2t)\sigma(t)$.

4.7. Consider a simple feedback system as shown in Figure 4.42 with

$$P(s) = \frac{2}{(s+1)(s+2)}$$

and $d(t) = \sin 20t$. Design a PI controller $C(s) = K_P + K_I/s$ so that the magnitude of the steady-state error with respect to this disturbance is no more than 0.1.

4.8. Consider the feedback system as shown in Figure 4.43.

1. Let $r(t) = t\sigma(t)$ (unit ramp function) and $d(t) = 0$. Find the range of K so that the magnitude of the steady-state error $e_{ss}(t)$, where $e(t) = r(t) - y(t)$, is less than 1, i.e., $|e_{ss}(t)| < 1$.
2. Let $r(t) = 0$ and $d(t) = (\sin t)\sigma(t)$. Find the range of K so that the magnitude of the steady-state error $e_{ss}(t)$ is less than 1, i.e., $|e_{ss}(t)| < 1$.

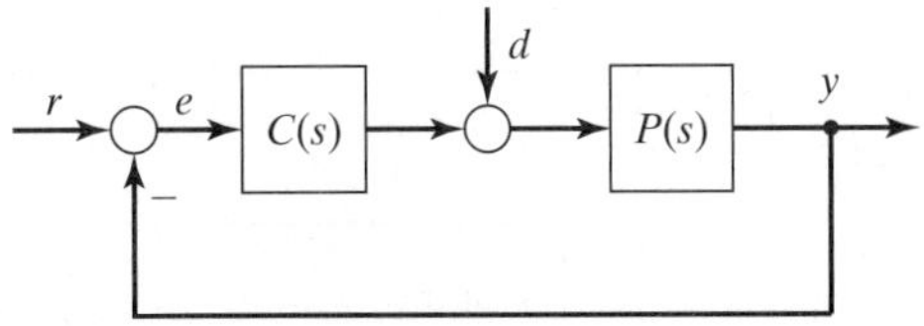

FIGURE 4.42: A unity feedback system.

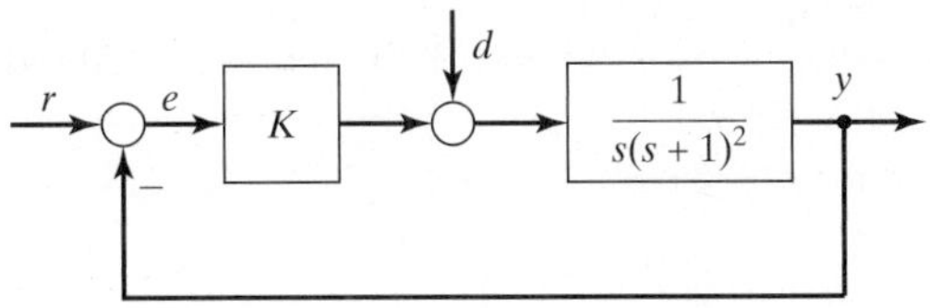

FIGURE 4.43: A unity feedback system.

4.9. Consider a unity feedback control system shown in Figure 4.41 with closed-loop transfer function

$$\frac{Y(s)}{R(s)} = \frac{Ks+b}{s^2+as+b}.$$

1. What is the loop transfer function $L(s)$?
2. What is the type of the loop transfer function $L(s)$?
3. What are the position, velocity, and acceleration error constants?

4.10. Assume that the closed-loop transfer function of an internally stable unity feedback control system, as shown in Figure 4.23, is

$$\frac{Y(s)}{R(s)} = \frac{L(s)}{1+L(s)} = \frac{b_1 s^{n-1} + b_2 s^{n-2} + \cdots + b_n}{s^n + a_1 s^{n-1} + a_2 s^{n-2} + \cdots + a_n}.$$

Prove that the open-loop system $L(s)$ is of type k $(k \leq n)$ if and only if $b_n = a_n$, $b_{n-1} = a_{n-1}, \ldots, b_{n-k+1} = a_{n-k+1}$, but $b_{n-k} \neq a_{n-k}$.

4.11. Answer the following questions and explain your answers briefly:

1. Does the step response of the closed-loop system $\frac{1}{s^2+s+1}$ have overshoot?
2. Does the step response of the closed-loop system $\frac{s+1}{s^2+s+1}$ have overshoot?
3. What is the type of open-loop plant $\frac{s+1}{s^2}$? What are its position, velocity, and acceleration error constants?
4. What is the steady-state value of the step response of system

$$\frac{s^2+3s+8}{(s+1)(s+2)(s+3)(s+4)}?$$

4.12. Consider the unity feedback system shown in Figure 4.44. Here $P(s) = \frac{1}{s(s+2)}$ and K is a pure gain.

1. Design K such that the velocity error constant is 5.
2. Design K such that the system is critically damped.
3. Design K such that the overshoot is $e^{-\pi}$.

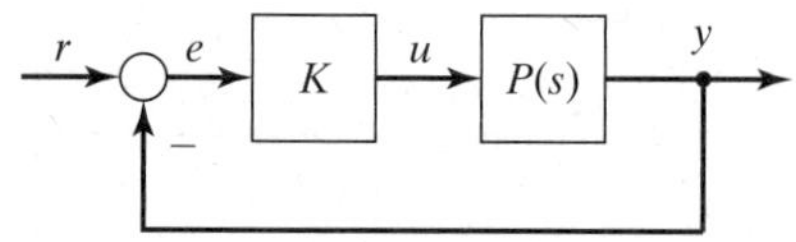

FIGURE 4.44: A unity feedback system.

4. Design K such that the real and imaginary parts of the closed-loop poles have equal magnitude.

4.13. Simulate the step responses of the following (closed-loop) systems:

1. $\dfrac{s^2 - 2s + 1.1}{s^3 + 3s^2 + 3s + 1}$;

2. $\dfrac{s^2 - 2s + 100}{s^3 + 3s^2 + 3s + 1}$.

Observe that a system with a pair of complex zeros with positive real part may or may not exhibit undershoot.

4.14. Let $y(t)$ be the response of a stable system with transfer function $G(s)$ to step input $\alpha\sigma(t)$. Prove that $\lim_{t\to\infty} y(t) = G(0)\alpha$. (This is why $G(0)$ is called the DC gain of the system. Hint: if you use the final value theorem, you need to be careful with its conditions.)

4.15. Consider the unity feedback system shown in Figure 4.23. Here $L(s)$ is at least type 1 and the transfer function from r to y has poles $p_1, p_2, \ldots, p_n$ and zeros $z_1, z_2, \cdots, z_m$. Show that

$$\frac{1}{K_v} = \sum_{i=1}^{m} \frac{1}{z_i} - \sum_{i=1}^{n} \frac{1}{p_i}.$$

4.16. Consider transfer function

$$G(s) = \frac{s^2 - 3s + 2}{s^3 + 3s^2 + 3s + 1}.$$

Does its step response exhibit type A undershoot? How about other types of undershoot? Simulate its step response. Observe what type of undershoot occurs.

4.17. 1. Assume that the relative degree of $G(s)$ is at least 2 and that its step response has overshoot. Show that the step response of $F(s) = (1 - \alpha s)G(s)$, where $\alpha > 0$, has both overshoot and undershoot. Furthermore, show that the PO of the step response of $F(s)$ is greater than or equal to that of $G(s)$.

2. Use the result of 1 to show that for all $\alpha > 0$, $\omega_n > 0$, and $\zeta \in (0, 1)$, the step response of the system

$$F(s) = \frac{\omega_n^2(1 - \alpha s)}{s^2 + 2\zeta\omega_n s + \omega_n^2}$$

has both overshoot and undershoot.

3. Simulate and plot the step response of

$$F(s) = \frac{1 - s}{s^2 + s + 1}$$

and compare it with that of

$$G(s) = \frac{1}{s^2 + s + 1}.$$

4.18. **1.** A stable system is said to have monotonic step response if its step response is either a monotonically increasing or monotonically decreasing function of time. Show that if a system has monotonic step response, then its step response has neither overshoot nor undershoot.

2. Show that a system has monotonic step response if and only if it has a sign-invariant impulse response, i.e., its impulse response is either nonnegative for all time or nonpositive for all time.

3. Show that if systems $G_i(s)$ have monotonic step responses, then the cascaded system $\prod_{i=1}^{m} G_i(s)$ has monotonic step response.

4. Show that the system

$$G(s) = \prod_{i=1}^{m} \frac{\alpha_i}{s + \alpha_i}$$

where $\alpha_i > 0$, has a monotonic step response.

4.19. Problem 4.18 says that a stable system with all real poles and without zeros has monotonic step response. This problem concerns a stable system with all real poles and negative real zeros.

1. Simulate the step responses of

$$G_1(s) = \frac{8(s+1)}{(s+2)(s+4)}, G_2(s) = \frac{8(s+3)}{3(s+2)(s+4)}, G_2(s) = \frac{8(s+5)}{5(s+2)(s+4)}$$

and observe the occurrence of overshoot.

2. What can you say about the occurrence of overshoot in the step response of

$$G(s) = \frac{a_1 a_2 (s+b)}{b(s+a_1)(s+a_2)}$$

where $a_1, a_2, b > 0$?

3. What can you say about the occurrence of overshoot in the step response of

$$G(s) = \frac{a_1 \cdots a_n (s+b_1) \cdots (s+b_m)}{b_1 \cdots b_n (s+a_1) \cdots (s+a_n)}$$

where $m < n$, $a_1, \ldots, a_n, b_1, \ldots, b_m > 0$?

4.20. Let $G_1(s)$ and $G_2(s)$ be the transfer functions of two stable systems. For the first system, assume $G_1(0) > 0$ and its impulse $g_1(t)$ response is nonnegative, i.e., $g_1(t) \geq 0$ for all $t \geq 0$. For the second system, assume $G_2(0) > 0$ and its step response $y_2(t)$ does not have overshoot, i.e., $y_2(t) \leq G_2(0)$ for all $t \geq 0$.

1. Show that the step response of the cascaded system $G_2(s)G_1(s)$ does not have overshoot.

2. Determine whether the step response of

$$G(s) = \frac{\alpha_0(s-z)}{-z(s+\alpha_0)} \prod_{i=1}^{n} \frac{\alpha_i}{s+\alpha_i}$$

where $\alpha_i > 0, i = 0, 1, 2, \ldots, n$, has overshoot.

3. Determine whether the step response of

$$G(s) = \frac{s^2 - 2\zeta\omega_n s + \omega_n^2}{(s+\omega_n)^2} \prod_{i=1}^{n} \frac{\alpha_i}{s+\alpha_i},$$

where $\zeta \geq 0$, $\omega_n > 0$, and $\alpha_i > 0, i = 1, 2, \ldots, n$, has overshoot.

4.21. Consider a plant $P(s) = \frac{1}{s^2}$.

1. Is it possible to design a unity feedback controller so that the closed-loop step response is free of overshoot?
2. Design a first-order 2DOF controller as shown in Figure 4.45 so that the closed-loop characteristic polynomial is $(s+3)^3$ and the transfer function from r to y is $\frac{9}{(s+3)^2}$.
3. With this design, does the closed-loop step response have overshoot?

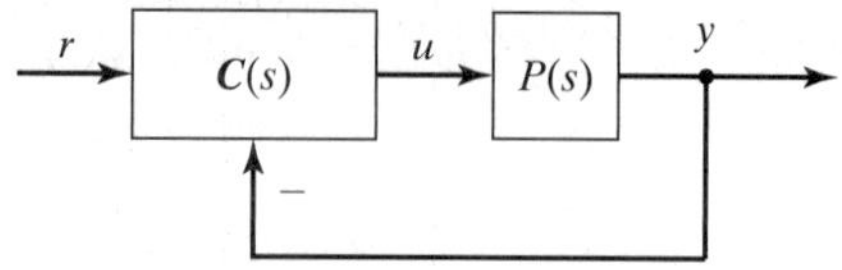

FIGURE 4.45: A 2DOF control system.

4.22. Consider the plant $P(s) = \frac{s-1}{s(s-2)}$.

1. Is it possible to design a unity feedback controller so that the closed-loop step response is free of overshoot?
2. Is it possible to design a 2DOF controller as in Figure 4.45 so that the closed-loop step response is free of undershoot?
3. Design a first-order 2DOF controller as in Figure 4.45 so that the closed-loop characteristic polynomial is $(s+1)(s+2)(s+3)$, the output $y(t)$ tracks a step in the reference $r(t)$, and the transfer function from $r(t)$ to $y(t)$ has poles at -1 and -2.
4. With this design, does the closed-loop step response have overshoot?

4.23. Consider a plant $P(s) = \frac{1}{1-s^2}$. Design a unity feedback controller $C(s)$, as shown in Figure 4.46, such that position error constant of the loop transfer function $L(s) = P(s)C(s)$ is infinity and the closed-loop poles are $-1, -2, -3, -4$. What is the velocity error constant of the loop transfer function?

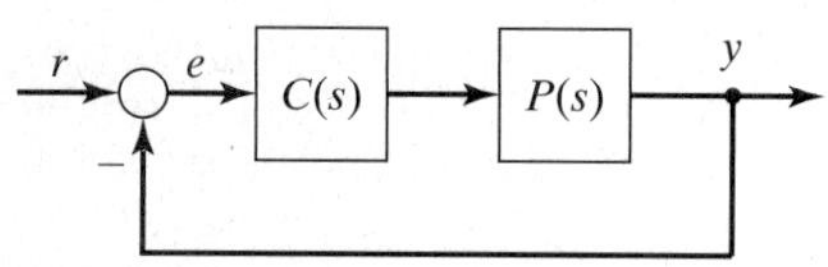

FIGURE 4.46: A unity feedback system.

4.24. For the feedback control system shown in Figure 4.47, assume that $r(t)$ is a unit step. Find the optimal gain K so that the performance criterion

$$J = \int_0^{\infty} [e^2(t) + 4u^2(t)]dt$$

is minimized.

4.25. Verify the equalities (4.5) and (4.6).

4.26. Verify the equalities in (4.8).

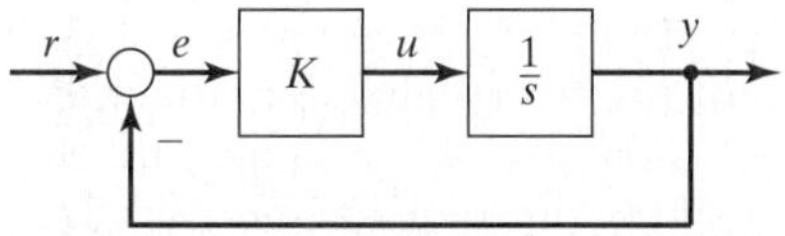

FIGURE 4.47: For Problem 4.24.

4.27. Compute the time-domain 2-norms of the following systems:

1. $$G(s) = \frac{s^3 + 2s^2 + 3s + 4}{s^4 + s^3 + 10s^2 + 6s + 8};$$

2. $$G(s) = \frac{s^4 + s^3 + 13s^2 + 9s + 21}{s^5 + 2s^4 + 16s^3 + 24s^2 + 48s + 32};$$

3. $$G(s) = \frac{2s^3 + 4s^2 + 24s + 32}{s^4 + 2s^3 + 12s^2 + 16s + 16};$$

4. $$G(s) = \frac{s^3 - s^2 + 18s - 9}{s^4 + 3s^3 + 27s^2 + 54s + 81}.$$

4.28. Verify equalities (4.11)–(4.13).

4.29. Derive an explicit formula for the time-domain 2-norm of the system with the transfer function

$$G(s) = \frac{b_1 s^3 + b_2 s^2 + b_3 s + b_4}{a_0 s^4 + a_1 s^3 + a_2 s^2 + a_3 s + a_4}.$$

4.30. Develop an algorithm to compute the time-domain inner product of two systems $F(s)$ and $G(s)$.

MATLAB PROBLEMS

4.31. Write a MATLAB program to compute the time-domain 2-norm of a system with a given transfer function. Let us name it "norm2" and it can call "routh" developed in Problem 3.24 as a subroutine. Test the program on the systems given in Problem 4.27.

4.32. Use MATLAB to solve this problem. For the system in Problem 2.14, controller

$$C(s) = \frac{0.8745s^3 + 4.0787s^2 + 2.4574s + 0.6105}{s^3 + 3.7897s^2 + 5.9143s}$$

is used to control the system using unity feedback.

1. Find the closed-loop transfer function.

2. Is the closed-loop system stable?

3. Plot the responses of the closed-loop system to unit impulse, unit step, and sinusoidal signal $\cos t$ for the time period 0 to 30 sec. (Use impulse, step, or lsim.)

4. For the step response, record the rising time, peak time, PO, and settling time.

5. Use the MATLAB program developed in Problem 4.31 to compute the 2-norm of the error signal due to unit step reference.

4.33. Write a MATLAB program to implement the algorithm in Problem 4.30 for the computation of the time-domain inner product of two systems with given transfer functions. Let us name it "iprod" and it can call "routh" developed in Problem 3.24 or/and "norm2" developed in Problem 4.31 as subroutines.

EXTRA CREDIT PROBLEMS

4.34. Let

$$G(s) = \frac{b_1 s^{n-1} + b_2 s^{n-2} + \cdots + b_n}{a_0 s^n + a_1 s^{n-1} + \cdots + a_n}$$

be a strictly proper stable transfer function. Let $\alpha_i, \beta_i, i = 1, 2, \ldots, n$, be the numbers obtained in Step 2 of Algorithm 4.31. Show that

$$\boldsymbol{A} = \begin{bmatrix} -\frac{1}{\alpha_1} & \frac{1}{\sqrt{\alpha_1\alpha_2}} & & & \\ -\frac{1}{\sqrt{\alpha_1\alpha_2}} & 0 & \ddots & & \\ & \ddots & \ddots & \ddots & \\ & & \ddots & 0 & \frac{1}{\sqrt{\alpha_{n-1}\alpha_n}} \\ & & & -\frac{1}{\sqrt{\alpha_{n-1}\alpha_n}} & 0 \end{bmatrix}, \quad \boldsymbol{b} = \begin{bmatrix} \beta_1 \\ \beta_2 \\ \vdots \\ \beta_n \end{bmatrix}$$

$$\boldsymbol{c} = \begin{bmatrix} \frac{2}{\alpha_1} & 0 & \cdots & 0 \end{bmatrix}, \qquad d = 0.$$

gives a realization of $G(s)$. (This realization is called the **Routh realization**) Draw the schematic op-amp circuit of the realization.

4.35. Show the following inequalities

$$\|g(t) * u(t)\|_\infty \le \|g(t)\|_1 \|u(t)\|_\infty$$
$$\|g(t) * u(t)\|_1 \le \|g(t)\|_1 \|u(t)\|_1$$
$$\|g(t) * u(t)\|_2 \le \|g(t)\|_1 \|u(t)\|_2.$$

The proofs of these three inequalities are progressively more difficult.

NOTES AND REFERENCES

The term internal model principle was coined in

B. F. Francis and W. M. Wonham, "The internal model principle of control theory," *Automatica*, vol. 12, pp. 457–465, 1976.

Related contributions can be found, for example, in

> E. J. Davison, "The robust control of a servomechanism problem for linear time-invariant multivariable systems," *IEEE Transactions on Automatic Control*, vol. AC-21, pp. 25–34, 1976.

Theorems 4.14 and 4.16 on undershoot and Theorem 4.18 on overshoot seem to have first appeared in

> M. M. Seron, J. H. Braslavsky, and G. C. Goodwin, *Fundamental Limitations in Filtering and Control*, Springer, London, 1997.

Theorem 4.17 on type A undershoot first appeared in

> T. Norimatsu and M. Ito, "On the zero non-regular control system," *Journal of Institute of Electrical Engineering of Japan*, vol. 81, pp. 566–575, 1961.

And it also appeared in

> T. Mita and H. Yoshida, "Undershooting phenomenon and its control in linear multivariable servomechanisms," *IEEE Transactions on Automatic Control*, vol. AC-26, pp. 402–407, 1981.
>
> M. Vidyasagar, "On undershoot and nonminimum phase zeros," *IEEE Transactions on Automatic Control*, vol. AC-31, p. 440, 1985.

Theorem 4.22 is due to

> S. Darbha and S. P. Bhattacharyya, "On the synthesis of controllers for a nonovershooting step response," *IEEE Transactions on Automatic Control*, vol. 48, pp. 797–799, 2003.

The proof of Theorem 4.22 given in Section 4.7 is a modified version of the one in the above paper.
The effort to find a simple method of computing the 2-norm of a signal or a system was initiated in the late 1940s by a group at the Massachusetts Institute of Technology (MIT). The initial effort ended up with formulas for a Laplace transform or transfer function up to 7th order, reported in

> H. M. James, N. B. Nichols, and R. S. Philips, *Theory of Servomechanisms*, McGraw-Hill, New York, 1947.

Another team effort was carried out in the 1950s by another group in MIT. This effort, documented in

> G. C. Newton, L. A. Gould, and J. F. Kaiser, *Analytical Design of Linear Feedback Control*, John Wiley & Sons, Inc., New York, 1957,

led to an algorithm based on matrix equations for system of arbitrarily high order and corrections to two formulas obtained in the first MIT effort. Algorithm 4.31 first appeared in

K. J. Åström, *Introduction to Stochastic Control Theory*, Academic Press, New York, 1970,

but was derived in a different way. Based on this algorithm, one can easily prove Theorem 4.29, as first observed in

C. P. Therapos, "Balanced minimal realization of SISO systems," *Electronics Letters*, vol. 19, pp. 424–426, 1983.

We adopt an opposite route here, deriving Algorithm 4.31 from Theorem 4.29 and proving Theorem 4.29 in a simple way after we build up more background, which will be done in Chapter 8.

CHAPTER

5

Root-Locus Method

It is now known from the analysis in the preceding chapters that the performance of a system is largely determined by the locations of the system's poles. If all system parameters are known, then determining the pole locations is rather easily done by using any standard computer-aided tools. However, it is generally more important to know how closed-loop poles vary when some parameters in the system are changed. There are at least two reasons for studying this kind of problem. First of all, many parameters in a system may vary with the working environment or cannot be determined exactly. It is then important to ensure that the system can still operate in the desired state. Thus, it is important to know how the closed-loop poles change with some system parameters. This, of course, can also be accomplished by extensive computer simulations, but in many cases a brief sketch of the pole variation trend is more insightful. Secondly, and perhaps more importantly, control engineers need to know how certain controller parameters affect system performance such that the desired controller parameters can be chosen to achieve the design objectives.

One classical technique in determining pole variations with parameters is known as the **root-locus method**, invented by W. R. Evens, which will be introduced in this chapter. It should be pointed out that in most cases *it is more important to know how to quickly sketch a root locus, which would indicate the trend of the root loci rather than the exact root-locus plot*, which can be done by using a computer if necessary. Thus, attention is not focused on the exact details of root-locus construction in this chapter. Instead, we shall concentrate on how we can use root locus as a tool for system analysis and controller design.

Walter R. Evans (1920–1999) earned his BS degree in Electrical Engineering from Washington University in 1941 and his MS Degree in Electrical Engineering from the University of California, Los Angeles, in 1951. He taught as an instructor in the Department of Electrical Engineering at Washington University from 1946 to 1948. In 1948, Mr Evans moved to Autonetics, a division of North American Aviation, now known as Rockwell International. It was during his lectures to his colleagues on analysis of servo-mechanisms in August 1948 that he finally came up with the root-locus techniques. That same year, he developed the Spirule, a tool used in conjunction with the application of the root-locus method, and The Spirule Company (formed by him) sold in the next few decades over 100,000 copies of the Spirule over 75 countries around the world. His root-locus method was published in the paper "Graphical analysis of control systems," *Transactions of the American Institute of Electrical Engineers,* vol. 67, pp. 547–551, 1948, and in the paper "Control system synthesis by root-locus method," *Transactions of the American Institute of Electrical Engineers,* vol. 69, pp. 66–69, 1950. Mr Evans worked with the technical staff of the Guidance and Control Department of the Re-Entry Systems Operation of the Ford Aeronautic Company from 1959 to 1971. He rejoined Autonetics where he worked with the technical staff of the Strategic Systems Division until his retirement in 1980. Mr Evans was awarded the prestigious Rufus Oldenburger Medal by the American Society of Mechanical Engineers in 1987 and the Richard E. Bellman Control Heritage Award of the American Automatic Control Council in 1988.

5.1 ROOT-LOCUS TECHNIQUES

Consider a standard feedback system for stabilization shown in Figure 5.1 or a unity feedback system shown in Figure 5.2.

The closed-loop poles are given by the roots of the following equation:

$$1 + P(s)C(s) = 0.$$

For simplicity of presentation, we shall denote

$$L(s) := P(s)C(s)$$

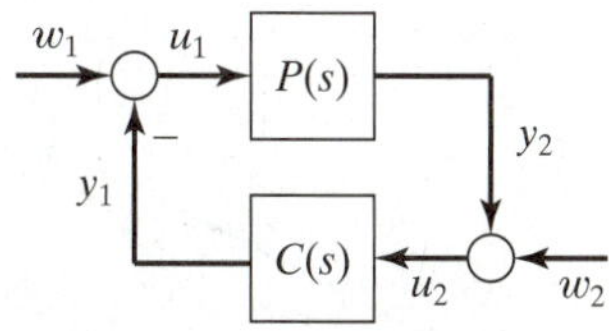

FIGURE 5.1: Feedback system for stabilization.

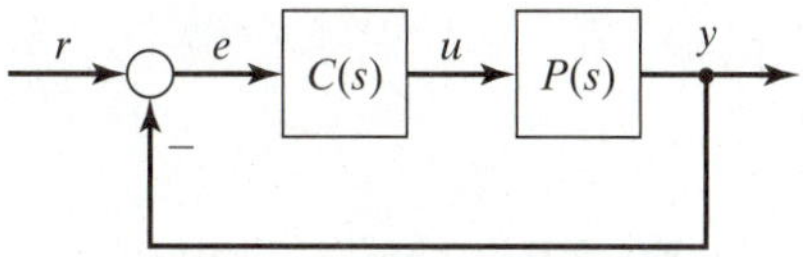

FIGURE 5.2: A unity feedback system.

and assume that $L(s)$ has the following form:

$$L(s) = \frac{K(s - z_1)(s - z_2)\cdots(s - z_m)}{(s - p_1)(s - p_2)\cdots(s - p_n)}$$

where $z_1, \ldots, z_m$ are the open-loop zeros; $p_1, \ldots, p_n$ are the open-loop poles; and K is a variable gain. Our objective is to study how the closed-loop poles change when K varies from 0 to ∞. We shall show later how problems involving other system parameters may be converted into problems like this.

Thus, our goal is to find all points that satisfy

$$L(s) = -1.$$

This equation can be equivalently written into two equations:

Magnitude condition:	$\lvert L(s)\rvert = 1$	(5.1)
Phase condition:	$\angle L(s) = (2k+1)180^\circ, \; k = 0, \pm 1, \ldots$	(5.2)

It is easy to see that the magnitude condition can always be satisfied by a suitable $K \geq 0$. On the other hand, the phase condition does not depend on the value of K (but depends on the sign of K):

$$\angle L(s) = \sum_{i=1}^{m} \angle(s - z_i) - \sum_{j=1}^{n} \angle(s - p_j) = (2k+1)180^\circ.$$

Thus, the key is to find all those points that satisfy the phase condition. Consider, for example, a system with open-loop transfer function

$$L(s) = \frac{K(s - z_1)(s - z_2)}{(s - p_1)(s - p_2)(s - p_3)}, \qquad p_2 = \overline{p}_1.$$

The open-loop poles and zeros of the system are shown in Figure 5.3 where a pole is represented by "×" and a zero is represented by "∘." The phase of $L(s)$ at a

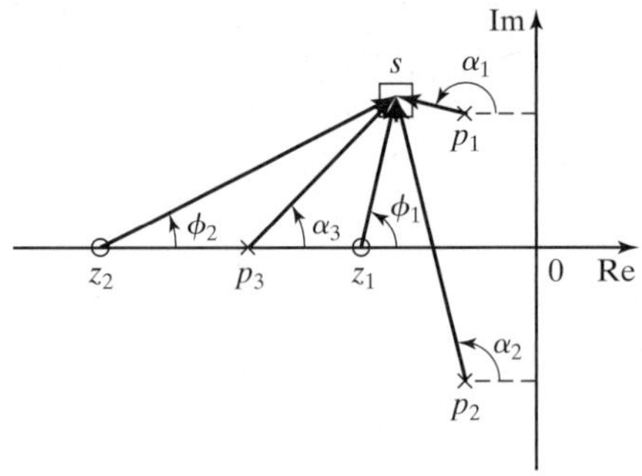

FIGURE 5.3: The phase of $L(s)$ at a point s.

point s in the complex plane is computed as

$$\begin{aligned}\angle L(s) &= \angle(s - z_1) + \angle(s - z_2) - \angle(s - p_1) - \angle(s - p_2) - \angle(s - p_3) \\ &= \phi_1 + \phi_2 - \alpha_1 - \alpha_2 - \alpha_3.\end{aligned}$$

Several basic rules can be derived from the phase condition, which will facilitate the sketching of the root locus. These rules are summarized in Table 5.1. The terminologies used in the table such as "asymptotes," "breakaway points," "angle of departure," and "angle of arrival" are illustrated in the following two examples.

EXAMPLE 5.1

In this example, asymptotes and breakaway points are illustrated. Consider an open-loop transfer function

$$L(s) = \frac{K}{s(s+4)(s+5)}.$$

The system has three poles and no zero, so the angles of the three asymptotes can be calculated as

$$\theta = \frac{(2k+1)180^\circ}{3} = 60^\circ, -60^\circ, 180^\circ$$

for $k = 0, -1$, and 1, and the intersection of the asymptotes with the real axis is given by

$$\kappa = \frac{0-4-5}{3} = -3.$$

Note that we could have set $k = 0, 1, 2$ to get $\theta = 60^\circ, 180^\circ, 300^\circ$, which are the same angles. The three asymptotes are shown in Figure 5.4.

The asymptotes clearly indicate that the system will become unstable when the gain is sufficiently large. It is quite easy to see by Rule 4 that this is always the case if the relative degree of $L(s)$ is at least 3, since in that case there will be at least one asymptote with an angle less than 90°.

1. The root locus is symmetric with respect to the real axis.

2. The root loci start from n poles p_i (when $K = 0$) and approach the n zeros (m finite zeros z_i and $n - m$ infinite zeros when $K \to \infty$).

3. The root locus includes all points on the real axis to the left of an odd number of open-loop real poles and zeros.

4. As $K \to \infty$, $n - m$ branches of the root-locus approach asymptotically $n - m$ straight lines (called **asymptotes**) with angles

$$\theta = \frac{(2k+1)180^\circ}{n-m}, \quad k = 0, \pm 1, \pm 2, \ldots$$

and the starting point of all asymptotes is on the real axis at

$$\kappa = \frac{\sum_{i=1}^{n} p_i - \sum_{j=1}^{m} z_j}{n-m} = \frac{\sum \text{poles} - \sum \text{zeros}}{n-m}.$$

5. The **breakaway points** (where the root loci meet and split away, usually on real axis) and the **breakin points** (where the root loci meet and enter the real axis) are among the roots of the equation: $\dfrac{dL(s)}{ds} = 0$. (On the real axis, only those roots that satisfy Rule 3 are breakaway or breakin points.)

6. The **departure angle** ϕ_k (from a pole, p_k) is given by

$$\phi_k = \sum_{i=1}^{m} \angle(p_k - z_i) - \sum_{j=1, j\neq k}^{n} \angle(p_k - p_j) \pm 180^\circ.$$

(In the case p_k is l repeated poles, the departure angle becomes ϕ_k/ℓ.)

The **arrival angle** ψ_k (at a zero, z_k) is given by

$$\psi_k = -\sum_{i=1, i\neq k}^{m} \angle(z_k - z_i) + \sum_{j=1}^{n} \angle(z_k - p_j) \pm 180^\circ.$$

(In the case z_k is l repeated zeros, the arrival angle becomes ψ_k/ℓ.)

TABLE 5.1: Root locus rules: $0 \leq K \leq \infty$.

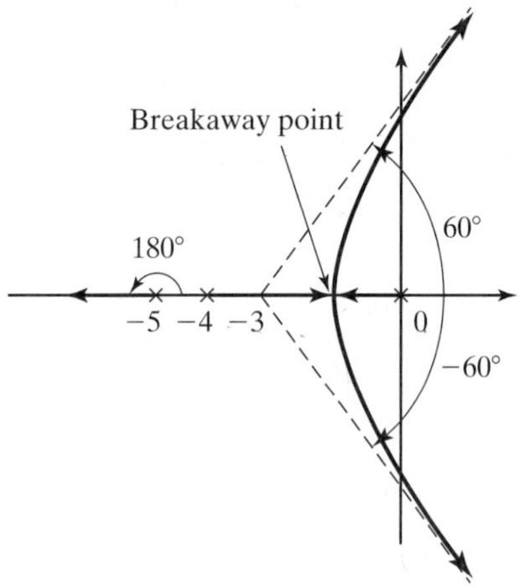

FIGURE 5.4: Breakaway point and asymptotes.

EXAMPLE 5.2

In this example, departure and arrival angles are illustrated. Consider an open-loop transfer function

$$L(s) = \frac{K(s^2+4s+8)}{(s+3)(s^2+2s+2)} = \frac{K(s-z_1)(s-z_2)}{(s-p_1)(s-p_2)(s-p_3)}$$

with zeros at $z_1 = -2+j2$ and $z_2 = -2-j2$ and poles at $p_1 = -3$, $p_2 = -1+j$, and $p_3 = -1-j$.

The departure angle ϕ at $p_2 = -1+j$ satisfies the following equation

$$\angle(p_2-z_1) + \angle(p_2-z_2) - \angle(p_2-p_1) - \phi - \angle(p_2-p_3) = -180°,$$

i.e.,

$$-45° + \tan^{-1} 3 - 90° - \phi - \tan^{-1}\frac{1}{2} = -180°$$

which gives $\phi = 90°$, while the arrival angle at the zero $z_1 = -2+j2$ can be calculated from

$$\psi + \angle(z_1-z_2) - \angle(z_1-p_1) - \angle(z_1-p_2) - \angle(z_1-p_3) = -180°,$$

i.e.,

$$\psi + 90° - \tan^{-1} 2 - 135° - (180° - \tan^{-1} 3) = -180°$$

which gives $\psi = 36.87°$.

The departure angle at $p_3 = -1-j$ and the arrival angle at $z_2 = -2-j2$ are $-\phi$ and $-\psi$, respectively. All these angles are shown in Figure 5.5.

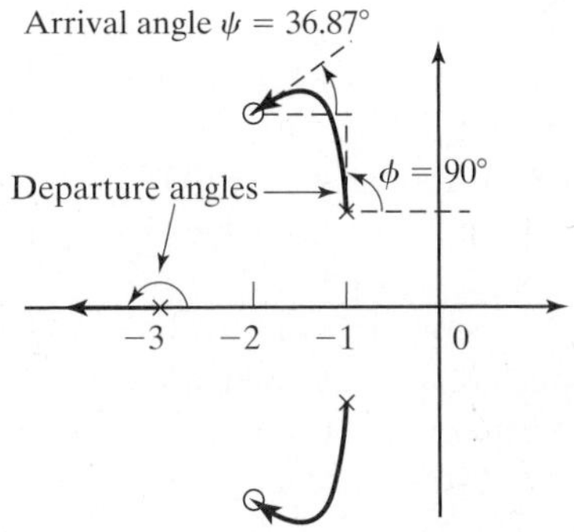

FIGURE 5.5: Departure and arrival angles.

The detailed derivations of these rules in Table 5.1 are given in the next section. We shall only look at Rule 3 here. Take s to be a point on the real axis as shown in Figure 5.6. It is easy to see that any complex pair of poles or zeros will contribute totally zero degree of phase for any test point on the real axis. Thus, we only need to look at real zeros and poles. It is also clear that any real pole or real zero on the left-hand side of the testing point s (real number) will also contribute zero degree of phase, and any pole or zero on the right-hand side will contribute 180 degree of phase. Hence, Rule 3 is verified.

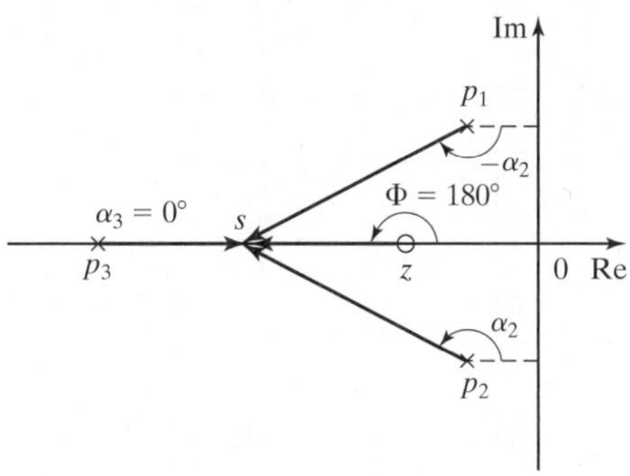

FIGURE 5.6: The phase of $L(s)$ at a point s on the real axis.

Normally, a quick sketch of the root-locus can be obtained by using only Rules 1–4. Rules 5 and 6 are rarely used nowadays since the exact root-locus can be easily generated using a computer program.

EXAMPLE 5.3

Consider a feedback system shown in Figure 5.2 with

$$L(s) = P(s)C(s) = \frac{8K(s+2)}{(s+1)(s+5)(s+10)}.$$

We would like to study how the closed-loop poles change when K varies from 0 to $+\infty$. This is done by constructing a root-locus plot of the system with respect to K. From Rule 3, we know there are root loci on the real axis in two intervals: $[-2,-1]$ and $[-10,-5]$. By Rule 2, one root locus starts from -1, a pole, and ends at -2, a zero, and other two root loci start from -10 and -5, respectively, and then approach two infinite zeros with two asymptotes. The two asymptotes start at

$$\kappa = \frac{(-1-5-10)-(-2)}{3-1} = -7$$

on the real axis with $90°$ and $-90°$, respectively. With the above information, a rough root locus can be sketched.

`z=-2;`	vector of zeros
`p=[-1,-5,-10];`	vector of poles
`k=8;`	gain (not the variable gain K)
`L=zpk(z,p,k)`	form the transfer function of the system
`rlocus(L)`	generate a root-locus plot with an automatically chosen range of gain K

An accurate root-locus plot can be generated by using the sequence of MATLAB commands listed in the box. The root-locus plot is shown in Figure 5.7.

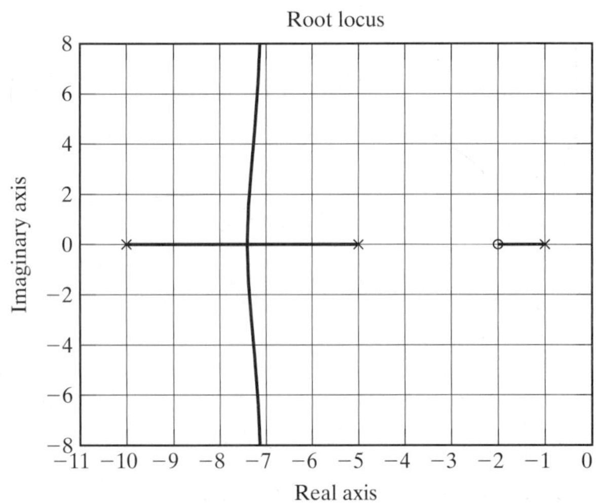

FIGURE 5.7: Root locus for Example 5.3.

Table 5.2 shows some typical root-locus plots with the given open-loop pole and zero patterns.

In many applications, the varying parameters do not necessarily appear as gains and they can, in general, appear anywhere in the transfer functions. The following example shows how to convert a nonstandard root-locus problem into a standard one.

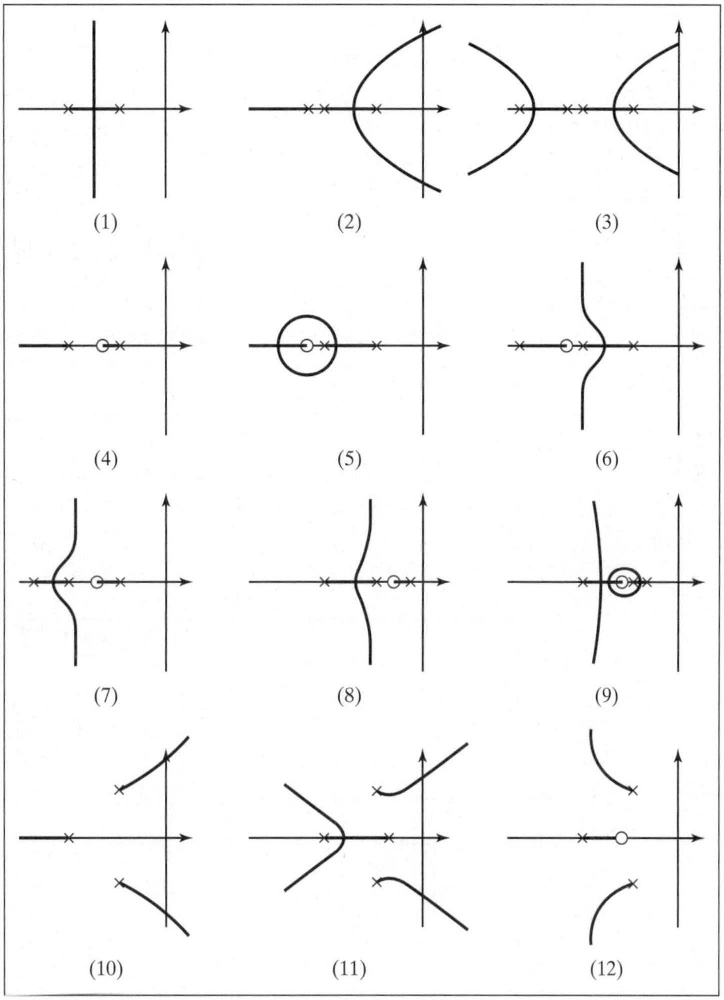

TABLE 5.2: Some typical root-locus plots.

EXAMPLE 5.4

Consider a feedback system with

$$P(s)C(s) = \frac{4(s+3)}{s(s+1)(s+K)}$$

where K is a variable pole position. We would like to analyze how K affects the system stability and performance. This problem is obviously not in the standard root-locus format. Nevertheless, the closed-loop poles are given by the roots of

$$1 + P(s)C(s) = 0$$

or

$$s(s+1)(s+K)+4(s+3)=0$$

which can be written as

$$(s+2)(s^2-s+6)+Ks(s+1)=0$$

or

$$1+\frac{Ks(s+1)}{(s+2)(s^2-s+6)}=0.$$

Let

$$L(s)=\frac{Ks(s+1)}{(s+2)(s^2-s+6)}.$$

`num=[1,1,0];` numerator coefficients of $L(s)$ excluding K

`den=conv([1,2], [1,-1,6]);` denominator coefficients of $L(s)$

`L=tf(num,den)` create the transfer function

`rlocus(L)` generate a root-locus plot

We can then construct the root locus for this system as usual. We can also use the above MATLAB commands for the task. The root locus of this system is shown in Figure 5.8.

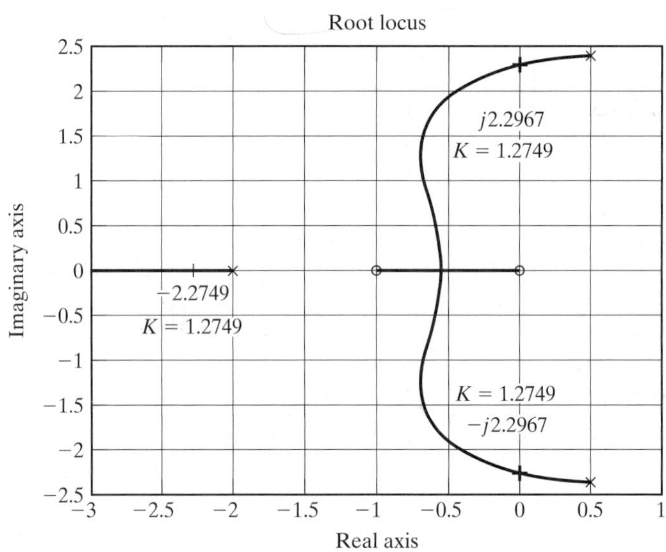

FIGURE 5.8: Root locus for Example 5.4 (The root locations for $K = 1.2749$ are shown with "+").

We can also get the numerical value of any point on the root-locus plot and the corresponding gain value by using `rlocfind`:

```
>> [K, poles]=rlocfind(L)
```
(just point and click on any desired point on the plot after entering this command)

For example, to get the critical value of K where the root locus enters the left half plane, simply enter the above command and then click on the intersection of the root locus and the imaginary axis. We get

$$K = 1.2749, \quad \text{poles} = -2.2749, \pm j2.2967.$$

(The actual numerical value may be slightly different from the above depending on how accurate the point was actually clicked.) This shows that the closed-loop system will only be stable for $K > 1.2749$.

This critical value of K can also be determined exactly by applying the Routh–Horwitz stability test in Chapter 3 to the characteristic polynomial:

$$s(s+1)(s+K) + 4(s+3) = s^3 + (1+K)s^2 + (K+4)s + 12.$$

The stability criterion then shows that the system is stable if and only if

$$(K+1)(K+4) - 12 > 0,$$

i.e.,

$$K > \frac{\sqrt{57}-5}{2} = 1.2749.$$

5.2 DERIVATIONS OF ROOT-LOCUS RULES*

Here we shall revisit the root-locus rules. Note that the closed-loop characteristic equation is

$$c(s) := (s-p_1)(s-p_2)\cdots(s-p_n) + K(s-z_1)(s-z_2)\cdots(s-z_m) = 0. \quad (5.3)$$

Rule 1. This is obvious. Since (5.3) has real coefficients, all roots of the equation appear in complex conjugate pairs.

Rule 2. When $K = 0$, (5.3) reduces to

$$(s-p_1)(s-p_2)\cdots(s-p_n) = 0$$

such that the solutions are $s = p_i, i = 1, 2, \ldots, n$. When K increases, the roots are moving continuously away from p_i.

When $K \to \infty$, we can rewrite (5.3) as

$$\frac{1}{K}(s-p_1)(s-p_2)\cdots(s-p_n) + (s-z_1)(s-z_2)\cdots(s-z_m) = 0.$$

If $|s|$ is finite, then the equation approaches

$$(s-z_1)(s-z_2)\cdots(s-z_m) = 0,$$

i.e., m roots approach $z_i, i = 1, 2, \ldots, m$. However, it is possible that $|s|$ is going to infinity as $K \to \infty$. In this case, (5.3) can be approximated as

$$1 + \frac{K}{(s - \kappa)^{n-m}} = 0$$

for some κ. To find κ, we note that

$$\begin{aligned} 0 &= 1 + \frac{K(s - z_1)(s - z_2)\cdots(s - z_m)}{(s - p_1)(s - p_2)\cdots(s - p_n)} \\ &= 1 + \frac{K}{s^{n-m} - \left(\sum_{i=1}^{n} p_i - \sum_{i=1}^{m} z_i\right) s^{n-m-1} + \cdots} \end{aligned}$$

by power series expansion and also note that

$$0 = 1 + \frac{K}{(s - \kappa)^{n-m}} = 1 + \frac{K}{s^{n-m} - (n - m)\kappa s^{n-m-1} + \cdots}.$$

Comparing coefficients, we get

$$(n - m)\kappa = \sum_{i=1}^{n} p_i - \sum_{i=1}^{m} z_i,$$

i.e.,

$$\kappa = \frac{\sum_{i=1}^{n} p_i - \sum_{i=1}^{m} z_i}{n - m} = \frac{\sum_{i=1}^{n} \text{poles} - \sum_{i=1}^{m} \text{zeros}}{n - m}.$$

Finally, the $n - m$ solutions of the equation

$$1 + \frac{K}{(s - \kappa)^{n-m}} = 0$$

are given by

$$s = \kappa + K^{\frac{1}{n-m}} e^{j\frac{(2k+1)\pi}{n-m}}, \quad k = 0, \pm 1, \pm 2, \ldots.$$

These are the $n - m$ straight lines described in Rule 4.

Rule 5. Breakaway and breakin points are those points where the characteristic equation has repeated roots. Suppose z is a breakaway or breakin point. Then

$$c(s) = (s - z)^r f(s)$$

where $r \geq 2$ and $f(s)$ is a polynomial of s. It is easy to see that

$$c(z) = 0, \quad \left.\frac{dc(s)}{ds}\right|_{s=z} = 0,$$

i.e.,

$$\prod_{i=1}^{n}(z-p_i)+K\prod_{i=1}^{m}(z-z_i)=0$$

$$\frac{d}{ds}\prod_{i=1}^{n}(s-p_i)\bigg|_{s=z}+K\frac{d}{ds}\prod_{i=1}^{m}(s-z_i)\bigg|_{s=z}=0.$$

Eliminating K, we get

$$\prod_{i=1}^{m}(z-z_i)\;\frac{d}{ds}\prod_{i=1}^{n}(s-p_i)\bigg|_{s=z}-\prod_{i=1}^{n}(z-p_i)\;\frac{d}{ds}\prod_{i=1}^{m}(s-z_i)\bigg|_{s=z}=0$$

which is equivalent to

$$\frac{dL(s)}{ds}\bigg|_{s=z}=0.$$

Note that this condition is only necessary, i.e., a point satisfying this condition is not always a breakaway or breakin point; we must check if it is on the root locus. The reason is that this condition is satisfied for both positive and negative K. Hence, the repeated roots for both positive and negative K are included in the solutions of this equation. Therefore, those points satisfying the above equation but not on the root locus for $K>0$ are the candidates for breakaway or breakin points of the root locus for $K<0$, i.e., the complementary root locus to be considered at the end of the chapter.

Rule 6. The departure angle from a pole can be found by testing a point very close to the pole. Suppose s is a point on the root locus and is very close to p_k. Then the departure angle from p_k is given by $\angle(s-p_k)$ when $s\to p_k$. By the phase condition, we have

$$\sum_{i=1}^{m}\angle(s-z_i)-\sum_{j=1,j\neq k}^{n}\angle(s-p_j)-\angle(s-p_k)=\pm180°.$$

Thus

$$\phi_k=\lim_{s\to p_k}\angle(s-p_k)=\sum_{i=1}^{m}\angle(p_k-z_i)-\sum_{j=1,j\neq k}^{n}\angle(p_k-p_j)\pm180°.$$

In the case p_k is l repeated poles, we have

$$\sum_{i=1}^{m}\angle(s-z_i)-\sum_{j\neq k}\angle(s-p_j)-l\,\angle(s-p_k)=\pm180°$$

such that the departure angle becomes ϕ_k/l.

The formulas for arrival angles can be derived similarly.

5.3 EFFECTS OF ADDING POLES AND ZEROS

One of the most useful applications of root-locus techniques is to predict how the changes of system parameters or structures affect system performance. In particular, we shall discuss in this section how additional open-loop poles or zeros may affect the locations of closed-loop poles. We shall start our analysis from Table 5.2.

- **Adding open-loop poles:** Plot (2) in Table 5.2 can be obtained from Plot (1) by adding one more open-loop pole. It is seen that two root loci in Plot (2) go to the right half plane along two asymptotes of 60° and −60°.

  Similarly, Plot (3) can be obtained by adding two more poles to Plot (1) or one more pole to Plot (2). In this case, two root loci in Plot (3) go to the right half plane along two asymptotes of 45° and −45°, meaning these two root loci will go to the right half plane faster than the two in Plot (2) as the parameter is increased. The same conclusions can be drawn by comparing Plot (10) with Plot (11).

  Conclusions: Adding open-loop poles tend to push the root loci to the right half plane.

- **Adding open-loop zeros (PD controllers):** We shall now compare Plot (1) with Plot (4) and Plot (5). Indeed, Plot (4) and Plot (5) can be obtained by adding one zero to Plot (1). It is not hard to see that the additional zero tends to attract the root loci. Thus, by suitably placing the additional zero, one can move the root loci to the desired region. This is how the PD controller works. Similar conclusions can be reached by comparing Plot (10) with Plot (12).

  Conclusions: Zeros attract root loci, and thus suitably placed zeros can move root loci to the desired region.

- **Adding a pair of pole and zero with zero on the right of the pole (Lead controllers):** We now compare Plot (1) with Plot (6) and Plot (7). We can think that Plot (6) and Plot (7) are obtained by adding a pair of a pole and a zero to Plot (1) with the zero on the right of the pole. In both cases, two root loci tend to move left toward the stable region while another root-locus approaches the open-loop zero. When a closed-loop pole is close enough to a zero, their effects on the closed-loop performance become negligible since this closed-loop pole will almost be canceled by this zero. Thus, the other two root loci play the dominant roles. Since these two root loci are in the more stable region, it is expected that the closed-loop performance with this lead (or PD) controller will be improved when the controller parameters are suitably chosen.

  Conclusions: A lead (or PD) controller can be used to improve the transient performance if it is suitably designed.

- **Adding a pair of a pole and a zero with a pole on the right of the zero (Lag controllers):** We shall now compare Plot (1) with Plot (8) and Plot (9). In contrast to the lead controller case, Plot (8) and Plot (9) are obtained from Plot (1) by adding a pair of a pole and a zero with the pole on the right of the zero. It is clear that the total effect is somewhat similar to adding a pole. Thus, part of the root-locus plot is pushed to the right to a certain degree. It is also important to note that the root-locus plot will not change much (except the local part near the pair of a pole and a zero) if the pair of a pole and a zero are much closer than other poles and zeros. This is indeed how a lag or PI controller works. By choosing the pair of a pole and a zero much closer to one another (usually also close to origin), this controller will essentially not change the dominant pole locations. Thus, it will not change the transient response much if it is suitably designed, but will improve steady-state tracking by increasing the system gain with the ratio of zero over the pole. (This is why the pole and zero are sometimes chosen to be close to the origin so that a large ratio can be obtained.)

Note that a PI controller can be regarded as a special lag controller with the pole placed at the origin. Thus, the discussion for lag controllers also applies to PI controllers.

Conclusions: A lag (or PI) controller can be used to improve steady-state tracking.

In the next few sections, we shall see how the above analysis can be used to design controllers.

5.4 PHASE-LAG CONTROLLER

We shall consider a first-order phase-lag controller in the following form:

$$C(s) = \frac{K(s+b)}{s+a}, \quad b > a > 0.$$

Then $C(0) = Kb/a$. Thus, a phase-lag controller has the potential to increase the steady-state gain constant by b/a times compared to a pure gain controller.

The pole-zero configuration of the controller is shown in Figure 5.9. It is clear that $\angle C(s) < 0$ for any s on the upper half of the complex plane, i.e., it always contributes a negative phase (or phase lag) to the root-locus equation. This can also be regarded as one of the reasons why these controllers are called phase-lag controllers. Another natural interpretation of this name is from the frequency responses

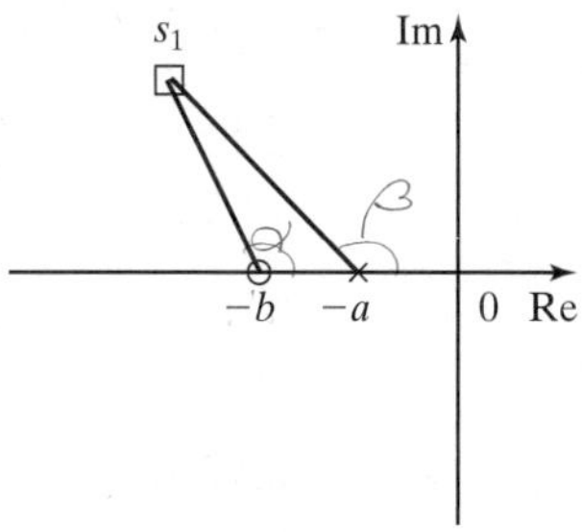

FIGURE 5.9: Pole/zero configuration of a phase-lag controller.

of these controllers, which will be presented in the next chapter. These controllers will always generate negative phase frequency responses, as will be discussed later.

A phase-lag controller is designed in such a way that $C(s)$ contributes very little phase at the desired closed-loop pole locations while providing substantial gain increase (b/a times) to reduce the steady-state error. Thus, a phase-lag controller is essentially a gain compensation controller. More specifically, suppose s_1 is the desired closed-loop (dominant) pole location, then typically b and a are chosen so that

$$\frac{s_1+b}{s_1+a} \approx 1,$$

i.e.,

$$C(s_1) \approx K$$

and

$$P(s_1)C(s_1) \approx KP(s_1) = -1.$$

This is usually achieved by choosing the zero $-b$ and the pole $-a$ far from s_1 but relatively close to the origin.

Note that a and b need not be small in the absolute sense. For example, if $s_1 = -1000 + j1000$, $b = 20$, and $a = 5$, then

$$C(s_1) = \frac{K(s_1+b)}{s_1+a} = \frac{K(-980+j1000)}{-995+j1000} \approx K$$

and $C(0) = Kb/a = 4K$.

Algorithm 5.5 (Phase-lag controller design).

Step 1: Construct a root-locus plot of $KP(s)$.

Step 2: Find the closed-loop (dominant) poles s_1 and $\bar{s}_1$ on the root-locus plot that will give the desired transient response and find the corresponding K value, say K_0.

Step 3: Calculate the value of K required to yield the desired steady-state response and denote this K as K_s.

Step 4: Pick a number b ($> a$) that is much smaller than $|s_1|$ (so that $\dfrac{s_1+b}{s_1+a} \approx 1$) and let $a = K_0 b/K_s$.

Step 5: Verify the controller design by simulation with

$$C(s) = \frac{K_0(s+b)}{s+a}.$$

Note that $C(0) = K_s$.

Note that this design procedure can also be used to design a PI controller by simply taking $K_s = \infty$.

We shall now illustrate the above design procedure through an example.

EXAMPLE 5.6

Consider a unity feedback system with a plant model

$$P(s) = \frac{10}{s(s+5)(s+10)}.$$

We would like to design a controller so that

- the response of the closed-loop system with respect to a step input has no more than 20% overshoot and the settling time is no greater than 4 [sec];
- the steady-state error with respect to a ramp input is no more than 0.05.

First of all, we shall find the desired dominant pole locations. Suppose that the dominant poles are a pair of complex poles, s_1 and $\bar{s}_1$, given as the poles of a standard second-order system

$$\frac{\omega_n^2}{s^2 + 2\zeta\omega_n s + \omega_n^2},$$

i.e.,

$$s_1 = -\zeta\omega_n + j\omega_n\sqrt{1-\zeta^2}.$$

Then, to guarantee that the overshoot is no more than 20%, we need the damping ratio to satisfy

$$\zeta \geq \frac{-\ln(20/100)}{\sqrt{\pi^2 + (\ln 20/100)^2}} = 0.456.$$

To guarantee that the settling time is no more than 4 [sec] with 2% tolerance, we need $\zeta\omega_n \geq 4/t_s = 1$. We shall pick s_1 in the region described above in the following design process.

We shall now follow the phase-lag design procedure to see whether a phase-lag controller can be designed to satisfy the above design specifications.

- We shall first construct a root-locus plot for

$$KP(s) = \frac{10K}{s(s+5)(s+10)}.$$

```
P=zpk([],[0,-5,-10],10);

rlocus(P);

sgrid    draw damping ratio lines
```

Then point and click on the desired point to find K_0 and s_1

```
[K0,s1]=rlocfind(P)
```

The root locus is shown in Figure 5.10. Now, choose the dominant poles on the root-locus plot so that their real parts are greater than 1 and they have approximately the damping ratio 0.5 by using the following MATLAB command

```
>> [K0,s1]=rlocfind(P)
```

We then get

$$K_0 = 13,\ s_1 = -1.665 + j2.8927,$$

$$\bar{s}_1 = -1.665 - j2.8927,\ s_2 = -11.67.$$

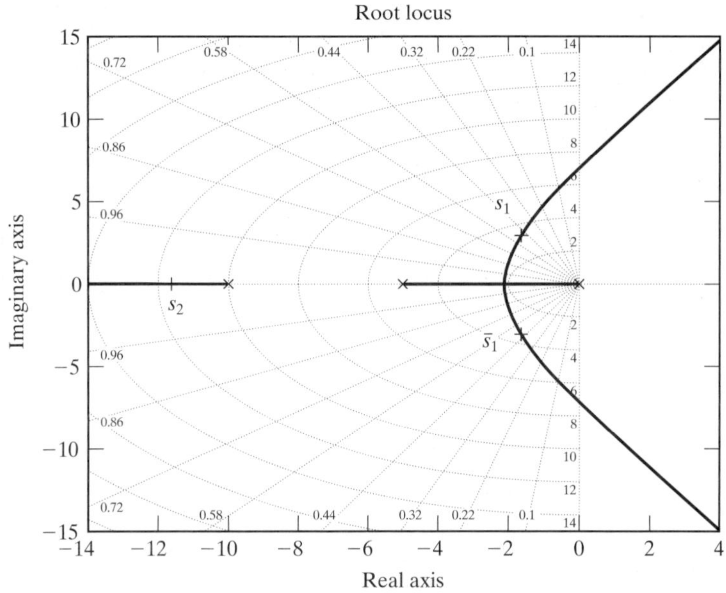

FIGURE 5.10: Uncompensated root locus of Example 5.6.

- We shall now find the required gain to satisfy the desired steady-state error. Note that $K_s = C(0)$ and

$$K_v = \lim_{s\to 0} sP(s)C(s) = \frac{C(0)}{5}.$$

Thus

$$e_{ss} = \frac{1}{K_v} = \frac{5}{C(0)} \leq 0.05$$

gives $K_s = C(0) \geq 100$. Take $K_s = 100$.

- Take $b = 0.05$, which is much smaller than $|s_1|$. Then $a = K_0 b/K_s = 0.0065$ and we have a controller

```
b=0.05;  K0=13;  Ks=100;
a=K0*b/Ks;
C=zpk(-b,-a,K0); controller
T=feedback(P*C,1); closed-loop
pole(T) closed-loop poles
step(T) step response
```

$$C(s) = \frac{13(s+0.05)}{s+0.0065}$$

which gives the closed-loop poles at

$$-11.6656, \; -0.0509,$$
$$-1.6450 \pm j2.8724$$

and a closed-loop zero at -0.05, which approximately cancels the closed-loop pole at -0.0509.

Thus, we can see that the closed-loop system has two dominant poles at $-1.6450 \pm j2.8724$, which are very close to s_1 and $\bar{s}_1$ as expected. The step response and the ramp response are shown in Figures 5.11 and 5.12, respectively. The design specifications are met with a settling time of 1.89 [sec] and an overshoot of 17.7%.

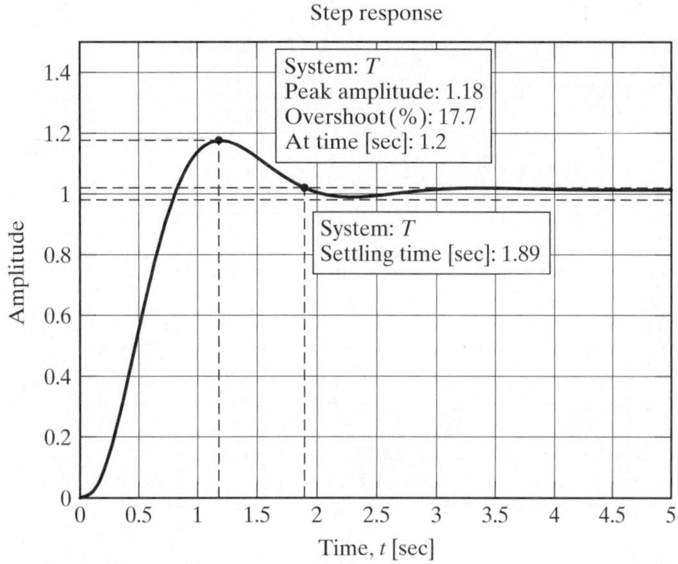

FIGURE 5.11: Step response of Example 5.6.

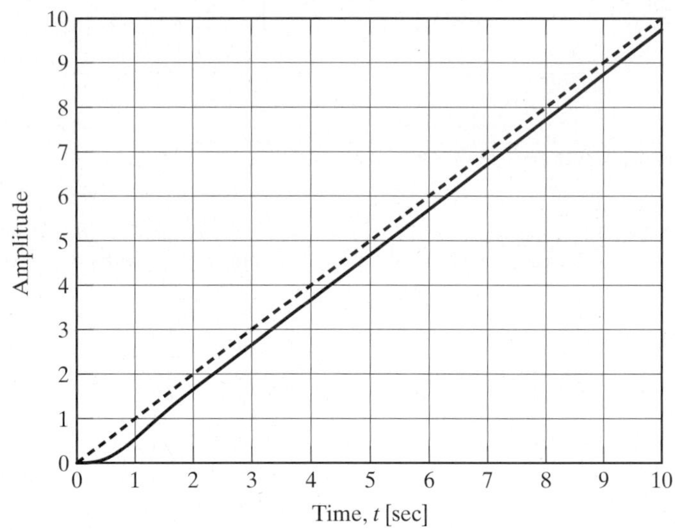

FIGURE 5.12: Ramp response of Example 5.6.

Note that the entire design process can also be done interactively using the GUI tool: `sisotool`. For example, one can start with

```
>> sisotool(P)
```

and then interactively change the controller pole and zero positions, see Figure 5.13.

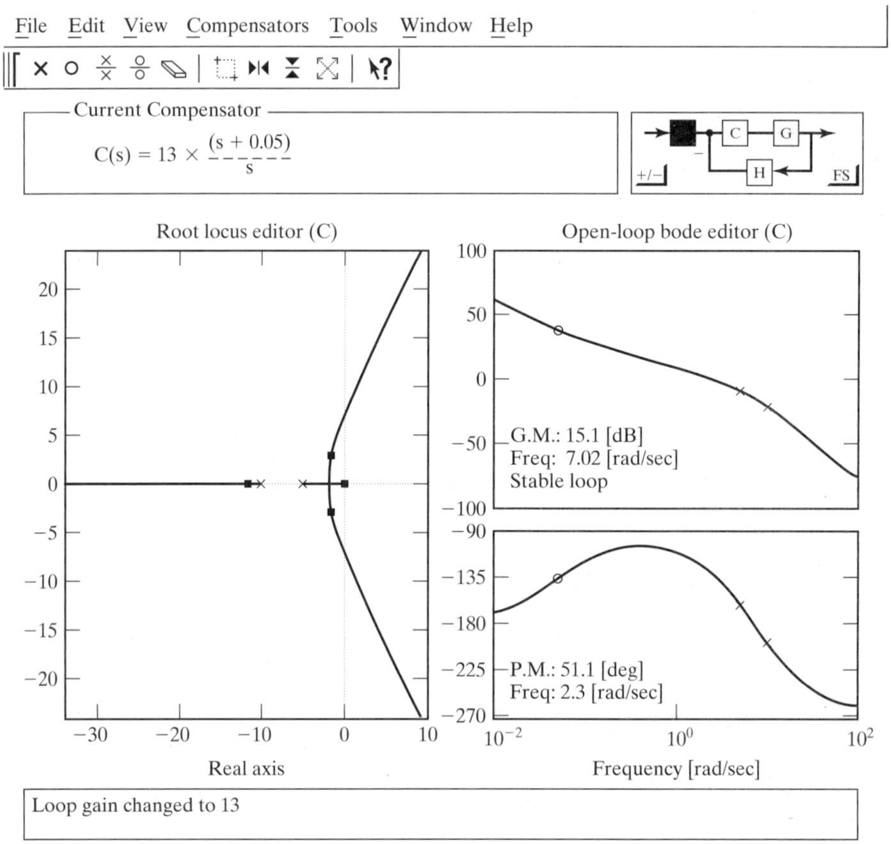

FIGURE 5.13: Using the GUI tool `sisotool` for Example 5.6: the left-hand side shows the root locus and right-hand side shows the Bode diagram, which will be discussed in the next chapter.

5.5 PI CONTROLLER

A PI controller has the following transfer function

$$C(s) = \frac{K(s+b)}{s}, \quad b > 0.$$

Thus $C(0) = \infty$. Hence, a PI controller increases the system type by 1. Since a PI controller can be considered as the extreme case of a phase-lag controller, a design procedure similar to the phase-lag case can be applied.

Algorithm 5.7 (PI controller design).

Step 1: Construct a root-locus plot of $KP(s)$.

Step 2: Find the closed-loop (dominant) poles s_1 and $\bar{s}_1$ on the root-locus plot that will give the desired transient response and find the corresponding K value, say K_0.

Step 3: Pick a b that is much smaller than $|s_1|$ (so that $\frac{s_1+b}{s_1} \approx 1$).

Step 4: Verify the controller design by simulation with

$$C(s) = \frac{K_0(s+b)}{s}.$$

EXAMPLE 5.8

Consider the design problem in Example 5.6. We can pick the same s_1 and b. Then

$$C(s) = \frac{13(s+0.05)}{s}$$

which gives the closed-loop poles at

$$-11.6649, \quad -1.6420 \pm j2.8693, \quad -0.0510.$$

The time responses with respect to step and ramp are almost the same as the ones shown in Figures 5.11 and 5.12 for the phase-lag controller.

5.6 PHASE-LEAD CONTROLLER

Suppose that it is required in Example 5.6 that the settling time be less than 1 [sec]. Then it is fairly easy to see that no phase-lag (or PI) controller can satisfy this specification since it requires that the dominant poles satisfy $\mathrm{Re}\, s_1 \leq -4$. In this case, a phase-lead (or PD) controller is needed to move the dominant poles a little further away from the imaginary axis.

A first-order phase-lead controller has the following general form:

$$C(s) = \frac{K(s+b)}{s+a}, \quad a > b > 0.$$

Since $\angle C(s) > 0$ for any s on the upper half of the complex plane, it contributes a positive angle (or phase-lead). In contrast to the phase-lag controller, a phase-lead controller is a phase compensator, i.e., the compensation of the controller is achieved by providing a positive phase in the root-locus phase equation so that the desired poles can be moved to the further left half plane or other desired locations. This can be illustrated through an example. Consider a system with three poles p_1, p_2, and p_3 shown in Figure 5.14 and suppose s_0 is a point on the root locus, i.e.,

$$-\phi_1 - \phi_2 - \phi_3 = -180^\circ.$$

Now, suppose that we need to move the closed-loop pole from s_0 to s_1; then we would need the following phase lead to guarantee that s_1 point satisfies the phase condition:

$$\theta := \alpha_1 + \alpha_2 + \alpha_3 - \phi_1 - \phi_2 - \phi_3 > 0.$$

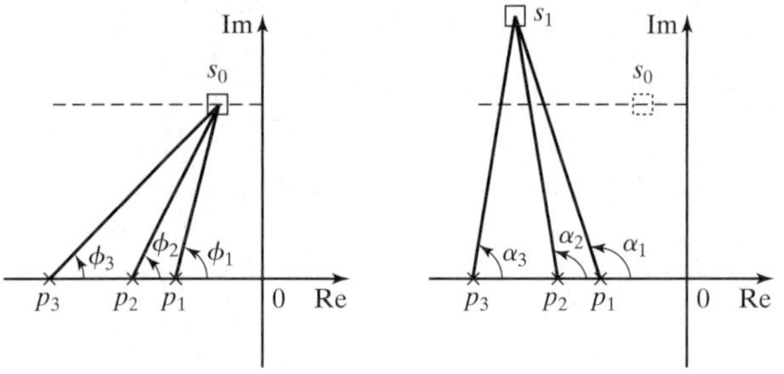

FIGURE 5.14: Phase needed to move from s_0 to s_1.

Algorithm 5.9 (Phase-lead controller design – one possible method).

Step 1: Construct a root-locus plot of $KP(s)$.

Step 2: Determine the desired closed-loop (dominant) poles s_1 and $\overline{s}_1$ that will give the desired transient response.

Step 3: Calculate the angle required so that s_1 is on the root-locus plot

$$\angle C(s_1) + \angle P(s_1) = (2k + 1)\ 180°,$$

i.e.,

$$\theta = \angle C(s_1) = (2k + 1)\ 180° - \angle P(s_1) > 0$$

for some suitable integer k.

Step 4: Find b and a so that

$$\angle(s_1 + b) - \angle(s_1 + a) = \theta$$

and make sure that s_1 is the dominant pole (Note that there are infinitely many choices as long as the angle between the two vectors shown in Figure 5.15 is θ).

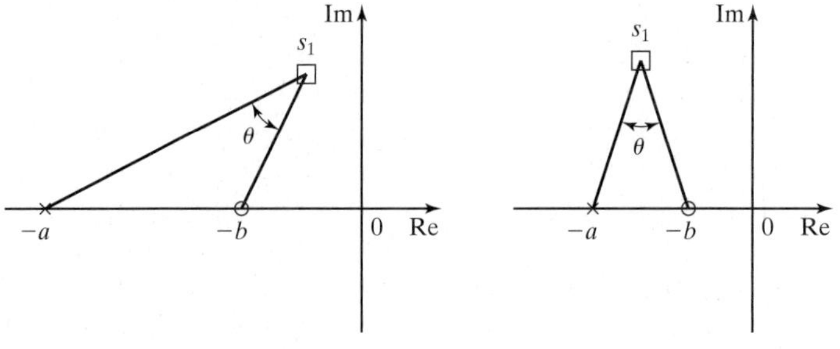

FIGURE 5.15: Two configurations of phase-lead controllers with the same phase.

Step 5: Find K_0 so that

$$\frac{K_0 |s_1 + b|}{|s_1 + a|} |P(s_1)| = 1.$$

Step 6: Verify the controller design by simulation with

$$C(s) = \frac{K_0(s+b)}{s+a}.$$

EXAMPLE 5.10

Consider the same unity feedback system as in Example 5.6 with

$$P(s) = \frac{10}{s(s+5)(s+10)}.$$

We would like to design a controller so that

- the response of the closed-loop system with respect to a step input has no more than 20% overshoot and the settling time is no more than 1 [sec].

As in Example 5.6, we can find the desired dominant pole locations:

$$\zeta \geq 0.456, \quad \zeta\omega_n \geq 4.$$

From the root-locus plot in Figure 5.10, it is clear that a phase-lag or PI controller cannot satisfy these design specifications. We shall try a phase-lead controller. Taking into consideration the effects by less dominant poles and zeros, we shall pick the dominant poles conservatively at

$$s_1 = -5 + j5, \quad \bar{s}_1 = -5 - j5.$$

Then $P(s_1) = j0.04$, which is calculated using `evalfr(P,s1)`, and

$$\angle P(s_1) = 90°.$$

```
P=zpk([],[0,-5,-10],10);

s1=-5+j*5;

Ps1=evalfr(P,s1);

theta=angle(Ps1);

b=4;

a=-real(s1)+imag(s1)
    /tan(angle(s1+b)-theta);

K0=abs(s1+a)/abs(Ps1)
    /abs(s1+b);

C=zpk(-b,-a,K0);

T=feedback(P*C,1);

step(T)
```

Thus, we need

$$\theta = \angle C(s_1) = 180° - 90° = 90°.$$

We shall now pick b and a. Note that if $b \geq 5$, then a must be ∞ from Figure 5.15 since $\theta = 90°$. As a first try, we shall pick $b = 2$. Then, a can be solved from

$$\angle(s_1 + b) - \angle(s_1 + a) = \theta$$

which gives $a = 13.3333$. Now K_0 can be found as

$$K_0 = 41.667.$$

Thus, we have a phase-lead controller

$$C_0(s) = \frac{41.667(s+2)}{s+13.333}.$$

The root locus of $1 + KP(s)C_0(s) = 0$ is shown in Figure 5.16.

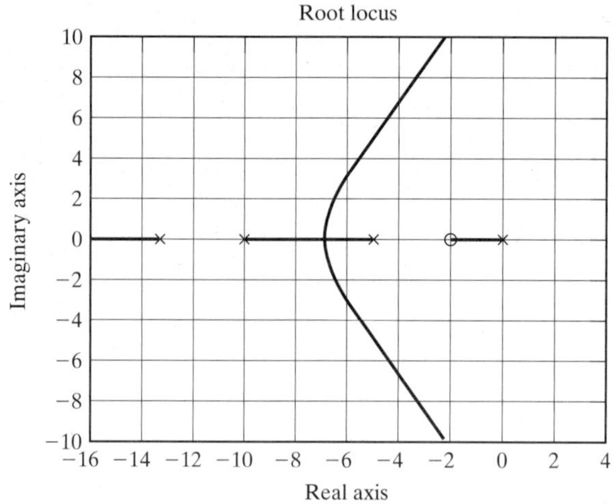

FIGURE 5.16: Root locus of Example 5.10 with $C_0(s)$.

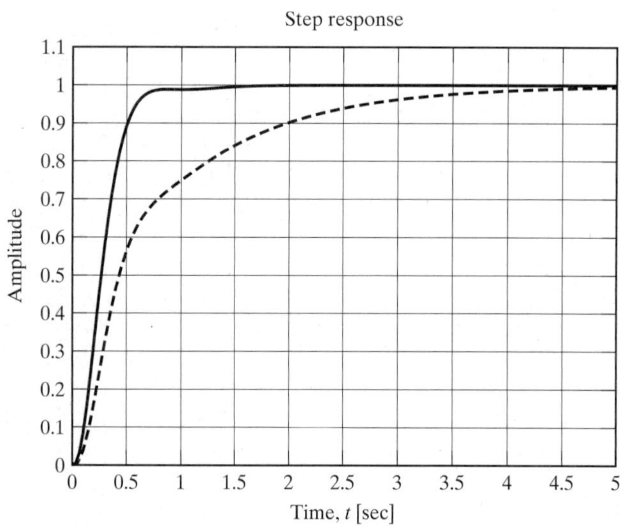

FIGURE 5.17: Step response of Example 5.10 with $C_0(s)$ (dashed) and $C(s)$ (solid).

Unfortunately, the numerical simulation shown in Figure 5.17 (dashed line) clearly indicates that the design specifications are not satisfied since the closed-loop poles are at

$$-17.3739, \quad -5 \pm j5, \quad -0.9593$$

and the pair of poles at $-5 \pm j5$ are actually not dominant poles. The pole at -0.95934, which is not close to any zero, turns out to be the dominant pole.

We shall now try with $b = 4$. Then, we have $a = 30$ and $K_0 = 125$, which gives

$$C(s) = \frac{125(s+4)}{s+30}$$

and the closed-loop poles at

$$-31.8614, -5 \pm j5, -3.1386.$$

The root locus of the closed-loop system with $C(s)$ is plotted in Figure 5.18.

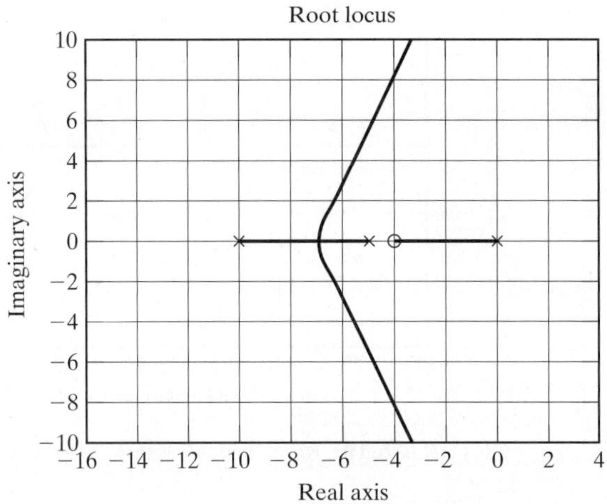

FIGURE 5.18: A part of the root locus for Example 5.10 with $C(s)$ while the other part of the root locus from the pole at -30 to negative infinity is not shown.

The step response shown in Figure 5.17 (solid line) indicates that the design specifications are met even though $-5 \pm j5$ are not the only dominant poles (since the closed-loop pole at -3.1386 is not very close to the closed-loop zero at -4). The velocity constant can be calculated as $K_v = 10/3$ with $C(s)$. Thus, the steady-state error with respect to a ramp input is $e_{ss} = 1/K_v = 3/10 = 0.3$.

Instead of picking the desired closed-loop poles first, it is sometimes more convenient to first pick some controller parameters. For example, in many cases, one can first choose the desired controller zero and pole, and then determine an appropriate gain.

EXAMPLE 5.11

Consider the same design problem as in the last example and note that the open-loop pole at -5 in the transfer function $P(s)$ tends to pull the closed-loop poles to the right half plane. Hence, it is appropriate to place the zero of the lead controller at the same location, i.e., $b = 5$, to cancel this undesirable pole and then adjust the pole location a and the gain K so that the desired performance specifications are achieved. For example, take

$$C(s) = \frac{K(s+5)}{s+20}.$$

Then the root locus with

$$L(s) = C(s)P(s) = \frac{10K}{s(s+10)(s+20)}$$

is shown in Figure 5.19.

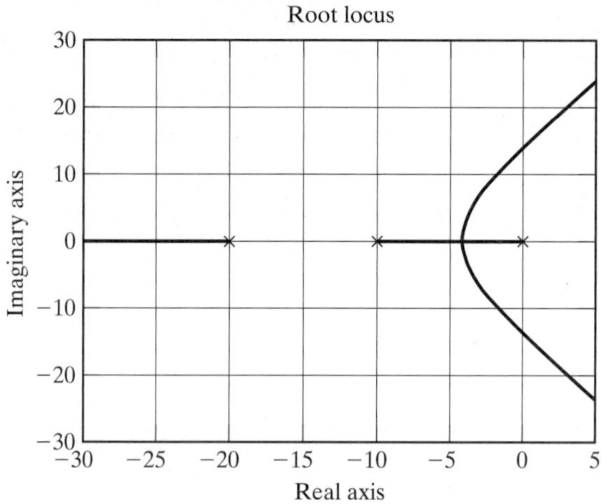

FIGURE 5.19: Root locus of Example 5.11 with $C_1(s)$.

Now, after suitably choosing the desired closed-loop pole locations on the root locus, i.e., choosing a suitable gain K results in a lead controller,

$$C(s) = \frac{90(s+5)}{s+20}.$$

The closed-loop poles with this controller are given by

$$-23.0074, -3.4963 \pm j5.1859.$$

The step response with this controller shown in Figure 5.20 indicates that the design specifications are met.

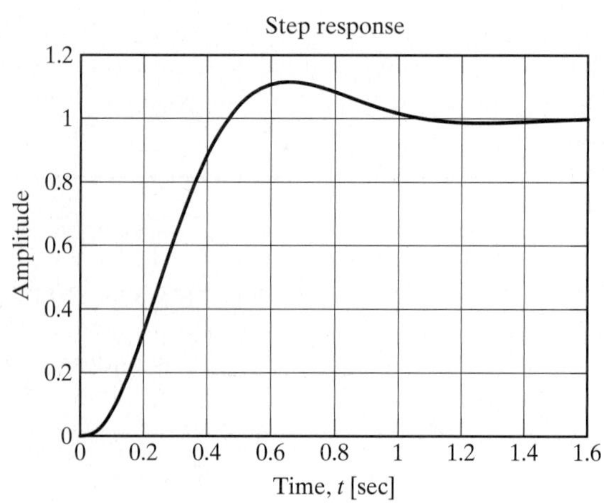

FIGURE 5.20: Step response of Example 5.11 with $C(s)$.

The above design process can be most effectively done using `sisotool`. If the initial choices of controller zero and pole do not result in satisfactory performance, they can be easily adjusted interactively by either dragging the controller zero and pole to new locations or by manually entering new values of the zero and pole locations. In addition, various closed-loop time/frequency responses such as step response can be generated from the pull-down menu of the interactive window. The best way to learn how this process works is by trying a simple example, which we strongly encourage to do now.

5.7 PD CONTROLLER

A PD controller takes the following form:

$$C(s) = K(s+b), \qquad b > 0.$$

These controllers are not physically realizable and are usually implemented approximately as

$$C(s) = K\left(\frac{s}{Ts+1} + b\right)$$

for some small $T > 0$. Nevertheless, it is included here for historical reasons.

This controller can be regarded as a special case of the phase-lead controller by taking $a \to \infty$ and keeping K/a constant. Thus, the design procedure for phase-lead controller can essentially be applied.

Algorithm 5.12 (PD controller design).

Step 1: Construct a root-locus plot of $KP(s)$.

Step 2: Determine the desired closed-loop (dominant) poles s_1 and $\bar{s}_1$ that will give the desired transient response.

Step 3: Calculate the angle required so that s_1 is on the root-locus plot

$$\angle C(s_1) + \angle P(s_1) = (2k+1)\ 180°,$$

i.e.,

$$\theta = \angle C(s_1) = (2k+1)\ 180° - \angle P(s_1) > 0$$

as shown in Figure 5.21.

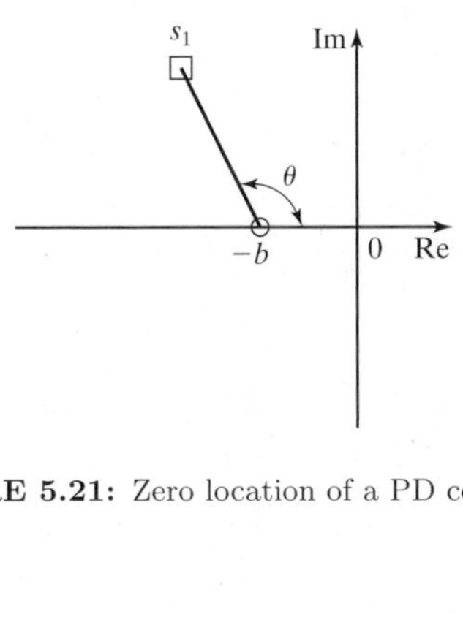

FIGURE 5.21: Zero location of a PD controller.

Step 4: Find b so that

$$\angle(s_1 + b) = \theta$$

and make sure that s_1 is the dominant pole;

Step 5: Find K_0 so that

$$K_0|s_1 + b|\,|P(s_1)| = 1;$$

Step 6: Verify the controller design by simulation with

$$C(s) = K_0(s + b).$$

EXAMPLE 5.13

Continuing from Example 5.10, we can also design a PD controller to satisfy the design specifications. If we again take $s_1 = -5 \pm j5$ and $b = 5$, then we need $K_0 = 5$, which gives a PD controller

$$C(s) = 5(s + 5).$$

The closed-loop poles with this PD controller are at

$$-5, -5 \pm j5.$$

The step response shown in Figure 5.22 indicates that the design specifications are met.

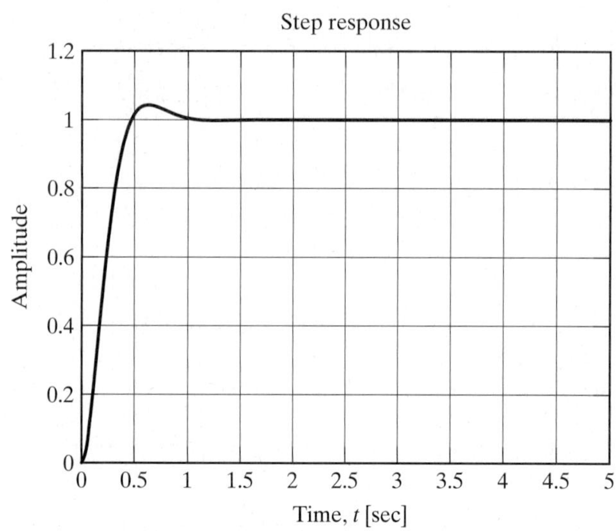

FIGURE 5.22: Step response of Example 5.13.

Of course, one can also start by picking the zero position $-b$ and varying K using the root locus to find the desirable gain.

EXAMPLE 5.14

Consider a double integrator plant

$$P(s) = \frac{1}{s^2}.$$

It is clear that a PD or a phase-lead controller is needed to stabilize this double integral system. It is also easy to see that almost any PD or phase-lead type controller will stabilize this system. However, it is critical in practical implementation that the derivative or phase lead should not be too large in order to reduce noise effect and to avoid saturation of the actuators. On the other hand, one would need enough phase leads or derivatives in order to get a fast response. To analyze the effects of the controller parameters on the closed-loop behavior, let

$$C(s) = \frac{K(s+b)}{s+10} = \frac{K_P(1+T_D s)}{0.1s+1}$$

with

$$K_P = 0.1Kb, \quad T_D = 1/b.$$

The root locus of the system $L(s) = P(s)C(s)$ with $b = 0.5$ (i.e., strong derivative control) is shown in Figure 5.23, and the closed-loop step responses of the system with different controller gains are shown in Figure 5.24.

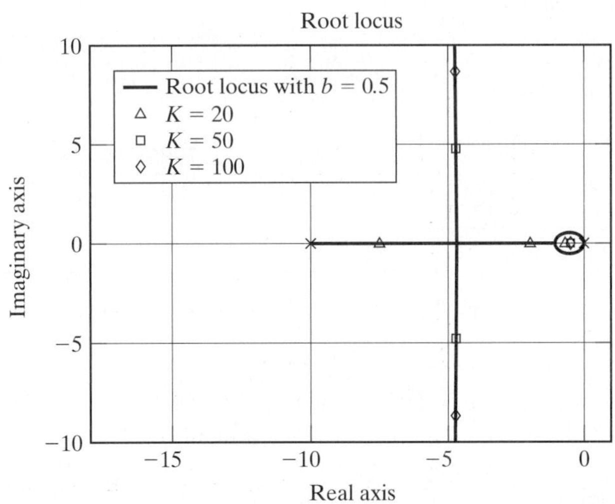

FIGURE 5.23: Root locus of Example 5.14 with $b = 0.5$.

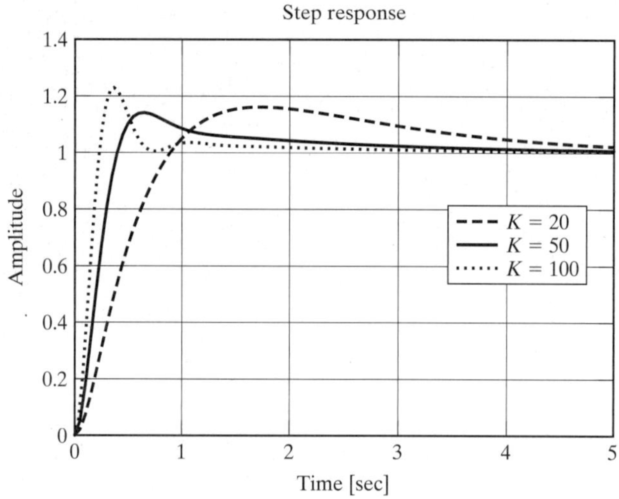

FIGURE 5.24: Step responses of system in Example 5.14 with $b = 0.5$.

It is clear from the root locus that the closed-loop system behaves very much like a second-order system for a sufficiently large controller gain K since one of the closed-loop poles will almost be canceled by the closed-loop zero from the controller. Hence a good transient performance can be achieved by choosing a suitable controller gain. However, a too large controller gain, say $K \gg 100$, would result in too much overshoot. On the other hand, if the controller gain is very small, say $K \ll 20$, the closed-loop behavior will be dominated by those closed-loop poles close to the imaginary axis thus resulting in slow transient response.

The closed-loop behavior of the system with different gain K are summarized in Table 5.3.

K	20 (small)	50 (medium)	100 (large)
Closed-loop poles	Δ	$\square$	$\Diamond$
	−7.5160	−4.7205+j4.7370	−4.7369+j8.5192
	−1.7024	−4.7205−j4.7370	−4.7369−j8.5192
	−0.7815	−0.5590	−0.5262
Closed-loop zeros	−0.5	−0.5	−0.5
Time response	Slow	Reasonable	Larger overshoot

TABLE 5.3: Closed-loop poles on the root locus with $b = 0.5$.

The root locus of the system $L(s) = P(s)C(s)$ with $b = 6$ (i.e., relatively weak derivative control) is shown in Figure 5.25. In this case, the closed-loop system behavior is dominated by the pair of complex poles close to the imaginary axis. It is not hard to see that the closed-loop system will respond much slower with significant overshoot than that when b is small, as shown in Figure 5.26. However, the closed-loop system with this controller will be much less sensitive to the sensor noise since the derivative term is much smaller.

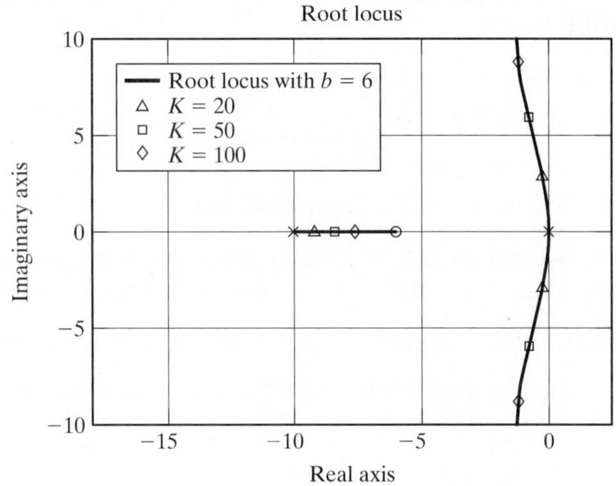

FIGURE 5.25: Root locus of Example 5.14 with $b = 6$.

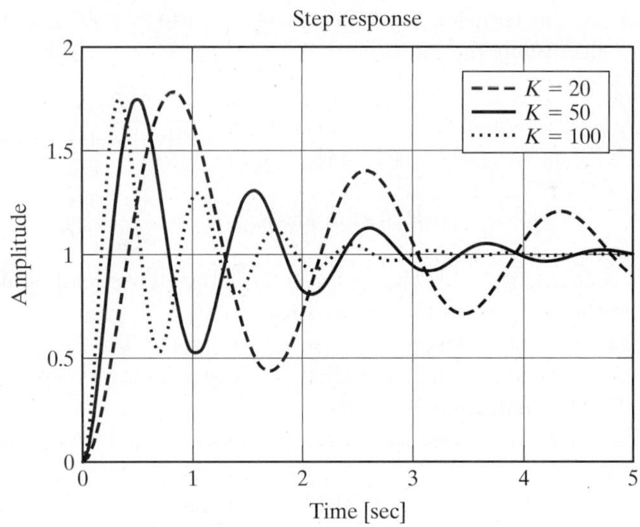

FIGURE 5.26: Step responses of system in Example 5.14 with $b = 6$.

5.8 LEAD-LAG OR PID CONTROLLER

Sometimes, a pure phase-lag or phase-lead controller may not provide enough design freedom to satisfy design specifications. In this case, a combination of phase-lag and phase-lead or PID controller is necessary. The basic idea is to use a phase lead to satisfy the transient response and a phase lag to satisfy the steady-state requirement. We shall now illustrate this design procedure through an example.

EXAMPLE 5.15

Consider a unity feedback system with

$$P(s) = \frac{10}{s(s+5)(s+10)}.$$

We would like to design a controller so that

- the response of the closed-loop system with respect to a step input has no more than 20% overshoot and the settling time is no greater than 1 [sec];
- the steady-state error with respect to a ramp input is no more than 0.05.

It is seen from Examples 5.10 and 5.13 that a phase-lead controller

$$C_{\text{lead}}(s) = \frac{125(s+4)}{s+30}$$

or a PD controller

$$C_{\text{pd}}(s) = 5(s+5)$$

will satisfy the transient specifications, i.e., PO $\leq 20\%$ and $t_s \leq 1$ [sec]. Now we define a new plant model

$$P_{\text{lead}}(s) = P(s)C_{\text{lead}}(s) = \frac{1250(s+4)}{s(s+5)(s+10)(s+30)}$$

or

$$P_{\text{pd}}(s) = P(s)C_{\text{pd}}(s) = \frac{50}{s(s+10)}.$$

The root locus of $1 + KP_{\text{lead}}(s) = 0$ with the phase-lead controller $C_{\text{lead}}(s)$ is the same as the one shown in Figure 5.18.

We shall now design a phase-lag controller for $P_{\text{lead}}(s)$ so that the steady-state error specification is satisfied. The steady-state error specification requires $K_v \geq 20$. We shall take $K_v = 20$.

Following the phase-lag controller design procedure, we pick the desired closed-loop poles at

$$-31.8614, \quad -5 \pm j5, \quad -3.1386$$

which are the closed-loop poles with only the phase-lead controller $C_{\text{lead}}(s)$. Recall from Example 5.10 that we have $K_v = 10/3$ when the phase-lead controller is used. Hence, we need to increase the system gain by

$$K_s = \frac{20}{10/3} = 6.$$

After a few trials, we pick $b = 0.08$ (at least 10 times smaller than the dominant poles). Then $a = b/K_s = 0.08/6 = 0.01333$ (with $K_0 = 1$). Thus we have

$$C_{\text{lag}}(s) = \frac{s+b}{s+a} = \frac{s+0.08}{s+0.01333}$$

and finally the desired lead-lag controller is

$$C(s) = C_{\text{lead}}(s)C_{\text{lag}}(s) = \frac{125(s+4)}{s+30}\,\frac{s+0.08}{s+0.01333}.$$

The closed-loop poles with this lead-lag controller are at

$$-31.858, \quad -4.9821 \pm j4.9624, \quad -3.1094, \quad -0.0817.$$

Note that the closed-loop pole at -0.0817 is almost canceled by a closed-loop zero at -0.08, and therefore they will have almost no effect on system response. The step response shown in Figure 5.27 indicates that the design specifications are met.

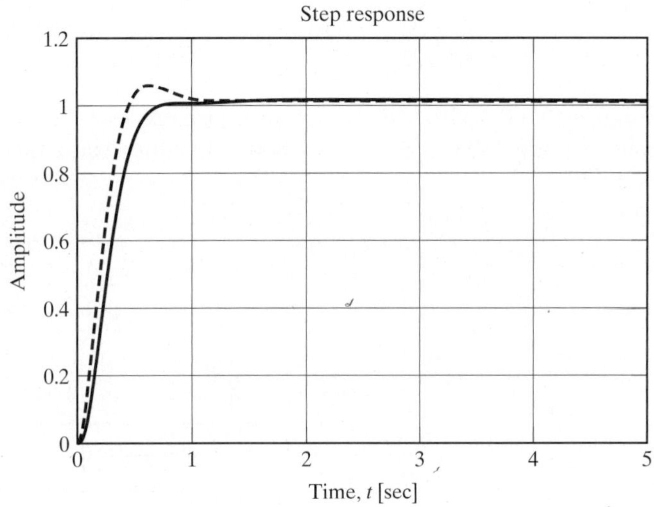

FIGURE 5.27: Step responses of Example 5.15 with a lead-lag controller (solid line) and a PID controller (dashed line).

Similarly, we can also design a PI controller for $P_{\text{pd}}(s)$ so that the steady-state error requirement is satisfied. Since a PI controller increases the system type by 1, the steady-state error requirement is automatically satisfied. We can simply pick a PI controller

$$C_{\text{pi}}(s) = \frac{s+0.08}{s}.$$

We then have a special PID controller as

$$C_{\text{pid}}(s) = C_{\text{pd}}(s)C_{\text{pi}}(s) = \frac{5(s+5)(s+0.08)}{s}.$$

The closed-loop poles with this PID controller are at

$$-4.9593 \pm j4.9597, \quad -0.0813.$$

The step response shown in Figure 5.27 indicates that the design specifications are also met.

5.9 2DOF CONTROLLERS

All design procedures discussed so far in this chapter are for one degree of freedom (1DOF) controllers. The design of a 2DOF controller can be carried out in two steps: first, a 1DOF controller, $C_2(s)$, is designed using the method described in previous sections; then the $C_1(s)$ part of the 2DOF controller $\boldsymbol{C}(s) = \begin{bmatrix} C_1(s) & C_2(s) \end{bmatrix}$ is chosen appropriately to satisfy other performance criteria. We shall now illustrate this design method through an example.

EXAMPLE 5.16

Consider the system and the 1DOF controller designed in Example 5.10:

$$P(s) = \frac{10}{s(s+5)(s+10)}, \quad C_2(s) = \frac{125(s+4)}{s+30}.$$

To design a 2DOF controller as shown in Figure 5.28, we can choose the 2DOF controller in the following form without changing the closed-loop characteristic polynomial:

$$\boldsymbol{C}(s) = \begin{bmatrix} C_1(s) & C_2(s) \end{bmatrix} = \frac{1}{s+30} \begin{bmatrix} as+b & 125(s+4) \end{bmatrix}$$

where the parameter a and b can be chosen to satisfy other performance criteria.

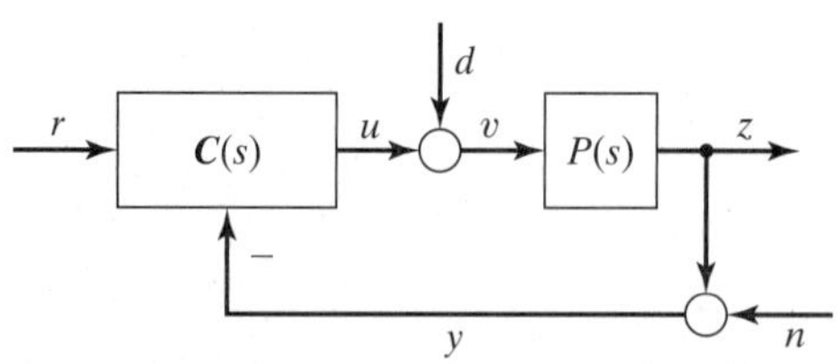

FIGURE 5.28: 2DOF feedback control system.

From Theorem 4.12, it is clear that we need to choose

$$b = 125 \times 4 = 500$$

in order for the system to track a step input.

With this b, the transfer function from r to z is given by

$$T_{zr}(s) = \frac{10(as+500)}{(s+31.8614)(s^2+10s+50)(s+3.1386)}.$$

If a fast step response is desirable, then a can be chosen to cancel the slow pole at -3.1386, i.e.,

$$a = 500/3.1386 = 159.3067.$$

Hence, we get a 2DOF controller

$$\boldsymbol{C}(s) = \frac{1}{s+30} \begin{bmatrix} 159.3067(s+3.1386) & 125(s+4) \end{bmatrix}$$

which gives the closed-loop transfer from r to z as

$$T_{zr}(s) = \frac{1593.067}{(s + 31.8614)(s^2 + 10s + 50)}.$$

Of course, the free parameters in controller $C_1(s)$ can be chosen to satisfy other performance criteria, and $C_1(s)$ can also take a much more complex form, in general.

5.10 GENERAL GUIDELINES IN ROOT-LOCUS DESIGN

One natural question we could raise is how one decides what type of controller should be used for a given control design problem. We shall now present some simple guidelines and procedures:

- Find the desired dominant pole regions from the transient requirements;
- Find the desired steady-state error constants from the steady-state tracking requirements;
- Decide whether you need to increase the system type in order to satisfy the design specifications: if the steady-state accuracy demands the increase of the system type, then PI or PID controllers must be used;
- Plot the root locus of $KP(s)$;
- Check whether the desired dominant poles can be selected on the root-locus plot: if yes, phase-lag or PI controllers can be used; if not, phase-lead, PD, PID, or lead-lag controllers must be used.

5.11 COMPLEMENTARY ROOT LOCUS

In all the preceding sections, the root-locus plots are constructed under the assumption that the varying parameters are positive. There are some cases where the parameters can be negative, in which cases the phase condition becomes

$$\angle L(s) = \angle K + \sum_{i=1}^{m} \angle(s - z_i) - \sum_{j=1}^{n} \angle(s - p_j) = (2k + 1)180^\circ, \quad k = 0, \pm 1, \pm 2, \ldots,$$

i.e.,

$$\boxed{\sum_{i=1}^{m} \angle(s - z_i) - \sum_{j=1}^{n} \angle(s - p_j) = k360^\circ, \qquad k = 0, \pm 1, \pm 2, \ldots}$$

since $\angle K = 180^\circ$ for $K < 0$.

Thus, the root-locus rules can be suitably modified as shown in Table 5.4. The root locus for a negative gain variation is called the **complementary root locus**.

1. The complementary root locus is symmetric with respect to the real axis.

2. The complementary root loci start from n poles p_i (when $K = 0$) and approach the n zeros (m finite zeros z_i and $n-m$ infinite zeros when $K \to -\infty$).

3. The complementary root locus includes all points on the real axis to the left of an **even** number of open-loop real poles and zeros.

4. As $K \to -\infty$, $n-m$ branches of the complementary root locus approach asymptotically $n-m$ straight lines (called asymptotes) with angles

$$\theta = \frac{k360^\circ}{n-m}, \quad k = 0, \pm 1, \pm 2, \ldots$$

and the starting point of all asymptotes is on the real axis at

$$\kappa = \frac{\sum_{i=1}^{n} p_i - \sum_{j=1}^{m} z_j}{n-m} = \frac{\sum \text{poles} - \sum \text{zeros}}{n-m}.$$

5. The breakaway points (where the root loci meet and split away, usually on the real axis) and the breakin points (where the root loci meet and enter the real axis) are among the roots of the equation: $\dfrac{dL(s)}{ds} = 0$. (On the real axis, only those roots satisfying Rule 3 are breakaway or breakin points.)

6. The departure angle ϕ_k (from a pole, p_k) is given by

$$\phi_k = \sum_{i=1}^{m} \angle(p_k - z_i) - \sum_{j=1, j\neq k}^{n} \angle(p_k - p_j).$$

(In the case p_k is l repeated poles, the departure angle becomes ϕ_k/l.)

The arrival angle ψ_k (at a zero, z_k) is given by

$$\psi_k = -\sum_{i=1, i\neq k}^{m} \angle(z_k - z_i) + \sum_{j=1}^{n} \angle(z_k - p_j).$$

(In the case z_k is l repeated zeros, the arrival angle becomes ψ_k/l.)

TABLE 5.4: Complementary root locus rules: $-\infty \leq K \leq 0$.

EXAMPLE 5.17

Consider the feedback system shown in Figure 5.2 with

$$L(s) = P(s)C(s) = \frac{2K(s+3)}{s(s+1)(s+2)}$$

and $K < 0$. We would like to see how the closed-loop poles change when K varies from 0 to $-\infty$.

From the complementary root-locus Rule 3, we know that there are root loci on the real axis in three intervals: $(-\infty, -3]$, $[-2, -1]$, and $[0, +\infty)$. By Rule 2, one root locus goes to zero at -3, and the other two root loci approach two infinite zeros with two asymptotes. The two asymptotes have angles

$$\theta = \frac{k360^\circ}{3-1} = 0^\circ, 180^\circ, \text{ for } k = 0, 1.$$

`z=-3;` vector of zeros

`p=[0,-1,-2];` vector of poles

`k=2;` gain (not the var iable gain K)

`L=zpk(z,p,k)` form the transfer function of the system

`rlocus(-L)` generate a complementary root-locus plot with an automatically chosen range of gain $K < 0$

With the above information, a rough root locus can be sketched. An accurate root-locus plot can be generated by using the MATLAB commands above. The root-locus plot is shown in Figure 5.29.

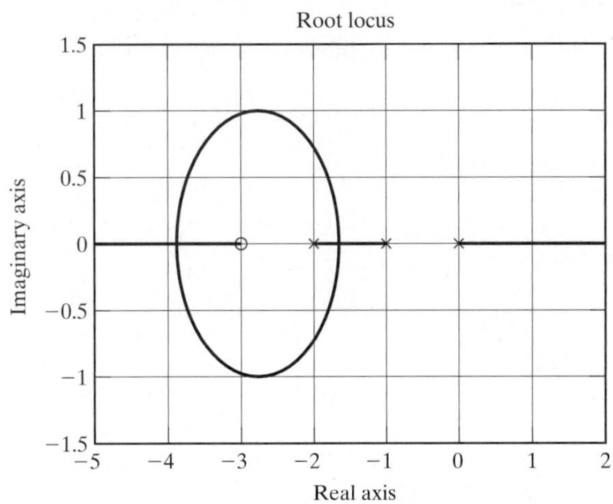

FIGURE 5.29: Complementary root locus for Example 5.17.

All controller design procedures described in the previous sections can also be applied in the same way as in the regular root-locus method.

5.12 STRONG STABILIZATION

Almost all controllers we have discussed so far are either stable, like lead, lag, and lead-lag controllers, or they may have poles at origin but otherwise stable like PI and PID controllers.

A natural question we would like to ask is, can any control system always be stabilized by a stable controller (or an almost stable controller with possible poles at origin)? The problem of stabilizing a system with a stable controller is called a **strong stabilization** problem. Since an unstable controller is generally undesirable in most practical applications, it is critically important to limit our controllers to be stable ones. Unfortunately, it is not always possible to stabilize an unstable system with a stable controller. Fortunately, there is a clear answer to the existence of a stable stabilizing controller for a given plant. Moreover, the root locus can be used to explain the nonexistence of such controllers.

Consider a plant $P(s)$ with real right half plane zeros at

$$0 \leq z_1 < \cdots < z_m$$

where $z_m = \infty$ is counted as a zero if $P(s)$ is strictly proper. Note that the complex unstable zeros of the plant, if any, are not counted.

Parity interlacing property (p.i.p.): A plant $P(s)$ is said to satisfy the p.i.p. if the number of unstable poles between every pair of real right half plane zeros of $P(s)$ is even.

EXAMPLE 5.18

Consider the following set of plants

$$P_1(s) = \frac{K(s-1)(s-4)}{(s-2)(s+5)}$$

$$P_2(s) = \frac{K(s-2)(s+5)}{(s-1)(s-4)}$$

$$P_3(s) = \frac{K(s-2)(s+5)}{(s-1)(s-4)(s+3)}$$

$$P_4(s) = \frac{K(s-1)(s+5)}{(s-2)(s-3)(s+4)}$$

$$P_5(s) = \frac{K(s-1)(s-4)(s+5)}{(s-2)(s-3)(s+4)(s+6)}.$$

Their pole-zero configurations are shown in Figure 5.30.

- $P_1(s)$ does not satisfy p.i.p.: $P_1(s)$ has two real right half plane zeros at 1 and 4, and there is one right half plane pole at 2, which is between these two real right half plane zeros.

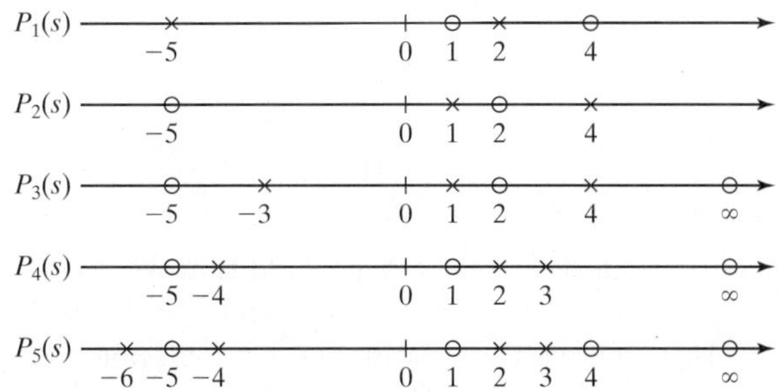

FIGURE 5.30: Pole-zero configurations.

- $P_2(s)$ satisfies p.i.p.: $P_2(s)$ has only one real right plane zero at 2 and there is no zero at $+\infty$ since $P_2(s)$ is not strictly proper. Hence, the p.i.p. is satisfied since there is no pair of real right half plane zeros.
- $P_3(s)$ does not satisfy p.i.p.: $P_3(s)$ has one finite real right plane zero at 2 and a zero at $+\infty$ since $P_3(s)$ is strictly proper. Since the real right half plane pole at 4 falls between 2 and $+\infty$, the p.i.p. is violated.
- $P_4(s)$ satisfies p.i.p.: $P_4(s)$ has one finite real right plane zero at 1 and a zero at $+\infty$ since $P_4(s)$ is strictly proper. Since there are two real right half plane poles at 2 and 3 between 1 and $+\infty$, the p.i.p. is satisfied.
- $P_5(s)$ satisfies p.i.p.: $P_5(s)$ has two finite real right plane zeros at 1 and 4 and a zero at $+\infty$ since $P_5(s)$ is strictly proper. Since there are two real right half plane poles at 2 and 3 between 1 and 4, the p.i.p. is satisfied.

EXAMPLE 5.19

Consider an unstable plant

$$P_1(s) = \frac{(s-1)(s-4)}{(s-2)(s+5)}$$

This plant cannot be stabilized by any stable controller. That is to say, we must use an unstable controller to stabilize this plant. The reason is that if the controller is stable, then the root-locus (or the complementary root-locus) method shows that there must be a closed-loop pole in either the interval $[1, 2]$ or the interval $[2, 4]$, with either negative feedback or positive feedback. This is true no matter where the controller zeros are located as long as the controller poles are in the closed left half plane. A simple sketch of the root locus or complementary root locus for this system with any controller poles/zeros configuration, as long as the controller poles are stable or marginally stable, will show that there is always an entire branch of root locus in either the interval $[1, 2]$ or the interval $[2, 4]$. Hence a stable (or a marginally stable) controller cannot stabilize this system.

To find a stabilizing controller for this system, we assume that the stabilizing controller takes the following form:

$$C_1(s) = \frac{q_0 s + q_1}{p_0 s + p_1}.$$

It is again clear from the root-locus method that we need to choose a controller pole in the interval $(1, 4)$ with the hope of stabilizing the system with this controller. This is because if there is an additional open-loop pole in the interval $(1, 4)$, then on the complementary root locus there is the possibility that none of the closed-loop roots is completely restricted on the positive real axis. It remains to determine the exact ranges of these controller parameters. This can be done by applying the pole placement design procedure in Chapter 3.

The closed-loop characteristic polynomial with this controller is given by

$$\begin{aligned} c(s) &= (s-2)(s+5)(p_0 s + p_1) + (s-1)(s-4)(q_0 s + q_1) \\ &= (p_0 + q_0)s^3 + (3p_0 + p_1 - 5q_0 + q_1)s^2 \\ &\quad + (-10p_0 + 3p_1 + 4q_0 - 5q_1)s + (-10p_1 + 4q_1). \end{aligned}$$

Now suppose that it is desirable to place the closed-loop poles at $-1, -1, -1$. We then need

$$c(s) = (s+1)^3 = s^3 + 3s^2 + 3s + 1.$$

Comparing coefficients, we get

$$\begin{bmatrix} 1 & 0 & 1 & 0 \\ 3 & 1 & -5 & 1 \\ -10 & 3 & 4 & -5 \\ 0 & -10 & 0 & 4 \end{bmatrix} \begin{bmatrix} p_0 \\ p_1 \\ q_0 \\ q_1 \end{bmatrix} = \begin{bmatrix} 1 \\ 3 \\ 3 \\ 1 \end{bmatrix}$$

which gives

$$\begin{bmatrix} p_0 \\ p_1 \\ q_0 \\ q_1 \end{bmatrix} = \begin{bmatrix} 2.7593 \\ -4.0926 \\ -1.7593 \\ -9.9815 \end{bmatrix}$$

and

$$C_1(s) = \frac{q_0 s + q_1}{p_0 s + p_1} = \frac{-1.7593s - 9.9815}{2.7593s - 4.0926} = \frac{-0.6376(s + 5.673)}{s - 1.4832}.$$

This controller $C_1(s)$ is obviously unstable itself.

The above observation can be generalized to a much more general case.

Theorem 5.20. A plant can be stabilized by a stable controller if and only if the plant satisfies the p.i.p. condition.

While the root-locus method can be used to judge whether the plant cannot be stabilized by a stable controller from its unstable pole-zero configuration, it is much more difficult to show that a plant can, indeed, be stabilized by a stable controller when the p.i.p. condition is satisfied, which typically requires the construction of a stable stabilizing controller.

EXAMPLE 5.21

Here we consider another plant

$$P_2(s) = \frac{(s-2)(s+5)}{(s-1)(s-4)}.$$

It has been shown in Example 5.18 that this system satisfies the p.i.p. Hence, it is possible to stabilize this system with a stable controller. It is also easy to see by the root-locus method that we need a controller zero in the interval $[1, 4]$ in order to stabilize the system. Therefore, we shall assume that the controller takes the following form:

$$C_2(s) = \frac{-K(s-z)}{s+5}$$

with $1 < z < 4$. The root locus of the system with $z = 2.5$ is shown in Figure 5.31 and shows that the closed-loop system is stable for $K > 1.1116$. On the other hand, the root locus of the system with $z = 3.5$ in Figure 5.32 shows that the closed-loop system is not stable for any controller gain. In fact, it can be shown that the system can always be stabilized for $1 < z < 3$ for a sufficiently large K and that the system cannot be stabilized if $z \geq 3$.

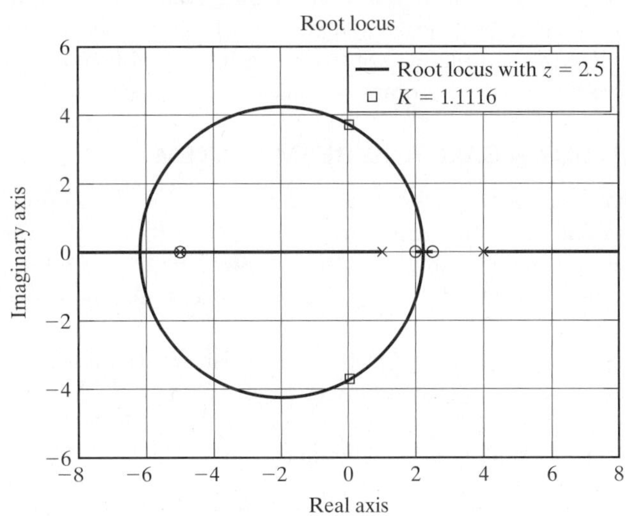

FIGURE 5.31: Root-locus for $L_2(s) = P_2(s)C_2(s)$ with $z = 2.5$.

Of course, we could have also designed a stable stabilizing controller using the pole placement design procedure in Chapter 3 as in the last example. Comparing

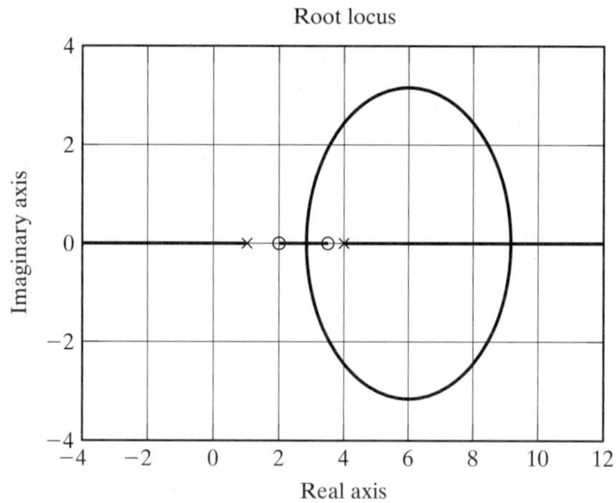

FIGURE 5.32: Root-locus for $L_2(s) = P_2(s)C_2(s)$ with $z = 3.5$.

with Example 5.19 and noting that $P_2(s) = 1/P_1(s)$, it is not difficult to see that a controller

$$C_{21}(s) = 1/C_1(s) = \frac{2.7593s - 4.0926}{-1.7593s - 9.9815} = \frac{-1.5684(s - 1.4832)}{s + 5.673}$$

will stabilize this plant with closed-loop poles at $-1, -1, -1$.

In general, the order of a stabilizing controller, whether it is itself stable or not, may need to be at least as high as $n-1$ where n is the order of the plant. Hence, finding a stable stabilizing controller for an unstable and high-order plant with the p.i.p. condition may be quite complicated.

5.13 CASE STUDY – BALL AND BEAM SYSTEM

An advantage of the root-locus design method over the pole placement method in Chapter 3 is that the former often leads to controllers of lower orders. Let us revisit the ball and beam system modeled in Section 2.10.1 and studied in Section 3.9.1. We will first consider the stabilization problem and then consider the step tracking problem.

For the stabilization problem, we wish to design a DISO controller, as shown in Figure 3.19 and redrawn here as Figure 5.33, such that the ball on the beam can be brought back to the origin automatically whenever it has deviated away from the origin for some reason. The mathematical model of the system is given by

$$\begin{bmatrix} \Theta(s) \\ X(s) \end{bmatrix} = \begin{bmatrix} P_\theta(s) \\ P_x(s) \end{bmatrix} U(s) = \frac{1}{s^3(s+50)} \begin{bmatrix} s^2/75 \\ g/750 \end{bmatrix} U(s).$$

As we have done in Section 3.9.1, we will first design $C_\theta(s)$ to stabilize $P_\theta(s)$ and then design $C_x(s)$ to stabilize the partially controlled system with the plant and $C_\theta(s)$. The root locus of $P_\theta(s)$ is shown in Figure 5.34. Clearly, any proportional

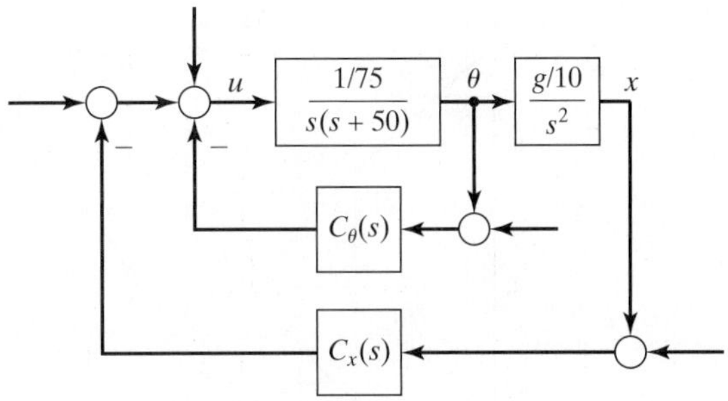

FIGURE 5.33: The ball and beam stabilization control system.

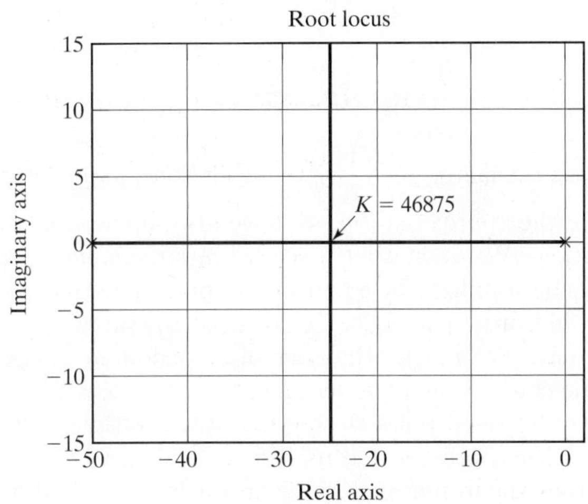

FIGURE 5.34: The root locus of $P_\theta(s)$.

controller $C_\theta(s) = K$ with $K > 0$ stabilizes the system. We choose $K = 46875$ to push the closed-loop poles away from the imaginary axis as much as possible, i.e., to maximize the speed of the response. The same proportional feedback was obtained using the pole placement method in Section 3.9.1 rather accidentally, whereas it comes out naturally here.

With the controller $C_\theta(s) = 46875$ connected, the partially controlled system, named $P_{\text{tmp}}(s)$, becomes

$$P_{\text{tmp}}(s) = \frac{g/750}{s^2(s+25)^2}.$$

Its root locus is shown in Figure 5.35. It is seen that a proportional controller is not able to stabilize this system. Since we are interested in the simplest of controllers, the next attempt would be to see whether a first-order controller is able to stabilize the system. Observe that this system has two dominant poles at the origin. Hence it can be approximated by a double integrator system

$$\tilde{P}_{\text{tmp}}(s) = \frac{1/(75 \times 25^2)}{s^2}$$

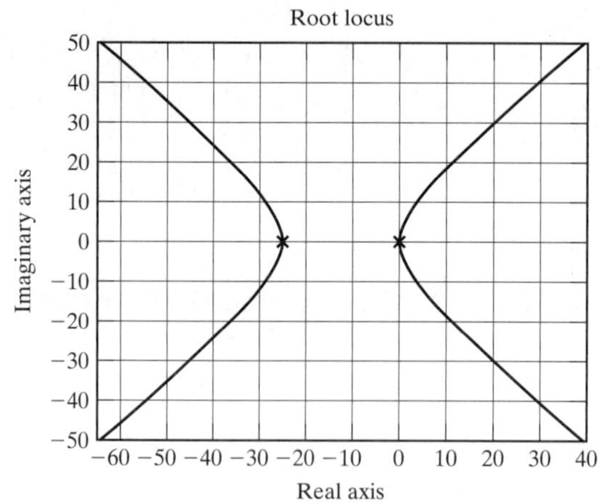

FIGURE 5.35: The root locus of $P_{\text{tmp}}(s)$.

by removing the factor $\dfrac{25^2}{(s+25)^2}$, which is considered being close to 1 for reasonably small s. In the approximation, we have also replaced g by 10 to simplify numerical manipulation. We know that a second-order system, such as a double integrator, can always be stabilized by a first-order controller. Hence we would also expect that the same first-order controller can be used to stabilize the system it approximates.

We have dealt with the control of a double integrator system quite a few times. One can use the pole placement method to design a first-order controller to place the closed-loop poles to the desirable positions. One can also guess the pole and zero of the controller and use the root locus to decide the gain. We would like to argue that the former method is an analytic method whereas the latter method is a trial-and-error method. Whenever both methods are available, the analytic method is preferred.

To use the pole placement method, let us specify that the closed-loop poles to be $-1 \pm j$ and -1 and assume that the controller has the form

$$C_x(s) = \frac{q_0 s + q_1}{p_0 s + p_1}.$$

Solving the Diophantine equation with respect to $\tilde{P}_{\text{tmp}}(s)$ and $C_x(s)$

$$s^2(p_0 s + p_1) + \frac{1}{75 \times 25^2}(q_0 s + q_1) = (s+1)(s^2 + 2s + 2),$$

we obtain

$$p_0 = 1, \quad p_1 = 3, \quad q_0 = 187500, \quad q_1 = 93750.$$

It is now crucial to see whether the designed controller $C_x(s)$ stabilizes the "true" plant $P_{\text{tmp}}(s)$. Simple checking shows that it does with the "true" closed-loop poles $-25.1779 \pm j2.0583$, $-0.8671 \pm j1.1653$, and -0.9099. Hence a stabilizing controller is designed as

$$\begin{bmatrix} C_\theta(s) & C_x(s) \end{bmatrix} = \begin{bmatrix} 46875 & \dfrac{187500s + 93750}{s+3} \end{bmatrix} = 46875 \begin{bmatrix} 1 & \dfrac{4s+2}{s+3} \end{bmatrix}.$$

Next, let us solve the step tracking problem. We wish to design a controller so that the ball rolls to a position given by a step reference $r_x(t)$ from whatever initial condition without overshoot. We are not quite interested in the tracking of the variable $\theta(t)$. Hence there is no need to modify the motor controller $C_\theta(s)$. Since the partially controlled plant with the inner loop connected $P_{\text{tmp}}(s)$ is of type 2, a 2DOF controller for the $x(t)$ variable has to be used to avoid overshoot. The block diagram of the closed-loop system is shown in Figure 5.36. Again, we are interested in controllers with the lowest order. Hence we start with a 2DOF controller $\boldsymbol{C}_x(s)$ having the same order as the stabilization controller $C_x(s)$

$$\boldsymbol{C}_x(s) = \begin{bmatrix} C_{x1}(s) & C_{x2}(s) \end{bmatrix} = 46875 \frac{\begin{bmatrix} qs+2 & 4s+2 \end{bmatrix}}{s+3}.$$

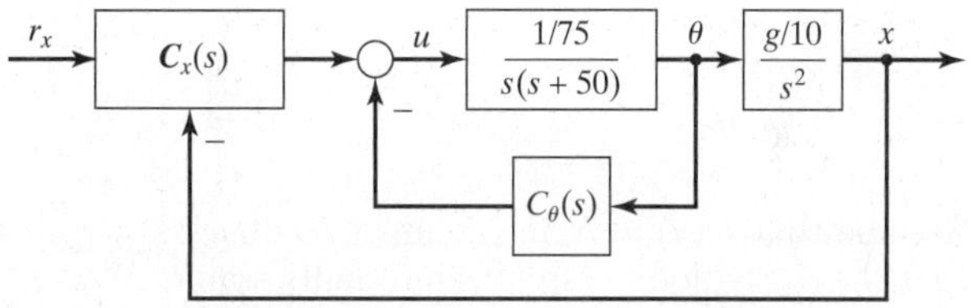

FIGURE 5.36: The ball and beam step tracking control system.

Here, $C_{x2}(s)$ is simply taken as the stabilization controller $C_x(s)$. The constant term of the numerator polynomial of $C_{x1}(s)$ is set to be the same as that of $C_{x2}(s)$ because of the step tracking requirement, prescribed by Theorem 4.12. Figure 5.37 shows the closed-loop step responses for several different values of q. The simple choice of $q = 0$ appears to be a good one. As a result, the final design of the controller for step tracking is

$$\begin{bmatrix} C_\theta(s) & \boldsymbol{C}_x(s) \end{bmatrix} = 46875 \begin{bmatrix} 1 & \dfrac{2}{s+3} & \dfrac{4s+2}{s+3} \end{bmatrix}.$$

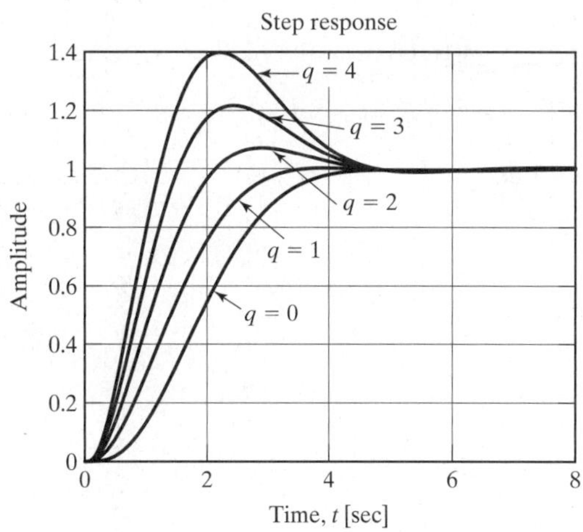

FIGURE 5.37: Closed-loop step responses for various values of q.

PROBLEMS

5.1. Sketch the root loci for a unity feedback system with an open-loop transfer function

1. $L(s) = \dfrac{K(s+1)}{s(s+5)(s+10)}$

2. $L(s) = \dfrac{10(Ts+1)}{s^2(s+2)}$

3. $L(s) = \dfrac{5(s-1)(s+1)}{(s+a)(s+10)(s+20)}$

where K, T, and a are nonnegative parameters.

5.2. Plot the root locus of $L(s) = \dfrac{K}{s(s^2+6s+12)}$. Show that the root locus consists of straight lines.

5.3. Sketch the root locus of

$$P(s) = \frac{1}{s(s+4)(s^2+4s+8)}.$$

Assume that $C(s) = K$ in the unity feedback system. Find the range of K such that the closed-loop system is internally stable. What is the value of K such that persistent oscillation occurs in the impulse response? What is the frequency of the persistent oscillation?

5.4. Plot the root locus of $P(s) = \dfrac{s+2}{s(s+1)}$. Now, let $C(s) = K$ and consider the feedback system shown in Figure 5.38. Design K so that the closed-loop poles have the largest ratio of the imaginary part over the real part. What are the corresponding closed-loop poles?

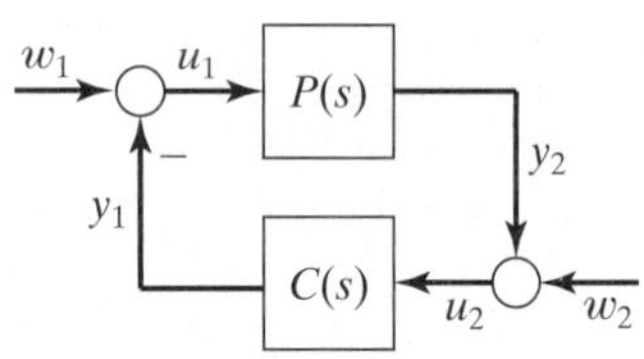

FIGURE 5.38: Feedback system for stabilization.

5.5. Prove that the root locus of

$$L(s) = \frac{s-z}{(s-p_1)(s-p_2)}$$

where z, p_1, p_2 are real numbers with $z < p_1$ and $z < p_2$, contains a circle centered at z with radius $\sqrt{|p_1 - z||p_2 - z|}$.

5.6. A feedback control system shown in Figure 5.39 is said to have infinite gain margin if the system is stable for all $K \in [1, \infty)$.

1. Design a simple controller for

$$P(s) = \frac{1}{s(s-1)}$$

such that the closed-loop system has infinite gain margin.

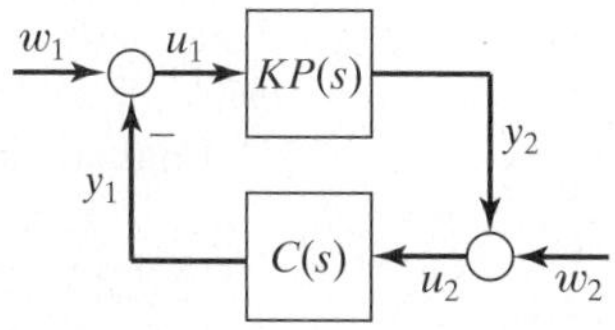

FIGURE 5.39: Feedback system with gain uncertainty.

2. Using root-locus arguments, show that if $P(s)$ has a zero on the open right-hand side of the complex plane, then there exists no nonzero controller $C(s)$ such that the closed-loop system has infinite gain margin.

5.7. In Figure 5.38, let

$$P(s) = \frac{1}{(s+2)(s+3)} \quad \text{and} \quad C(s) = K\frac{s+a}{s+b}$$

Using root-locus techniques, determine a and b that will produce closed-loop poles at $s = -1 \pm 1j$.

5.8. Given a double integrator plant $P(s) = \frac{1}{s^2}$, design a lead compensator so that the closed system has dominant poles at $-1 \pm j\sqrt{3}$.

5.9. Consider a unity feedback system with a plant model given by

$$P(s) = \frac{10(s-5)}{s(s^2+s+4)}$$

and a controller given by

$$C(s) = \frac{K(s+b)}{s+a}$$

for $K > 0$ and some real b and a.

- Use the root-locus technique to determine the signs of b and a so that the closed-loop system is stable for all $K \in (0, K_u)$ for some $K_u > 0$.
- Sketch the possible forms of the root locus in terms of the pole and zero locations of $C(s)$.

5.10. Let $P(s) = \frac{1}{s-1}$. Design a PI controller so that the velocity error constant is 5 and the closed-loop system has two repeated poles. Where is the location of the closed-loop poles? Is the closed-loop system stable?

5.11. Let $P(s) = \frac{1}{s^2}$. Design a PID controller so that the closed-loop system has poles at $-1 \pm j$ and -2.

5.12. Consider a feedback system with a plant model

$$P(s) = \frac{(s-1)(s-3)}{(s-2)(s+1)^2}.$$

Use the root-locus technique to show that the system cannot be stabilized by a stable controller.

5.13. Find a stable controller that will stabilize the following plant:

$$P_4(s) = \frac{K(s-1)(s+5)}{(s-2)(s-3)(s+4)}$$

5.14. One way to design a controller is to cancel some stable but undesirable poles and zeros. However, it is, in general, impossible to get an exact cancelation. This can be very critical in the case of lightly damped poles. Assume we have a plant model given by

$$P(s) = \frac{K(s+1)(s+2)}{s[(s+0.1)^2 + a^2]}, \quad a \in [3, 5].$$

Now let $C_1(s)$ and $C_2(s)$ be two controllers given by

$$C_1(s) = \frac{(s+0.1)^2 + 5^2}{(s+1)(s+2)}$$

$$C_2(s) = \frac{(s+0.1)^2 + 3^2}{(s+1)(s+2)}.$$

Using the root-locus technique to show that $C_1(s)$ may not be able to stabilize the plant for all $a \in [3, 5]$, whereas $C_2(s)$ stabilizes the plant for all $a \in [3, 5]$.

MATLAB PROBLEMS

5.15. Use MATLAB to solve this problem.

1. Plot the root locus of the system

$$\frac{K}{s(s+1)(s+a)}$$

for $a = 0$, $\frac{1}{2}$, 1, and 2. Compare it with the root locus of

$$\frac{K}{s(s+1)}.$$

Use the same coordinate ranges in all plots. Observe that an extra pole tends to shift the root locus to the right and the larger the extra pole, the greater the effect.

2. Plot the root locus of system

$$\frac{K(s+b)}{s(s+1)^2}$$

for $b = 10$, 5, 1, and 0.5. Compare it with the root locus of

$$\frac{K}{s(s+1)^2}.$$

Use the same coordinate ranges in all plots. Observe that an extra zero tends to shift the root locus to the left and the larger the zero, the greater the effect.

3. Plot the root locus of system

$$\frac{K(s+b)}{s(s+1)(s+a)}$$

for $a = 0.5$, $b = 2$ and $a = 2$, $b = 0.5$. Compare it with the root locus of

$$\frac{K}{s(s+1)}.$$

Observe

- if $a < b$, the effect of the pole at $-a$ is stronger than the effect of the zero at $-b$, which makes the root locus shift to the right;
- if $a > b$, the effect of the zero at $-b$ is stronger than the effect of the pole at $-a$, which makes the root locus shift to the left.

5.16. Sketch the root locus of the system

$$L(s) = \frac{K(s^2 + a^2)}{s(s^2 + 10^2)}$$

for $a = 1, 10/3, 8$, and 10, respectively. Compute angles of departure from complex poles, angles of arrival at complex zeros, and breakaway and breakin points (if applicable). Use MATLAB to verify your results.

NOTES AND REFERENCES

The root-locus method was invented by Walter R. Evans in 1948 while he was lecturing his colleagues on the analysis of servo-mechanisms.

W. R. Evans, "Graphical analysis of control systems," *Transactions of the American Institute of Electrical Engineers,* vol. 67, pp. 547–551, 1948.

W. R. Evans, "Control system synthesis by root locus method," *Transactions of the American Institute of Electrical Engineers,* vol. 69, pp. 66–69, 1950.

W. R. Evans, *Control-System Dynamics*, McGraw-Hill, New York, 1954.

The root-locus method has since been included in almost every control textbook. The root-locus method, together with the frequency-domain method by Bode and Nyquist to be presented in the subsequent chapters, form the core of what is called the classical control method.

The parity interlacing property for strong stabilization was first shown in

D. C. Youla, J. J. Bongiorno Jr., and C. N. Lu, "Single-loop feedback stabilization of linear multivariable plants," *Automatica*, vol. 10, pp. 159–173, 1974.

CHAPTER 6

Frequency-Domain Analysis

We have seen from Chapter 4 that the steady-state response of a stable system with respect to a sinusoidal input is still a sinusoid with the magnitude amplified by a factor and the phase shifted by a certain amount. The amplification factor and the phase shift vary with the frequency of the sinusoidal input and are respectively called **the magnitude frequency response** and **the phase frequency response** of the system. The frequency response of a system contains all information about the system and the controller design can be done on the basis of only the frequency response. One might wonder the advantage of using the frequency response method instead of other methods. The fact is that in many cases the frequency response method is more convenient. First of all, frequency response model of a system can often be obtained experimentally for a stable system even though an analytic model may not be available. Second, controller design can be done completely on the basis of frequency response without using a transfer function description of the system; this method will be explored in detail in the next chapter. Third, there are good correlations between time-domain specifications and frequency-domain specifications. And finally, some design specifications are explicitly stated in terms of frequency response, such as bandwidth and tolerance of phase delays.

6.1 FREQUENCY RESPONSE

Recall from Chapter 4 that the steady-state response of a stable dynamical system $P(s)$ with respect to a sinusoidal input $r(t) = (\sin \omega t)\sigma(t)$ is

$$y_{ss}(t) = |P(j\omega)| \sin(\omega t + \angle P(j\omega))\sigma(t)$$

where $\angle P(j\omega) = \tan^{-1} \dfrac{\text{Im}P(j\omega)}{\text{Re}P(j\omega)}$.

We shall now consider some specific examples.

EXAMPLE 6.1

Let a plant model be given by

$$P(s) = \frac{10}{(s+1)(s+2)}$$

and let the input to the plant be a sinusoidal signal $r(t) = (\sin \omega t)\sigma(t)$ for some given frequency ω. Then

$$Y(s) = \frac{10\omega/(1+\omega^2)}{s+1} - \frac{10\omega/(4+\omega^2)}{s+2} + \frac{s\text{Im}P(j\omega) + \omega\text{Re}P(j\omega)}{s^2+\omega^2}$$

and the transient response and the steady-state response are respectively

$$y_t(t) = \left(\frac{10\omega}{1+\omega^2}e^{-t} - \frac{10\omega}{4+\omega^2}e^{-2t}\right)\sigma(t)$$

and

$$\begin{aligned} y_{ss}(t) &= [\text{Im}\, P(j\omega) \cos \omega t + \text{Re}\, P(j\omega) \sin \omega t]\, \sigma(t) \\ &= |P(j\omega)| \sin(\omega t + \angle P(j\omega))\sigma(t). \end{aligned}$$

Some typical time responses for $\omega = 1, 2, 5, 10$ are shown in Figure 6.1. It is clear that the transient part goes away very quickly and what one can see are the sustained sinusoidal steady-state responses. Table 6.1 gives the magnitudes and phase shifts of the steady-state responses with different frequencies.

However, if the system is not stable, for example, let

$$P(s) = \frac{1}{s-1},$$

then

$$y(t) = \left[\frac{10\omega}{1+\omega^2}e^{t} + |P(j\omega)| \sin(\omega t + \angle P(j\omega))\right]\sigma(t).$$

It is impossible to observe the sinusoidal response $|P(j\omega)| \sin(\omega t + \angle P(j\omega))$ experimentally from the time response of $y(t)$ since the e^t term does not go to zero with time.

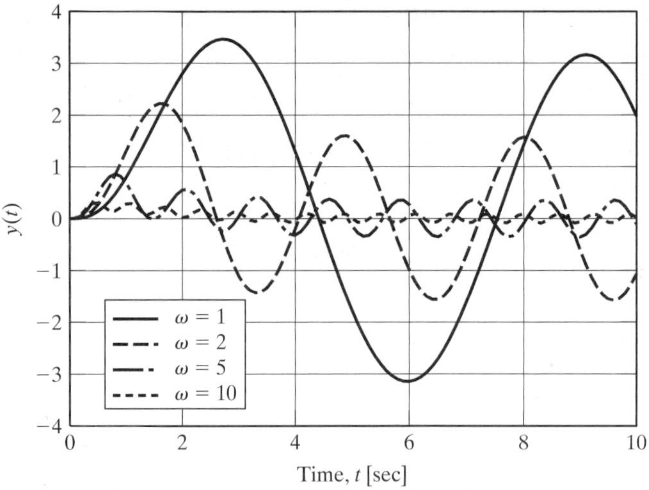

FIGURE 6.1: Time response to sinusoidal signals in Example 6.1.

ω	1	2	5	10
$\lvert P(j\omega)\rvert$	3.1623	1.5811	0.3642	0.0976
$\angle P(j\omega)$	−71.5651	−108.4349	−146.8887	−162.9795

TABLE 6.1: Frequency response.

The above observation has very practical and important implications in system identification of stable systems. Suppose that a stable dynamical system has the transfer function $P(s)$ which is not analytically available. We shall apply a sequence of sinusoidal signals $r(t) = A(\omega)\sin\omega t$ with a range of frequencies ω to the system. Next, we shall measure the steady-state response of the output, which is $y(t) = B(\omega)\sin(\omega t + \phi(\omega))$ for some $B(\omega)$ and $\phi(\omega)$. We can then obtain the following data:

$$|P(j\omega)| = \frac{B(\omega)}{A(\omega)}, \quad \angle P(j\omega) = \phi(\omega).$$

We shall call $|P(j\omega)|$ the **magnitude frequency response** and $\angle P(j\omega)$ the **phase frequency response**. We shall see in the subsequent sections and chapters that most control system analyses and designs can be accomplished on the basis of only frequency response without the analytic model. This method can be used to generate an approximate frequency response model in an interval of frequencies even if the original system is not linear.

Theoretically, $A(\omega)$ can be taken to be a constant for all frequencies without changing the results. However, there are some practical reasons why the magnitude of the input signal $A(\omega)$ should not be taken constant over all frequencies. One factor to be considered is measurement noise, which is usually significant in the high-frequency range. Since most dynamical systems have significantly small gain in the high-frequency range and the output measurement to high-frequency signals can be very noisy, the output measurement $B(\omega)$ would be very small and inaccurate if $A(\omega)$ is small. Hence, the frequency response $B(\omega)/A(\omega)$ and $\phi(\omega)$ would be far less accurate. Therefore, it is desirable to increase the input signal magnitude (i.e., the signal-to-noise ratio) in the high-frequency range to reduce the modeling error.

Next, we consider the system response to a periodic, but nonsinusoidal signal. Let $f(t)$ be a bounded periodic signal with period $2T$ and suppose it is piecewise differentiable with $f(t) = [f(t^+) + f(t^-)]/2$ at jumps. Then the signal admits a Fourier series expansion and the frequency response discussed in the previous example can be applied to each component of the Fourier series.

EXAMPLE 6.2

Let $f(t)$ be a square wave given by

$$f(t) = \begin{cases} 1, & 2kT < t < (2k+1)T, \\ -1, & (2k+1)T < t < 2(k+1)T, \\ 0, & t = kT \end{cases}$$

for $k = 0, 1, 2, \ldots$. Note that we have defined the function at the jumps to be zero so that there is an exact Fourier series representation. Thus, this function has the following Fourier series expansion:

$$f(t) = \sum_{n=0}^{\infty} \frac{4}{(2n+1)\pi} \sin \frac{(2n+1)\pi}{T} t.$$

We now consider the response of a stable system $P(s)$ with the input signal $r(t) = f(t)$. The steady-state response can then be written as

$$y_{ss}(t) = \sum_{n=0}^{\infty} \frac{4}{(2n+1)\pi} \left| P(j\frac{(2n+1)\pi}{T}) \right| \sin \left(\frac{(2n+1)\pi}{T} t + \angle P(j\frac{(2n+1)\pi}{T}) \right).$$

On the other hand, one can take the first few terms of the Fourier series of $f(t)$ as a rough approximation. For example, take

$$f_a(t) = \sum_{n=0}^{N} \frac{4}{(2n+1)\pi} \sin \frac{(2n+1)\pi}{T} t.$$

The steady-state response of the system $P(s)$ with respect to $f_a(t)$ is then

$$y_{ass}(t) = \sum_{n=0}^{N} \frac{4}{(2n+1)\pi} \left| P(j\frac{(2n+1)\pi}{T}) \right| \sin \left(\frac{(2n+1)\pi}{T} t + \angle P(j\frac{(2n+1)\pi}{T}) \right).$$

The function $f(t)$ and the approximation $f_a(t)$ with $T = 5$ and $N = 4$ are shown in Figure 6.2.

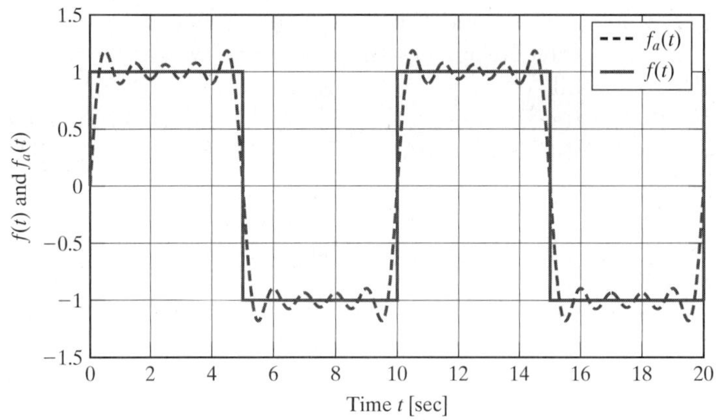

FIGURE 6.2: $f(t)$ and an approximation $f_a(t)$ with $T = 5$ and $N = 4$ in Example 6.2.

Now let $P(s)$ again be given by

$$P(s) = \frac{10}{(s+1)(s+2)}.$$

The steady-state responses of the system $P(s)$ with respect to $f(t)$ and $f_a(t)$ with $N = 0$ and $N = 1$ are then shown in Figure 6.3.

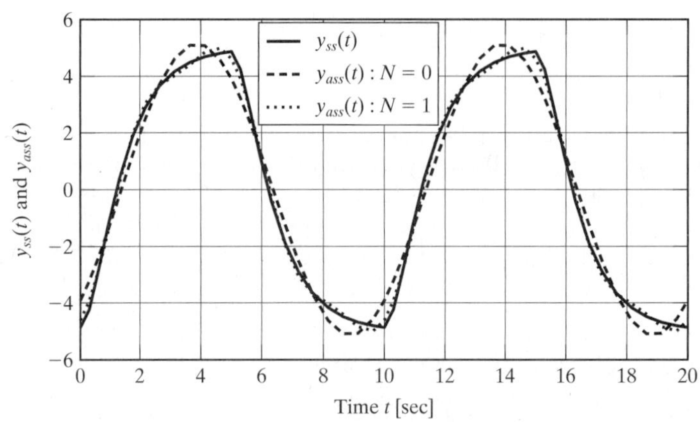

FIGURE 6.3: $y_{ss}(t)$ and $y_{ass}(t)$ with $N = 0$ and $N = 1$ in Example 6.2.

It can then be seen that the first few frequency components play the most important role in the steady-state response where the signal has the most strength and the frequency response of the plant has significantly larger magnitude than at the higher frequencies. Hence, even though the input signal $r(t) = f(t)$ itself cannot be accurately approximated by the first few frequency components, the steady-state response can be fairly accurately approximated by considering only the first few frequency components if the transfer function has much smaller magnitude frequency response at higher frequencies.

Observations:

(**1.**) The steady-state response of a system to a signal depends on two factors: (i) the signal strength at the frequency components, and (ii) the magnitude frequency response of the system at those frequency components.

(**2.**) The step response can be regarded as a limiting case when $T \to \infty$. Hence it is clear that the frequency response near 0 frequency plays the most critical role in this response since $\frac{(2n+1)\pi}{T} \to 0$ as $T \to \infty$ for any fixed n.

When a signal $f(t)$ is not periodic, it cannot, in general, be expanded into discrete frequency components. However, the Fourier (or Laplace) transform may be applied under some mild conditions so that the signal can be regarded as having continuous frequency components. Thus, all the frequency response concepts discussed above can be applied.

EXAMPLE 6.3

Consider the unity feedback system shown in Figure 6.4, where $P(s)$ denotes the plant transfer function, $C(s)$ the controller transfer function, $r(t)$ the reference input, $z(t)$ the output of the system, $y(t)$ the measurement, $d(t)$ the disturbance, and $n(t)$ the sensor noise.

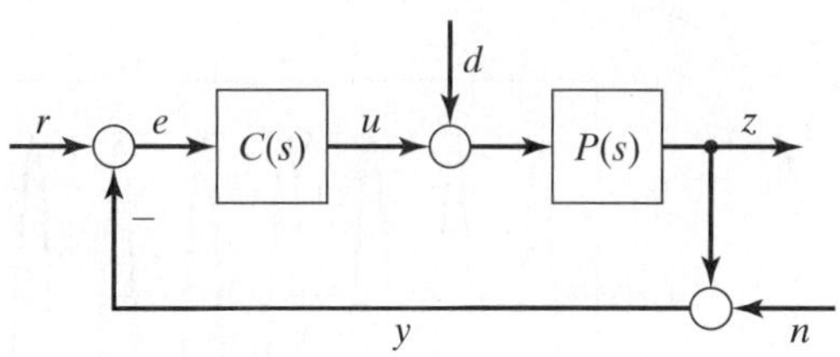

FIGURE 6.4: Standard unity feedback system.

Now, denote the transfer function from d to z as $T_{zd}(s)$. Then

$$T_{zd}(s) = \frac{P(s)}{1 + P(s)C(s)}.$$

Let $P(s)$ and $C(s)$ be given by

$$P(s) = \frac{10}{s(s+5)(s+10)}, \quad C(s) = \frac{90(s+5)}{s+20}.$$

Then

$$T_{zd}(s) = \frac{P(s)}{1 + P(s)C(s)} = \frac{10(s+20)}{s(s+10)(s+20)+900}.$$

If the disturbance $d(t) = f(t)$ is a square wave as given in Example 6.2, the steady-state response of the system with respect to $d(t)$ is given by

$$z_{ss}(t) = \sum_{n=0}^{\infty} \frac{4}{(2n+1)\pi} \left| T_{zd}(j\frac{(2n+1)\pi}{T}) \right| \sin\left(\frac{(2n+1)\pi}{T} t + \angle T_{zd}(j\frac{(2n+1)\pi}{T}) \right).$$

Figure 6.5 shows the steady-state response to a relatively slow square wave with $T = 5$. The corresponding frequencies of the signal components in this case are

$$\frac{(2n+1)\pi}{T} = \frac{\pi}{5}, \frac{3\pi}{5}, \frac{5\pi}{5}, \ldots, \quad n = 0, 1, 2 \ldots.$$

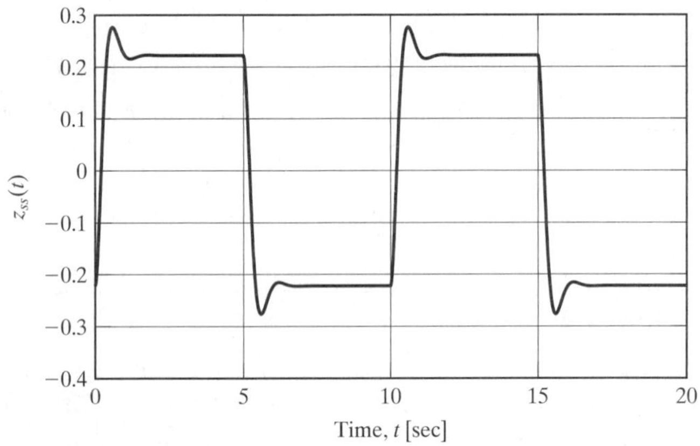

FIGURE 6.5: $z_{ss}(t)$ with $r(t) = f(t)$ for $T = 5$ in Example 6.3.

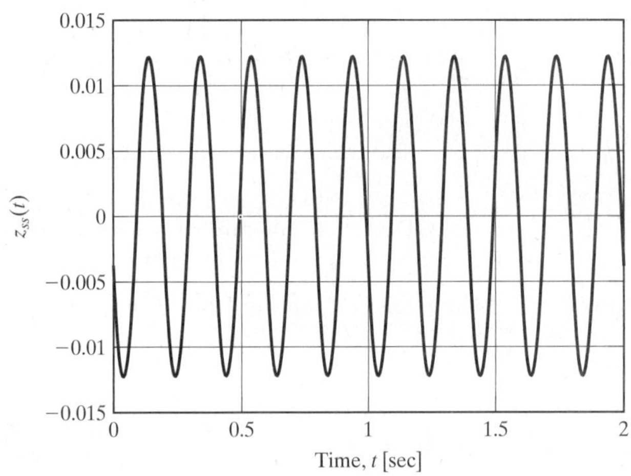

FIGURE 6.6: $z_{ss}(t)$ with $r(t) = f(t)$ for $T = 0.1$ in Example 6.3.

Figure 6.6 shows the steady-state response to a relatively fast square wave with $T = 0.1$. The corresponding frequencies of the signal components in this case are

$$\frac{(2n+1)\pi}{T} = 10\pi, 30\pi, 50\pi, \ldots, \quad n = 0, 1, 2 \ldots.$$

It is clear that the magnitude of a steady-state response in this case is much smaller than the previous case since $|T_{zd}(j\omega)|$ is much smaller at these high frequencies.

We can conclude from the above that this closed-loop system can significantly reject the disturbance $d(t)$ if the magnitude of $T_{zd}(j\omega)$ is small at those frequencies ω at which $d(t)$ has significant strength.

Similarly, let the transfer function from $n(t)$ to $z(t)$ be denoted by $T_{zn}(s)$ and note that

$$T_{zn}(s) = \frac{-P(s)C(s)}{1 + P(s)C(s)}.$$

The effects of the sensor noise $n(t)$ on the system output can then be effectively reduced if the magnitude of the frequency response $|T_{zn}(j\omega)|$ is small in the frequency range where $n(t)$ has significant strength.

On the other hand, the transfer function from $r(t)$ to $z(t)$ is

$$T(s) = \frac{P(s)C(s)}{1 + P(s)C(s)}$$

and thus it is desirable to have $T(j\omega) \approx 1$ in the frequency range where $r(t)$ has significant components. This requirement seems to be in conflict with the requirement for sensor noise reduction where it is desirable to have $|T(j\omega)|$ as small as possible. Fortunately, this may not necessarily be the case since $r(t)$ is typically a low-frequency signal, i.e., the most significant components of $r(t)$ are in the low-frequency range, and $n(t)$ is typically a high-frequency signal, i.e., its most significant components are in the high-frequency range.

The conclusions from Example 6.3 are applicable in a much more general setting. To that end, again consider the system shown in Figure 6.4. Recall that the open-loop transfer function is defined as

$$L(s) = C(s)P(s),$$

the **sensitivity function** as

$$S(s) = \frac{1}{1 + L(s)}$$

which is the transfer function from $n(t)$ to $y(t)$ or from $r(t)$ to $e(t)$, and the **complementary sensitivity function** as

$$T(s) = 1 - S(s) = \frac{L(s)}{1 + L(s)}$$

which is the transfer function from $r(t)$ to $z(t)$. (The word *complementary* is used to signify the fact that $T(s)$ is the complement of $S(s)$, $S(s) + T(s) = 1$.)

The closed-loop system satisfies the following equations:

$$Z(s) = T(s)(R(s) - N(s)) + S(s)P(s)D(s) \tag{6.1}$$

$$R(s) - Z(s) = S(s)R(s) + T(s)N(s) - S(s)P(s)D(s) \tag{6.2}$$

$$U(s) = C(s)S(s)(R(s) - N(s)) - T(s)D(s). \tag{6.3}$$

These three equations show the fundamental benefits and design objectives inherent in feedback loops, which will be explained in detail below. For example, (6.1) shows that the effects of disturbance $d(t)$ on the plant output can be made small by making the frequency response of $S(s)P(s)$ small over a corresponding frequency range as discussed in Example 6.3.

In the following, we shall use "$\gg$" and "$\ll$" to mean "much bigger" and "much smaller." Then the following facts will be useful in the subsequent discussions.

Proposition 6.4. Let ω be any frequency.

- Suppose $|L(j\omega)| \gg 1$. Then

$$|S(j\omega)| \ll 1, \quad |S(j\omega)P(j\omega)| \approx \frac{1}{|C(j\omega)|},$$

$$|T(j\omega)| \approx 1, \quad |C(j\omega)S(j\omega)| \approx \frac{1}{|P(j\omega)|}.$$

- Suppose $|L(j\omega)| \ll 1$. Then

$$|S(j\omega)| \approx 1, \quad |S(j\omega)P(j\omega)| \approx |P(j\omega)|,$$

$$|T(j\omega)| \approx |L(j\omega)|, \quad |C(j\omega)S(j\omega)| \approx |C(j\omega)|.$$

Proof. Note that

$$|L(j\omega)| - 1 \leq |1 + L(j\omega)| \leq 1 + |L(j\omega)|.$$

Then, for any frequency ω such that $|L(j\omega)| > 1$, we have

$$\frac{1}{|L(j\omega)|+1} \leq |S(j\omega)| \leq \frac{1}{|L(j\omega)|-1}$$

$$\frac{|L(j\omega)|}{|L(j\omega)|+1} \leq |T(j\omega)| \leq \frac{|L(j\omega)|}{|L(j\omega)|-1}$$

$$\frac{1}{|C(j\omega)|}\frac{|L(j\omega)|}{|L(j\omega)|+1} \leq |S(j\omega)P(j\omega)| \leq \frac{1}{|C(j\omega)|}\frac{|L(j\omega)|}{|L(j\omega)|-1}$$

$$\frac{1}{|P(j\omega)|}\frac{|L(j\omega)|}{|L(j\omega)|+1} \leq |C(j\omega)S(j\omega)| \leq \frac{1}{|P(j\omega)|}\frac{|L(j\omega)|}{|L(j\omega)|-1}.$$

Thus, if $|L(j\omega)| \gg 1$, we have

$$|S(j\omega)| \ll 1,\ |T(j\omega)| \approx 1,\ |S(j\omega)P(j\omega)| \approx \frac{1}{|C(j\omega)|},\ |C(j\omega)S(j\omega)| \approx \frac{1}{|P(j\omega)|}.$$

The rest follows similarly. □

Typically, the reference signal $r(t)$ is a low-frequency signal: for example, a step signal, a relatively slow sinusoidal signal, etc. A typical disturbance signal may be a low-frequency signal such as a constant 60 [Hz] electrical signal from power source or a slowly varying load applied to a motor shaft. Disturbance may also appear as other frequency signals depending on the system setup.

Since the transfer function from $r(t)$ to the tracking error $r(t) - z(t)$ is $S(s)$, it is clear that good tracking requires $|S(j\omega)|$ be made small over the frequency range where $r(t)$ is significant, typically in the low-frequency range. Similarly, good disturbance rejection at the plant output $z(t)$ would require that $|S(j\omega)|$ and $|S(j\omega)P(j\omega)|$ be made small, particularly in the low-frequency range where $d(t)$ is usually significant.

By Proposition 6.4, we conclude that *good tracking and disturbance rejection at plant output* $z(t)$ *in general requires large loop gain* $|L(j\omega)|$ *in the frequency range where* $r(t)$ *is significant and large enough controller gain* $|C(j\omega)| \gg 1$ *in the frequency range where* $d(t)$ *is significant for desensitizing* $d(t)$.

EXAMPLE 6.5

It is known from the internal model principle that to reject a step disturbance at the plant output the loop transfer function $L(s) = C(s)P(s)$ must have a pole at the origin so that $|L(0)| = \infty$.

Similarly, to reject a step disturbance at the plant input the controller $C(s)$ must have a pole at origin so that $|C(0)| = \infty$.

To prevent the control signal $u(t)$ from becoming too large to saturate the actuator, it is seen from (6.3) that $|C(j\omega)S(j\omega)|$ must be kept within a reasonable range. The magnitude of $|C(j\omega)S(j\omega)|$ in the low-frequency range is essentially limited by the allowable cost of control effort and saturation limit of the actuators; hence, in general, the maximum gain of $C(s)S(s)$ can be fairly large, while the high-frequency gain is essentially limited by the controller bandwidth and the (sensor) noise frequencies. Ideally, one would like to roll off as fast as possible beyond the desired control bandwidth so that the high-frequency noises are attenuated as much as possible.

Another performance trade-off is with the sensor noise error reduction. The conflict between the disturbance rejection and the sensor noise reduction is evident in (6.1). Large $|C(j\omega)|$ and $|L(j\omega)|$ values over a large frequency range make errors due to $d(t)$ small. However, they also make errors due to $n(t)$ large because this noise is "passed through" over the same frequency range; that is,

$$Z(s) = T(s)[R(s) - N(s)] + S(s)P(s)D(s) \approx R(s) - N(s).$$

Note that $n(t)$ is typically significant in the high-frequency range.

Worse still, large loop gains outside the bandwidth of $P(s)$, i.e., $|L(j\omega)| \gg 1$ while $|P(j\omega)| \ll 1$, can make the control activity $u(t)$ quite unacceptable, which may cause the saturation of actuators. This follows from

$$U(s) = C(s)S(s)[R(s) - N(s)] - T(s)D(s) \approx \frac{1}{P(s)}[R(s) - N(s)] - D(s)$$

since the resulting equation shows that sensor noise is actually amplified at $u(t)$ whenever the frequency range significantly exceeds the bandwidth of $P(s)$.

Similarly, the controller gain $|C(j\omega)|$ should also be kept not too large in the frequency range where the loop gain is small in order not to saturate the actuators. This is because for small loop gain $|L(j\omega)| \ll 1$

$$U(s) = C(s)S(s)[R(s) - N(s)] - T(s)D(s) \approx C(s)[R(s) - N(s)].$$

Therefore, it is desirable to keep $|C(j\omega)|$ not too large when the loop gain is small. (That is why a pure PD controller is not desirable since a pure PD controller has a growing gain as the frequency increases.)

The rest of this chapter will discuss the frequency-domain tools for analysis of the dynamic system.

6.2 BODE DIAGRAMS

As we have pointed out in the last section, the frequency response of a system contains all information about the system and control system analysis and design can be done completely on the basis of such information. Thus, it is important to introduce some tools and techniques for system analysis and design based on the frequency response.

Although the frequency response technique can most naturally be studied for stable systems, it can be generalized to even unstable systems although the frequency response of an unstable system cannot be obtained directly through experimental data as above. In general, for any given transfer function $P(s)$, we shall call $|P(j\omega)|$ the magnitude frequency response and $\angle P(j\omega)$ the phase frequency response even if $P(s)$ is not stable.

Thus, the frequency response of a system can be obtained directly from its transfer function. The frequency response of a system is conventionally represented using the magnitude plot in the log scale for both frequency and magnitude, and the phase plot in the log scale for the frequency and the linear scale for the phase. These magnitude and phase plots are called **Bode diagrams** of the system.

The magnitude Bode diagrams are sometimes plotted in decibels (dB). Let $P(s)$ be a transfer function, then the unit in $20\log_{10}|P(j\omega)|$ is [dB]. Thus, if $|P(j\omega)| = 0.1, 1, 10$, we have $20\log_{10}|P(j\omega)| = -20, 0, 20$ [dB], respectively. As a convention, the $\log_{10}$ will simply be written as log.

Hendrik Wade Bode (1905–1982) was born in Madison, Wisconsin. He received his B.A. degree in 1924 and M.A. Degree in 1926, both from the Ohio State University. While employed at Bell Laboratories, he attended Columbia University Graduate School, and received his Ph.D. degree in 1935. While working on the design of a variable equalizer, he discovered the relationship between gain and phase of a stable and minimum phase transfer function. He also developed the construction techniques for drawing the so-called Bode diagram. His main contributions were summarized in the paper "Relations between attenuation and phase in feedback amplifier design," *Bell System Technical Journal*, pp. 421–454, July 1940, and the book *Network Analysis and Feedback Amplifier Design*, Van Nostrand, 1945. Bode retired from Bell Telephone Laboratories in October 1967 at the age of 61, after 41 years of distinguished service in his career. He was immediately elected Gordon McKay Professor of Systems Engineering at Harvard University. Bode is a fellow of a number of scientific and engineering societies, including the IEEE, the American Physical Society, and the American Academy of Arts and Sciences. He has also achieved further accolade by being elected to the National Academy of Sciences and the National Academy of Engineering. He was awarded the 1969 IEEE Edison Medal "For fundamental contributions to the arts of communication, computation, and control; for leadership in bringing mathematical science to bear on engineering problems; and for guidance and creative counsel in systems engineering."

EXAMPLE 6.6

Let a first-order system be

$$P(s) = \frac{1}{Ts+1}.$$

Then its magnitude frequency response and phase frequency response are given by

$$|P(j\omega)| = \frac{1}{\sqrt{T^2\omega^2+1}}$$

$$\angle P(j\omega) = -\tan^{-1} T\omega.$$

Note that at the **corner frequency** $\omega = 1/T$, we have

$$|P(j1/T)| = \frac{1}{\sqrt{2}} = 3[\text{dB}], \quad \angle P(j1/T) = -\tan^{-1} 1 = -45°.$$

In the log scale, we have

$$\log|P(j\omega)| = -\frac{1}{2}\log(T^2\omega^2+1).$$

If $T\omega \ll 1$, then

$$\log|P(j\omega)| \approx 0.$$

On the other hand, if $T\omega \gg 1$, then

$$\log|P(j\omega)| \approx -\log(T\omega).$$

Thus, $\log|P(j\omega)|$ can be approximated by two straight lines as shown in Figure 6.7. $\omega = 1/T$ is the intersection (or corner) of these two straight lines.

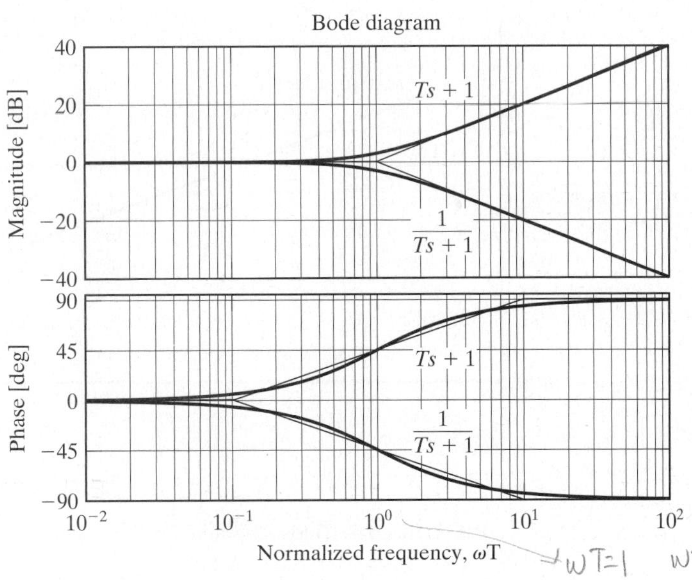

FIGURE 6.7: Bode diagrams for Example 6.6.

Similarly, if $T\omega \leq 0.1$, then

$$\angle P(j\omega) \approx 0$$

and if $T\omega \geq 10$, then

$$\angle P(j\omega) \approx -90°.$$

Hence $\angle P(j\omega)$ can also be approximated using three straight lines as shown in Figure 6.7.

Since $\log|1/P(j\omega)| = -\log|P(j\omega)|$ and $\angle(1/P(j\omega)) = -\angle P(j\omega)$, we can also plot the frequency response of $Ts + 1$ easily as shown in Figure 6.7. The exact frequency response of $P(s)$ and $1/P(s)$ are also shown in Figure 6.7 as smooth curves.

EXAMPLE 6.7

Let m be an integer and

$$P(s) = \frac{K}{s^m}.$$

Then

$$|P(j\omega)| = \frac{K}{\omega^m}, \quad \angle P(j\omega) = -m \times 90°.$$

The Bode diagrams for this system are illustrated in Figure 6.8. Note that if m is a negative integer, say $m = -1$, then the slope of the magnitude plot becomes +20[dB/decade] and the phase becomes +90°.

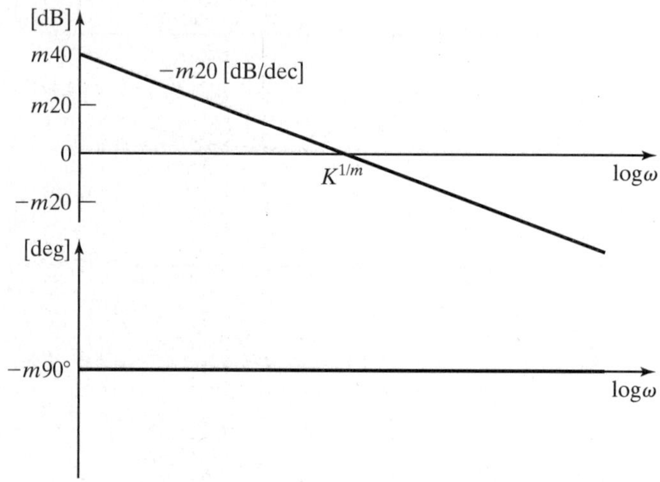

FIGURE 6.8: Bode diagrams for Example 6.7.

EXAMPLE 6.8

The Bode diagrams for transfer functions with complex poles are much more complicated to construct since they depend critically on their damping ratios. Figure 6.9 illustrates the Bode diagrams for a standard second-order system

$$P(s) = \frac{\omega_n^2}{s^2 + 2\zeta\omega_n s + \omega_n^2}, \quad 0 < \zeta \leq 1$$

with various damping ratios. In general, one needs to do some detailed calculations to get a fairly accurate Bode diagram for a system with lightly damped poles. Nevertheless, knowing the frequency responses of the system at some key frequencies can be very helpful in sketching a rough Bode diagram.

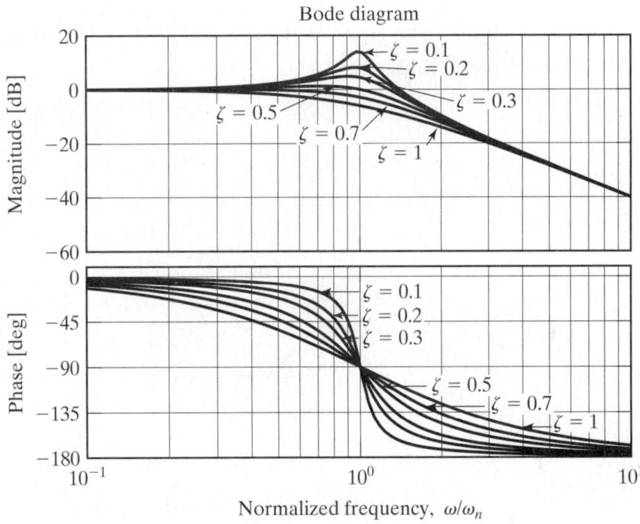

FIGURE 6.9: Frequency response of a second-order system.

Table 6.2 gives some key values or approximations of the frequency response at various frequencies.

ω	$\lvert P(j\omega)\rvert$	$\angle P(j\omega)$
0	1	$0°$
$0.1\omega_n$	≈ 1	$(0, -12°)$
ω_n	$\frac{1}{2\zeta}$	$-90°$
ω_p (for $0 < \zeta < \frac{1}{\sqrt{2}}$)	$\frac{1}{2\zeta\sqrt{1-\zeta^2}}$	$-\tan^{-1}\frac{\sqrt{1-2\zeta^2}}{\zeta}$
$10\omega_n$	≈ 0.01	$(-168°, -180°)$
∞	0	$-180°$

TABLE 6.2: Frequency response of a second-order system at some key frequencies.

It is noted that the magnitude frequency response has a peak value if $0 < \zeta < \frac{1}{\sqrt{2}}$, and the peak frequency ω_p (to be derived in Section 6.5) is given by

$$\omega_p = \omega_n \sqrt{1 - 2\zeta^2}.$$

On the other hand, the magnitude frequency response is monotonically decreasing for $\zeta \geq \frac{1}{\sqrt{2}}$.

EXAMPLE 6.9

Let

$$P_1(s) = \frac{1}{Ts - 1}, \quad P_2(s) = \frac{1}{Ts + 1}.$$

Then it is clear that $|P_1(j\omega)| = |P_2(j\omega)|$ and thus $P_1(s)$ and $P_2(s)$ have the same magnitude Bode diagrams. However,

$$\begin{aligned} \angle P_1(j\omega) &= -\angle(jT\omega - 1) = -(\pi - \angle(jT\omega + 1)) = -\pi + \angle(jT\omega + 1) \\ &= -\pi - \angle P_2(j\omega) \end{aligned}$$

as shown in Figure 6.10.

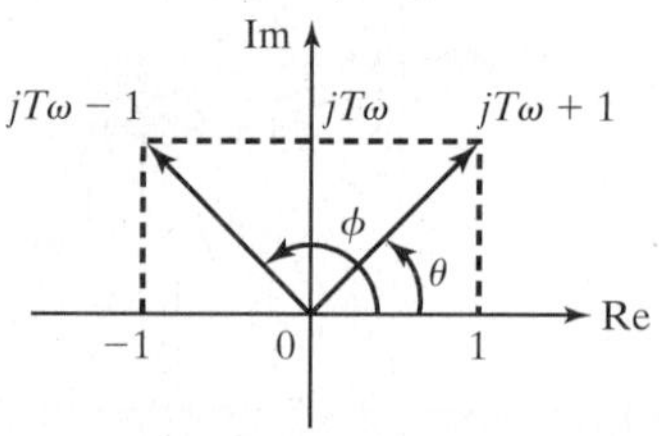

FIGURE 6.10: Example 6.9: $\phi = \pi - \theta$.

The frequency responses of $P_1(s)$ and $P_2(s)$ are shown in Figure 6.11. Thus an unstable pole results in a much larger negative phase than a stable pole with the same magnitude.

It is not difficult to draw the Bode diagrams for $Ts - 1$ since $\angle(jT\omega - 1) = \pi - \angle(jT\omega + 1)$. Again, it is clear that an unstable zero gives a much larger phase than a stable zero. That is why an unstable zero is called a **nonminimum phase** zero.

In general, let $P(s) = P_1(s)P_2(s)P_3(s)$. Then

$$\begin{aligned} \log|P(j\omega)| &= \log|P_1(j\omega)| + \log|P_2(j\omega)| + \log|P_3(j\omega)|, \\ \angle P(j\omega) &= \angle P_1(j\omega) + \angle P_2(j\omega) + \angle P_3(j\omega). \end{aligned}$$

Thus, the Bode diagrams of a system $P(s)$ can be obtained by graphically adding the Bode diagrams of $P_1(s)$, $P_2(s)$, and $P_3(s)$.

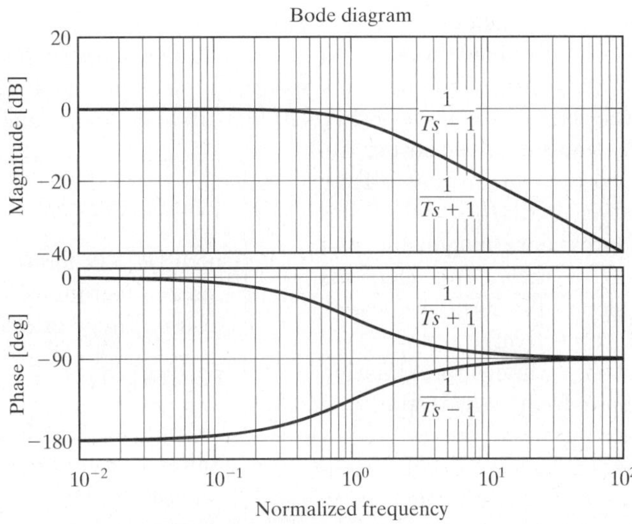

FIGURE 6.11: Bode diagrams for Example 6.9.

EXAMPLE 6.10

Let two transfer functions be given by

$$P_1(s) = \frac{10(s+1)}{s+4},$$

$$P_2(s) = \frac{0.2(s+0.5)}{s+0.1}.$$

Then the approximate Bode diagrams can be obtained easily by graphically adding the straight lines for a simple real pole and zero as shown in Figure 6.12 for the

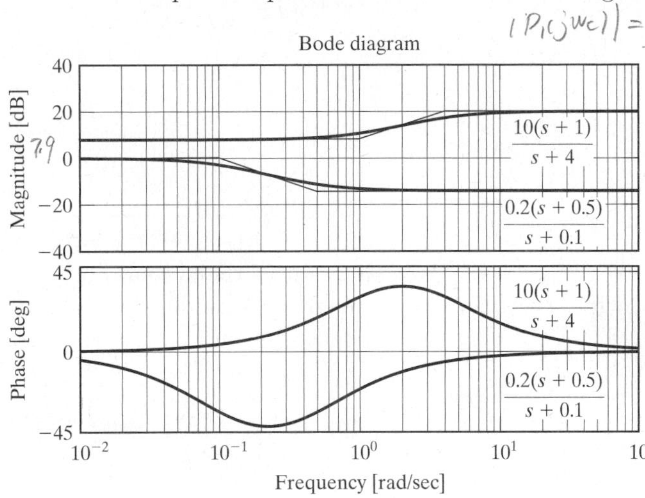

FIGURE 6.12: Bode diagrams for Example 6.10.

magnitude frequency response. For example, the approximate magnitude plot of $P_1(s)$ is obtained by starting a straight line with 0 slope at a magnitude of $P_1(0) = 2.5$, then changing the slope of the straight line at the corner frequency 1 (i.e., the zero at -1) by a positive 20 [dB/decade], and changing again the slope of the straight line at the corner frequency 4 (i.e., the pole at -4) by a negative 20 [dB/decade]. Hence the approximate magnitude plot consists of three line segments with slopes 0, 20 [dB/decade], and 0, respectively.
The straight line approximation for phase frequency response is less accurate and hence is not commonly used. The accurate Bode diagrams shown in Figure 6.12 can be obtained by using the MATLAB commands on the right.

`P1=tf(10*[1,1],[1,4]);`

`bode(P1)` generate Bode magnitude and phase diagrams of P_1 with automatically chosen range of frequencies.

`P2=zpk(-0.5,-0.1,0.2);`

`w=logspace(-2,2,200);` set the frequency range from 10^{-2} to 10^{2} with 200 sample points.

`bode(P2,w)` or use `ltiview` below

`ltiview('bode',P2)`

EXAMPLE 6.11

Transportation delays are very common in process control. The following transfer function models a system with input or output time delay

$$P(s) = \frac{e^{-Ts}}{s(s+1)}.$$

Note that

$$|P(j\omega)| = \left|\frac{e^{-jT\omega}}{j\omega(j\omega+1)}\right| = \left|\frac{1}{j\omega(j\omega+1)}\right|$$

and

$$\angle P(j\omega) = -T\omega + \angle\frac{1}{j\omega(j\omega+1)}.$$

Thus, the magnitude Bode diagram of the system does not depend on the time delay, while the phase Bode diagram is modified by an ever-increasing negative phase. The Bode diagram of a delay system can also be generated using MATLAB routines.

```
>> sys=tf([1],[1,1,0],'inputdelay', 2);
```
take $T = 2$.

```
>> bode(sys)
```

The Bode diagrams of the system with $T = 0$ and $T = 2$ are shown in Figure 6.13.

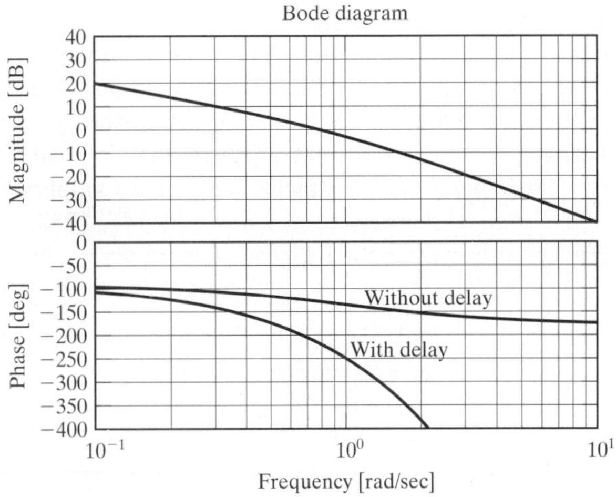

FIGURE 6.13: Bode diagrams for Example 6.11 with $T = 0$ and $T = 2$.

It should be emphasized that the Bode diagrams of a system contain all the information about the system very much like its transfer function does and the system analysis and design can be done solely on the basis of Bode diagrams. Controller designs based on Bode diagrams will be described in the next chapter. Here, we shall look at how the steady-state error constants can be calculated from Bode diagrams.

Suppose the Bode diagram of an open-loop transfer function $L(s) = C(s)P(s)$ is given. The steady-state error constants K_p, K_v, and K_a can then all be found from the open-loop Bode magnitude diagram as shown in Figure 6.14. The idea is very simple since the transfer function $L(s)$ at a very low frequency can be approximated by K_p if $L(s)$ has no pole at origin, by K_v/s if $L(s)$ has one pole at origin, and by K_a/s^2 if $L(s)$ has two poles at origin.

Harry Nyquist (1889–1976) was born in Nilsby, Sweden. He received his B.S. and M.S. degrees in electrical engineering in 1914 and 1915, respectively, from the University of North Dakota, Grand Forks. He received his Ph.D. degree in physics from Yale University in 1917. He worked at AT&T's Department of Development and Research from 1917 to 1934, and continued when it became Bell Telephone Laboratories in that year, until his retirement in 1954. During his 37 years of service with the Bell System, he received 138 U.S. patents and published 12 technical articles. He contributed important efforts to the study of thermal noise ("Johnson–Nyquist noise"), the stability of feedback amplifiers, telegraphy, facsimile, television, and other important communications problems. With Herbert E. Ives he helped develop AT&T's first facsimile machines that were made public in 1924. In a paper published in 1924: "Certain factors affecting telegraph speed", *Bell System Technical Journal*, vol. 3, pp. 324–346, 1924, he determined the bandwidth requirements for transmitting information, which laid the foundations for later advances on information theory by Claude Shannon. In 1927, Nyquist determined that the number of independent pulses that could be put through a telegraph channel per unit time is limited to twice the bandwidth of the channel. This rule is essentially a dual of what is now known as the Nyquist–Shannon sampling theorem. In 1932, he published a classic paper on the stability of feedback amplifiers: "Regeneration theory," *Bell System Technical Journal*, vol. 11, pp. 126–147, 1932.

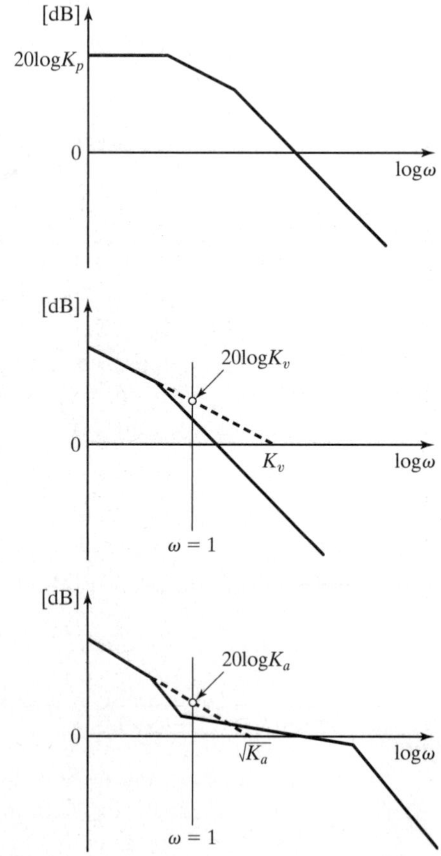

FIGURE 6.14: Determination of steady-state error constants from open-loop magnitude Bode diagrams.

6.3 NYQUIST STABILITY CRITERION

Let Γ be a closed contour and let

$$F(s) = K\frac{\prod_{i=1}^{m}(s - z_i)}{\prod_{j=1}^{n}(s - p_i)}$$

where p_i and z_j are (real or complex) poles and zeros of $F(s)$, respectively.

Now, evaluate $F(s)$ along the contour of Γ; then the trajectory of $F(s)$ is a closed curve. As illustrated in Figure 6.15, for any p (pole or zero) outside the contour, the trajectory of $s - p$ evaluated along Γ will not encircle the origin. On the other hand, for any p inside of the contour, the trajectory of $s - p$ evaluated along Γ will encircle the origin once in the same direction as Γ.

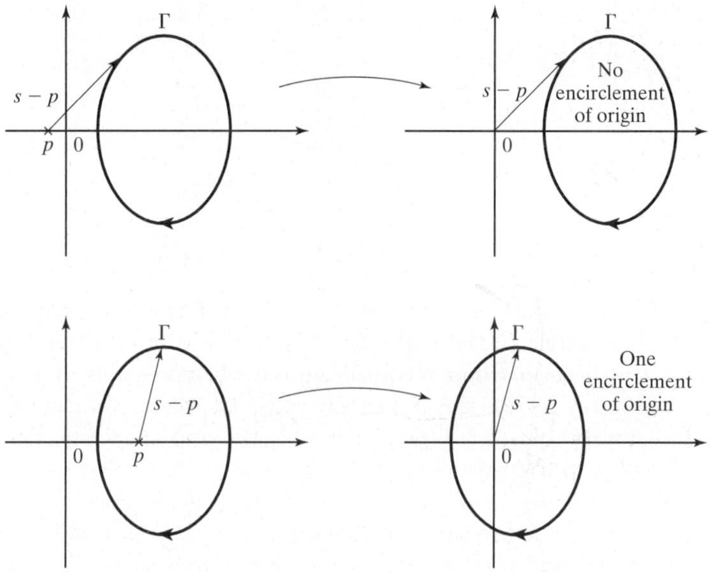

FIGURE 6.15: Encirclements of a single pole.

Define

$N_Z =$ the number of zeros of $F(s)$ inside Γ;

$N_P =$ the number of poles of $F(s)$ inside Γ;

$N =$ the number of encirclements of $F(s)$ around the origin. (An encirclement in the same direction as the curve Γ counts as a positive encirclement; otherwise, it counts as a negative encirclement.)

It then follows that

$$N = N_Z - N_P.$$

In our application, we shall take

$$F(s) = 1 + C(s)P(s)$$

and a special contour Γ that includes the entire right half plane. This contour is called the **Nyquist contour** as shown in Figure 6.16. The Nyquist contour is chosen so that it is indented around all possible poles of $F(s)$ on the imaginary axis.

Clearly, the number of encirclements of $F(s)$ around the origin is the number of encirclements of $L(s) = C(s)P(s)$ around $(-1, 0)$. Note that $L(s)$ evaluated along the semicircle of the Nyquist contour in the right half plane will only result in an infinitesimal variation at origin. Thus, we only need to evaluate $L(s)$ on the Nyquist contour along the imaginary axis.

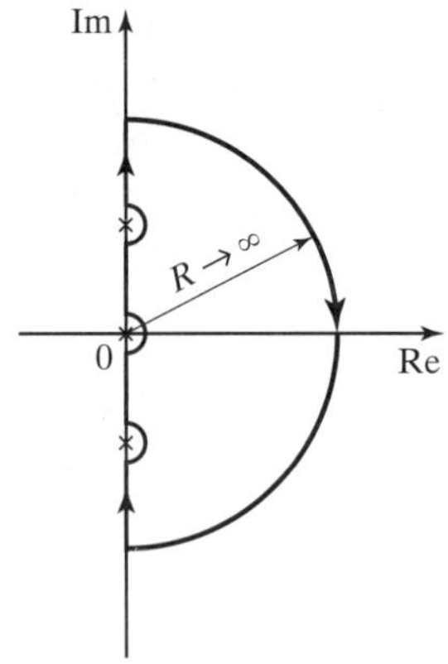

FIGURE 6.16: Nyquist contour.

Since the contour is symmetric to the real axis, the Nyquist plot of $F(s)$ will also be symmetric to the real axis. Thus, it is only necessary to plot the Nyquist plot along the upper half Nyquist contour. In particular, if $F(s)$ or, equivalently, $L(s)$ has no poles on the imaginary axis, then the Nyquist plot of $L(s)$ can be constructed by plotting $L(j\omega)$, $0 \leq \omega \leq \infty$, and then reflecting it symmetrically with respect to the real axis.

Note that the number of the right half plane poles of $F(s)$ enclosed by the Nyquist contour is the same as the number of the right half plane poles of $L(s) = C(s)P(s)$ enclosed by the Nyquist contour. Furthermore, the number of right half plane zeros of $F(s)$ enclosed by the Nyquist contour is the number of closed-loop right half plane poles. Thus the closed-loop system is stable if and only if $N_Z = 0$, i.e., $N = -N_P$. That is, the Nyquist plot of $L(s) = C(s)P(s)$ must encircle the $(-1, 0)$ point N_P times in the anticlockwise direction.

Hence we have the following stability criterion:

Theorem 6.12 (Nyquist Stability Criterion). Let N_P be the number of right half plane poles of $L(s) = C(s)P(s)$ enclosed by the Nyquist contour. Then, the closed-loop system is stable if and only if the Nyquist plot of $L(s) = C(s)P(s)$ encircles the $(-1, 0)$ point N_P times in the anticlockwise direction.

In particular, if $L(s)$ has no open right half plane poles (i.e., $N_P = 0$), then the closed-loop system is stable if and only if the Nyquist plot of $L(s)$ does not encircle the $(-1, 0)$ point.

EXAMPLE 6.13

Let $L(s) = \dfrac{K}{s+1}$, $K > 0$. Then $L(s)$ has no poles on the imaginary axis. Thus, the Nyquist plot can be obtained by plotting $L(j\omega), -\infty \leq \omega \leq \infty$:

$$L(j\omega) = \frac{K}{j\omega + 1} = \frac{K(1 - j\omega)}{\omega^2 + 1}.$$

The Nyquist plot for this system is shown in Figure 6.17. Since the Nyquist plot does not encircle the $(-1, 0)$ point, the closed-loop system is stable.

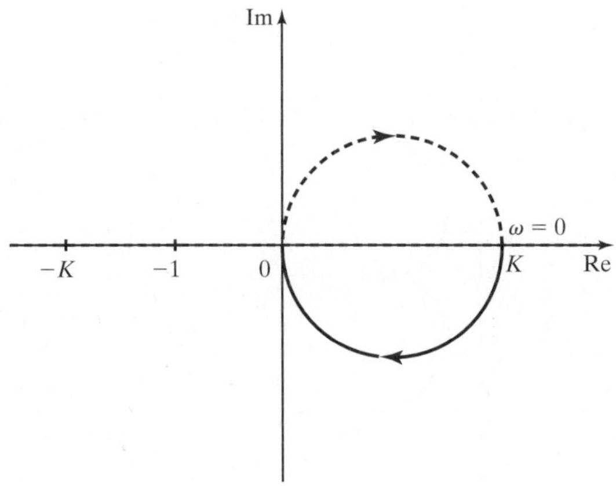

FIGURE 6.17: Nyquist plot for Example 6.13.

EXAMPLE 6.14

Let

$$L(s) = \frac{K}{(s+1)(s+2)(s+3)}.$$

Then, $L(s)$ has no unstable poles. Thus, $N_P = 0$. The Nyquist plot can be generated by using a MATLAB command

```
>> Nyquist(L) or ltiview('nyquist',L)
```

The Nyquist plots for $K = 10$ and $K = 100$ are shown in Figures 6.18 and 6.19, respectively. It is clear from the Nyquist plots that the closed-loop system is stable for $K = 10$ since the Nyquist plot in Figure 6.18 does not encircle $(-1, 0)$, and is

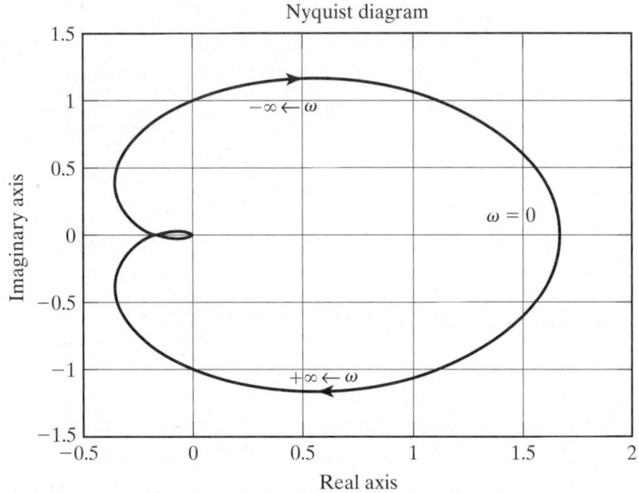

FIGURE 6.18: Nyquist plot for Example 6.14 with $K = 10$.

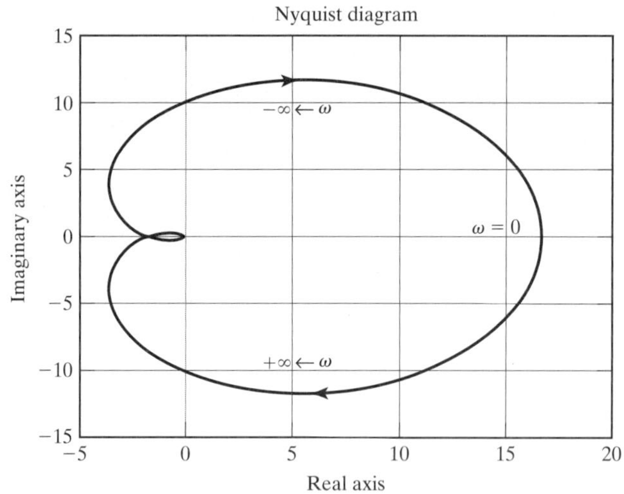

FIGURE 6.19: Nyquist plot for Example 6.14 with $K = 100$.

unstable for $K = 100$ since the Nyquist plot in Figure 6.19 encircles $(-1, 0)$ twice. Thus, the closed-loop system with $K = 100$ has two unstable poles. It is also fairly easy to see that the Nyquist plot passes through the $(-1, 0)$ point when $K = 60$, which means that the closed-loop system with $K = 60$ has two poles on the imaginary axis.

EXAMPLE 6.15

Let

$$L(s) = \frac{K}{s(s+1)}$$

with $K > 0$. Since $L(s)$ has a pole at the origin, the Nyquist contour has an indented semicircle around it. Let the radius of the semicircle be $\epsilon > 0$. Then $s = \epsilon e^{j\theta}$ for $-\pi/2 \le \theta \le \pi/2$, and

$$L(s) = \frac{K}{\epsilon e^{j\theta}(\epsilon e^{j\theta} + 1)} \approx \frac{K}{\epsilon} e^{-j\theta}.$$

On the imaginary axis, we have

$$L(j\omega) = \frac{K}{j\omega(j\omega + 1)} = -\frac{K}{1+\omega^2} - j\frac{K}{\omega(1+\omega^2)}.$$

The Nyquist plot of $L(s)$ with $K = 10$ is illustrated in Figure 6.20 and a MATLAB-generated Nyquist plot is shown in Figure 6.21. It is clear that we should be very careful in interpreting the Nyquist plot near $\omega = 0$ since MATLAB cannot easily show how the phase changes with large magnitude. Since the Nyquist plot does not encircle the $(-1, 0)$ point, the closed-loop system is stable.

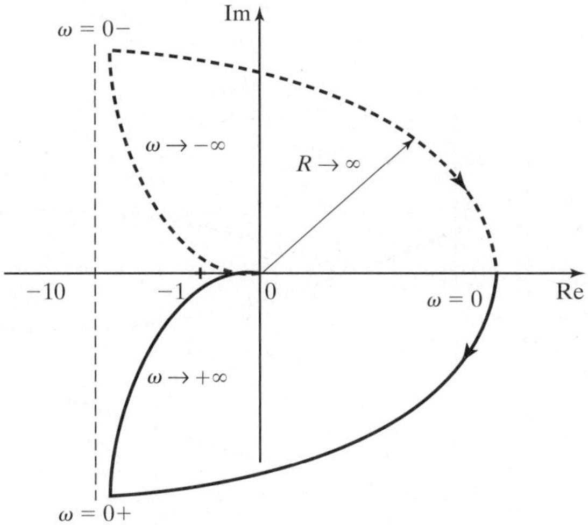

FIGURE 6.20: Nyquist plot for $L(s)$ in Example 6.15 with $K = 10$.

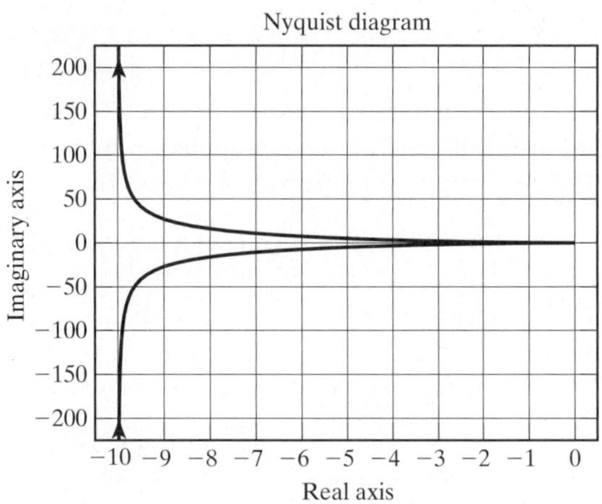

FIGURE 6.21: MATLAB-generated Nyquist plot for $L(s)$ in Example 6.15 with $K = 10$.

We can also approximately plot the Nyquist diagram by making a small perturbation on the poles on the imaginary axis. For example, we could have reached the same conclusion by plotting

$$\frac{K}{(s+\epsilon)(s+1)}$$

for a small $\epsilon > 0$.

The approximation method can be especially useful for plants with multiple poles on the imaginary axis. For example, a MATLAB-generated Nyquist diagram for a system

$$L_1(s) = \frac{K}{s^2(s+1)}$$

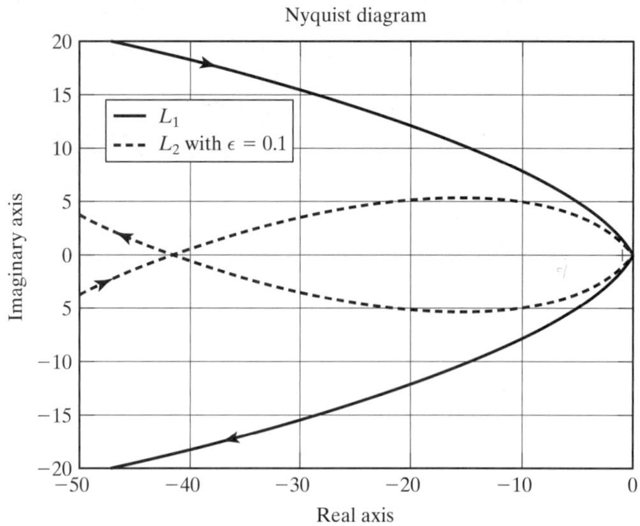

FIGURE 6.22: (Partial) Nyquist plot for $L_1(s)$ and $L_2(s)$ in Example 6.15 with $K = 10$ and $\epsilon = 0.1$.

is shown in Figure 6.22 (solid line). It is not obvious how many times the Nyquist diagram of $L_1(s)$ has actually encircled the $(-1, 0)$ point without knowing how the Nyquist diagram changes from the frequency range of $\omega = 0^-$ to 0^+.

Now define an approximate transfer function

$$L_2(s) = \frac{K}{(s+\epsilon)^2(s+1)}, \quad \epsilon > 0.$$

Figure 6.23 shows the (complete) Nyquist diagram of $L_2(s)$ with $K = 10$ and $\epsilon = 0.1$. Since this approximate Nyquist plot encircles the $(-1, 0)$ point twice (see also Figure 6.22), the closed-loop system has two unstable poles.

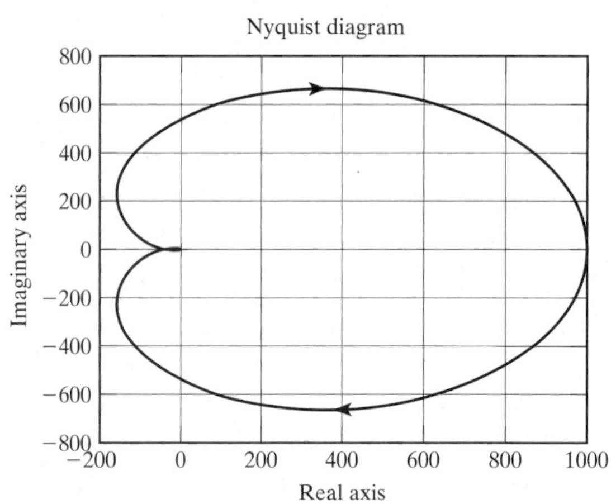

FIGURE 6.23: Complete Nyquist plot for $L_2(s)$ with $K = 10$ and $\epsilon = 0.1$.

EXAMPLE 6.16

Let

$$L(s) = \frac{K(s+1)}{s(s-4)}.$$

Note that

$$L(j\omega) = -\frac{5K}{\omega^2+16} + j\frac{K(4-\omega^2)}{\omega(\omega^2+16)}$$

and

$$L(j2) = -K/4.$$

Since

$$L(s) \approx \frac{-K/4}{s}$$

for the frequencies near 0 and

$$L(s) \approx \frac{K}{s}$$

for the frequencies near ∞, we can sketch the Nyquist plot as shown in Figure 6.24.

Note that $N_P = 1$ and $N = -1$ if $K > 4$ and $N = 1$ if $K < 4$; we conclude that the closed-loop system is stable if $K > 4$ and is unstable with two unstable closed-loop poles if $K < 4$. Furthermore, it is clear that the closed-loop system has two imaginary poles when $K = 4$.

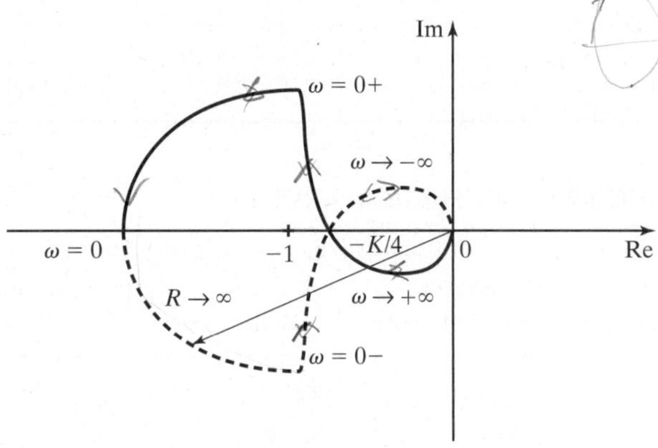

FIGURE 6.24: Nyquist diagram for $L(s)$ in Example 6.16.

EXAMPLE 6.17

Let

$$L(s) = \frac{K(s-1)}{s(s+1)(s-2)}.$$

The Nyquist plot of this system is somewhat similar to that of

$$\frac{K}{s(s+1)}.$$

Note that

$$L(j\omega) = -\frac{K(\omega^2+3)}{(\omega^2+1)(\omega^2+4)} - j\frac{2K}{\omega(\omega^2+1)(\omega^2+4)}$$

and

$$\operatorname{Re} L(j0) = -3K/4.$$

A sketch of the Nyquist plot is shown in Figure 6.25. Note that since $N_P = 1$ and $N = 0$, the closed-loop system is unstable for all $K > 0$.

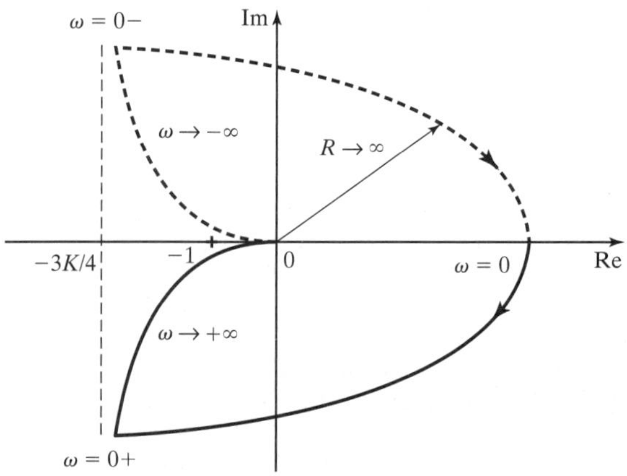

FIGURE 6.25: Nyquist plot for $L(s)$ in Example 6.17.

6.4 GAIN MARGIN AND PHASE MARGIN

One of the advantages of the Nyquist stability criterion is that it not only decides the stability of the system but also, and perhaps more importantly, gives some quantitative measures of relative stability, i.e., the stability margins. The stability margins in the frequency domain are conventionally given in two quantities: gain margin (GM) and phase margin (PM).

Gain margin: The amount of system gain that can be increased without jeopardizing the closed-loop stability.

Suppose the closed-loop system with the open-loop transfer function $L(s)$ is stable. Let ω_g be the only **phase crossover frequency**, i.e.,

$$\angle L(j\omega_g) = -180^\circ$$

and let $|L(j\omega_g)| < 1$. Then $L(j\omega_g) = -|L(j\omega_g)|$, and the closed-loop system with an open-loop transfer function $KL(s)$ will still be stable if $K < 1/|L(j\omega_g)|$ since $KL(j\omega_g) > -1$ and the Nyquist plot of $KL(j\omega)$ will have the same number of encirclements as $L(j\omega)$. On the other hand, the closed-loop system with an open-loop transfer function $KL(s)$ will become unstable if $K \geq 1/|L(j\omega_g)|$ since $KL(j\omega_g) \leq -1$ and the Nyquist plot of $KL(j\omega)$ will have a different number of encirclements compared to the Nyquist plot of $L(j\omega)$.

Therefore, the GM is given by

$$\textbf{Gain margin (GM)} = \frac{1}{|L(j\omega_g)|}.$$

In the case when the loop transfer function $L(s)$ has more than one phase crossover frequency, then the above argument carries over if ω_g is taken as the one so that $|L(j\omega_g)|$ is less than 1 but closest to 1.

Phase margin: The amount of phase lag that can be tolerated without jeopardizing the closed-loop stability.

Again suppose that the closed-loop system with the open-loop transfer function $L(s)$ is stable. Let ω_c be the only **gain crossover frequency**, i.e.,

$$|L(j\omega_c)| = 1.$$

Then $L(j\omega_c) = e^{j\angle L(j\omega_c)}$ and the closed-loop system with an open-loop transfer function $e^{-j\phi}L(s)$ will still be stable if $\angle L(j\omega_c) - \phi > -180°$ since $e^{-j\phi}$ $L(j\omega_c) = e^{j(\angle L(j\omega_c)-\phi)}$ and the Nyquist plot of $e^{-j\phi}L(j\omega)$ will have the same number of encirclements as $L(j\omega)$. On the other hand, the closed-loop system with an open-loop transfer function $e^{-j\phi}L(s)$ will become unstable if $\angle L(j\omega_c) - \phi \leq -180°$ since $e^{-j\phi}L(j\omega_c) = e^{j(\angle L(j\omega_c)-\phi)}$ and the Nyquist plot of $e^{-j\phi}L(j\omega)$ will have a different number of encirclements compared to the Nyquist plot of $L(j\omega)$.

Therefore, the PM is given by

$$\textbf{Phase margin (PM)} = 180° + \angle L(j\omega_c).$$

The GM and PM are illustrated in Figure 6.26 on a typical Nyquist plot.

The stability margins of a system are often more conveniently computed from its Bode diagrams as shown in Figure 6.27.

Note that, in general, a large GM does not necessarily imply a large PM, and a large PM does not necessarily imply a large GM as shown in Figure 6.28(a) and (b).

There are also cases in which both GM and PM are large but the Nyquist plots are very close to the $(-1, 0)$ point as illustrated in Figure 6.28(c). In those cases, the closed-loop systems can be destabilized if there are simultaneous gain and phase perturbations. Thus, a more sensible measure of stability margin is the smallest distance r from $(-1, 0)$ point to the Nyquist plot as shown in Figure 6.28(c).

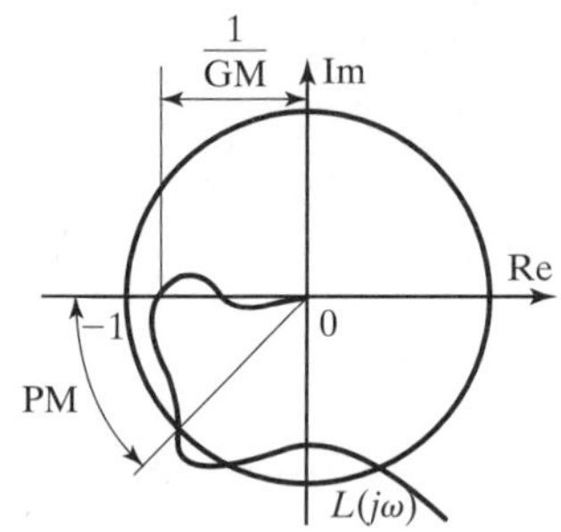

FIGURE 6.26: Stability margins on the Nyquist plot. (Only the part $0 \leq \omega \leq \infty$ is shown. The other part is symmetric to the real axis.)

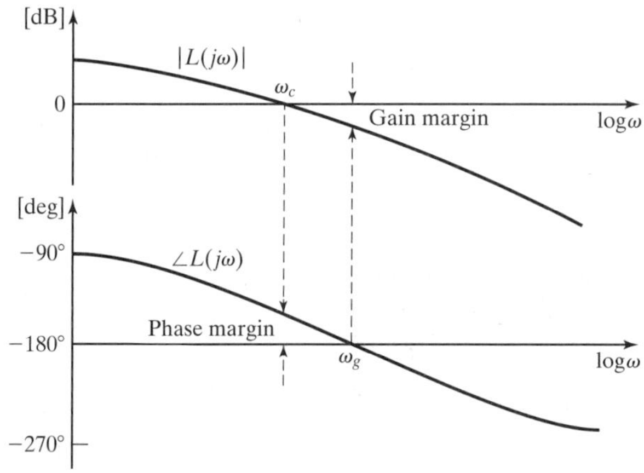

FIGURE 6.27: Stability margins on Bode diagrams.

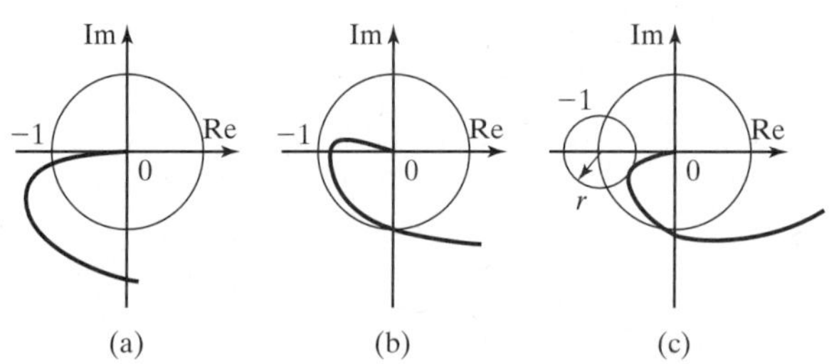

FIGURE 6.28: Comparison of stability margins.

EXAMPLE 6.18

Let a first-order open-loop transfer function be given by

$$L(s) = C(s)P(s) = \frac{K}{s+a}$$

with $K > a > 0$. Then it is clear that the GM is ∞ since the Nyquist diagram of $L(s)$ never crosses the negative real axis. The gain crossover frequency ω_c can be obtained by setting $|L(j\omega_c)| = 1$ and is given by $\omega_c = \sqrt{K^2 - a^2}$. Hence, the PM is given by

$$\mathrm{PM} = 180^\circ + \angle L(j\omega_c) = 180^\circ - \tan^{-1}\frac{\sqrt{K^2 - a^2}}{a} > 90^\circ.$$

EXAMPLE 6.19

Consider a second-order system with

$$L(s) = C(s)P(s) = \frac{\omega_n^2}{s(s + 2\zeta\omega_n)}.$$

It is then easy to see that the GM is ∞. Now let $|L(j\omega_c)| = 1$, and we get

$$\omega_c = \omega_n\sqrt{\sqrt{1 + 4\zeta^4} - 2\zeta^2}.$$

Thus, the PM is given by

$$\mathrm{PM} = 180^\circ + \angle L(j\omega_c) = \tan^{-1}\frac{2\zeta}{\sqrt{\sqrt{1 + 4\zeta^4} - 2\zeta^2}}\ \text{[rad]}.$$

The relationship between PM and ζ is shown in Figure 6.29. For $0 \leq \zeta \leq 0.6$, the damping ratio ζ and the PM have the following approximate relation:

$$\zeta \approx \frac{\mathrm{PM[deg]}}{100}.$$

This formula is often used in frequency-domain controller design to estimate system damping and overshoot.

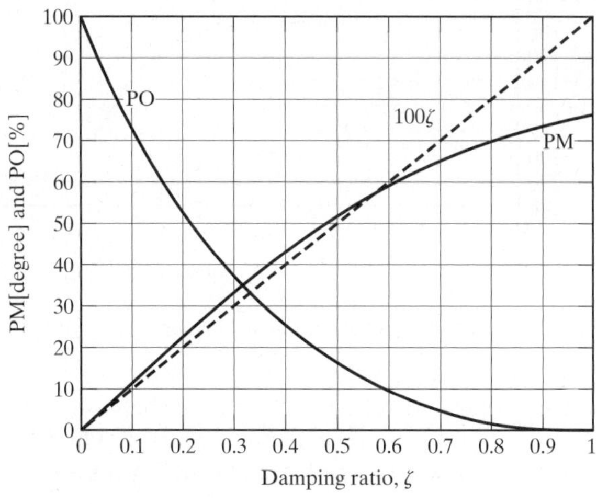

FIGURE 6.29: The relationships among phase margin, damping ratio, and percentage overshoot for a second-order system.

EXAMPLE 6.20

Let $L(s)$ be given by

$$L(s) = \frac{K(s+10)}{s(s^2+2s+4)}.$$

We shall calculate the stability margins of the system. To analytically find the range of $K > 0$ so that the system is stable, note that

$$L(j\omega) = \frac{-K(16+\omega^2)}{(4-\omega^2)^2+4\omega^2} - j\frac{K(40-8\omega^2)}{\omega((4-\omega^2)^2+4\omega^2)}.$$

Hence $\operatorname{Im} L(j\omega) = 0$ if $\omega = \sqrt{5}$. Thus $\omega_g = \sqrt{5}$ and we have $L(j\omega_g) = -K$. Therefore, the GM is given by $1/K$ or $20\log(1/K)$ [dB].

To calculate the PM analytically, we must find ω_c so that $|L(j\omega_c)| = 1$. This is usually much more difficult when the order of $L(s)$ is higher than 2. However, approximate solutions can be found very easily from Bode diagrams. The Bode diagrams of this system with $K = 10$ and $K = 0.1$ are shown in Figure 6.30.

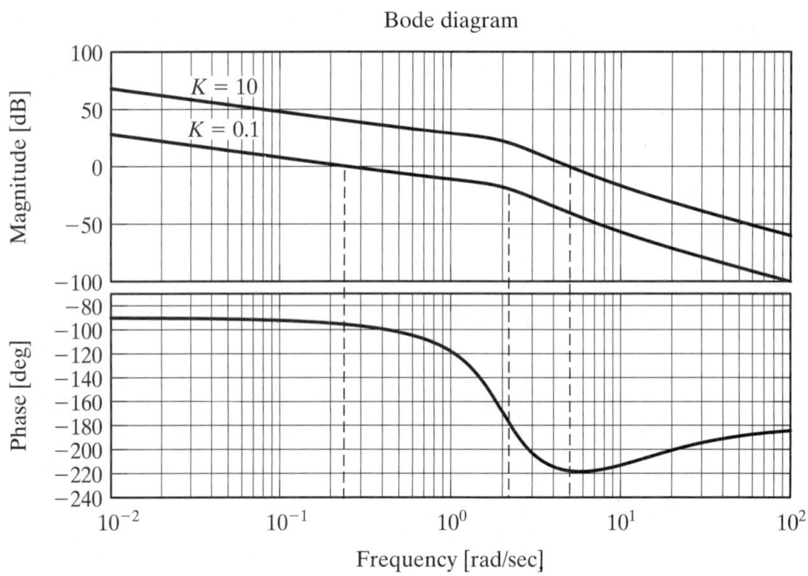

FIGURE 6.30: Bode diagrams for Example 6.20 with $K = 10$ and $K = 0.1$.

The Bode diagrams clearly show that the closed-loop system is unstable for $K = 10$ since the stability margins are negative. On the other hand, the stability margins for $K = 0.1$ can be read from the diagrams as

$$\text{PM} = 84^\circ, \quad \text{GM} = 20\,[\text{dB}].$$

EXAMPLE 6.21

Continuing from the last example, let $L(s)$ be given by

$$L(s) = L_0(s)e^{-Ts}, \qquad L_0(s) = \frac{0.1(s+10)}{s(s^2+2s+4)}.$$

We shall find the largest possible time delay T so that the closed-loop system is stable. Since the Bode diagram in Figure 6.30 shows that the system has a PM of 84° with crossover frequency $\omega_c = 0.25$, the largest possible delay is given by the solution of the following equation:

$$\angle L(j\omega_c) = \angle L_0(j\omega_c) - T_{\max}\omega_c = -180^\circ,$$

i.e.,

$$T_{\max}\omega_c = 84^\circ \times \frac{\pi}{180}$$

or $T_{\max} = 5.86$. Hence the system is stable for $T < T_{\max}$.

6.5 CLOSED-LOOP FREQUENCY RESPONSE

Note that the closed-loop transfer function from r to y in Figure 6.4 is

$$T(s) = \frac{C(s)P(s)}{1 + C(s)P(s)}.$$

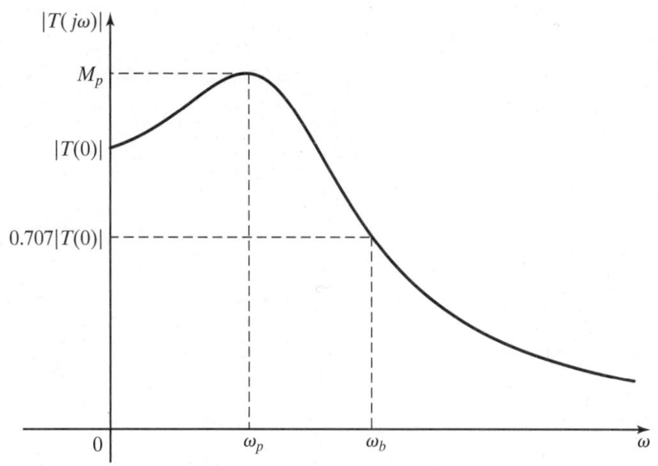

FIGURE 6.31: Closed-loop frequency response.

Figure 6.31 illustrates a typical closed-loop frequency response $T(j\omega)$ where M_p, ω_p, and ω_b are defined as follows:

Peak resonance M_p: The maximum value of $|T(j\omega)|$. This is an indicator of relative stability (or damping) of the system. A relatively large M_p indicates a lightly damped system.

Peak frequency ω_p: $M_p = |T(j\omega_p)|$.

Bandwidth ω_b: The smallest frequency such that $|T(j\omega_b)| = \frac{1}{\sqrt{2}}|T(j0)|$. This frequency indicates that the system will pass essentially all signals in the band $[0, \omega_b]$ without much attenuation. This is usually used as an indicator of rise time (as well as settling time which also strongly depends on M_p). It also gives an indication of the system's ability to reject high-frequency noise.

Cutoff rate: The slope of $|T(j\omega)|$ beyond the bandwidth. A large (negative) slope indicates a good noise rejection capability. Ideally, an infinite slope is desired, indicating a total rejection of signals in the high-frequency range.

We shall now look at some typical closed-loop frequency responses. As an example, let a first-order transfer function be

$$T(s) = \frac{1}{\tau s + 1}.$$

Then $\omega_b = 1/\tau$. There is no peak, and the cutoff rate is -20 [dB/decade] (or equivalently, it has a slope of -1).

Now, consider a standard second-order system

$$T(s) = \frac{\omega_n^2}{s^2 + 2\zeta\omega_n s + \omega_n^2}.$$

Then

$$|T(j\omega)| = \frac{1}{\sqrt{\left(1 - \left(\frac{\omega}{\omega_n}\right)^2\right)^2 + 4\zeta^2\left(\frac{\omega}{\omega_n}\right)^2}}.$$

To find M_p and ω_p, set

$$\frac{d|T(j\omega)|}{d\omega} = 0$$

which gives

$$\omega_p = \omega_n\sqrt{1 - 2\zeta^2}, \quad 0 < \zeta \leq \frac{1}{\sqrt{2}}$$

and

$$M_p = \frac{1}{2\zeta\sqrt{1 - \zeta^2}}.$$

Since the percentage overshoot is a function of ζ, M_p also gives the information about the overshoot. This relationship is shown in Figure 6.32.

Thus $|T(j\omega)|$ has a peak at $\omega_p = \omega_n\sqrt{1 - 2\zeta^2}$ if $0 < \zeta \leq 1/\sqrt{2}$ and has no peak if $\zeta > 1/\sqrt{2}$.

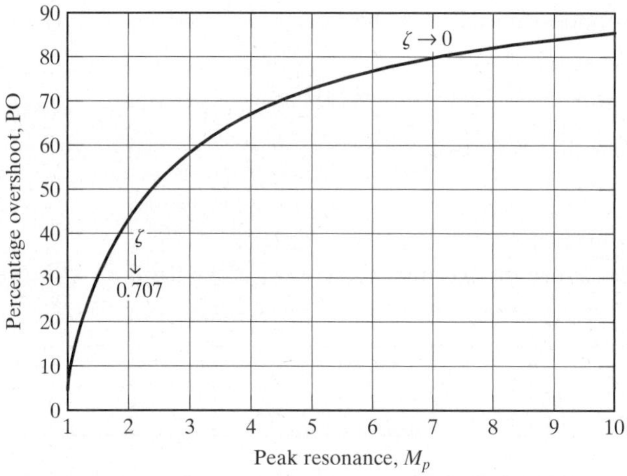

FIGURE 6.32: The relationship between M_p and PO for a standard second-order system with $0 < \zeta \leq 0.707$.

To find the bandwidth ω_b, let

$$|T(j\omega_b)| = \frac{1}{\sqrt{2}}.$$

Then ω_b can be found as

$$\omega_b = \omega_n\sqrt{1 - 2\zeta^2 + \sqrt{4\zeta^4 - 4\zeta^2 + 2}}, \quad 0 \leq \zeta < \infty.$$

The relationship between ω_b and ζ is shown in Figure 6.33.

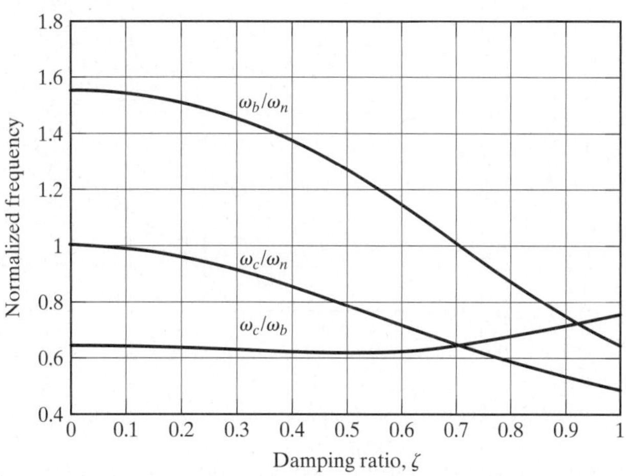

FIGURE 6.33: The relationships among ω_b/ω_n, ω_c/ω_n, ω_c/ω_b, and ζ for a standard second-order system.

Suppose that $T(s)$ is the closed-loop transfer function of an open-loop system with

$$C(s)P(s) = \frac{\omega_n^2}{s(s + 2\zeta\omega_n)}.$$

Then, the crossover frequency ω_c, i.e., the frequency such that $|C(j\omega_c)P(j\omega_c)| = 1$, is given by

$$\omega_c = \omega_n \sqrt{\sqrt{1 + 4\zeta^4} - 2\zeta^2}.$$

Its dependence on ζ is also shown in Figure 6.33. It is interesting to note that the ratio of ω_c/ω_b is almost constant for a fairly large range of ζ:

$$\frac{\omega_c}{\omega_b} \approx 0.64.$$

This is useful in estimating a closed-loop system bandwidth from its open-loop Bode diagram in control system design.

6.6 NICHOLS CHART

Another useful alternative representation of a system's frequency response is the gain–phase diagram. This diagram is generated by plotting the magnitude [dB] versus the phase [deg] of the open-loop frequency response as shown in Figure 6.34.

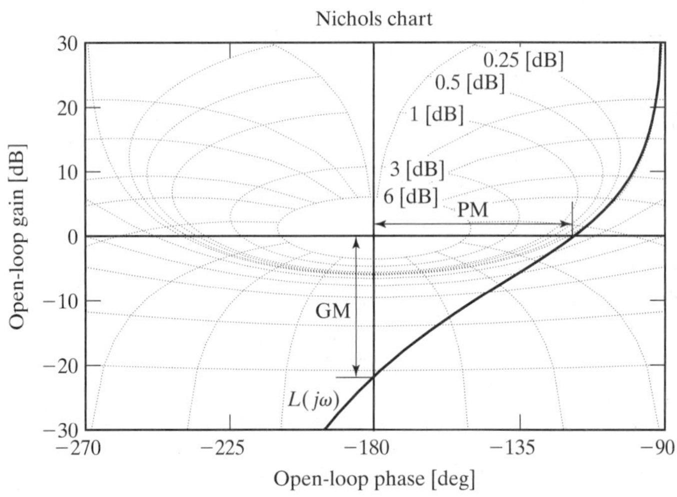

FIGURE 6.34: Gain margin and phase margin on a Nichols chart.

It is seen that the system's stability margins can be found very easily from this diagram. Like the Bode diagram, another advantage of this gain–phase diagram is that changing the system gain only moves the diagram in the vertical direction but does not change the shape of the diagram. On the other hand, adding a phase only moves the diagram in the horizontal direction. This can be very useful in control system design. A disadvantage of this diagram is that it does not explicitly show the dependence of the magnitude and phase on frequency, just like the Nyquist plot.

Assume that the system has the unity feedback. Then the closed-loop transfer function $T(s)$ is given by

$$T(s) = \frac{L(s)}{1 + L(s)}, \quad L(s) = C(s)P(s).$$

One can easily derive the relationship between the magnitude and phase of the closed-loop frequency response $T(j\omega)$ and those of the open-loop frequency response $L(j\omega)$. The loci of $L(j\omega)$ for a constant $|T(j\omega)|$ and a constant $\angle T(j\omega)$ are usually plotted in the gain–phase diagram. This diagram is usually referred to as the **Nichols chart** as shown in Figure 6.34.

Nathaniel B. Nichols (1914–1997) was born in Michigan. He received his B.S. degree from Central Michigan University in 1936, M.S. degree from the University of Michigan in 1937, an Honorary Doctor of Science degree from Central Michigan University in 1964, and an Honorary Doctor of Science degree from Case Western Reserve University in 1968. He worked with the Aerospace Corporation, Dow Chemical Company, MIT, Taylor Instrument Companies, University of Minnesota, and Raytheon Manufacturing Company. His professional experience included automatic control, automatic radar tracking and fire control computers, power-driven servomechanisms, industrial process controllers, recording and controlling instruments, spacecraft attitude controls, and space experiment controls and instruments. He is well known for the Ziegler and Nichols PID tuning rules published in two papers with John G. Ziegler: "Optimum setting for automatic controllers," *Transactions of the ASME*, vol. 64, pp. 759–768, 1942, and "Process lags in automatic control circuits," *Transactions of the ASME*, vol. 65, pp. 433–444, 1943. While working for the Aerospace Corporation, Nichols and his colleagues created the gain–phase diagram (now known as the Nichols chart) to facilitate the calculations of closed-loop frequency response. The Nichols chart is a tool for the designer to read off closed-loop gain and phase directly from a plot of open-loop logarithmic gain and phase, parameterized by frequency. The chart was brought to the control engineering community primarily by its inclusion in the 1947 book *Theory of Servomechanisms* by James, Nichols, and Phillips. The Nichols chart is one contribution among several for which Nichols was honored by the creation of the Nichols Medal of the International Federation of Automatic Control.

EXAMPLE 6.22

Let $L(s)$ be the open-loop transfer function of a unity feedback system

$$L(s) = \frac{s+1}{s(s+5)(s^2+0.8s+4)}.$$

The Nichols chart can then be generated by the following MATLAB command:

```
>> L=tf([1,1],conv([1,5, 0],[1,0.8,4]))

>> nichols(L)

>> ngrid

>> axis([-360, 0, -40,40]) change the displaying axis
```

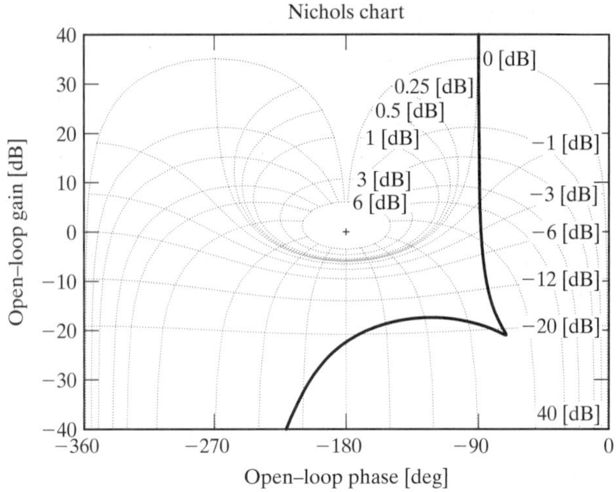

FIGURE 6.35: Nichols chart for Example 6.22.

The Nichols chart for this system is shown in Figure 6.35. The closed-loop frequency response of this system can also be obtained from this diagram by inspecting the intersections of the $L(j\omega)$ curve and the constant contours of the closed-loop frequency response. For example, it can be seen from Figure 6.35 that the closed-loop frequency response changes from 0, −1, −3, −6, −12, −20 [dB], and so on. The exact closed-loop frequency can of course be obtained by plotting $T(j\omega)$ as shown in Figure 6.36. The corresponding MATLAB commands are

```
>> T=feedback(L,1)    % find the closed-loop transfer function

>> bode(T,0.01,10)
```

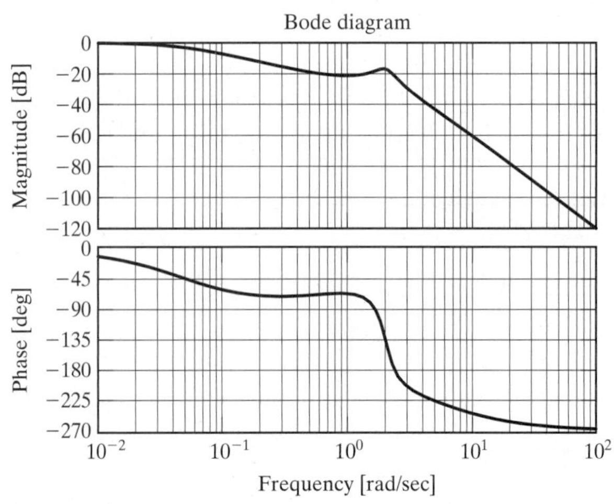

FIGURE 6.36: Closed-loop frequency response for Example 6.22.

6.7 RIEMANN PLOT

A disadvantage of the Nyquist plot is its difficulty in handling the infinity. For example, if $L(s)$ has one or more poles on the imaginary axis, then part of its Nyquist plot is at infinity and it is troublesome to figure out how this part of the Nyquist plot behaves, especially by numerical computation. We have seen from the Nyquist stability criteria that the information of the Nyquist plot at infinity is actually very important in determining closed-loop stability. An alternative to the Nyquist plot, which overcomes the difficulty at infinity, is given by the Riemann plot. It turns out that the Riemann plot not only serves as an alternative to the Nyquist plot because of its easy handling of the infinity but also laying the foundation of a comprehensive robust control theory which is to be covered in this book.

Georg Friedrich Bernhard Riemann (1826–1866) was born in Breselenz, a village near Dannenberg in the Kingdom of Hanover in what is today Germany. His name was most frequently used in mathematics to name various concepts, subjects, and results, such as the Riemann integral, Riemannian geometry, Riemann hypothesis, etc. In 1847 he entered the University of Gottingen, where he first met Carl Friedrich Gauss and attended his lectures on the method of least squares and became Gauss's student. In 1853, Gauss asked Riemann to prepare a Habilitationsschrift on the foundations of geometry. Over many months, Riemann developed his theory of higher dimensions. Riemann held his first lectures in 1854, which not only founded the field of Riemannian geometry but set the stage for Einstein's general relativity. Riemann was arguably the most influential mathematician of the middle of the nineteenth century. His published works constitute only a small volume, but it opened up research areas combining analysis with geometry. He made some of the most famous contributions to modern analytic number theory. In a single, short paper (the only one he published on the subject of number theory), he introduced the Riemann zeta function and established its importance for understanding the distribution of prime numbers. He made a series of conjectures about properties of the zeta function, one of which is the well-known Riemann hypothesis. The Riemann hypothesis today remains one of the most important unsolved problems in mathematics. The first person who solves it will be awarded one million US dollars by the Clay Institute of Mathematics.

Let us place a sphere with unit diameter at the origin of the complex plane, as shown in Figure 6.37. This sphere is called the **Riemann sphere** and is denoted by $\mathbb{S}$. The origin of the complex plane is its south pole S, and the pole at the top is its north pole N. The equator of the Riemann sphere is obviously the horizontal circle at the same level as its center. For a point c in the complex plane, connect it and the north pole by a straight line. This straight line will then intersect the Riemann sphere at one and only one point. This point is called the stereographic

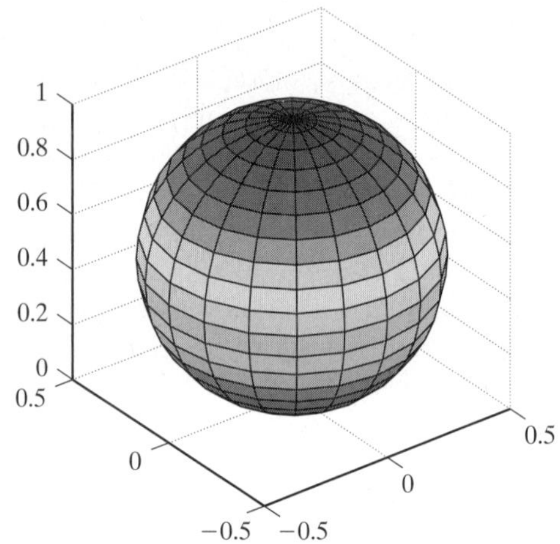

FIGURE 6.37: The Riemann sphere.

projection of c on the Riemann sphere and is denoted by $\phi(c)$. It is, of course, reasonable to define the stereographic projection of ∞ to be the north pole N. Figure 6.38 graphically shows the stereographic projection process.

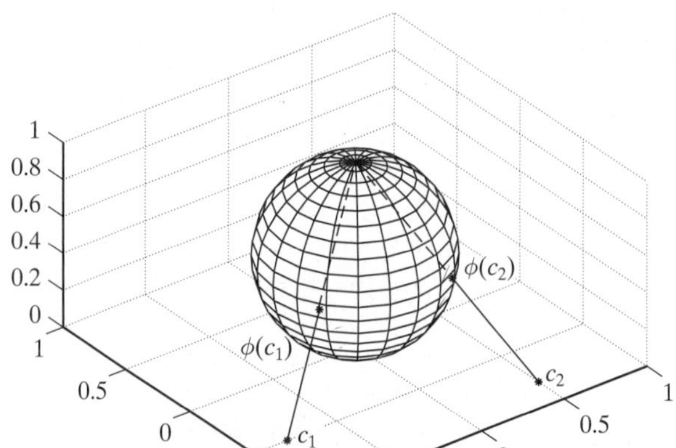

FIGURE 6.38: The stereographic projection.

Let us identify the complex plane with $\mathbb{R}^2$ and identify its real and imaginary axes with the x and y axes, respectively. We also call the vertical axis the z axis. The Riemann sphere then sits in $\mathbb{R}^3$ with axis x-y-z. If a complex number $c = \mathbb{C} \cup \{\infty\}$ is given, what are the x-y-z coordinates of $\phi(c)$? First let us look down from the top. The whole z axis then becomes a point and we only see the complex plane, i.e., the $x - y$ plane. The segment connecting c and N looks like a segment connecting c and the origin. Since $\phi(c)$ is somewhere in the segment, it follows that

$$x = \alpha \operatorname{Re} c, \qquad y = \alpha \operatorname{Im} c$$

for some number $\alpha \in [0,1]$. Let us now cut the Riemann sphere with a vertical plane passing through the z axis and the point c. The intersection of this plane and the Riemann sphere is shown in Figure 6.39. From this figure it can be shown that

$$z = \frac{|c|^2}{1+|c|^2}$$

and

$$x^2 + y^2 = \alpha^2|c|^2 = \frac{|c|^2}{1+|c|^2} - z^2 = \frac{|c|^2}{(1+|c|^2)^2}.$$

Hence

$$\alpha = \frac{1}{1+|c|^2}.$$

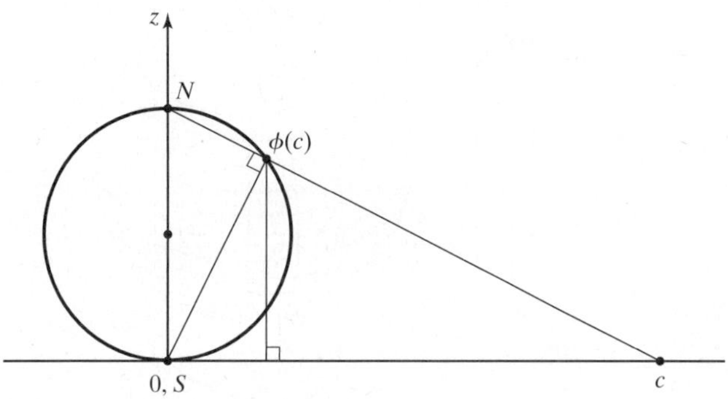

FIGURE 6.39: Relation of c and $\phi(c)$.

In summary, we obtain

$$\phi(c) = \left(\frac{\text{Re } c}{1+|c|^2}, \frac{\text{Im } c}{1+|c|^2}, \frac{|c|^2}{1+|c|^2}\right).$$

For example,

$$\phi(0) = (0,0,0), \quad \phi(1) = \left(\frac{1}{2}, 0, \frac{1}{2}\right), \quad \phi(\infty) = (0,0,1), \quad \phi(1+j1) = \left(\frac{1}{3}, \frac{1}{3}, \frac{2}{3}\right).$$

It is easy to see that the projection of the unit circle in the complex plane is the equator of the Riemann sphere.

Notice that different complex numbers definitely have different projections. This shows that the stereographic projection defines a one-to-one correspondence between $\mathbb{C} \cup \{\infty\}$ and the Riemann sphere $\mathbb{S}$.

The following sequence of MATLAB commands generates the Riemann sphere:

```
>> [X,Y,Z]=sphere;
>> X=X/2; Y=Y/2; Z=Z/2+ones(size(Z))/2;
>> surf(X,Y,Z)
>> axis equal
```

The first line of command generates the data on the x-y-z axes for plotting a unit sphere, i.e., a sphere centered at the origin with unit radius. The data for plotting the Riemann sphere can then be obtained by modifying the data for the unit sphere. The commands from the second line are just for this purpose. The third line of command is for the plotting. The last command is to make the sphere round, instead of being fat. If we change the fifth command to

```
>> mesh(X,Y,Z),
```

we get a different appearance of the Riemann sphere, as shown in Figure 6.40.

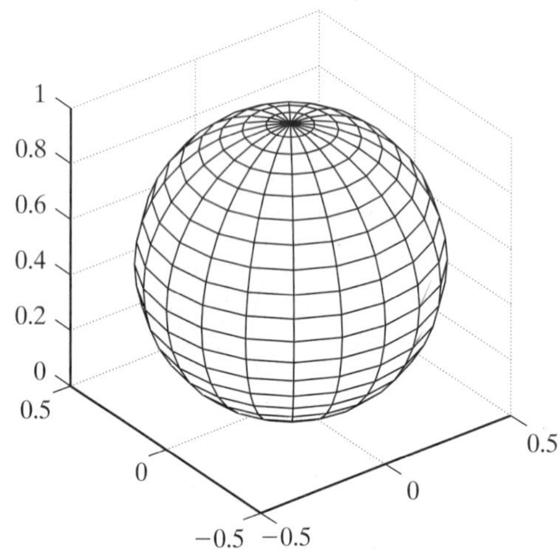

FIGURE 6.40: A different appearance of the Riemann sphere.

It is more convenient and sometimes more natural to represent a number in $\mathbb{C} \cup \{\infty\}$ by b/a for some $a, b \in \mathbb{C}$ which are not simultaneously zero. In this case

$$\boxed{\phi\left(\frac{b}{a}\right) = \left(\frac{\text{Re } a^*b}{|a|^2 + |b|^2}, \frac{\text{Im } a^*b}{|a|^2 + |b|^2}, \frac{|b|^2}{|a|^2 + |b|^2}\right).}$$

Now, suppose we have a frequency response

$$L(j\omega) = \frac{b(j\omega)}{a(j\omega)}.$$

The Riemann plot is then simply the stereographic projection of its Nyquist plot on the Riemann sphere. Clearly, it is given by the curve

$$\left(\frac{\text{Re } a(-j\omega)b(j\omega)}{|a(j\omega)|^2 + |b(j\omega)|^2}, \frac{\text{Im } a(-j\omega)b(j\omega)}{|a(j\omega)|^2 + |b(j\omega)|^2}, \frac{|b(j\omega)|^2}{|a(j\omega)|^2 + |b(j\omega)|^2}\right)$$

as ω goes from $-\infty$ to ∞.

EXAMPLE 6.23

Figure 6.41 gives the Riemann plot of transfer function

$$G(s) = \frac{s+1.5}{s^2+1.5s+2}.$$

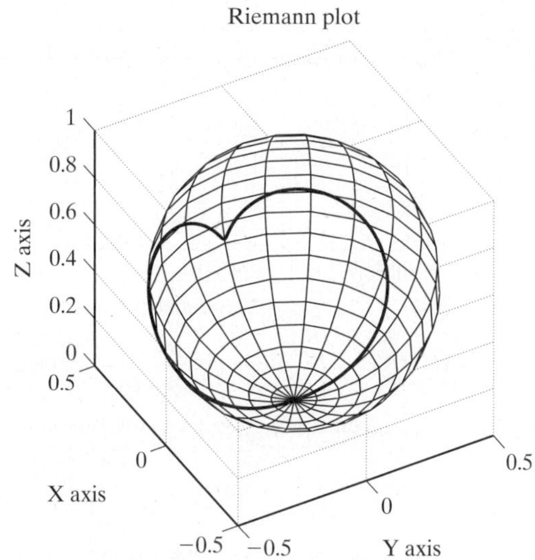

FIGURE 6.41: The Riemann plot of the system in Example 6.23.

PROBLEMS

6.1. The Bode plot for a stable system $P(s)$ is given in Figure 6.42 for the frequency range $0.01 \le \omega \le 10$. Assuming the Bode plot has captured all low-frequency information, find the steady-state response of $P(s)$ due to the input

$$u(t) = 10 - \sin t, \qquad t \ge 0.$$

6.2. Show that the Nyquist plot of the system

$$K\frac{s-z}{s-p}, \qquad p \ne 0,$$

is a circle. Find the intersections of the circle with the real axis. Now assume $K = 1$. From the Nyquist stability criterion, determine the condition for the stability of the closed-loop system under unit negative feedback.

6.3. Consider a unity feedback system with a loop transfer function

$$L(s) = P(s)C(s) = \frac{K(s+2)}{s(s-1)}.$$

Use the Nyquist stability criterion to determine the range of K ($K > 0$) so that the closed-loop system is stable. Then use Routh or Hurwitz criterion to check the result.

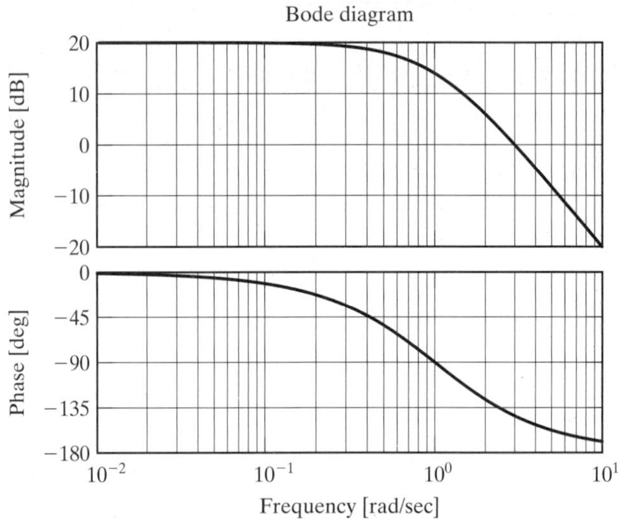

FIGURE 6.42: Bode diagram of a stable system $P(s)$.

6.4. For the following given open-loop transfer function

$$L_1(s) = \frac{K}{s^2(\tau s+1)},\ L_2(s) = \frac{K(Ts+1)}{s^2(\tau s+1)},\ L_3(s) = \frac{K(Ts+1)}{s(\tau s+1)^2}$$

sketch the Nyquist plots for all possible closed-loop stability situations (e.g., asymptotically stable, marginally stable, and unstable) and determine the ranges of K for stability.

6.5. Let the loop transfer function of a unity feedback system be

$$L(s) = \frac{K(\tau s+1)}{s(s+1)(Ts+1)}$$

where $\tau \geq 0$ and $T \geq 0$. Discuss the closed-loop stability of the system using the Nyquist stability criterion.

6.6. Consider a unity feedback system with loop transfer function

$$L(s) = D + K\frac{\prod_{i=1}^{n}(s-a_i)}{\prod_{i=1}^{n}(s+a_i)}$$

where D is a real number and K, a_i, $i = 1, 2, \ldots, n$, are nonzero real numbers.

1. Show that the Nyquist plot of $L(s)$ is a circle.

2. If a_i, $i = 1, 2, \ldots, n$, are all positive, give the condition for the closed-loop stability.

3. If m of a_i, $i = 1, 2, \ldots, n$, are negative, give the condition for the closed-loop stability.

6.7. Suppose that system

$$\frac{1}{s(s^2+6s+10)}$$

is controlled by a pure proportional feedback controller with gain K. We are interested in finding the value of K such that persistent oscillation occurs when the unity feedback system is given a step input. Plot its root locus and find such

K using the information provided by the root locus. Plot its Nyquist plot and find such K using the information provided by the Nyquist plot.

MATLAB PROBLEMS

6.8. Sketch the Bode diagrams using MATLAB for the following systems:

$$P_1(s) = \frac{100}{(s+1)(s+10)}, \qquad P_2(s) = \frac{1000(s+1)}{s(s+20)(s+30)},$$

$$P_3(s) = \frac{10^4(s+1)}{(s+10)(s^2+2s+100)}, \qquad P_4(s) = \frac{2000e^{-5s}}{s(s+1)(s+50)},$$

$$P_5(s) = \frac{10}{(s-1)(s+5)}, \qquad P_6(s) = \frac{10(s-1)}{s(s-2)}.$$

6.9. Plot the Nyquist plots for the systems in Problem 6.8 and find the ranges of K, if possible, so that the closed-loop systems with open-loop transfer functions $KP_i(s)$ for each i are stable.

6.10. Find the closed-loop GMs and PMs of unity feedback systems with open-loop transfer functions given in Problem 6.8.

6.11. Obtain the Nyquist plots of the following systems using the nyquist command. Determine whether the closed-loop systems formed by these open-loop systems under unity negative feedback are stable. For those giving stable closed-loop systems, find their GM and PM.

1. $P_1(s) = \dfrac{1}{s(s^2+s+0.5)}$.
2. $P_2(s) = \dfrac{20(s+1)}{s(s+5)(s^2+2s+10)}$.
3. $P_3(s) = \dfrac{s+0.5}{s^3+s^2+1}$.
4. $P_4(s) = \dfrac{100(s+0.5)}{s^2(s+2)(s+10)}$.

6.12. Obtain the Bode plots of the following systems. For those giving stable closed-loop systems under unity feedback, find their GMs and PMs. (The following commands are useful: bode, logspace, loglog, semilogx.)

1. $P_1(s) = \dfrac{10(s+1)}{(s+2)(s+5)}$.
2. $P_2(s) = \dfrac{1}{s^2+1}$.
3. $P_3(s) = \dfrac{20(s^2+s+0.5)}{s(s+1)(s+10)}$.
4. $P_4(s) = \dfrac{s+0.5}{s^3+s^2+1}$.

6.13. Consider a unity feedback system with an open-loop transfer function

$$L(s) = P(s)C(s) = \frac{100e^{-\tau s}(s+1)}{s(s+10)(s^2+2s+10)}.$$

Find the largest possible time delay τ so that the closed-loop system is stable.

6.14. Write a MATLAB program to plot the Riemann plot of a transfer function. Use this program to plot the Riemann plot of the following transfer functions:

1. $G(s) = \dfrac{1}{s(s+1)}$.
2. $G(s) = \dfrac{s+1}{s^2(s-1)}$.

EXTRA CREDIT PROBLEMS

6.15. Show that the sterographic projection of a straight line in the complex plane onto the Riemann sphere is a circle.

6.16. Show that the stereographic projection of a circle in complex plane onto the Riemann sphere is a circle.

NOTES AND REFERENCES

The use of the Bode diagram, Nyquist plot, and Nichols chart in feedback control has a long history. Together with the root locus, they constitute the main tools in the so-called classical control theory.

H. W. Bode, *Network Analysis and Feedback Amplifier Design*, Van Nostrand, New York, 1945.

H. Nyquist, "Regeneration theory," *Bell System Technical Journal*, vol. 11, pp. 126–147, 1932.

H. M. James, N. B. Nichols, and R. S. Phillips, *Theory of Servomechanisms*, McGraw-Hill, New York, 1947.

The use of the Riemann plot, however, is quite recent. Its first appearance is probably in

A. K. El-Sakkary, "Estimating robustness on the Riemann sphere," *International Journal of Control*, vol. 42, pp. 561–567, 1989.

A more systematic book on the use of Riemann plot and a complete feedback system analysis and synthesis theory based on it is

G. Vinnicombe, *Uncertainty and Feedback: $\mathcal{H}_\infty$ Loop-Shaping and the ν-Gap Metric*, Imperial College Press, London, 2001.

Classical Design in Frequency Domain

CHAPTER 7

Controller design in frequency domain can be very convenient in many applications. In this chapter, we shall explore some simple design techniques using Bode diagrams. Our focus will be on the simple lead-lag or PID type of controllers since they are widely used in industry, particularly in process industry. The main reason for their popularity is perhaps their simplicity in design and tuning. Over the years, many PID controller design and tuning techniques have been proposed. The root-locus method discussed in Chapter 5 and the frequency-response method to be discussed in this chapter are both model based, i.e., their designs are based on the availability of plant mathematical models or their frequency responses. However, plant models are not generally available for many control systems. It is then important to develop techniques or rules to design or tune controllers so that the desired performance can be achieved. Two widely used classical tuning techniques are due to Ziegler and Nichols and are commonly known as the Ziegler and Nichols tuning rules. In the later part of this chapter we will discuss these tuning techniques and give some justifications for these rules using frequency-domain analysis. We shall also point out the limitations of these techniques through examples. Some practical considerations such as saturation and sensor noise rejection are also considered in this chapter. We then look at a loopshaping controller design method without putting any structure restrictions on the controller. Finally, performance limitations of a closed-loop system due to the right half plane poles, zeros, and bandwidth constraints are explored using the classical Bode integral relations.

7.1 PHASE-LAG CONTROLLER

We shall consider a first order phase-lag controller in the following form:

$$C(s) = \frac{K_c(s/b+1)}{s/a+1}, \quad b > a.$$

Note that $C(0) = K_c$ and $C(\infty) = K_c a/b$. The Bode diagram of this controller with $K_c = 1$ is shown in Figure 7.1. In order to better illustrate how the controller parameters are selected on the Bode diagrams, we have drawn controller magnitude Bode plots using straight line approximations throughout this chapter.

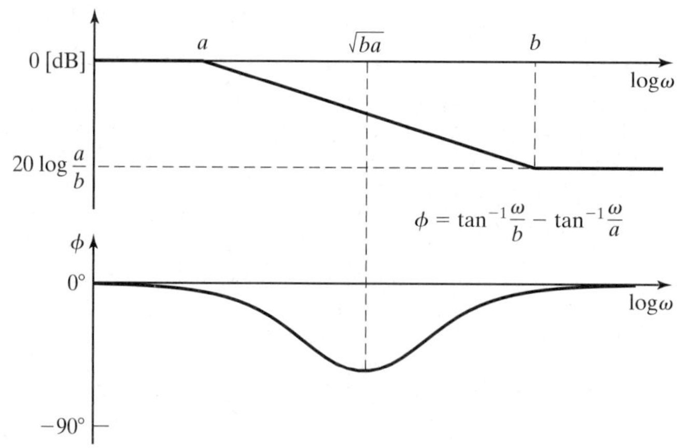

FIGURE 7.1: Bode diagram of a phase-lag controller $C(s) = \dfrac{s/b+1}{s/a+1}$.

As we have pointed out in our discussion of the root-locus design, a phase-lag controller is a gain compensator. It is used mainly to increase low-frequency gain without affecting the frequency response near the crossover region. The following procedure can be used to design a phase-lag controller.

Algorithm 7.1 (Phase-lag controller design). (See Figure 7.2)

Step 1: Find K_c so that the DC gain requirements of the open-loop system $L(s) = C(s)P(s)$ are satisfied. For example, K_c has to be chosen to satisfy steady-state error requirements on tracking, disturbance rejection, and so on.

Step 2: Determine the desired crossover frequency ω_c and PM ϕ_c. (Recall that the crossover frequency is related to the rise time and settling time, and the PM is related to the overshoot of the system response.)

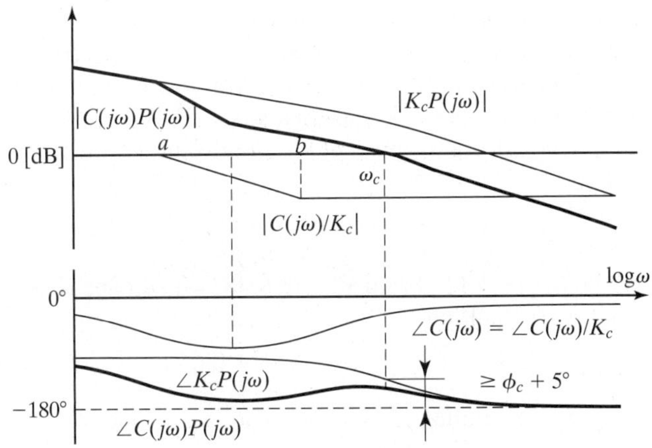

FIGURE 7.2: Phase-lag controller design.

Step 3: Plot the Bode diagram of $K_cP(s)$.

Step 4: Check the phase plot of $K_cP(s)$ at the desired crossover frequency ω_c. If $180° + \angle K_cP(j\omega_c) \geq \phi_c + 5°$, then a phase-lag controller can be designed to satisfy the design specifications. (The reason for adding 5° is that we expect controller $C(s)$ to contribute about $-5°$ at ω_c when the controller parameters are appropriately chosen as below.)

Step 5: Choose

$$b \approx 0.1\omega_c, \quad a = \frac{b}{K_c|P(j\omega_c)|}.$$

Step 6: Finally, a phase-lag controller is given by

$$C(s) = \frac{K_c(s/b+1)}{s/a+1}.$$

We shall now illustrate the above design procedure through an example.

EXAMPLE 7.2

Consider a unity feedback system with

$$P(s) = \frac{8(s+10)}{s(s+3)(s+20)}.$$

We are required to design a feedback controller so that the following specifications are satisfied:

- The step response of the system has no more than 20% maximum overshoot and 8 [sec] of maximum settling time (2% tolerance).
- The steady-state error with respect to a ramp input is no greater than 0.05.

To design a controller satisfying the above design specifications, we first need to translate the above time-domain specifications into frequency-domain specifications. We shall use the formula for the standard second-order system to find the equivalent frequency-domain specifications. From Figure 6.29, we can see that a PO $\leq 20\%$ overshoot is equivalent to a damping $\zeta \geq 0.45$ or PM $\geq 47°$. Now, the settling time

$$t_s = \frac{4}{\zeta \omega_n} \leq 8$$

gives $\omega_n \geq 0.5/\zeta = 1.11$. From Figure 6.33, we can see that we need $\omega_c \geq 0.8\omega_n = 0.89$ when $\zeta \approx 0.45$. Finally,

$$\frac{1}{K_v} \leq 0.05$$

gives $K_v \geq 20$. In summary, we need to design a controller so that the loop transfer function $L(s) = C(s)P(s)$ has

- PM $\geq 47°$ and $\omega_c \geq 0.89$;
- $K_v \geq 20$.

Since

$$K_v = \lim_{s \to 0} sC(s)P(s) = 4C(0)/3,$$

we need $C(0) \geq 15$. Now take $K_c = 15$ and plot the Bode diagram of $K_cP(s)$ as shown in Figure 7.3 (dashed lines). The Bode diagram shows that we can choose $\omega_c = 2$ and a lag controller can be designed to satisfy the PM and the K_v condition since

$$180° + \angle P(j2) = 62° > 47° + 5°.$$

```
P=zpk(-10,[0, -3, -20],8)
Pat2=bode(P,2)     computing |P(j2)|
C=tf(15*[5,1],[42,1]);
L=P*C;
bode(L)
[Gm,Pm,wg,wp]=margin(L)   computing margins
```

We now pick $b = 0.1\omega_c = 0.2$. Then

$$a = \frac{b}{K_c|P(j2)|} = \frac{0.2}{8.4432} = 0.0237$$

which gives a phase-lag controller

$$C(s) = \frac{K_c(s/b+1)}{s/a+1} = \frac{15(s/0.2+1)}{s/0.0237+1} = \frac{15(5s+1)}{42s+1}.$$

The compensated loop transfer function is shown in Figure 7.3 as solid lines with 56.75° PM. The step response of the closed-loop system is shown in Figure 7.4 and it is clear that the design specifications are satisfied. In fact, the overshoot is 16% and the settling time is about 6.94 [sec].

A controller design process is an iterative process and usually involves some (or many) trial-and-error steps. For example, if one chooses $\omega_c = 3$, then the above design process will produce a controller

$$C_1(s) = \frac{50s+15}{16.23s+1}$$

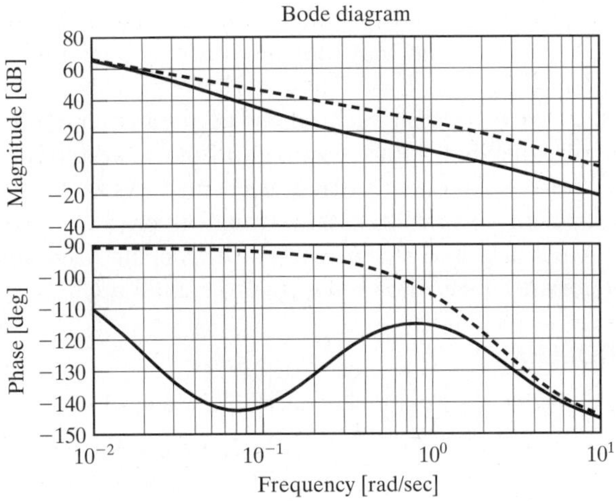

FIGURE 7.3: Example 7.2 – $K_cP(s)$ (dashed) and $C(s)P(s)$ (solid).

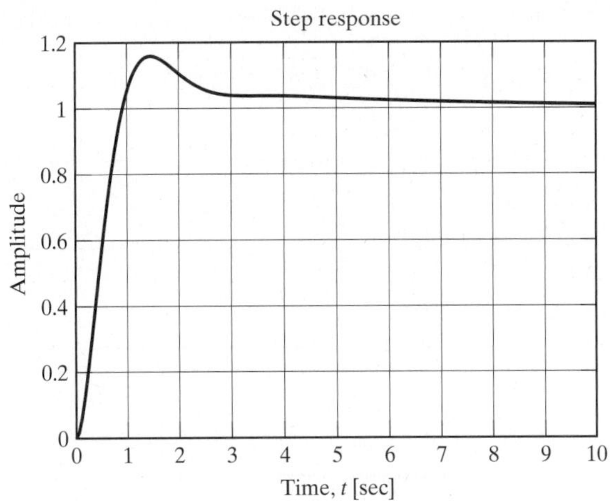

FIGURE 7.4: Example 7.2 – Step response.

with a PM of 48.58°. However, this controller design will result in a 23% overshoot to a step response and therefore the design specification will not be satisfied. On the other hand, one may get much better transient response by carrying out some minor tuning on the controller $C(s)$. For example, picking b and a a few times smaller, say $b = 0.2/5 = 0.04$ and $a = 0.0237/5 = 0.00474$, we get

$$C_2(s) = \frac{15(25s + 1)}{210s + 1}$$

which will result in a closed-loop response with 10% overshoot and 2.4 [sec] settling time.

It is not hard to see from the above design process that a controller satisfying the design specifications may also be obtained by taking even a smaller b, for example, $b = 0.01\,\omega_c$ or $b = 0.001\,\omega_c$. However, this is generally undesirable since a controller designed with such choice of parameters will significantly reduce the magnitude frequency response (loop gain) of the system in the low-frequency range in a wider range of frequencies than necessary. As we have discussed in Section 6.1, the loop gain in the low-frequency range is critically related to the disturbance rejection capability and the steady-state tracking accuracy of the control system. It is therefore necessary not to choose too small a b.

7.2 PI CONTROLLER

We shall consider a PI controller in the following form:

$$C(s) = \frac{K_c(s/b+1)}{s}.$$

Note that a PI controller increases the system type by 1 with $C(0) = \infty$ and $C(\infty) = \dfrac{K_c}{b}$. The Bode diagram of this controller with $K_c = 1$ is shown in Figure 7.5.

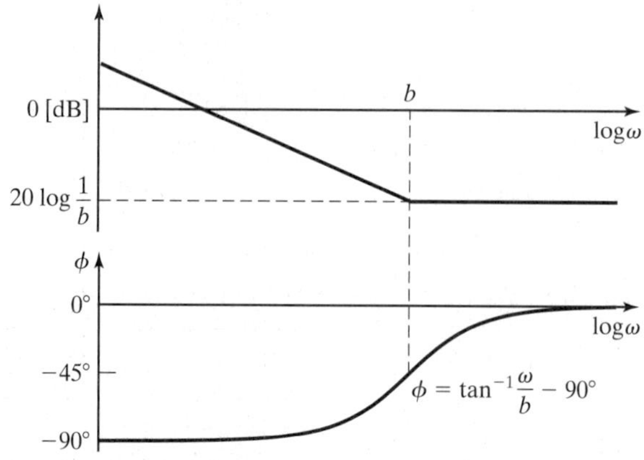

FIGURE 7.5: Bode diagram of a PI controller $C(s) = \dfrac{s/b+1}{s}$.

Algorithm 7.3 (PI controller design).

Step 1: Determine the desired crossover frequency ω_c and PM ϕ_c. (Recall that the crossover frequency is related to the rise time and settling time, and the PM is related to the overshoot of the system response.)

Step 2: Plot the Bode diagram of $KP(s)$ for any given K (for example, $K = 1$).

Step 3: Check the phase plot of $KP(s)$ at the desired crossover frequency ω_c. If $180° + \angle KP(j\omega_c) \geq \phi_c + 5°$, then a PI controller can be designed to satisfy the design specifications. (The reason for adding 5° is that we expect controller $C(s)$ to contribute about $-5°$ at ω_c when the controller parameters are appropriately chosen as below.)

Step 4: Choose

$$b \approx 0.1\omega_c, \quad K_c = \frac{b}{|P(j\omega_c)|}.$$

Step 5: Finally, a PI controller is given by

$$C(s) = \frac{K_c(s/b+1)}{s}.$$

We shall now illustrate the above design procedure through an example.

EXAMPLE 7.4

Assume that a plant transfer function is given by

$$P(s) = \frac{30}{(s+5)(s^2+4s+8)}$$

and we desire to design a feedback controller so that the following specifications are satisfied:

- The system has a PM of at least 40° and a crossover frequency of at least 2.
- The steady-state error with respect to a step input is no more than 0.01.

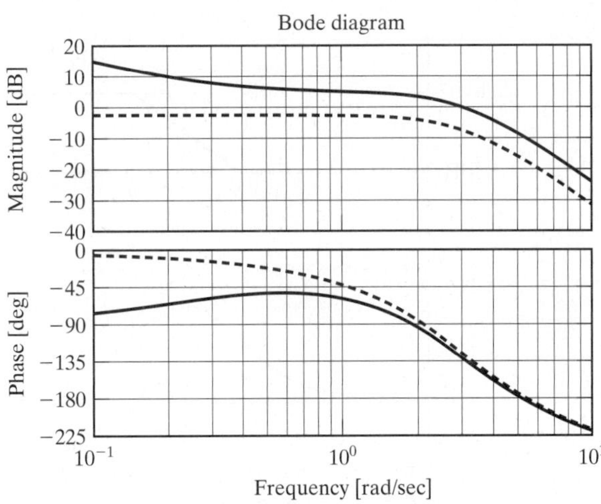

FIGURE 7.6: Example 7.4 – $P(s)$ (dashed) and $C(s)P(s)$ (solid).

We first plot the Bode diagram of $P(s)$ as shown in Figure 7.6. The Bode diagram shows that the phase of $P(s)$ at $\omega = 2$ is about $-85°$. Thus a phase-lag or PI controller can be designed to satisfy the design specifications. We choose

$\omega_c = 3$ for fast response. Then $\angle P(j3) \approx -126°$ and $|P(j3)| = 0.4273$. Take $b = 0.1\omega_c = 0.3$ and

$$K_c = \frac{b}{|P(j\omega_c)|} = \frac{0.3}{0.4273} = 0.7021.$$

We then have a PI controller

$$C(s) = \frac{K_c(s/b+1)}{s} = \frac{0.7021(s/0.3+1)}{s}.$$

The compensated loop transfer function is shown in Figure 7.6 as solid lines. The compensated system is type 1 and has a PM of 48° with $\omega_c = 3$. Thus, all design specifications are met.

One can also design a phase-lag controller to satisfy the design specification. For example, a phase-lag controller

$$C(s) = \frac{133.3(s/0.3+1)}{s/0.0053+1}$$

satisfies the design specifications.

7.3 PHASE-LEAD CONTROLLER

We shall consider a first-order phase-lead controller in the following form:

$$C(s) = \frac{K_c(s/b+1)}{s/a+1}, \quad b < a.$$

The Bode diagram of this controller with $K_c = 1$ is shown in Figure 7.7.

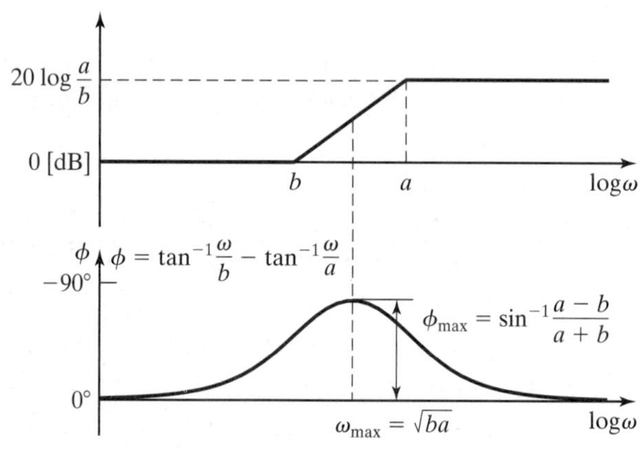

FIGURE 7.7: Phase-lead controller $C(s) = \dfrac{s/b+1}{s/a+1}$ $(b < a)$.

Note that the maximum phase of this compensator (see Figure 7.8) is given by

$$\phi_{\max} = \sin^{-1}\frac{a-b}{a+b} = \sin^{-1}\frac{a/b-1}{a/b+1}$$

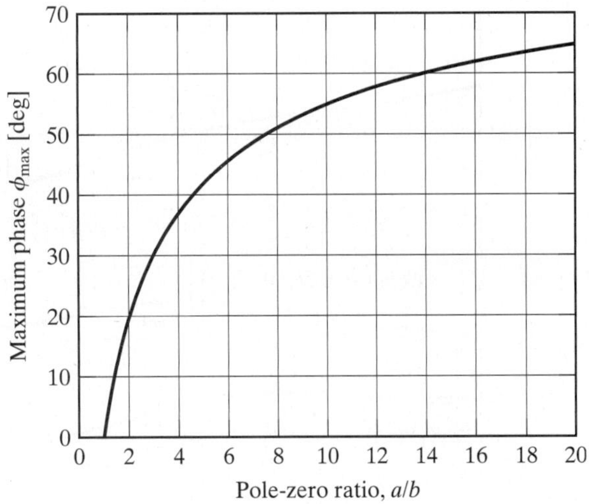

FIGURE 7.8: Maximum phase angle for $C(s) = \dfrac{s/b+1}{s/a+1}$: $\phi_{\text{max}} = \sin^{-1}\dfrac{a-b}{a+b}$.

or

$$\frac{a}{b} = \frac{1+\sin\phi_{\text{max}}}{1-\sin\phi_{\text{max}}},$$

and the frequency where the maximum is achieved is

$$\omega_{\text{max}} = \sqrt{ba}.$$

As we have pointed out in the root-locus design, a phase-lead controller is a phase compensator. It is used mainly to increase PM without reducing the crossover frequency. The following procedure can be used to design a phase-lead controller.

Algorithm 7.5 (Phase-lead controller design). (See Figure 7.9)

Step 1: Find K_1 so that the DC gain requirements of the open-loop system $L(s) = K_1P(s)$ are satisfied. For example, K_1 has to be chosen to satisfy steady-state error requirements on tracking, disturbance rejection, and so on.

Step 2: Determine the desired crossover frequency ω_{desired}, PM ϕ_{desired}, and so on. (Recall that the crossover frequency is related to the rise time and settling time, and the PM is related to the overshoot of the system response.)

Step 3: Plot the Bode diagram of $K_1P(s)$ and calculate the crossover frequency ω_1.

Step 4: If $\omega_1 \ll \omega_{\text{desired}}$, let $\omega_c = \omega_{\text{desired}}$ and let

$$\phi_{\text{max}} = \phi_{\text{desired}} - \angle K_1P(j\omega_c) - 180°.$$

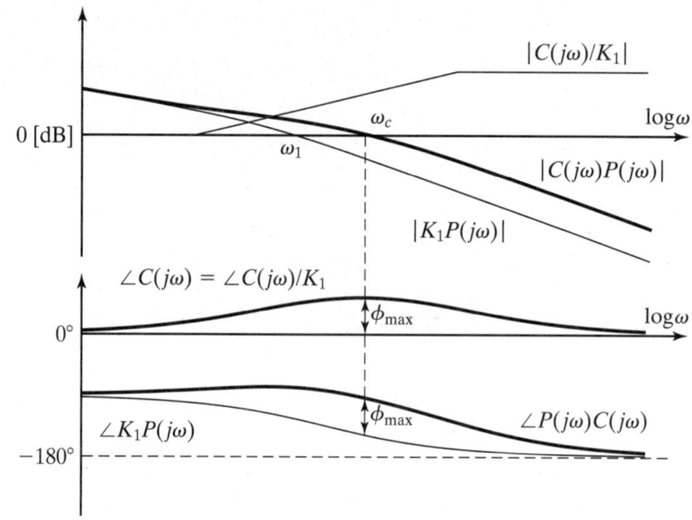

FIGURE 7.9: Phase-lead controller design.

Find a and b from the following equations:

$$\frac{a}{b} = \frac{1 + \sin\phi_{\max}}{1 - \sin\phi_{\max}}, \quad \omega_c = \sqrt{ba}.$$

Find K_c such that

$$\left|\frac{K_c(j\omega_c/b + 1)}{j\omega_c/a + 1}\right| |P(j\omega_c)| = 1.$$

If $K_c \geq K_1$, go to step 6. Otherwise, go to step 5.

Step 5: If $\omega_1 \approx \omega_{\text{desired}}$ or $\omega_1 \geq \omega_{\text{desired}}$, let $K_c = K_1$. Estimate the phase $\phi_{\max}$ needed by examining the Bode diagram in the frequency range $\geq \omega_1$. (It is expected that the lead compensator will somewhat increase the crossover frequency.) Let

$$\frac{a}{b} = \frac{1 + \sin\phi_{\max}}{1 - \sin\phi_{\max}}$$

and let ω_c be such that

$$|K_1 P(j\omega_c)| = \sqrt{\frac{b}{a}}.$$

Find b and a by setting

$$\omega_c = \sqrt{ab}.$$

Step 6: Let a lead controller be given by

$$C(s) = \frac{K_c(s/b + 1)}{s/a + 1}.$$

Step 7: Plot the Bode diagram of $C(s)P(s)$ and check the design specifications. Adjust K_c, a, and b if necessary. If the design specifications cannot be satisfied, then a lead-lag controller may be needed.

We shall now illustrate the above design procedure through two examples.

EXAMPLE 7.6

A plant transfer function is given by

$$P(s) = \frac{100}{s(s+20)}.$$

We desire to design a feedback controller so that the following specifications are satisfied:

- The system has a PM of at least 50° and a crossover frequency of at least 40.
- The steady-state error with respect to a ramp input is no more than 0.1.

The steady-state error requirement is easily satisfied for any $K_c \geq K_1 := 2$. The Bode diagram for $K_1 P(s)$ is shown in Figure 7.10. The crossover frequency is $\omega_1 = 10$, which is much smaller than the desired crossover frequency $\omega_{\text{desired}} = 40$. Since $\angle K_1 P(j\omega_{\text{desired}}) \approx -154°$, the positive phase needed at $\omega_{\text{desired}} = 40$ to guarantee the PM is

$$\phi_{\max} = 50 + 154 - 180 = 24°.$$

Let

$$\frac{a}{b} = \frac{1+\sin\phi_{\max}}{1-\sin\phi_{\max}} = 2.3712,$$

$$\sqrt{ba} = \omega_{\text{desired}} = 40,$$

i.e.,

$$a \approx 61.6, \quad b \approx 26.$$

```
P=tf(100,[1,20,0]);

K1P=2*P;    bode(K1P)

[Mg40,Ph40]=bode(K1P,40)
```
computing phase at $j40$

```
C1=tf([1/26,1],[1/61.6,1])

C1P40=bode(C1*P,40)
```
computing $|C_1(j40)P(j40)|$

```
L=11.6*C1*P

hold, grid, bode(L)

[Gm,Pm,wg,wp]=margin(L)
```
computing margins

We shall now determine K_c so that the crossover frequency is $\omega_c = \omega_{\text{desired}} = 40$, i.e.,

$$|C(j40)P(j40)| = \left|\frac{K_c(j40/26+1)}{j40/61.6+1}P(j40)\right| = 0.086\ K_c = 1.$$

We thus need $K_c = 11.6$, which is greater than K_1. Hence the final controller is given by

$$C(s) = \frac{11.6(s/26+1)}{s/61.6+1}$$

and all design specifications are satisfied. The Bode diagram of the compensated loop transfer function is also shown in Figure 7.10, which shows that the system has 50° PM.

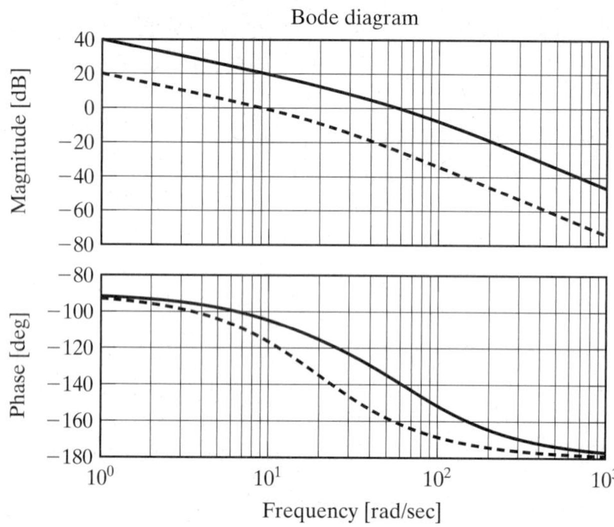

FIGURE 7.10: Example 7.6 – $K_1P(s)$ (dashed) and $C(s)P(s)$ (solid).

EXAMPLE 7.7

We shall now consider a system with the same transfer function as in Example 7.6,

$$P(s) = \frac{100}{s(s+20)}$$

and we desire to design a feedback controller so that the following specifications are satisfied:

- The system has a PM of at least 50° and a crossover frequency of at least 40.
- The steady-state error with respect to a ramp input is no more than 0.01.

The steady-state error requirement is satisfied for any $K_c \geq K_1 := 20$. Plot the Bode diagram for $K_1P(s)$ as shown in Figure 7.11. The crossover frequency is $\omega_1 = 42.5$, which is greater than the desired crossover frequency $\omega_{\text{desired}} = 40$. Thus, we need not increase the gain any further and we can take $K_c = K_1 = 20$. We also expect that the crossover frequency after the lead compensation will be larger than ω_1. Now, examining the phase diagram of $K_1P(s)$, we can see that the lead controller will probably need to provide a positive phase of about 30° degree. Let

$$\phi_{\max} = 40°$$

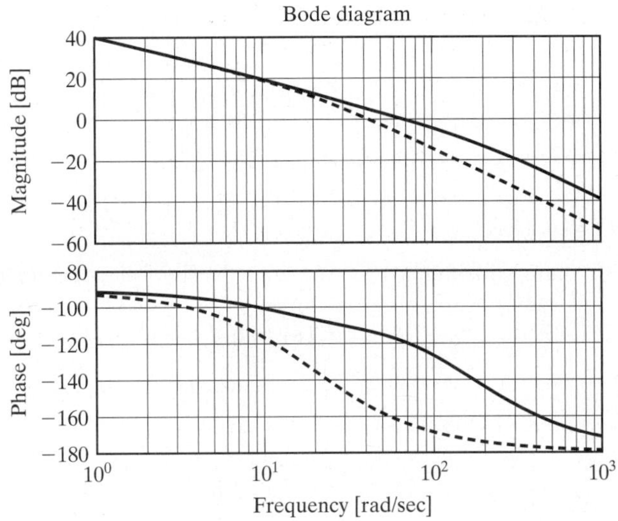

FIGURE 7.11: Example 7.7 – $K_1P(s)$ (dashed) and $C(s)P(s)$ (solid).

and let

$$\frac{a}{b} = \frac{1 + \sin\phi_{\max}}{1 - \sin\phi_{\max}} = 5.6254.$$

Let ω_c be such that

$$|K_1P(j\omega_c)| = \sqrt{\frac{1}{5.6254}} = 0.42.$$

From the Bode diagram, we can find $\omega_c \approx 67$. Let

$$\sqrt{ba} = \omega_c = 67,$$

i.e.,

$$a \approx 158.9, \quad b \approx 28.25.$$

Thus we have

$$C(s) = \frac{20(s/28.25 + 1)}{s/156.9 + 1}.$$

The actual PM achieved is 60.7°. The Bode diagrams of the compensated system are also shown in Figure 7.11.

It is not hard to see that in Examples 7.6 and 7.7 we could have selected a controller in the following form

$$C(s) = \frac{K_c(s/20 + 1)}{s/a + 1}$$

so that the undesirable pole at −20 is cancelled. This can significantly simplify the design process. Indeed, it is fairly easy to verify that a controller

$$C(s) = \frac{8(s/20 + 1)}{s/47.67 + 1}$$

will satisfy the design specification in Example 7.6 and a controller

$$C(s) = \frac{20(s/20+1)}{s/119.17+1}$$

will satisfy the design specification in Example 7.7.

7.4 PD CONTROLLER

We shall again consider a PD controller in the following form:

$$C(s) = K_c(s/b+1).$$

The Bode diagram of this controller with $K_c = 1$ is shown in Figure 7.12. Note that the phase of this compensator is given by

$$\phi = \tan^{-1}\frac{\omega}{b}.$$

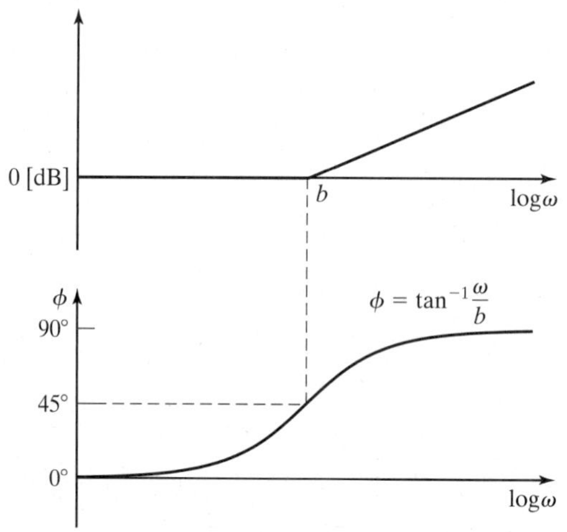

FIGURE 7.12: Bode diagram of a PD controller $C(s) = s/b + 1$.

Algorithm 7.8 (PD controller design).

Step 1: Find K_1 so that the DC gain requirements of the open-loop system $L(s) = K_1 P(s)$ are satisfied. For example, K_1 has to be chosen to satisfy steady-state error requirements on tracking, disturbance rejection, and so on.

Step 2: Determine the desired crossover frequency ω_{desired}, PM ϕ_{desired}, and so on. (Recall that the crossover frequency is related to the rise time and settling time, and the PM is related to the overshoot of the system response.)

Step 3: Plot the Bode diagram of $K_1 P(s)$ and calculate the crossover frequency ω_1.

Step 4: If $\omega_1 \ll \omega_{\text{desired}}$, let $\omega_c = \omega_{\text{desired}}$ and let

$$\phi_{\max} = \phi_{\text{desired}} - \angle K_1 P(j\omega_c) - 180^\circ.$$

Find b from the following equation

$$b = \frac{\omega_c}{\tan \phi_{\max}}.$$

Find K_c such that

$$K_c \left| j\omega_c/b + 1 \right| \left| P(j\omega_c) \right| = 1.$$

If $K_c \geq K_1$, go to step 6. Otherwise, go to step 5.

Step 5: If $\omega_1 \approx \omega_{\text{desired}}$ or $\omega_1 \geq \omega_{\text{desired}}$, let $K_c = K_1$. Estimate the phase $\phi_{\max}$ needed by examining the Bode diagram in the frequency range $\geq \omega_1$. (It is expected that the PD compensator will somewhat increase the crossover frequency.) Let ω_c and b be chosen such that

$$b = \frac{\omega_c}{\tan \phi_{\max}}$$

and

$$\left| K_c (j\omega_c/b + 1) P(j\omega_c) \right| \approx 1.$$

(This can be done graphically from the Bode diagram.)

Step 6: Let a PD controller be given by

$$C(s) = K_c (s/b + 1).$$

Step 7: Plot the Bode diagram of $C(s)P(s)$ and check the design specifications. Adjust b and K_c if necessary. If the design specifications cannot be satisfied, then a PID or lead-lag controller may be needed.

We shall now illustrate the above design procedure through an example.

EXAMPLE 7.9

Consider a feedback system with a plant transfer function

$$P(s) = \frac{50}{s^2(s+10)}.$$

Suppose we desire to design a PD feedback controller so that the system has a PM of at least 40° and a crossover frequency of at least 1. Since the plant has a double integral, we shall not pose any additional condition on the steady-state requirement.

To design a PD controller for this system, the Bode diagram of $P(s)$ is constructed as shown in Figure 7.13. Since $\angle P(j10) = -225°$ and the maximum phase of the PD controller is 90°, it is clear that the crossover frequency must be less than 10 in order to get the desired PM with a PD compensation. Since there is no specific requirement on the low-frequency gain, we can choose $K_c < 1$. Let $\omega_c = 1$ and note that $\angle P(j1) = -185.7°$. Hence, to achieve at least 40° PM at $\omega_c = 1$, we shall take $\phi_{\max} = 50°$. Then

$$b = \frac{\omega_c}{\tan \phi_{\max}} = 0.8391.$$

Let K_c be such that

$$K_c |j\omega_c/b + 1| \, |P(j\omega_c)| = 1.$$

Then $K_c = 0.1292$. Thus a PD controller is given by

$$C(s) = K_c(s/b + 1) = 0.1292(s/0.8391 + 1).$$

The compensated Bode diagram is shown in Figure 7.13 as solid lines.

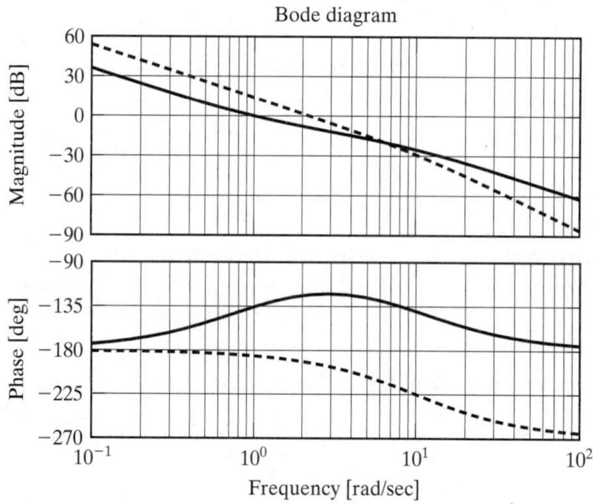

FIGURE 7.13: Example 7.9 – $P(s)$ (dashed) and $C(s)P(s)$ (solid).

Now suppose we desire to have the loop gain greater than 20 [dB] for all frequencies $\omega \leq 1$ with the same PM. This would be satisfied for $K_1 P(s)$ with $K_1 = 2.01$. Since the crossover frequency for $K_1 P(s)$ is greater than 3 from the Bode diagram in Figure 7.14, it is necessary to pick $\omega_c \geq 3$. Now plot the Bode diagram of $K_1 P(s)$ as shown in Figure 7.14. Let $\omega_c = 7$. We then need $\phi_{\max} = 75°$ and

$$b = \frac{\omega_c}{\tan \phi_{\max}} = \frac{7}{3.7321} = 1.8756.$$

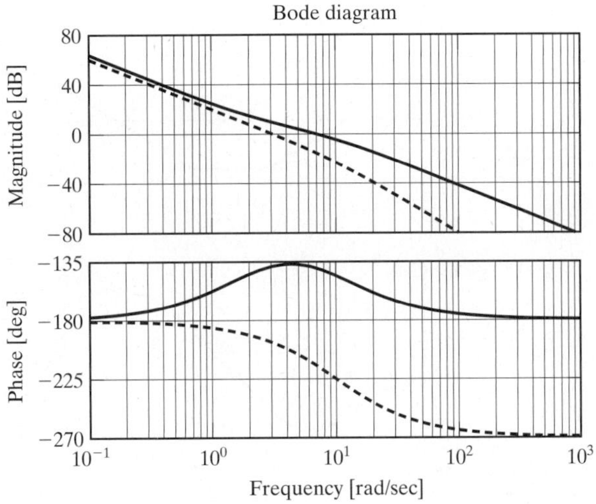

FIGURE 7.14: Example 7.9 – $K_1P(s)$ (dashed) and $C_1(s)P(s)$ (solid).

Let K_c be such that

$$K_c|j\omega_c/b+1|\,|P(j\omega_c)| = 1,$$

i.e., $K_c = 3.096 > K_1$. We then obtain a PD controller as

$$C_1(s) = 3.096(s/1.8756+1),$$

and the compensated Bode diagram is shown in Figure 7.14.

The design process described above can be regarded only as a starting point for any practical design. Subsequent adjustments to the design parameters are usually highly desirable to produce a satisfactory controller design.

7.5 LEAD-LAG OR PID CONTROLLER

A lead-lag controller can take the following general form:

$$C(s) = \frac{K_c(s/b_1+1)}{s/a_1+1}\,\frac{s/b_2+1}{s/a_2+1}, \quad a_1 < b_1 < b_2 < a_2.$$

The Bode diagram of this controller with $K_c = 1$ is shown in Figure 7.15.

The design of this controller essentially follows a combination of a lead controller design and a lag controller design although it is much more flexible.

The following procedure can be used to design a lead-lag controller.

Algorithm 7.10 (Lead-lag controller design).

Step 1: Find K_c so that the DC gain requirements of the open-loop system $L(s) = K_cP(s)$ are satisfied. For example, K_c has to be chosen to satisfy steady-state error requirements on tracking, disturbance rejection, and so on.

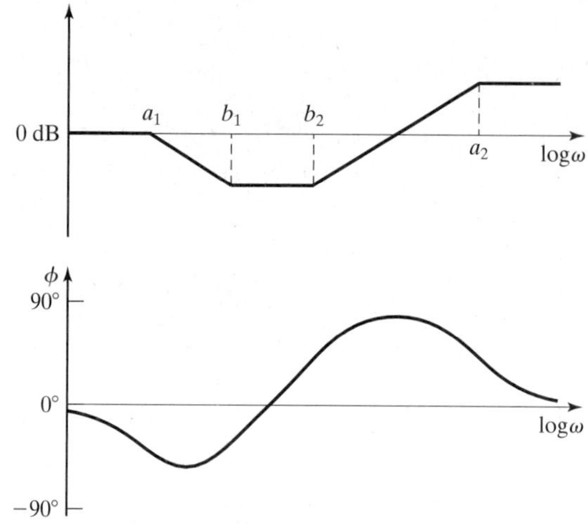

FIGURE 7.15: Lead-lag controller $C(s) = \dfrac{s/b_1 + 1}{s/a_1 + 1}\,\dfrac{s/b_2 + 1}{s/a_2 + 1}$.

Step 2: Determine the desired crossover frequency ω_c and PM ϕ_{desired}.

Step 3: Plot the Bode diagram of $K_c P(s)$ and calculate the phase $\phi_{\max}$ needed at ω_c in order to achieve the desired PM:

$$\phi_{\max} = \phi_{\text{desired}} - \angle K_c P(j\omega_c) - 180^\circ + 5^\circ.$$

Step 4: Choose a_2 and b_2 such that

$$\frac{a_2}{b_2} = \frac{1 + \sin\phi_{\max}}{1 - \sin\phi_{\max}}, \quad \omega_c = \sqrt{b_2 a_2}.$$

Let

$$C_{\text{lead}}(s) = \frac{K_c(s/b_2 + 1)}{s/a_2 + 1}.$$

Step 5: Choose

$$b_1 \approx 0.1\omega_c, \quad a_1 = \frac{b_1}{|C_{\text{lead}}(j\omega_c)P(j\omega_c)|}.$$

Step 6: Plot the Bode diagram of $C(s)P(s)$ and check the design specifications.

Similarly, a PID controller can be designed by essentially combining the design process for PD and PI controllers. The Bode diagram of a PID controller

$$C(s) = \frac{K_c(s/b_1 + 1)(s/b_2 + 1)}{s}$$

with $K_c = 1$ is shown in Figure 7.16.

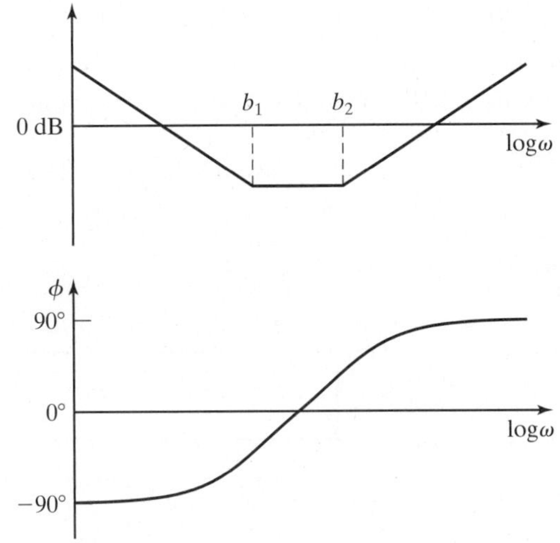

FIGURE 7.16: Bode diagram of a PID controller $C(s) = \dfrac{(s/b_1 + 1)(s/b_2 + 1)}{s}$.

Algorithm 7.11 (PID controller design).

Step 1: Determine the desired crossover frequency ω_c and PM ϕ_{desired}.

Step 2: Plot the Bode diagram of $P(s)$ and calculate the phase $\phi_{\max}$ needed at ω_c in order to achieve the desired PM:

$$\phi_{\max} = \phi_{\text{desired}} - \angle P(j\omega_c) - 180^\circ + 5^\circ.$$

Step 3: Choose b_2 such that

$$b_2 = \frac{\omega_c}{\tan \phi_{\max}}.$$

Let

$$C_{\text{pd}}(s) = s/b_2 + 1.$$

Step 4: Choose

$$b_1 \approx 0.1\omega_c, \quad K_c = \frac{b_1}{|C_{\text{pd}}(j\omega_c)P(j\omega_c)|}.$$

Step 5: A PID controller is given by

$$C(s) = \frac{K_c(s/b_1 + 1)(s/b_2 + 1)}{s}.$$

Step 6: Plot the Bode diagram of $C(s)P(s)$ and check the design specifications.

7.6 ZIEGLER AND NICHOLS TUNING RULES

We consider the unit feedback system shown in Figure 7.17. We assume that the controller $C(s)$ is a PID controller, which has the following general form:

$$C(s) = K_P \left(1 + \frac{1}{T_I s} + T_D s\right)$$

where K_P is the proportional gain, T_I is the integral constant, and T_D is the derivative constant.

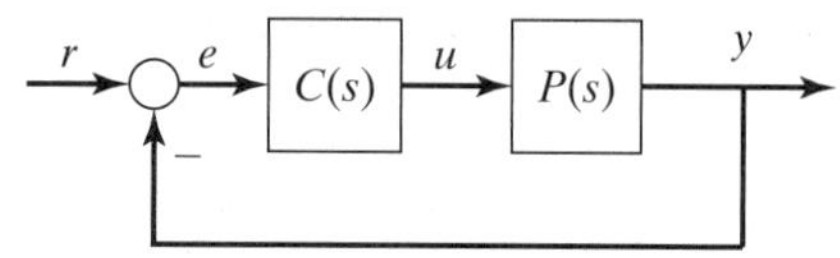

FIGURE 7.17: PID-controlled system.

7.6.1 Ziegler and Nichols first method

The Ziegler and Nichols first method uses the step response of the system. First of all, a step response of the system is obtained experimentally (or by simulation). If the step response has the S shape as shown in Figure 7.18, which means that the system is actually a high-order system, then it can be assumed that the open-loop system can be approximated by the following first-order system with a transport delay:

$$P(s) = \frac{Ke^{-Ls}}{\tau s + 1}$$

where L and τ are obtained by drawing a line with the largest slope on the step response.

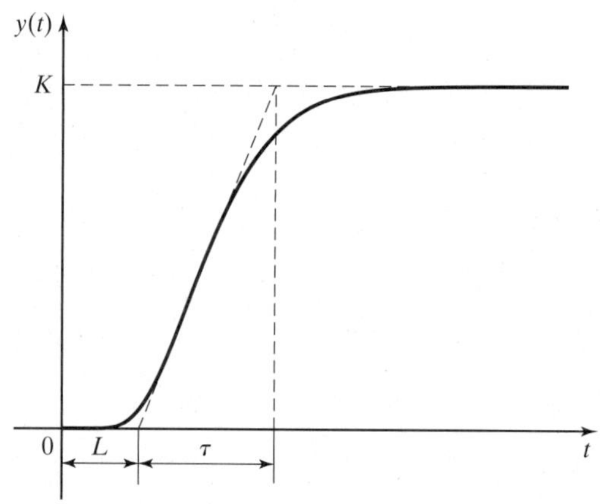

FIGURE 7.18: Step response of the process.

Ziegler and Nichols suggested that the PID controller parameters can be selected, as shown in Table 7.1, as a starting point for further fine tuning. In the following paragraphs, we shall give some detailed analysis using Bode diagrams explaining why these selections are reasonable.

Controller type	K_P	T_I	T_D
P	$\frac{\tau}{KL}$		
PI	$\frac{0.9\tau}{KL}$	$\frac{L}{0.3}$	
PID	$\frac{1.2\tau}{KL}$	$2L$	$0.5L$

TABLE 7.1: The Ziegler and Nichols first method.

It should be noted that the value of K does not play any role in the performance since it is canceled by the controller gain. It is the L and the ratio τ/L that will determine the system performance.

7.6.2 Frequency-response analysis of the Ziegler and Nichols tuning rules

It is noted that the Ziegler and Nichols first tuning method is based on the assumption that the plant can be approximately modeled by

$$P(s) = \frac{Ke^{-Ls}}{\tau s + 1}.$$

The Bode diagram of $P(s)$ can be approximately plotted as in Figure 7.19. Note that the phase of $P(s)$ is given by

$$\angle P(j\omega) = -L\omega - \tan^{-1}\tau\omega.$$

Also note that to achieve reasonable transient performance the open loop must have enough PM. For example, to achieve no more than 50% of the maximum overshoot, we must have a PM of at least 25°.

Proportional controller Let $C(s)$ be a proportional controller. Then

$$\angle P(j\omega)C(j\omega) = -L\omega - \tan^{-1}\tau\omega.$$

If we choose the crossover frequency of $P(s)C(s)$ close to $1/L$, i.e.,

$$\omega_c \approx \frac{1}{L},$$

then the PM is given by

$$\begin{aligned} \text{PM} &= \pi + \angle P(j\omega_c)C(j\omega_c) \approx \pi - 1 - \tan^{-1}\frac{\tau}{L} \text{ [rad]} \\ &= 123^\circ - \frac{180^\circ}{\pi}\tan^{-1}\frac{\tau}{L} > 33^\circ. \end{aligned}$$

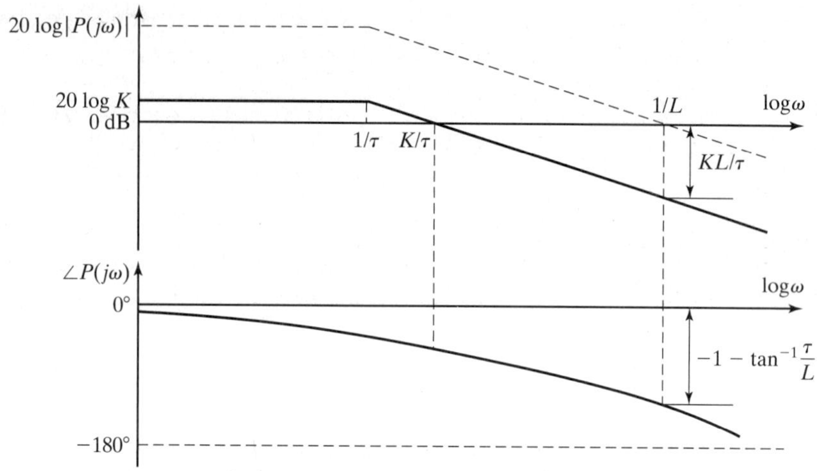

FIGURE 7.19: Frequency response of the process $P(s) = \dfrac{Ke^{-Ls}}{\tau s + 1}$.

Thus, if the crossover frequency is chosen to be $\omega_c \approx 1/L$, then we would have enough PM to guarantee that the maximum overshoot is no greater than 50%. Finally, note that

$$|P(j\omega_c)| \approx |P(j/L)| = \frac{K}{\sqrt{1 + (\tau/L)^2}} < \frac{KL}{\tau}.$$

Thus, if we take

$$C(s) = \frac{\tau}{KL},$$

then it is guaranteed that $\omega_c \leq 1/L$ and PM $> 33°$.

PI controller Let $C(s)$ be a PI controller with

$$C(s) = \frac{K_P(s + 1/T_I)}{s}.$$

Since we expect that $C(s)$ will contribute some negative phase at the crossover frequency, the crossover frequency for the system with a PI controller will be a bit smaller than the crossover frequency of the system with only a proportional controller. We shall choose

$$\omega_c \approx \frac{0.9}{L}.$$

Then, for $T_I = L/0.3$, we have

$$\begin{aligned}
\angle C(j\omega_c)P(j\omega_c) &= \angle(j0.9/L + 0.3/L) - 90° - \frac{180°}{\pi}(0.9 + \tan^{-1}\frac{0.9\tau}{L}) \\
&= \frac{180°}{\pi}\tan^{-1} 3 - 90° - \frac{180°}{\pi}(0.9 + \tan^{-1}\frac{0.9\tau}{L}) \\
&= -70° - \frac{180°}{\pi}\tan^{-1}\frac{0.9\tau}{L}.
\end{aligned}$$

Thus

$$\mathrm{PM} = 180^\circ + \angle C(j\omega_c)P(j\omega_c) = 110^\circ - \frac{180^\circ}{\pi}\tan^{-1}\frac{0.9\tau}{L} > 38^\circ$$

if $\tau/L \le 10/3$ and $\mathrm{PM} \ge 20^\circ$ for any τ/L. Also note that

$$|C(j\omega_c)P(j\omega_c)| = \frac{\sqrt{0.9}\;\tau/L}{\sqrt{1+0.9^2(\tau/L)^2}} \le 1$$

if $\tau/L \le 10/3$. That is, if $\tau/L \le 10/3$, we can guarantee that $\omega_c \le 0.9/L$ and $\mathrm{PM} \ge 38^\circ$.

PID controller Note that the Ziegler and Nichols tuning rule gives

$$C(s) = \frac{1.2\tau}{KL}\left(1 + \frac{1}{2Ls} + 0.5Ls\right) = \frac{1.2\tau}{KL}\,\frac{(1+Ls)^2}{2Ls}.$$

We have

$$C(j1.2/L)P(j1.2/L) = \frac{\tau/L(0.6399 - 1.0387j)}{1 + j1.2\tau/L}$$

and

$$|C(j1.2/L)P(j1.2/L)| \le 1$$

if $\tau/L \le 4.5455$. Furthermore,

$$\angle C(j1.2/L)P(j1.2/L) = -58.75^\circ - \frac{180^\circ}{\pi}\tan^{-1}\frac{1.2\tau}{L}$$

and

$$\mathrm{PM} = 180^\circ + \angle C(j\omega_c)P(j\omega_c) = 121.25^\circ - \frac{180^\circ}{\pi}\tan^{-1}\frac{1.2\tau}{L} > 31.25^\circ.$$

Hence we can guarantee that $\mathrm{PM} \ge 31.25^\circ$ with $\omega_c \le 1.2/L$ if $\tau/L \le 4.5455$.

EXAMPLE 7.12

We consider a time delay system

$$P(s) = \frac{Ke^{-Ls}}{\tau s + 1}$$

with $K = 10$ and $L = 2$. We shall examine how the controllers obtained using Ziegler and Nichols first method perform for various τ.

The step responses of the system with proportional controllers, PI controllers, and PID controllers obtained using Ziegler and Nichols first method are shown in Figures 7.20–7.22. It is quite clear from these simulations that the performance of these controllers depends critically on the values of τ (actually the ratio τ/L). In particular, the percentage overshoot can be very large when the

PI or the PID controller is used with a very large τ/L. Hence, we should not use these tuning formulas to calculate the final design parameters; rather they should be used only as a starting point of controller tuning. It is also clear that the performance of these controllers are consistent with our frequency-domain analysis.

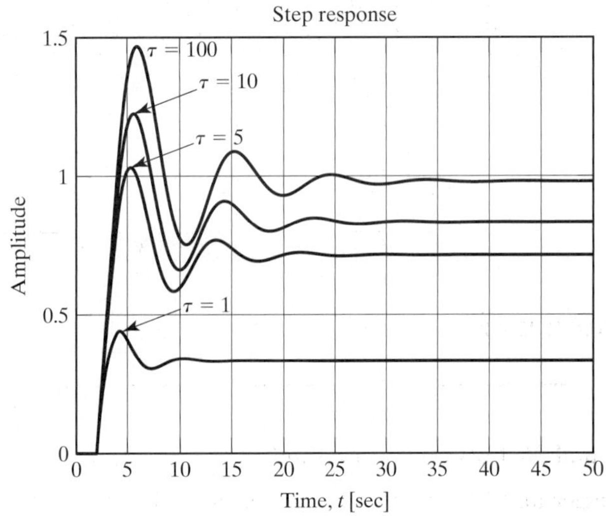

FIGURE 7.20: Example 7.12 – Proportional controller for $K = 10$, $L = 2$, and various τ.

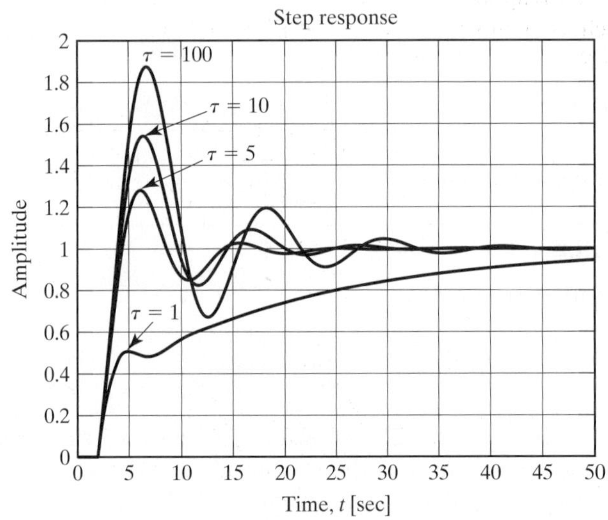

FIGURE 7.21: Example 7.12 – PI controller for $K = 10$, $L = 2$, and various τ.

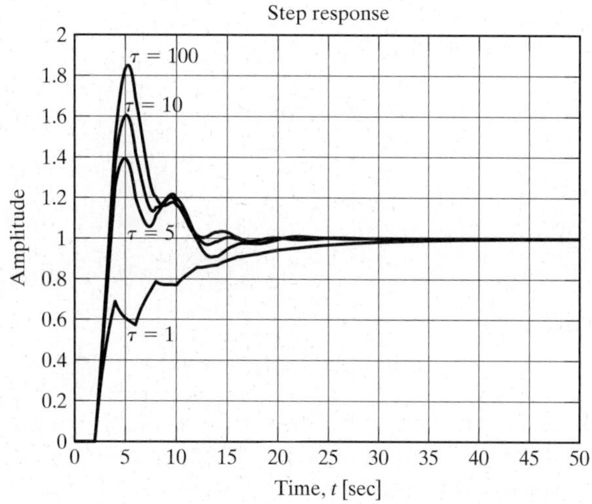

FIGURE 7.22: Example 7.12 – PID controller for $K = 10$, $L = 2$, and various τ.

7.6.3 Ziegler and Nichols second method

The Ziegler and Nichols second method uses the so-called critical gain and critical period (or frequency). First, an experiment (or simulation) is done with only a proportional controller shown in Figure 7.23. The proportional controller gain K is increased until a sustained oscillation is observed. Let this critical gain be K_{cr} and the corresponding oscillation period be P_{cr} as shown in Figure 7.24. Ziegler and Nichols then suggest that the PID controller parameters can be selected as in Table 7.2.

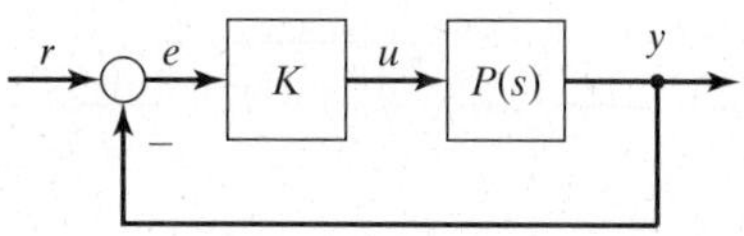

FIGURE 7.23: Tuning using a proportional controller.

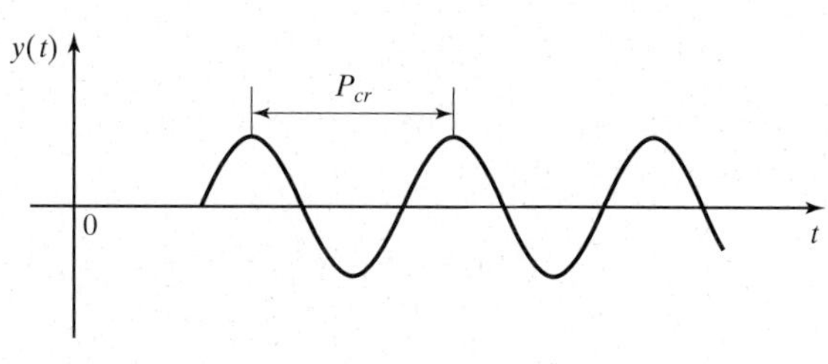

FIGURE 7.24: Sustained oscillation period P_{cr}.

Controller	K_P	T_I	T_D
P	$0.5K_{\rm cr}$		
PI	$0.45K_{\rm cr}$	$\frac{1}{1.2}P_{\rm cr}$	
PID	$0.6K_{\rm cr}$	$0.5P_{\rm cr}$	$0.125P_{\rm cr}$

TABLE 7.2: Ziegler and Nichols second method.

Again, the parameters in this table should be taken only as the starting point for further tuning. This method of choosing tuning parameters can probably be understood using the root-locus concept by noting the facts that K_{cr} is the gain where the root locus crosses the imaginary axis and the crossing point on the imaginary axis is given by $j\omega = j2\pi/P_{cr}$. We shall leave the detailed analysis to the readers.

7.7 DERIVATIVE CONTROL

We have seen that derivative control is sometimes needed in order to achieve the desired performance. However, special care must be taken when a derivative control law is implemented. In practice, a derivative control can never be implemented as it is. It is always implemented with a suitable low-pass filter. To gain further understanding of this issue, let us consider the simple feedback control system shown in Figure 7.25 with a PD controller

$$C(s) = K_P(1 + T_D s).$$

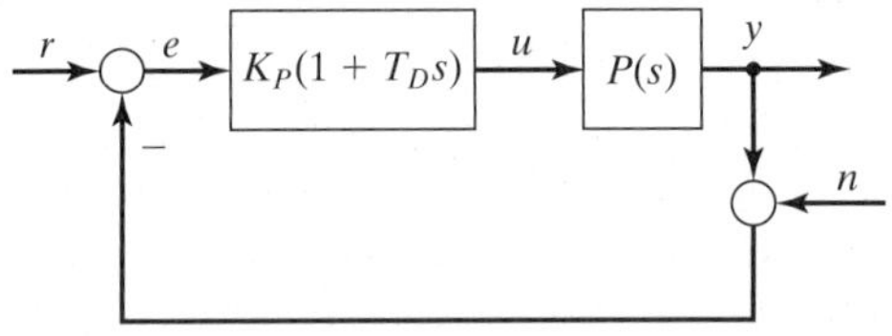

FIGURE 7.25: Standard implementation of a PD controller.

Since a pure derivative is not physically realizable, it is usually implemented with suitable filtering, for example,

$$C(s) = K_P \left(1 + \frac{T_D s}{Ts + 1}\right)$$

with a small $T > 0$ so that the derivative control is approximately realized. The choice of T is also necessary in order to filter out the sensor noise.

The standard implementation shown in Figure 7.25 may not always be desirable in practice since the output of any actuator is limited. Thus, when a command signal r has a jump, the output of the derivative controller will be extremely large (in fact, it is theoretically infinite), which will certainly saturate the actuator. Hence, the controller will not perform as we would expect. We shall now illustrate this with an example.

EXAMPLE 7.13

Consider the feedback system designed in Example 5.13 with

$$C(s) = 5(s+5) = 25(0.2s+1), \quad P(s) = \frac{10}{s(s+5)(s+10)}.$$

A SIMULINK diagram of this feedback system is shown in Figure 7.26, where we have implemented the PD controller as $C(s) = 25\left(1 + \frac{0.2s}{Ts+1}\right)$ with $T = 0.01$.

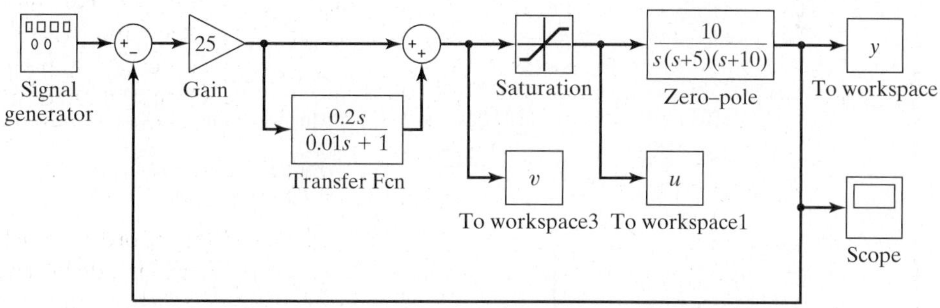

FIGURE 7.26: Example 7.13 – A PD-controlled system with actuator saturation.

The response of the feedback system without actuator saturation and with saturation limit $u_{\max} = 5, 10$, and 20 are shown in Figure 7.27. We have to note that the control signal due to the derivative control is very large when the command signal jumps. (In fact, $v(t) \to \infty$ at the moment of sudden change in command.) Thus, the actuator is easily saturated.

An alternative is to use a rate sensor to measure the derivative of the output directly or implement the derivative control in the feedback as shown in Figure 7.28. There are certain advantages of this inner-loop feedback over standard cascade implementation. For example, it can reduce the possibility of actuator saturation because of the abrupt change in input signals. The system responses with this implementation are shown in Figure 7.29, and it is clear that these responses are much smoother.

We should also note that the choice of T or a low-pass filter in the derivative feedback loop is also very important since there is inevitable measurement noise $n(t)$. Without proper filtering, the sensor noise $n(t)$ can be drastically amplified by the derivative feedback even if $n(t)$ is very small. The step responses of the system with $n(t) = 0.01 \sin 1000t$ for $T = 0$ and $T = 0.01$ are shown in Figure 7.30 for the case where there is no actuator saturation. The outputs for both cases are almost the same. However, the control signals are drastically different. The control signal

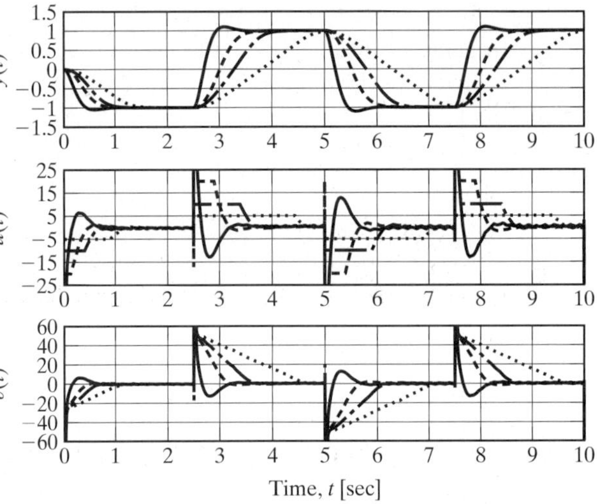

FIGURE 7.27: Example 7.13 – Responses to a square wave of 0.2 [Hz] with actuator saturation $u_{\max} = 5$ (dotted), $u_{\max} = 10$ (dash-dot), $u_{\max} = 20$ (dashed), and $u_{\max} = \infty$ (solid).

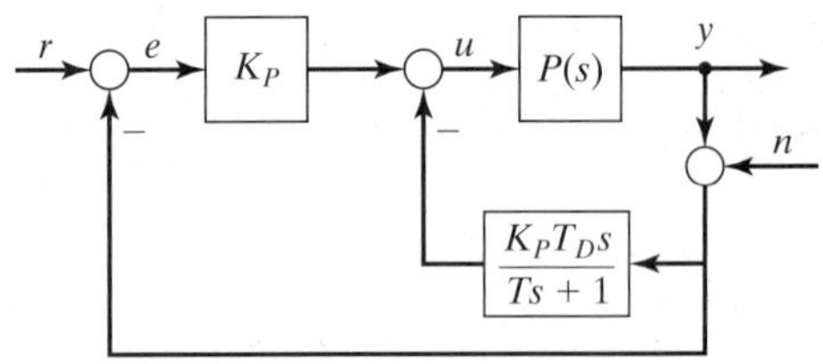

FIGURE 7.28: PD control using an inner loop feedback with a rate sensor or derivative.

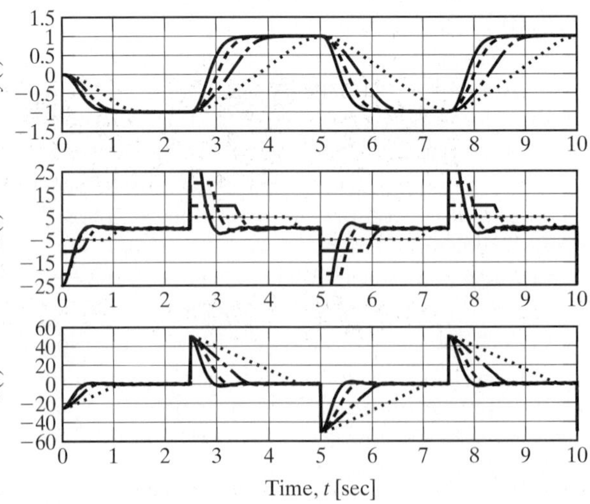

FIGURE 7.29: Example 7.13 with inner loop feedback – Responses to a square wave of 0.2 [Hz] with actuator saturation $u_{\max} = 5$ (dotted), $u_{\max} = 10$ (dash-dot), $u_{\max} = 20$ (dashed), and $u_{\max} = \infty$ (solid).

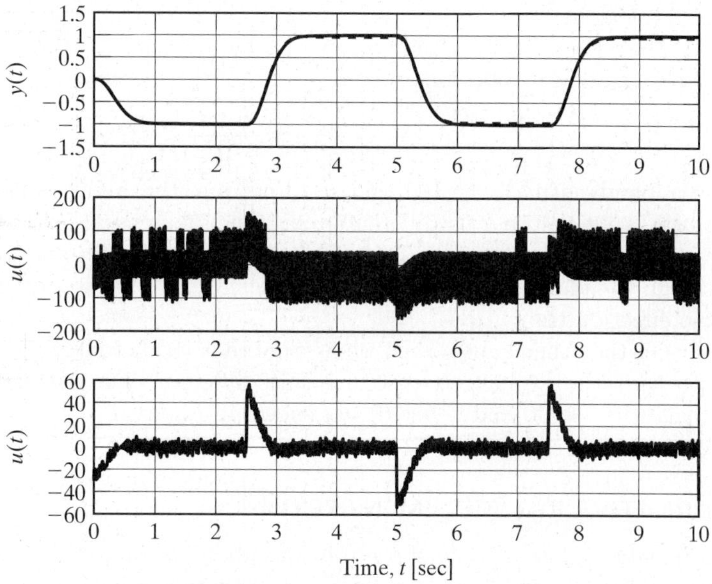

FIGURE 7.30: Example 7.13 – PD control using an inner-loop derivative feedback with a sensor noise $n(t) = 0.01 \sin 1000t$ and without actuator saturation. Top: output $y(t)$ for $T = 0.01$ (solid) and $T = 0$ (dashed); Middle: actuator output $u(t)$ for $T = 0$; Bottom: actuator output $u(t)$ for $T = 0.01$.

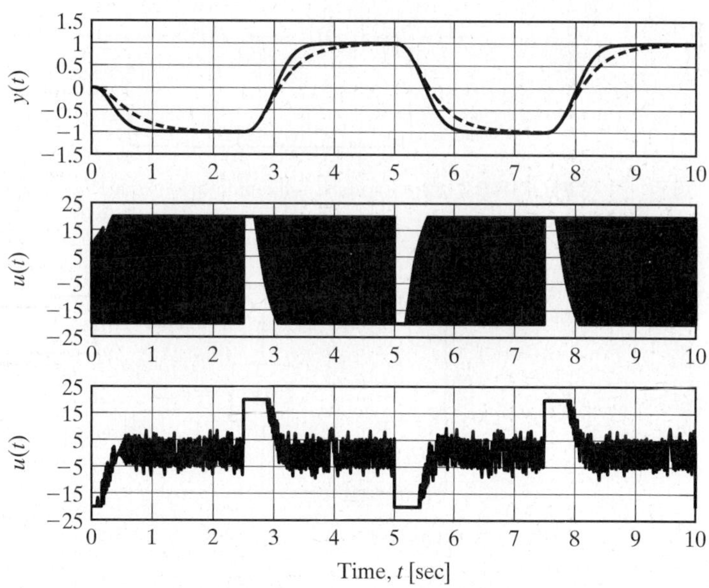

FIGURE 7.31: Example 7.13 – PD control using an inner-loop derivative feedback with a sensor noise $n(t) = 0.01 \sin 1000t$ and actuator saturation $u_{max} = 20$. Top: output $y(t)$ for $T = 0.01$ (solid) and $T = 0$ (dashed); middle: actuator output $u(t)$ for $T = 0$; bottom: actuator output $u(t)$ for $T = 0.01$.

without the filtering (i.e., $T = 0$) shown in the middle of Figure 7.30 is much more noisy than the control signal with a suitable filtering ($T = 0.01$) as shown in the bottom of Figure 7.30. This should not be surprising since the low-pass filter

$$\frac{1}{Ts+1} = \frac{1}{0.01s+1}$$

has a bandwidth $\omega_b = 100$ and it filters out the high-frequency noise $n(t) = 0.01 \sin 1000t$ before it is differentiated. Since it is assumed that there is no saturation, the high-frequency control signal is actually filtered out when it passes through the plant, which is a low-pass filter itself. That is why it seems that there is no effect on the output.

On the other hand, when we assume that the actuator has a saturation limit $u_{\max} = 20$, the system responses are quite different. The outputs and the actuator signals with $T = 0$ and $T = 0.01$ are shown in Figure 7.31.

7.8 ALTERNATIVE PID IMPLEMENTATION

We should note that some of the previous analyses for PD controllers also apply to PI (and PID) controllers. In general, when there is a jump in command signal, the output of the proportional controller may also generate a large control signal, which may saturate the actuator. It may therefore be also desirable to move the proportional controller as well as the derivative controller to the inner feedback loop as shown in Figures 7.32–7.34.

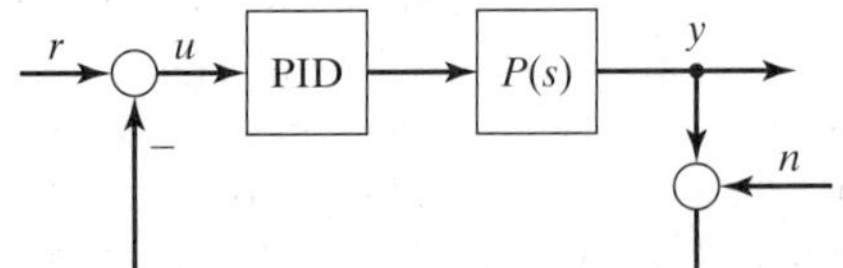

FIGURE 7.32: Standard implementation of a PID controller.

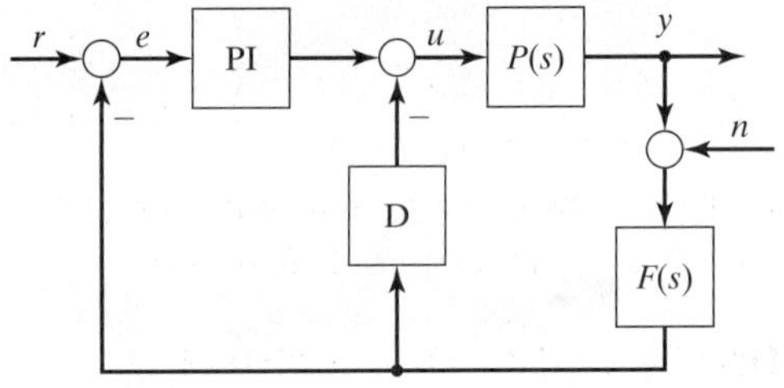

FIGURE 7.33: PID control using an inner-loop feedback with a rate sensor/output derivative.

Since there will inevitably be measurement noise, it is always necessary to prefilter the measurement signal before it is fed back. Thus, a suitably designed prefilter $F(s)$ is always needed.

We should be clear that these alternative implementations are not equivalent to the standard implementation. Although they all have the same closed-loop poles,

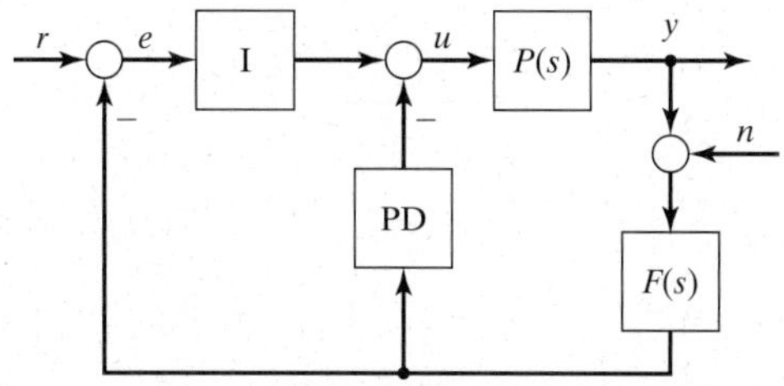

FIGURE 7.34: PID control using an inner-loop feedback with the output measurement and a rate sensor/output derivative.

they may have different zeros and steady-state error constants. To be more specific, let us assume that the PID controller is simply a PD controller:

$$\text{PID} = K_P(1 + T_D s).$$

Let us further assume that $F(s) = 1$ and that the plant has the following transfer function:

$$P(s) = \frac{b(s)}{a(s)}.$$

The open-loop transfer functions of the standard implementation and the alternative implementation then, shown in Figure 7.33, are respectively

$$L_s(s) = \frac{K_P(1 + T_D s)b(s)}{a(s)}, \quad L_a(s) = \frac{K_P P(s)}{1 + K_P T_D s P(s)} = \frac{K_P b(s)}{a(s) + K_P T_D s b(s)}.$$

It is now clear that both implementations have the same closed-loop poles given by the roots of the characteristic polynomial $a(s) + K_P(1 + T_D s)b(s)$ but the standard implementation has an extra zero, which means that the standard implementation may have a shorter rise time but a larger overshoot compared to the alternative implementation. Furthermore, if $P(s)$ has no poles at origin, the steady-state position error constants are the same since $L_s(0) = K_P P(0) = L_a(0)$. However, if $P(s)$ has a pole at origin, say $a(s) = s a_1(s)$, then the velocity error constants are different:

$$K_v = \frac{K_P b(0)}{a_1(0)}, \quad \text{standard implementation}$$

$$K_v = \frac{K_P b(0)}{a_1(0) + K_P T_D b(0)} \quad \text{alternative implementation.}$$

Thus, in general, the alternative implementation may result in a larger steady-state error compared to the standard implementation.

7.9 INTEGRAL CONTROL AND ANTIWINDUP

It is well understood that any actuator has an output limit. A common problem associated with actuator saturation and integral control is the so-called integrator windup. Consider the PID-controlled system shown in Figure 7.35.

Briefly, an integrator windup refers to the process wherein the output of the integral controller (u_c) keeps increasing even though the actuator has reached the saturation limit (i.e., $u_c > u_{\max}$ or $u_c < u_{\min}$). This can cause significant performance degradation since the increasing value of u_c does not help to reduce the

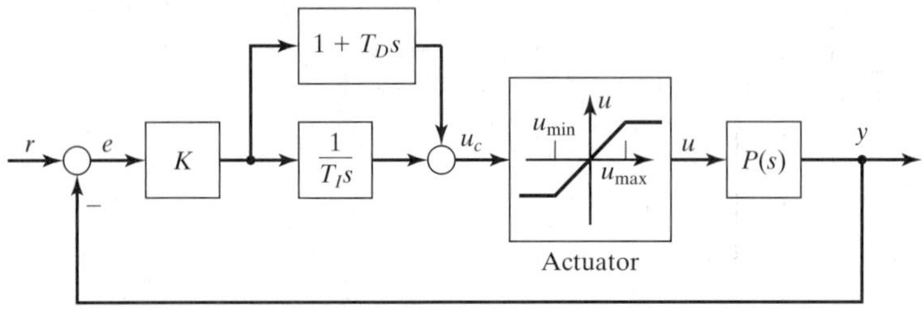

FIGURE 7.35: A PID-controlled system with actuator saturation.

system tracking error but will require considerable error e to discharge the integrator to its proper value. The method by which to avoid such integrator windup is to stop integrating as soon as u_c reaches the actuator limit. This can be done very easily in a digital controller where a suitable program can be set up for such purpose. Figures 7.36 and 7.37 show two antiwindup approaches, where K_a is a suitable parameter. In reality, it may be difficult to measure the output of the actuator, so the approach shown in Figure 7.37 is more suitable.

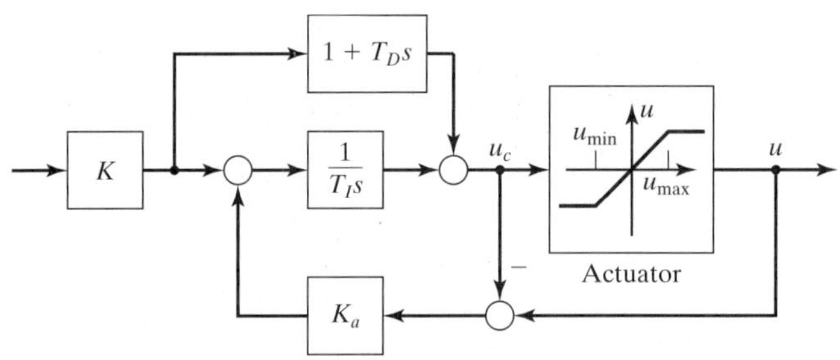

FIGURE 7.36: PID controller with antiwindup.

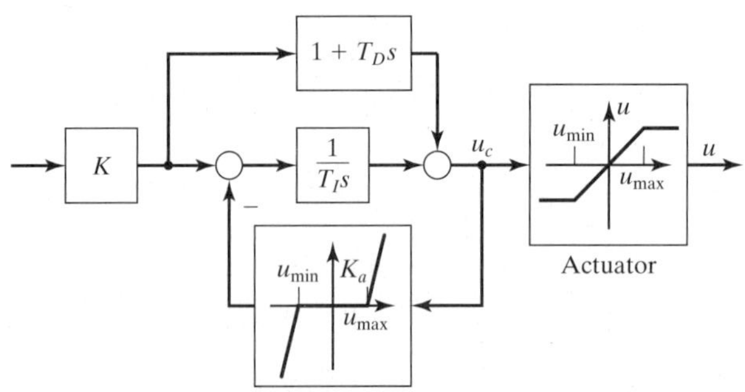

FIGURE 7.37: PID controller with a practical antiwindup mechanism.

EXAMPLE 7.14

Consider the system with an integral controller shown in Figure 7.38. The saturation limit is $|u_{\max}| = |u_{\min}| = 1$. The step response of the system with and without the antiwindup compensation are shown in Figures 7.39 and 7.40

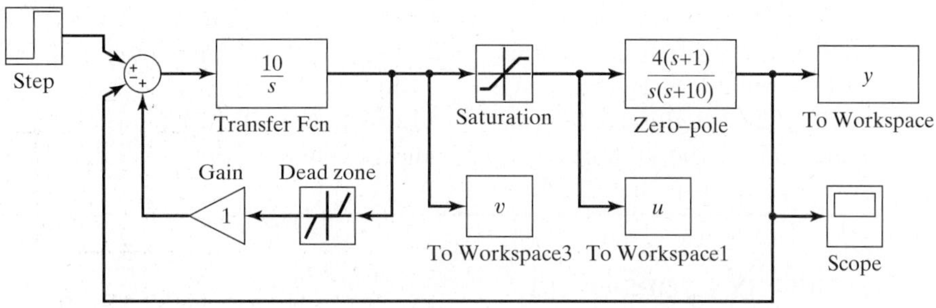

FIGURE 7.38: Example 7.14 – Integral control with antiwindup.

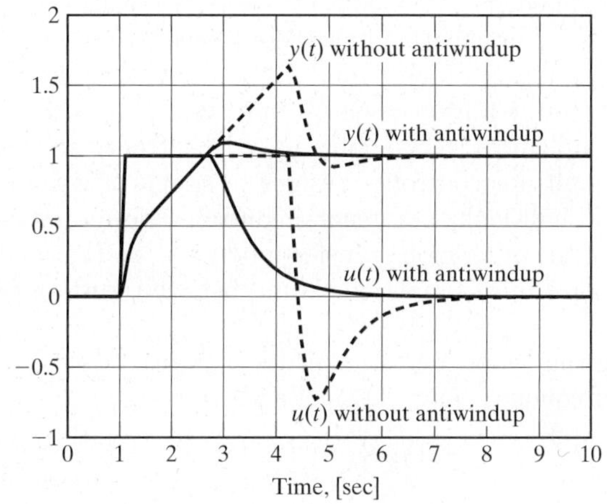

FIGURE 7.39: Example 7.14 – Integral control with antiwindup (solid) and without antiwindup (dashed).

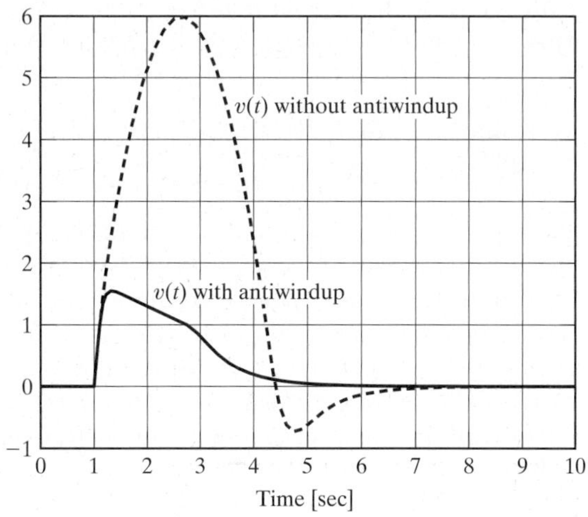

FIGURE 7.40: Example 7.14 – $v(t)$ with antiwindup (solid) and without antiwindup (dashed).

with $K_a = 1$. We have to note that the control signal u for the system with the antiwindup compensation starts to decrease almost immediately after $y(t)$ exceeds 1, the step reference signal. On the other hand, the control signal for the system without antiwindup compensation will stay in the saturation level for a much longer time. Thus, the system with antiwindup compensation exhibits a much smaller overshoot and settling time.

7.10 DESIGN BY LOOPSHAPING

We have so far restricted our controller structures to some simple forms: lead, lag, lead-lag, or PID. The greatest advantages of these simple controller structures are that effects of each controller parameters are relatively clear and that they can be tuned easily by application engineers. Nevertheless, these simple controller structures may put unnecessary limitations on achievable performance for a feedback system. With the advancement of modern digital technology, it is now simple and cheap to implement very sophisticated controller structures. Hence, there is no reason to limit our controller designs to simple lead-lag or PID controllers. In this section, we shall look at a general controller design procedure based on the **loopshaping** of the open-loop transfer functions. To that end, we shall now recall how an open-loop transfer function should behave if certain performance criteria are to be satisfied.

Note that good performance is required in some frequency range, typically some low-frequency range $[\omega_0, \omega_l]$, such that

$$|S(j\omega)|, \quad |P(j\omega)S(j\omega)|, \quad |C(j\omega)S(j\omega)|$$

be made small, and good robustness requires that

$$|T(j\omega)|$$

be made small typically over some high-frequency range $[\omega_h, \infty)$.

In terms of open-loop transfer function $L(s)$, the above conditions translate into

$$|L(j\omega)| \gg 1 \;\text{ and }\; |C(j\omega)| \gg 1 \;\; \forall \omega \in [\omega_0, \omega_l] \;\text{ where } |P(j\omega)| \;\text{ is not too small}$$

$$|L(j\omega)| \ll 1 \;\text{ and not too large } |C(j\omega)|, \; \forall \omega \in [\omega_h, \infty).$$

These design requirements are graphically shown in Figure 7.41.

The specific frequencies ω_0, ω_l, and ω_h depend on the specific applications and the knowledge we have on the disturbance characteristics, the modeling uncertainties, and the sensor noise levels. In most cases, $\omega_0 = 0$. (However, there are cases where having high loop gains at very low-frequency range is impossible, for example, systems with the output measurements taken from accelerometers, in which case the system transfer function has zero gain at 0 frequency, i.e., $P(j0) = 0$.)

We shall now describe a general loopshaping design approach that can be used as a starting point if the controller order is not limited. To start with, let $P(s)$ be factorized as

$$P(s) = P_u(s)P_m(s)$$

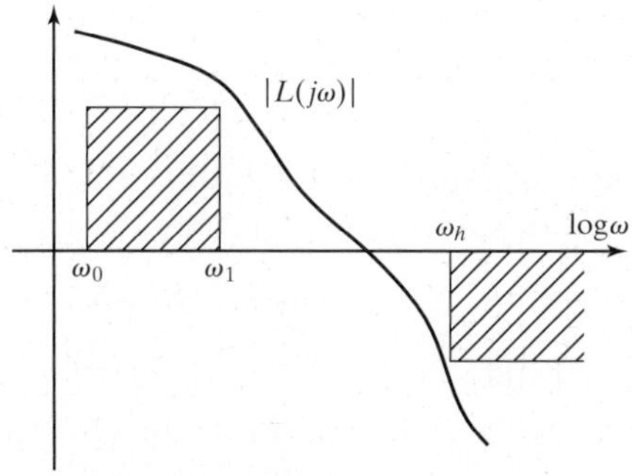

FIGURE 7.41: Desired loop gain.

where $P_u(s)$ includes all unstable (including imaginary axis) poles and zeros and $P_m(s)$ is strictly stable and minimum phase. To simplify the subsequent development, we shall limit the factorization to a special form so that $P_m(s)$ is biproper.

Let the controller $C(s)$ take the following form

$$C(s) = C_1(s)/P_m(s)$$

where $C_1(s)$ is a proper controller to be designed so that the closed-loop system with loop transfer function

$$L(s) = P_u(s)C_1(s)$$

and controller $C(s)$ satisfy all design specifications.

The above design procedure usually results in a high-order controller.

EXAMPLE 7.15

Let

$$P(s) = \frac{10(s+2)}{s(s+10)}.$$

Then

$$P_u(s) = \frac{10}{s}, \quad P_m(s) = \frac{s+2}{s+10}$$

is an appropriate factorization. Let

$$C_1(s) = 1$$

be a controller satisfying all design constraints for the plant $P_u(s)$. Then

$$C(s) = C_1(s)/P_m(s) = \frac{s+10}{s+2}$$

is the desired controller.

EXAMPLE 7.16

Let

$$P(s) = \frac{50(s+1)}{s(s^2+2s+10)}.$$

Then

$$P_u(s) = \frac{5}{s(s+a)}, \quad P_m(s) = \frac{10(s+1)(s+a)}{s^2+2s+10}$$

for any $a > 0$ form a factorization of $P(s) = P_u(s)P_m(s)$. Note that $P_u(s)$ contains a stable pole in order to make P_m biproper. Suppose

$$C_1(s) = \frac{20(s+a)}{s+10}$$

is a controller satisfying all design constraints for the plant $P_u(s)$. Then

$$C(s) = C_1(s)/P_m(s) = \frac{20(s+a)}{s+10}\,\frac{s^2+2s+10}{10(s+1)(s+a)} = \frac{2(s^2+2s+10)}{(s+1)(s+10)},$$

and the compensated loop transfer function is given by

$$L(s) = P(s)C(s) = \frac{100}{s(s+10)}.$$

These examples seem to suggest that we could shape the loop transfer function any way we want to if the transfer function is stable and minimum phase. Indeed, we could achieve almost any performance theoretically if there are no limits on the controller bandwidth and noise/disturbance effects on the system. For example, we could choose the following controller for the last example:

$$C(s) = \frac{K(s^2+2s+10)}{(s+1)(\tau s+1)}$$

for a small $\tau \to 0$ and a large $K \to \infty$. The resulting loop transfer function would be

$$L(s) = P(s)C(s) = \frac{50K}{s(\tau s+1)} \to \frac{50K}{s}.$$

Unfortunately, this controller is usually not practically implementable for $\tau \to 0$ and $K \to \infty$.

In the following sections and the next chapter, we shall consider in some detail the limitations imposed by model uncertainties, controller bandwidth, external disturbances, sensor noises, etc., such that $L(s)$ cannot be shaped arbitrarily even if $P(s)$ is stable and strictly minimum phase. In addition, there are further limitations if the plant has nonminimum phase zeros and unstable poles.

7.11 BODE'S GAIN AND PHASE RELATION

One important question that arises frequently is the level of performance that can be achieved in a feedback design. We have shown in the preceding sections that the feedback design goals are inherently conflicting, and a trade-off must be made among the different design objectives. It is also known that the fundamental requirements, such as stability and robustness, impose inherent limitations on the feedback properties irrespective of design methods, and the design limitations become more severe in the presence of right half plane zeros and poles in the open-loop transfer function.

In the classical feedback theory, Bode's gain–phase integral relation has been used as an important tool to express design constraints. This integral relation says that the phase of a stable and minimum phase transfer function is determined uniquely by the magnitude of the transfer function.

Theorem 7.17. Let $L(s)$ be a stable and minimum phase transfer function. Then

$$\angle L(j\omega_0) = \frac{1}{\pi}\int_{-\infty}^{\infty} \frac{d\ln|L(j\omega)|}{d\nu} \ln\coth\frac{|\nu|}{2}d\nu \qquad (7.1)$$

where $\nu := \ln(\omega/\omega_0)$.

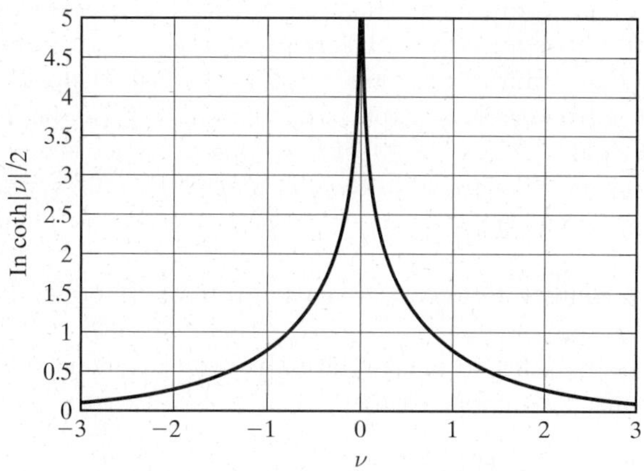

FIGURE 7.42: The function $\ln\coth\frac{|\nu|}{2}$ vs. ν.

The function $\ln\coth\frac{|\nu|}{2} = \ln\frac{e^{|\nu|/2}+e^{-|\nu|/2}}{e^{|\nu|/2}-e^{-|\nu|/2}}$ is plotted in Figure 7.42. Note that $\ln\coth\frac{|\nu|}{2}$ rapidly decreases as ω deviates from ω_0 and hence the integral depends mostly on the behavior of $\frac{d\ln|L(j\omega)|}{d\nu}$ near the frequency ω_0. This is clear from the following integration:

$$\frac{1}{\pi}\int_{-\alpha}^{\alpha} \ln\coth\frac{|\nu|}{2}d\nu = \begin{cases} 1.1406 \text{ (rad)}, & \alpha = \ln 3 \\ 1.3146 \text{ (rad)}, & \alpha = \ln 5 \\ 1.443 \text{ (rad)}, & \alpha = \ln 10 \end{cases} = \begin{cases} 65.3^\circ, & \alpha = \ln 3 \\ 75.3^\circ, & \alpha = \ln 5 \\ 82.7^\circ, & \alpha = \ln 10. \end{cases}$$

Note that $\frac{d\ln|L(j\omega)|}{d\nu}$ is the slope of the Bode plot, which is generally negative for almost all frequencies. It follows that $\angle L(j\omega_0)$ will be large (i.e., less negative) if the gain $L(j\omega)$ attenuates slowly near ω_0 and small (i.e., more negative) if it attenuates rapidly near ω_0. For example, suppose the slope $\dfrac{d\ln|L(j\omega)|}{d\nu} = -\ell$; that is, ($-20\ell$ dB per decade), in the neighborhood of ω_0; then it is reasonable to expect

$$\angle L(j\omega_0) < \begin{cases} -\ell \times 65.3^\circ, & \text{if the slope of } L(j\omega) = -\ell \text{ for } \frac{1}{3} \le \frac{\omega}{\omega_0} \le 3 \\ -\ell \times 75.3^\circ, & \text{if the slope of } L(j\omega) = -\ell \text{ for } \frac{1}{5} \le \frac{\omega}{\omega_0} \le 5 \\ -\ell \times 82.7^\circ, & \text{if the slope of } L(j\omega) = -\ell \text{ for } \frac{1}{10} \le \frac{\omega}{\omega_0} \le 10. \end{cases}$$

The behavior of $\angle L(j\omega)$ is particularly important near the crossover frequency ω_c, where $|L(j\omega_c)| = 1$ since $\pi + \angle L(j\omega_c)$ is the PM of the feedback system. Further,

$$|1 + L(j\omega_c)| = |1 + L^{-1}(j\omega_c)| = 2\left|\sin\frac{\pi + \angle L(j\omega_c)}{2}\right|$$

must not be too small for good stability robustness. If $\pi + \angle L(j\omega_c)$ is forced to be very small by rapid gain attenuation, the feedback system will amplify disturbances and exhibit little uncertainty tolerance at and near ω_c. Since it is generally required that the loop transfer function L roll off as fast as possible in the high-frequency range, it is reasonable to expect that $\angle L(j\omega_c)$ in the high-frequency range is close to or much less than $-\ell \times 90^\circ$ if the slope of $L(j\omega)$ is $-\ell$ near ω_c.

Thus, it is important to keep the slope of $L(j\omega)$ near ω_c not much smaller than -1 for a reasonably wide range of frequencies in order to guarantee some reasonable performance.

The conflict between attenuation rate and loop quality near crossover is thus evident.

Bode's gain and phase relation can be extended to stable and nonminimum phase transfer functions easily.

Theorem 7.18. Let $z_1, z_2, \ldots, z_k$ be the right half plane zeros of a stable $L(s)$. Then

$$\angle L(j\omega_0) = \frac{1}{\pi}\int_{-\infty}^{\infty} \frac{d\ln|L(j\omega)|}{d\nu} \ln\coth\frac{|\nu|}{2} d\nu + \sum_{i=1}^{k} \angle\frac{-j\omega_0 + z_i}{j\omega_0 + z_i}. \tag{7.2}$$

Proof. Note that $L(s)$ can be factorized as

$$L(s) = \frac{-s + z_1}{s + z_1}\frac{-s + z_2}{s + z_2}\cdots\frac{-s + z_k}{s + z_k}L_{\mathrm{mp}}(s),$$

where $L_{\mathrm{mp}}(s)$ is the stable and minimum phase and $|L(j\omega)| = |L_{\mathrm{mp}}(j\omega)|$. Hence

$$\begin{aligned} \angle L(j\omega_0) &= \angle L_{\mathrm{mp}}(j\omega_0) + \angle\prod_{i=1}^{k}\frac{-j\omega_0 + z_i}{j\omega_0 + z_i} \\ &= \frac{1}{\pi}\int_{-\infty}^{\infty} \frac{d\ln|L_{\mathrm{mp}}(j\omega)|}{d\nu} \ln\coth\frac{|\nu|}{2} d\nu + \sum_{i=1}^{k} \angle\frac{-j\omega_0 + z_i}{j\omega_0 + z_i}, \end{aligned}$$

which gives

$$\angle L(j\omega_0) = \frac{1}{\pi}\int_{-\infty}^{\infty} \frac{d\ln|L(j\omega)|}{d\nu}\ln\coth\frac{|\nu|}{2}d\nu + \sum_{i=1}^{k}\angle\frac{-j\omega_0 + z_i}{j\omega_0 + z_i}.$$

□

Since $\angle\frac{-j\omega_0 + z_i}{j\omega_0 + z_i} \leq 0$ for each i, a nonminimum phase zero contributes an additional phase lag and imposes limitations on the roll-off rate of the open-loop gain. For example, suppose $L(s)$ has a zero at $z > 0$; then

$$\phi_1(\omega_0/z) := \angle\frac{-j\omega_0 + z}{j\omega_0 + z}\bigg|_{\omega_0 = z, z/2, z/4} = -90^\circ, -53.13^\circ, -28^\circ,$$

as shown in Figure 7.43. Since the slope of $|L(s)|$ near the crossover frequency is, in general, no greater than -1, meaning that the phase due to the minimum phase part $L_{\text{mp}}(s)$ of $L(s)$ will in general be no greater than -90°, the crossover frequency (or the closed-loop bandwidth) must satisfy

$$\boxed{\omega_c < z/2} \qquad (7.3)$$

in order to guarantee closed-loop stability and some reasonable closed-loop performance.

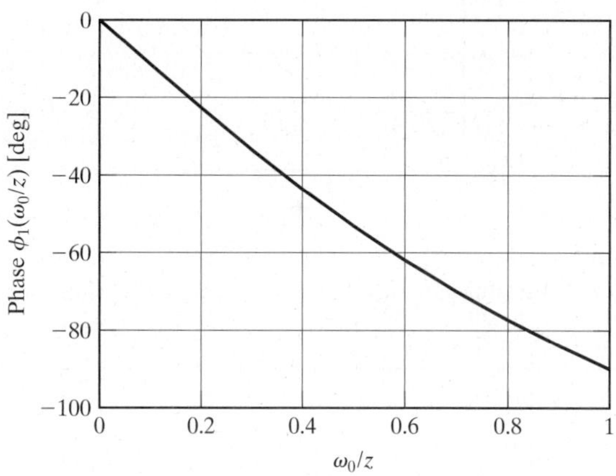

FIGURE 7.43: Phase $\phi_1(\omega_0/z)$ due to a real zero $z > 0$.

Next, suppose $L(s)$ has a pair of complex right half plane zeros at $z = x \pm jy$ with $x > 0$; then

$$\phi_2(\omega_0/|z|) := \angle\frac{-j\omega_0 + z}{j\omega_0 + z}\frac{-j\omega_0 + \bar{z}}{j\omega_0 + \bar{z}}\bigg|_{\omega_0 = |z|, |z|/2, |z|/3, |z|/4}$$

$$\approx \begin{cases} -180^\circ, & -106.26^\circ, & -73.7^\circ, & -56^\circ, & \text{Re}(z) \gg \text{Im}(z) \\ -180^\circ, & -86.7^\circ, & -55.9^\circ, & -41.3^\circ, & \text{Re}(z) \approx \text{Im}(z) \\ -360^\circ, & 0^\circ, & 0^\circ, & 0^\circ, & \text{Re}(z) \ll \text{Im}(z) \end{cases}$$

as shown in Figure 7.44. In this case we conclude that the crossover frequency must satisfy

$$\omega_c < \begin{cases} |z|/4, & \mathrm{Re}\,(z) \gg \mathrm{Im}\,(z) \\ |z|/3, & \mathrm{Re}\,(z) \approx \mathrm{Im}\,(z) \\ |z|, & \mathrm{Re}\,(z) \ll \mathrm{Im}\,(z) \end{cases} \tag{7.4}$$

in order to guarantee closed-loop stability and some reasonable closed-loop performance.

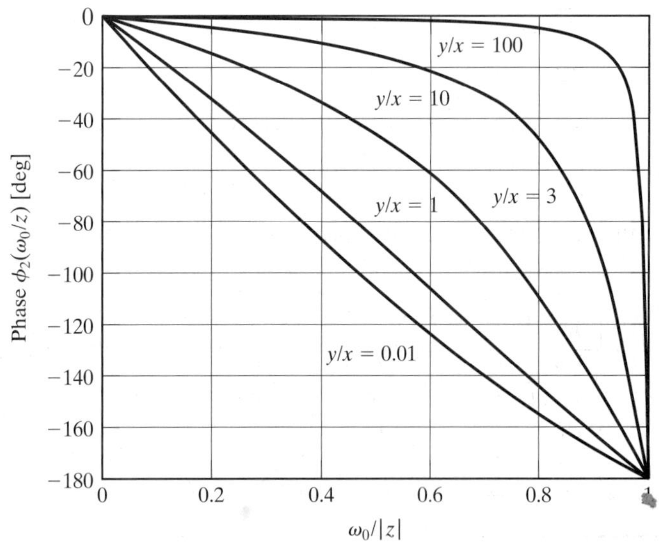

FIGURE 7.44: Phase $\phi_2(\omega_0/|z|)$ due to a pair of complex zeros: $z = x \pm jy$ and $x > 0$.

EXAMPLE 7.19

Let

$$L(s) = \frac{K(10-s)}{(s+10)(s+2)(s+3)} = \frac{10-s}{s+10}\,\frac{K}{(s+2)(s+3)}.$$

Then

$$L_m(s) = \frac{K}{(s+2)(s+3)}.$$

The Bode diagrams of $L(s)$ and $L_m(s)$ are shown in Figure 7.45. It is clear from Table 7.3 that the stability margin of the system with $L(s)$ is very small when the crossover frequency ω_c is approximately half of the right half plane zero $z = 10$. Indeed, the closed-loop system with the nonminimum phase system $L(s)$ in the loop becomes unstable when $\omega_c = z/2 = 5$.

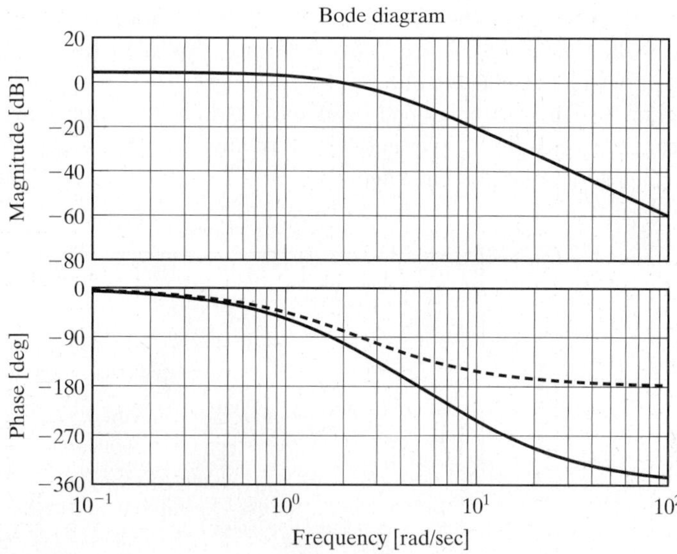

FIGURE 7.45: Bode diagrams of $L(s)$ (solid) and $L_m(s)$ (dashed) with $K = 10$.

K	10	15	20	25	30	40
PM of $L_m(s)$	102.67	79.6	67.5	59.67	54.07	46.41
ω_c of $L(s)$	1.95	2.95	3.7	4.3	4.86	4.98
PM of $L(s)$	80.6	46.7	26.9	13	2.24	−13.77

TABLE 7.3: Phase margins and ω_c.

7.12 BODE'S SENSITIVITY INTEGRAL

In this section we consider the design limitations imposed by bandwidth constraints and the right half plane poles using Bode's sensitivity integral.

Theorem 7.20. Let $L(s)$ be the open-loop transfer function with at least two more poles than zeros and let $p_1, p_2, \ldots, p_m$ be the open right half plane poles of $L(s)$. The following Bode's sensitivity integral then holds:

$$\int_0^{\infty} \ln |S(j\omega)| d\omega = \pi \sum_{i=1}^{m} p_i. \tag{7.5}$$

In the case where $L(s)$ is stable, the integral simplifies to

$$\int_0^{\infty} \ln |S(j\omega)| d\omega = 0. \tag{7.6}$$

These integrals show that there will exist a frequency range over which the magnitude of the sensitivity function exceeds 1 if it is to be kept below 1 at other frequencies. This is the so-called **water bed effect**. This water bed effect was vividly illustrated in Figure 7.46 by Gunter Stein in his 1989 Bode Lecture and 2003 paper in *IEEE Control Systems Magazine*. He described that a serious control designer is like a ditch digger who moves *dirt* from one place to another, using appropriate tools, and never getting rid of any of it. For every ditch dug somewhere, a mound is deposited somewhere else.

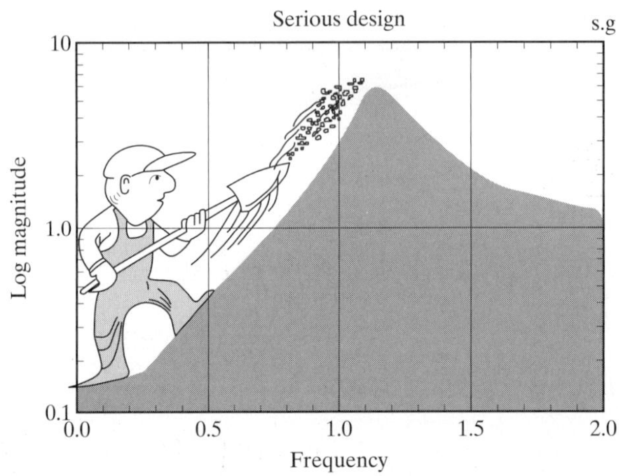

FIGURE 7.46: Gunter Stein's interpretation of the water bed effect: Sensitivity reduction at low frequency unavoidably leads to sensitivity increase at higher frequencies.

Bandwidth constraints in feedback design typically require that the open-loop transfer function be small above a specified frequency, and that it rolls off at a rate of more than one pole-zero excess above that frequency. In other words, we must have $|S(j\omega)| \approx 1, \forall\omega \in [\omega_h, \infty)$ for some high-frequency ω_h. These constraints are commonly needed to ensure stability robustness despite the presence of modeling uncertainty in the plant model particularly at high frequencies. This will be discussed in the next chapter. It is therefore easy to see that $\max_\omega |S(j\omega)|$ can be fairly large if we want to make $|S(j\omega)|$ small in some frequency range.

EXAMPLE 7.21

Suppose that the feedback system is designed such that the level of sensitivity reduction is given by

$$|S(j\omega)| \leq \epsilon < 1, \quad \forall\omega \in [0, \ \omega_l]$$

where $\epsilon > 0$ is a given constant.

Also suppose that

$$|L(j\omega)| \leq \frac{M_h}{\omega^2} \leq \tilde{\epsilon} < 1, \quad \forall\omega \in [\omega_h, \ \infty)$$

where $\omega_h > \omega_l$, and $M_h > 0$ is a given constant. Then for $\omega \geq \omega_h$,

$$|S(j\omega)| \leq \frac{1}{1 - |L(j\omega)|} \leq \frac{1}{1 - \frac{M_h}{\omega^2}}$$

and

$$\begin{aligned}
\int_{\omega_h}^{\infty} \ln|S(j\omega)|d\omega &\leq -\int_{\omega_h}^{\infty} \ln\left(1-\frac{M_h}{\omega^2}\right)d\omega = \sum_{i=1}^{\infty}\int_{\omega_h}^{\infty}\frac{1}{i}\left(\frac{M_h}{\omega^2}\right)^i d\omega \\
&= \sum_{i=1}^{\infty}\frac{1}{i}\frac{\omega_h}{2i-1}\left(\frac{M_h}{\omega_h^2}\right)^i \leq \omega_h \sum_{i=1}^{\infty}\frac{1}{i}\left(\frac{M_h}{\omega_h^2}\right)^i \\
&= -\omega_h \ln\left(1-\frac{M_h}{\omega_h^2}\right) \leq -\omega_h \ln(1-\tilde{\epsilon}).
\end{aligned}$$

Then

$$\begin{aligned}
\pi\sum_{i=1}^{m} p_i &= \int_0^{\infty} \ln|S(j\omega)|d\omega \\
&= \int_0^{\omega_l} \ln|S(j\omega)|d\omega + \int_{\omega_l}^{\omega_h} \ln|S(j\omega)|d\omega + \int_{\omega_h}^{\infty} \ln|S(j\omega)|d\omega \\
&\leq \omega_l \ln\epsilon + (\omega_h-\omega_l)\max_{\omega\in[\omega_l,\omega_h]} \ln|S(j\omega)| - \omega_h \ln(1-\tilde{\epsilon}),
\end{aligned}$$

which gives

$$\max_{\omega\in[\omega_l,\omega_h]} |S(j\omega)| \geq e^{\alpha}\left(\frac{1}{\epsilon}\right)^{\frac{\omega_l}{\omega_h-\omega_l}}(1-\tilde{\epsilon})^{\frac{\omega_h}{\omega_h-\omega_l}}$$

where

$$\alpha = \frac{\pi\sum_{i=1}^{m} p_i}{\omega_h-\omega_l} > 0.$$

The above lower bound shows that the sensitivity can be very significant in the transition band.

PROBLEMS

7.1. Let a plant be described by the following transfer function:

$$P(s) = \frac{50}{(s+1)(s+2)}.$$

Design a lag compensator so that the closed-loop PM is at least 50° and the steady-state error with respect to a step input is no more than 0.02. In addition, the loop transfer function should have the largest possible crossover frequency ω_c.

7.2. For the preceding problem, design a PI controller so that the design specifications are satisfied.

7.3. The transfer function model of a plant is given by

$$P(s) = \frac{20(s+2)}{s(s^2+3s+100)}.$$

Design a phase-lag (or PI) controller so that

- the steady-state error with respect to a unit ramp input is no more than 1;
- the percentage overshoot with respect to a step input is no more than 20%;
- the settling time to the step input is as short as possible.

7.4. Given a plant

$$P(s) = \frac{1000}{s(s+1)(s+5)(s^2+4s+100)},$$

design a phase-lag (or PI) controller so that

- the steady-state error with respect to a unit ramp input is no more than 0.1;
- the closed-loop PM is at least 45°;
- the open-loop crossover frequency ω_c is as large as possible.

7.5. Given a plant

$$P(s) = \frac{10}{s(s+4)},$$

design a lead compensator so that the closed-loop PM is at least 45° and the velocity error constant is at least 50.

7.6. Given a plant

$$P(s) = \frac{10}{s(s+1)(s+10)},$$

design a lead controller so that

- the steady-state error with respect to a unit ramp input is no more than 0.5;
- the closed-loop PM is at least 45°;
- the open-loop crossover frequency ω_c is at least 3.

7.7. Given a plant

$$P(s) = \frac{1000}{s(s+1)(s+5)(s^2+4s+100)},$$

design a lead compensator so that the closed-loop PM is at least 45° and the open-loop crossover frequency ω_c is at least 2.

7.8. Given a plant

$$P(s) = \frac{10}{s+1},$$

design a proper compensator (as simple as possible) so that the closed-loop PM is greater than 50° and the velocity error constant is 100.

7.9. Given a plant

$$P(s) = \frac{1000}{s(s+1)(s+5)(s^2+4s+100)},$$

design a suitable controller so that

- the steady-state error with respect to a unit ramp input is no more than 0.1;
- the closed-loop PM is at least 45°;
- the open-loop crossover frequency ω_c is at least 2.

7.10. Plot the root locus of

$$P(s) = \frac{1}{s(s+4)(s^2+4s+8)}.$$

Assume that $C(s) = K$ in the unity feedback system.

1. What is the value of K such that persistent oscillation occurs in the step response? What is the frequency of the persistent oscillation?

2. Choose K so that the gain margin of the feedback system is 2.
3. Choose K according to Ziegler–Nichols' rule #2.

NOTES AND REFERENCES

The PID controllers tuning rules were first proposed by Ziegler and Nichols in

J. G. Ziegler and N. B. Nichols "Optimum setting for automatic controllers," *Transactions of the ASME*, vol. 64, pp. 759–768, 1942.

J. G. Ziegler, N. B. Nichols, and N. Y. Rochester, "Process lags in automatic control circuits," *Transactions of the ASME*, vol. 65, pp. 433–444, 1943.

There have been constant efforts in devising new PID tuning techniques ever since. The following book contains an extensive summary of these efforts.

K. J. Åström and T. Hägglund, *Advanced PID Control*, the Instrumentation, Systems, and Automation Society, Research Triangle Park, NC, 2006.

The gain and phase relationship first appeared in

H. W. Bode, "Relations between attenuation and phase in feedback amplifier design," *Bell System Technical Journal*, vol. 19, pp. 421–454, 1940.

The gain and phase relationship together with the original Bode sensitivity integral and many other important works by Bode and his colleagues can be found in

H. W. Bode, *Network analysis and feedback amplifier design*, Van Nostrand, New York, 1945.

The Bode sensitivity integral has its mathematical origin from the so-called Jensen's formula in complex analysis.

J. L. Jensen, "Sur un nouvel et important thorme de la thorie des fonctions," *Acta Mathematica*, vol. 22, pp. 359–364, 1899.

L. V. Ahlfors, *Complex Analysis*, 3rd Edition, McGraw-Hill, New York, 1979.

The generalized Bode sensitivity integral formula given in this chapter is due to

J. S. Freudenberg and D. P. Looze, *Frequency Domain Properties of Scalar and Multivariable Feedback Systems,* Lecture notes in control and information sciences, Springer-Verlag, New York, 1988.

Gunter Stein gave the first Hendrik W. Bode Lecture at the IEEE Conference on Decision and Control in Tampa, Florida, in December 1989. The modified lecture was published as

G. Stein, "Respect the unstable," *IEEE Control Systems Magazine*, vol. 23(4), pp. 12–25, 2003.

CHAPTER 8

Performance and Robustness

In this chapter, we first introduce two frequency-domain norms for systems, namely, the 2-norm and the ∞-norm. These two system norms will be used to measure system performance and robustness under various disturbances and model uncertainties. We then introduce two types of commonly used generic model uncertainties, namely, the additive and multiplicative model uncertainties, and derive the respective robust stability criteria under such model uncertainties. The limitations of these model representations lead us to introduce a new distance measure between dynamical systems in terms of chordal and spherical distances. Robustness under this distance measure is then analyzed. Controller design under these system norms and uncertain measures will be considered in the next chapter.

8.1 FREQUENCY-DOMAIN 2-NORM OF SIGNALS AND SYSTEMS

In Chapter 4, we defined a set of signal and system norms in the time domain. There are also a few very useful signal and system norms in the frequency domain. In this section, we will study the frequency-domain 2-norm. In the next section, we will study the frequency-domain ∞-norm.

Definition 8.1. Let $X(s)$ be the Laplace transform of a signal. Its 2-norm is then defined as

$$\|X(s)\|_2 = \left[\frac{1}{2\pi}\int_{-\infty}^{\infty} |X(j\omega)|^2 d\omega\right]^{1/2}.$$

EXAMPLE 8.2

Let us revisit Example 4.24. Consider the unity feedback system shown in Figure 8.1.

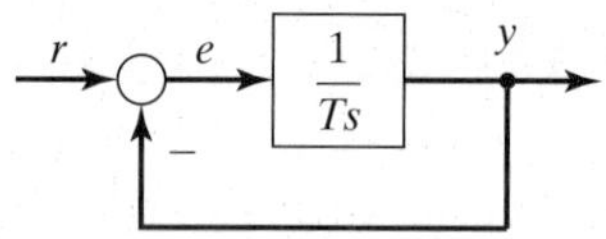

FIGURE 8.1: A first-order unity feedback system.

Assume that $r(t)$ is a unit step function $\sigma(t)$, i.e., $R(s) = 1/s$. We are interested in finding the frequency-domain 2-norm of $E(s)$, the Laplace transform of the error $e(t)$. We know that

$$E(s) = \frac{T}{Ts+1}.$$

Hence,

$$\begin{aligned}\|E(s)\|_2 &= \left[\frac{1}{2\pi}\int_{-\infty}^{\infty}\frac{T^2}{|j\omega T+1|^2}d\omega\right]^{1/2} = \left[\frac{1}{2\pi}\int_{-\infty}^{\infty}\frac{T^2}{\omega^2T^2+1}d\omega\right]^{1/2} \\ &= \left[\frac{T}{2\pi}\arctan T\omega\Big|_{-\infty}^{\infty}\right]^{1/2} = \sqrt{\frac{T}{2}}.\end{aligned}$$

EXAMPLE 8.3

Here we revisit Example 4.25. Consider the unity feedback system shown in Figure 8.2. We are interested in the frequency-domain 2-norm of the error signal when the reference is a unit step.

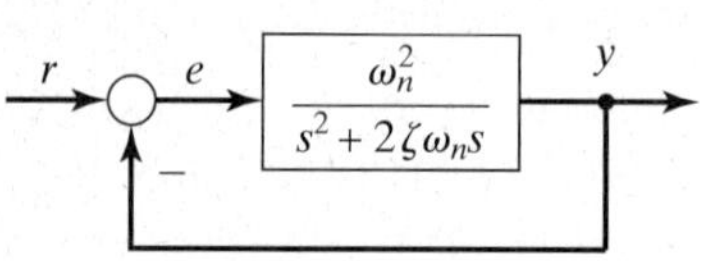

FIGURE 8.2: A second-order unity feedback system.

Analytically, we know

$$E(s) = \frac{s+2\zeta\omega_n}{s^2+2\zeta\omega_n s+\omega_n^2}.$$

Hence,

$$\|E(s)\|_2^2 = \frac{1}{2\pi}\int_{-\infty}^{\infty} E(j\omega)E(-j\omega)d\omega$$
$$= \frac{1}{2\pi}\int_{-\infty}^{\infty} \frac{j\omega + 2\zeta\omega_n}{(j\omega)^2 + 2\zeta\omega_n(j\omega) + \omega_n^2}\frac{-j\omega + 2\zeta\omega_n}{(-j\omega)^2 + 2\zeta\omega_n(-j\omega) + \omega_n^2}d\omega.$$

The integrant is a fourth-order rational function of ω, which is not easy to handle. Let us simplify it slightly. A partial fractional expansion of $E(s)E(-s)$ is given by

$$E(s)E(-s) = \frac{\frac{4\zeta^2+1}{4\zeta\omega_n}s + 2\zeta^2}{s^2 + 2\zeta\omega_n s + \omega_n^2} + \frac{-\frac{4\zeta^2+1}{4\zeta\omega_n}s + 2\zeta^2}{s^2 - 2\zeta\omega_n s + \omega_n^2}.$$

We name the first term on the right by $F(s)$. Then the second term is $F(-s)$. Since

$$\int_{-\infty}^{\infty} F(-j\omega)d\omega = \int_{-\infty}^{\infty} F(j\omega)d\omega,$$

it follows that

$$\|E(s)\|_2^2 = \frac{1}{\pi}\int_{-\infty}^{\infty} F(j\omega)d\omega.$$

The evaluation of such an integral is usually done by using a complex contour integral. Let $\mathcal{D}$ be the standard contour in the complex plane encircling the whole right half complex plane, as shown in Figure 8.3. Then

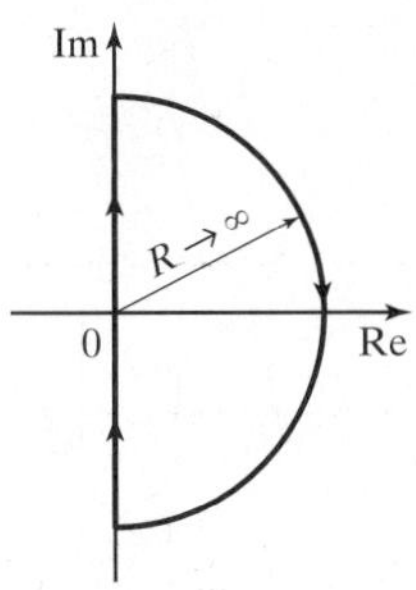

FIGURE 8.3: The $\mathcal{D}$-contour.

$$\oint_{\mathcal{D}} F(s)ds = \lim_{R\to\infty}\left[\int_{-jR}^{jR} F(j\omega)dj\omega + \int_{jR}^{-jR} F(Re^{j\theta})dRe^{j\theta}\right]$$
$$= j\int_{-\infty}^{\infty} F(j\omega)d\omega + j\int_{\pi/2}^{-\pi/2} \lim_{R\to\infty}\left[F(Re^{j\theta})Re^{j\theta}\right]d\theta$$
$$= j\int_{-\infty}^{\infty} F(j\omega)d\omega + j\int_{\pi/2}^{-\pi/2} \frac{4\zeta^2+1}{4\zeta\omega_n}d\theta$$
$$= j\pi\|E(s)\|_2^2 - j\pi\frac{4\zeta^2+1}{4\zeta\omega_n}.$$

Since the integrant in the far left is analytic in the right-hand side of the complex plane, the Cauchy integral theorem tells us that the contour integral is zero. Therefore,

$$\|E(s)\|_2^2 = \frac{4\zeta^2 + 1}{4\zeta\omega_n}$$

and

$$\|E(s)\|_2 = \sqrt{\left(\zeta + \frac{1}{4\zeta}\right)\frac{1}{\omega_n}}.$$

In the above two examples, we see that the frequency-domain 2-norms of $E(s)$ are exactly equal to the time-domain 2-norms of $e(t)$, computed in Examples 4.24 and 4.25, respectively. This is no coincidence.

Let $X(s)$ be the Laplace transform of $x(t)$. Assume that $\|x(t)\|_2 < \infty$. The region of convergence of $X(s)$ then contains the imaginary axis. Hence, the Laplace transform and the inverse Laplace transform relations give

$$X(j\omega) = \int_{0-}^{\infty} x(t)e^{-j\omega t}dt$$

$$x(t) = \frac{1}{2\pi}\int_{-\infty}^{\infty} X(j\omega)e^{j\omega t}d\omega.$$

Therefore,

$$\begin{aligned}
\|X(s)\|_2^2 &= \frac{1}{2\pi}\int_{-\infty}^{\infty} X(j\omega)X(-j\omega)d\omega \\
&= \frac{1}{2\pi}\int_{-\infty}^{\infty} X(j\omega)\left[\int_{0-}^{\infty} x(t)e^{j\omega t}dt\right]d\omega \\
&= \int_{0-}^{\infty} x(t)\left[\frac{1}{2\pi}\int_{-\infty}^{\infty} X(j\omega)e^{j\omega t}d\omega\right]dt \\
&= \int_{0-}^{\infty} x^2(t)dt \\
&= \|x(t)\|_2^2.
\end{aligned}$$

This gives the well-known Parseval's identity.

Theorem 8.4 (Parseval's Identity, version 1). Let $X(s)$ be the Laplace transform of $x(t)$ and assume $\|x(t)\|_2 < \infty$. Then

$$\|x(t)\|_2 = \|X(s)\|_2.$$

Because of this theorem, there is no need to distinguish between the time-domain 2-norm and frequency-domain 2-norm of a signal. The same notation $\|\cdot\|_2$ used for both norms will unlikely cause any confusion. The computation of the frequency-domain 2-norm can also be carried out by using Algorithm 4.31.

Note that the assumption on the finiteness of $\|x(t)\|_2$ in the theorem is important as evidenced in the following example.

EXAMPLE 8.5

Let $x(t) = e^t\sigma(t)$. Clearly, $\|x(t)\|_2 = \infty$ but

$$\|X(s)\|_2 = \left\|\frac{1}{s-1}\right\|_2 = \left[\frac{1}{2\pi}\int_{-\infty}^{\infty}\frac{1}{|j\omega-1|^2}d\omega\right]^{1/2} = \frac{1}{\sqrt{2}}.$$

For a system with transfer function $G(s)$, its frequency-domain 2-norm is simply the frequency-domain 2-norm of $G(s)$ considered as the Laplace transform of its impulse response. Again, owing to the Parseval's identity, the meaning of $\|G(s)\|_2$ has no ambiguity though it has two interpretations.

In Chapter 4, an algorithm was given to compute the time-domain 2-norm of a signal or a system. Because of Parseval's identity, the same algorithm can be used to compute the frequency-domain 2-norm. However, the algorithm was not proved in Chapter 4 since we were not able to prove the key Theorem 4.29 in the derivation of the algorithm. We are now in the position to prove this theorem. Before doing this, we first introduce the frequency-domain inner product and prove a more general version of Parseval's identity.

Definition 8.6. Let $X(s)$ and $Y(s)$ be the Laplace transforms of two signals. Their frequency-domain inner product is then defined as

$$\langle X(s), Y(s)\rangle = \frac{1}{2\pi}\int_{-\infty}^{\infty} X(j\omega)Y(-j\omega)d\omega.$$

Theorem 8.7 (Parseval's Identity, version 2). Let $X(s)$ and $Y(s)$ be the Laplace transforms of two signals $x(t)$ and $y(t)$. Assume that $\|x(t)\|_2 < \infty$ and $\|y(t)\|_2 < \infty$. Then

$$\langle X(s), Y(s)\rangle = \langle x(t), y(t)\rangle.$$

The proof of version 2 is not much harder than that of version 1. It simply follows from

$$\begin{aligned}\langle X(s), Y(s)\rangle &= \frac{1}{2\pi}\int_{-\infty}^{\infty} X(j\omega)Y(-j\omega)d\omega \\ &= \frac{1}{2\pi}\int_{-\infty}^{\infty} X(j\omega)\left[\int_{0^-}^{\infty} y(t)e^{j\omega t}dt\right]d\omega \\ &= \int_{0^-}^{\infty}\left[\frac{1}{2\pi}\int_{-\infty}^{\infty} X(j\omega)e^{j\omega t}d\omega\right]y(t)dt \\ &= \int_{0^-}^{\infty} x(t)y(t)dt \\ &= \langle x(t), y(t)\rangle.\end{aligned}$$

Now consider a strictly proper, stable, rational function

$$X(s) = \frac{b(s)}{a(s)} = \frac{b_1 s^{n-1} + \cdots + b_n}{a_0 s^n + a_1 s^{n-1} + \cdots + a_n}, \quad a_0 > 0. \tag{8.1}$$

Construct the Routh table of stable polynomial $a(s)$ as in Table 8.1.

s^n	$r_{00} = a_0$	$r_{01} = a_2$	$r_{02} = a_4$	$r_{03} = a_6$	$\cdots$
s^{n-1}	$r_{10} = a_1$	$r_{11} = a_3$	$r_{12} = a_5$	$r_{13} = a_7$	$\cdots$
s^{n-2}	r_{20}	r_{21}	r_{22}	r_{23}	$\cdots$
s^{n-3}	r_{30}	r_{31}	r_{32}	r_{33}	$\cdots$
$\vdots$	$\vdots$	$\vdots$	$\vdots$	$\vdots$	
s^2	$r_{(n-2)0}$	$r_{(n-2)1}$			
s^1	$r_{(n-1)0}$				
s^0	r_{n0}				

TABLE 8.1: Routh table.

Since $a(s)$ is stable, the Routh table can always be constructed to the end and all $r_{i0}, i = 0, 1, \ldots, n$, are positive. For each row (except the first one) of the Routh table, define a polynomial

$$\begin{aligned}
r_1(s) &= r_{10}s^{n-1} + r_{11}s^{n-3} + \cdots \\
r_2(s) &= r_{20}s^{n-2} + r_{21}s^{n-4} + \cdots \\
&\vdots \\
r_{n-1}(s) &= r_{(n-1)0}s \\
r_n(s) &= r_{n0}.
\end{aligned}$$

Also define

$$\alpha_i = \frac{r_{(i-1)0}}{r_{i0}}, \quad i = 1, 2, \ldots, n,$$

and functions

$$X_i(s) = \sqrt{2\alpha_i}\frac{r_i(s)}{a(s)}, \quad i = 1, 2, \ldots, n.$$

An equivalent version of Theorem 4.29 in the frequency domain can then be stated as follows.

Theorem 8.8.

$$\langle X_i(s), X_j(s) \rangle = \begin{cases} 1, & i = j; \\ 0, & i \neq j. \end{cases}$$

Proof. What we need to prove is

$$\left\langle \frac{r_i(s)}{a(s)}, \frac{r_j(s)}{a(s)} \right\rangle = \begin{cases} \dfrac{1}{2\alpha_i}, & \text{if } i = j \\ 0, & \text{otherwise} \end{cases}$$

where $i, j = 1, 2, \ldots, n$. By the definition of the inner product

$$\left\langle \frac{r_i(s)}{a(s)}, \frac{r_j(s)}{a(s)} \right\rangle = \frac{1}{2\pi}\int_{-\infty}^{\infty} \frac{r_i(j\omega)}{a(j\omega)}\frac{r_j(-j\omega)}{a(-j\omega)} d\omega. \quad (8.2)$$

The construction of the Routh table implies that

$$r_i(s) = r_{i-2}(s) - \alpha_{i-1} s r_{i-1}(s), \; i = 2, 3, \ldots, n. \quad (8.3)$$

It is easy to verify that

$$r_0(s) = \frac{1}{2}[c_0(s)a(s) + (-1)^n c_0(-s)a(-s)]$$

$$r_1(s) = \frac{1}{2}[c_1(s)a(s) + (-1)^{n-1} c_1(-s)a(-s)]$$

where $c_0(s) = 1$ and $c_1(s) = 1$. Therefore,

$$\begin{aligned} r_2(s) &= r_0(s) - \alpha_1 s r_1(s) \\ &= \frac{1}{2}[c_2(s)a(s) + (-1)^{n-2} c_2(-s)a(-s)] \end{aligned}$$

where

$$c_2(s) = c_0(s) - \alpha_1 s c_1(s).$$

By mathematical induction, we have, for $i = 2, 3, \ldots, n$

$$r_i(s) = \frac{1}{2}[c_i(s)a(s) + (-1)^{n-i} c_i(-s)a(-s)], \quad (8.4)$$

where $c_i(s)$ satisfies the same recursive relation as $r_i(s)$, namely,

$$c_i(s) = c_{i-2}(s) - \alpha_{i-1} s c_{i-1}(s).$$

This shows that

$$\begin{aligned} c_i(s) &= (-1)^{i-1}\alpha_{i-1}\ldots\alpha_1 s^{i-1} + \text{lower order terms} \\ &= (-1)^{i-1}\frac{r_{00}}{r_{(i-1)0}} s^{i-1} + \text{lower order terms.} \end{aligned}$$

Substituting (8.4) into (8.2) yields

$$\left\langle \frac{r_i(s)}{a(s)}, \frac{r_j(s)}{a(s)} \right\rangle = \frac{1}{4\pi}\int_{-\infty}^{\infty} \frac{c_i(j\omega) r_j(-j\omega)}{a(-j\omega)} d\omega + \frac{(-1)^{n-i}}{4\pi}\int_{-\infty}^{\infty} \frac{c_i(-j\omega) r_j(-j\omega)}{a(j\omega)} d\omega.$$

Notice that $r_j(-j\omega) = (-1)^{n-j} r_j(j\omega)$ and for each function $F(j\omega)$,

$$\int_{-\infty}^{\infty} F(-j\omega) d\omega = \int_{-\infty}^{\infty} F(j\omega) d\omega.$$

Then,

$$\int_{-\infty}^{\infty} \frac{c_i(j\omega)r_j(-j\omega)}{a(-j\omega)} d\omega = (-1)^{n-j} \int_{-\infty}^{\infty} \frac{c_i(j\omega)r_j(j\omega)}{a(-j\omega)} d\omega$$
$$= (-1)^{n-j} \int_{-\infty}^{\infty} \frac{c_i(-j\omega)r_j(-j\omega)}{a(j\omega)} d\omega.$$

Therefore,

$$\left\langle \frac{r_i(s)}{a(s)}, \frac{r_j(s)}{a(s)} \right\rangle = \frac{1}{4\pi} \int_{-\infty}^{\infty} \frac{c_i(-j\omega)r_j(-j\omega)}{a(j\omega)} d\omega \left[(-1)^{n-j} + (-1)^{n-i}\right].$$

For each

$$e(s) = e_1 s^{n-1} + e_2 s^{n-2} + \cdots + e_n,$$

let us compute

$$\int_{-\infty}^{\infty} \frac{e(j\omega)}{a(j\omega)} d\omega$$

using the complex contour integral idea in Example 8.3. Since $e(s)/a(s)$ is analytic in the right hand side of the complex plane, its integral along the $\mathcal{D}$-contour shown in Figure 8.3 is zero, i.e.,

$$0 = \oint_{\mathcal{D}} \frac{e(s)}{a(s)} ds = \lim_{R\to\infty} \int_{-jR}^{jR} \frac{e(j\omega)}{a(j\omega)} dj\omega + \int_{jR}^{-jR} \frac{e(Re^{j\theta})}{a(Re^{j\theta})} dRe^{j\theta}$$
$$= j\int_{-\infty}^{\infty} \frac{e(j\omega)}{a(j\omega)} d\omega + j\int_{\pi/2}^{-\pi/2} \lim_{R\to\infty} \frac{e(Re^{j\theta})Re^{j\theta}}{a(Re^{j\theta})} d\theta$$
$$= j\int_{-\infty}^{\infty} \frac{e(j\omega)}{a(j\omega)} d\omega + j\int_{\pi/2}^{-\pi/2} \frac{e_1}{a_0} d\theta$$
$$= j\int_{-\infty}^{\infty} \frac{e(j\omega)}{a(j\omega)} d\omega - j\pi\frac{e_1}{a_0}.$$

Consequently,

$$\int_{-\infty}^{\infty} \frac{e(j\omega)}{a(j\omega)} d\omega = \pi\frac{e_1}{a_0}.$$

If we set $e(j\omega) = c_i(-j\omega)r_j(-j\omega)$, then

$$e_1 = \begin{cases} (-1)^{n-i}\dfrac{r_{00}}{r_{(i-1)0}} r_{i0}, & i = j \\ 0, & i < j \end{cases} = \begin{cases} (-1)^{n-i}\dfrac{a_0}{\alpha_i}, & i = j \\ 0, & i < j. \end{cases}$$

This implies that

$$\left\langle \frac{r_i(s)}{a(s)}, \frac{r_j(s)}{a(s)} \right\rangle = \begin{cases} \dfrac{1}{2\alpha_i}, & i = j \\ 0, & i < j. \end{cases}$$

For $j > i$, the inner product is also zero since

$$\left\langle \frac{r_i(s)}{a(s)}, \frac{r_j(s)}{a(s)} \right\rangle = \left\langle \frac{r_j(s)}{a(s)}, \frac{r_i(s)}{a(s)} \right\rangle.$$

□

8.2 FREQUENCY-DOMAIN ∞-NORM OF SYSTEMS

For a system, there is a second frequency-domain norm, which is very useful.

Definition 8.9. For a system with transfer function $G(s)$, its (frequency-domain) ∞-norm is defined as

$$\|G(s)\|_\infty = \sup_{\omega \in \mathbb{R}} |G(j\omega)|.$$

In simpler words, the (frequency-domain) ∞-norm of a system is the peak value of its frequency magnitude response. It is also called the **resonance peak** of $G(s)$. The frequency-domain ∞-norm is only used for systems and not for signals, whereas the time-domain ∞-norm is only used for signals and not for systems. Therefore, the meaning of $\|\cdot\|_\infty$ can be determined from the context and from the nature of the object whose norm is to be considered. For example, $\|x(t)\|_\infty$ means the time-domain ∞-norm of signal $x(t)$, and $\|G(s)\|_\infty$ means the frequency-domain ∞-norm of system $G(s)$.

The frequency-domain ∞-norm of a stable system has interesting input/output interpretations. Since the norm is the peak frequency magnitude response, it is then the largest magnification in steady-state sinusoidal response. More specifically, if the input of the system is a sinusoidal signal with magnitude A and the steady-state output is a sinusoidal signal with magnitude B, then the frequency-domain ∞-norm is the maximum value of B/A over all possible frequencies. The largest magnification interpretation of the frequency-domain ∞-norm can be further extended. Let us notice that for a stable system $G(s)$ with input $U(s)$ and output $Y(s)$ we have

$$\begin{aligned}\|Y(s)\|_2^2 = \|G(s)U(s)\|_2^2 &= \frac{1}{2\pi}\int_{-\infty}^{\infty} |G(j\omega)U(j\omega)|^2 d\omega \\ &\le \|G(s)\|_\infty^2 \frac{1}{2\pi}\int_{-\infty}^{\infty} |U(j\omega)|^2 d\omega = \|G(s)\|_\infty^2 \|U(s)\|_2^2.\end{aligned}$$

Hence

$$\frac{\|Y(s)\|_2}{\|U(s)\|_2} \le \|G(s)\|_\infty,$$

i.e., the amplification in the output 2-norm over the input 2-norm is bounded by the ∞-norm of the system. Equivalently, the amplification of the output energy over the input energy is bounded by the square of the ∞-norm of the system. Furthermore, this bound is tight, i.e., for some special input/output pair, the amplification is exactly equal to (or arbitrarily close to) the ∞-norm of the system. This gives the following theorem.

Theorem 8.10. For a stable system $G(s)$, we have

$$\|G(s)\|_\infty = \sup_{0<\|u(t)\|_2<\infty} \frac{\|Y(s)\|_2}{\|U(s)\|_2} = \sup_{0<\|u(t)\|_2<\infty} \frac{\|y(t)\|_2}{\|u(t)\|_2}.$$

EXAMPLE 8.11

Consider a standard first-order system

$$G(s) = \frac{1}{Ts+1}.$$

Since

$$|G(j\omega)| = \frac{1}{\sqrt{T^2\omega^2+1}},$$

it is easy to see that its maximum is attained at $\omega = 0$ and the maximum value is 1. Hence $\|G(s)\|_\infty = 1$.

For a standard second-order system,

$$G(s) = \frac{\omega_n^2}{s^2 + 2\zeta\omega_n s + \omega_n^2}.$$

The material in Section 6.5 shows that

$$\|G(s)\|_\infty = \begin{cases} \dfrac{1}{2\zeta\sqrt{1-\zeta^2}} & 0 < \zeta < \dfrac{1}{\sqrt{2}} \\ 1 & \zeta \geq \dfrac{1}{\sqrt{2}}. \end{cases}$$

In the first case above ($0 < \zeta < 1/\sqrt{2}$), the frequency at which the peak of the frequency response is attained is $\omega = \omega_n\sqrt{1-2\zeta^2}$. In the second case ($\zeta \geq 1/\sqrt{2}$), the frequency reaches its maximum magnitude at $\omega = 0$.

There are several methods to compute the frequency-domain ∞-norm of a general system. The most straightforward and brute-force method is to just plot the Bode magnitude plot and read the maximum from the plot. This method is effective especially when accuracy and computational efficiency are not of major concern. The second method is to use the standard procedure for optimizing a univariate function: i.e., taking a derivative, finding the critical points where the derivative is zero, and comparing the values of the function at the critical points. Some simplification can be achieved if we apply this procedure to $|G(j\omega)|^2$ instead of $|G(j\omega)|$. Let a stable, proper system be represented by

$$G(s) = \frac{b(s)}{a(s)} = \frac{b_0 s^n + b_1 s^{n-1} + \cdots + b_n}{a_0 s^n + a_1 s^{n-1} + \cdots + a_n}$$

where $a_0 \neq 0$. Then

$$|G(j\omega)|^2 = \frac{b(-j\omega)b(j\omega)}{a(-j\omega)a(j\omega)}.$$

It can be seen that both $a(-j\omega)a(j\omega)$ and $b(-j\omega)b(j\omega)$ are real polynomials of ω^2, i.e.,

$$\begin{aligned} a(-j\omega)a(j\omega) &= p_1(\lambda) \\ b(-j\omega)b(j\omega) &= p_2(\lambda) \end{aligned}$$

where $\lambda = \omega^2$. Hence, we have

$$|G(j\omega)|^2 = \frac{b(-j\omega)b(j\omega)}{a(-j\omega)a(j\omega)} = \frac{p_2(\lambda)}{p_1(\lambda)}.$$

Since $p_1(\lambda)$ and $p_2(\lambda)$ have degree n, lower than that of $a(-j\omega)a(j\omega)$ and $b(-j\omega)$ $b(j\omega)$, it leads to some savings in computation and algebra if we apply the procedure to the maximization of $p_2(\lambda)/p_1(\lambda)$.

Yet there is a third method that is often used. It is easy to see that a lower bound of $\|G(s)\|_\infty$ is given by $|G(j\omega_0)|$ for any $\omega_0 \in [-\infty, \infty]$. Normally, we take $\omega_0 = 0$ or $\omega_0 = \infty$. This gives $|G(j\omega_0)| = |b_n/a_n|$ or $|G(j\omega_0)| = |b_0/a_0|$ respectively. Let $\gamma > |G(j\omega_0)|$. Then $\gamma \leq \|G(s)\|_\infty$ if and only if there is a frequency $\omega \in \mathbb{R}$ such that

$$|G(j\omega)|^2 = \frac{b(-j\omega)b(j\omega)}{a(-j\omega)a(j\omega)} = \gamma^2.$$

This happens if and only if

$$b(-s)b(s) - \gamma^2 a(-s)a(s)$$

has roots on the imaginary axis. On the basis of this, a bisection search algorithm can be implemented to find $\|G(s)\|_\infty$.

Algorithm 8.12.

Step 1 Let $\alpha = |G(j\omega_0)|$ for any $\omega_0 \in [-\infty, \infty]$. Let β be any upper bound of $\|G(s)\|_\infty$.

Step 2 Let $\gamma = (\alpha + \beta)/2$ and check if the polynomial

$$b(-s)b(s) - \gamma^2 a(-s)a(s)$$

has roots on the imaginary axis.

Step 3 If yes, then $\gamma \leq \|G(s)\|_\infty$ and let $\alpha = \gamma$; otherwise, $\gamma > \|G(s)\|_\infty$ and let $\beta = \gamma$. Go to Step 2 unless $|\alpha - \beta|$ is within error tolerance.

Step 4 Let $\|G(s)\|_\infty = (\alpha + \beta)/2$.

EXAMPLE 8.13

Consider a transfer function

$$G(s) = \frac{8(s+1)}{(s+2)(s+4)}.$$

The plot of $|G(j\omega)|$ versus the frequency ω is shown in Figure 8.4. It can be read from the plot that approximately $\|G(s)\|_\infty = 1.42$ and the peak frequency is approximately $\omega = 2.4$.

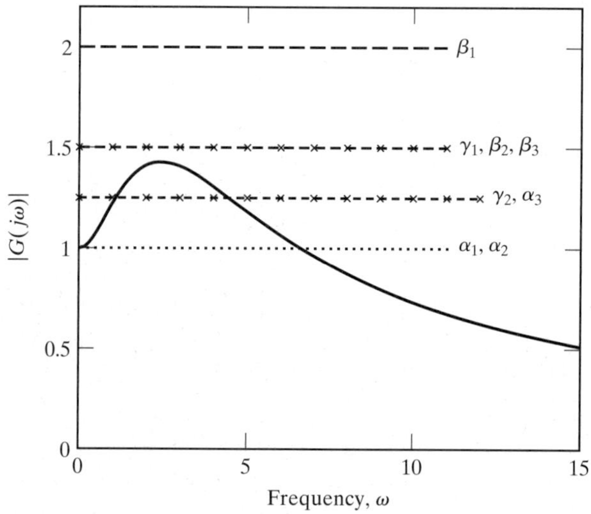

FIGURE 8.4: Magnitude plot of the transfer function in Example 8.13.

The second method to compute $\|G(s)\|_\infty$ is to maximize

$$|G(j\omega)|^2 = \frac{64(\omega^2+1)}{(\omega^2+4)(\omega^2+16)} = \frac{64(\lambda+1)}{(\lambda+4)(\lambda+16)}$$

using calculus. Taking the derivative with respect to λ, we have

$$\frac{d}{d\lambda}\frac{64(\lambda+1)}{(\lambda+4)(\lambda+16)} = \frac{64(\lambda+4)(\lambda+16) - 64(\lambda+1)(2\lambda+20)}{(\lambda+4)^2(\lambda+16)^2}.$$

Setting it to zero, we obtain $\lambda = 3\sqrt{5}-1$. This shows that $|G(j\omega)|^2$ is maximized at $\omega = \sqrt{3\sqrt{5}-1}$ and the maximum value is $(24-8\sqrt{5})/3$. This gives $\|G(s)\|_\infty = 1.4273$.

We can also use Algorithm 8.12 to compute $\|G(s)\|_\infty$. We can take $\alpha_1 = G(0) = 1$ as an initial lower bound and $\beta_1 = 2$ as a reasonable upper bound. Then $\gamma_1 = (\alpha_1 + \beta_1)/2 = 1.5$. In this case,

$$b(-s)b(s) - \gamma_1^2 a(-s)a(s) = 64(1-s^2) - 2.25(4-s^2)(16-s^2)$$

and it does not have roots on the imaginary axis. We then keep the old lower bound $\alpha_2 = \alpha_1 = 1$ and set the new upper bound as $\beta_2 = \gamma_1 = 1.5$. The new middle point of the lower and upper bounds is now $\gamma_2 = (\alpha_2 + \beta_2)/2 = 1.25$. In this case,

$$b(-s)b(s) - \gamma_2^2 a(-s)a(s) = 64(1 - s^2) - 1.5625(4 - s^2)(16 - s^2)$$

and it has roots on the imaginary axis. We then set the new lower bound as $\alpha_3 = \gamma_2 = 1.25$ and keep the previous upper bound $\beta_3 = \beta_2 = 1.5$. After two iterations, we have reduced our very original estimation $1 \leq \|G(s)\|_\infty \leq 2$ to a better estimation $1.25 \leq \|G(s)\|_\infty \leq 1.5$. If we keep the iteration going, the estimation will eventually be sufficiently accurate.

8.3 MODEL UNCERTAINTIES AND ROBUST STABILITY

Most controller design methods we have introduced in the previous chapters are based on the use of a mathematical model. A mathematical model is at best an approximation of the real system under certain working conditions. The quality of a mathematical model for a particular performance objective depends on how closely its responses match those of the true plant for that particular objective and application. One model may be better for one application than another that may be better for yet another application. A good model should be simple enough to facilitate design, yet complex enough to give the engineer the confidence that designs based on the model will work on the true plant.

The term **uncertainty** refers to the differences or errors between models and reality, and whatever mechanism is used to express these errors will be called a **representation of uncertainty**. Representations of uncertainty vary primarily in terms of the amount of structure they contain. This reflects both our knowledge of the physical mechanisms that cause differences between the model and the plant and our ability to represent these mechanisms in a way that facilitates convenient manipulation. The following discussions will be based on the feedback configuration in Figure 8.5.

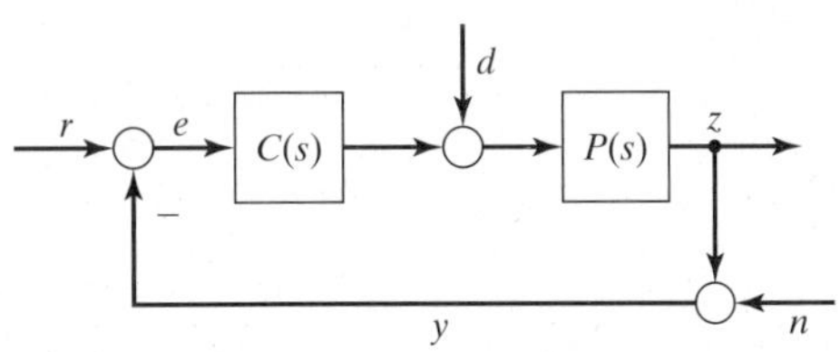

FIGURE 8.5: Standard unity feedback system.

Two commonly used uncertain model representations are **additive and multiplicative uncertain models**. For instance, an additive model representation,

$$P_\Delta(s) = P(s) + \Delta_a(s), \quad |\Delta_a(j\omega)| \leq |W_a(j\omega)|, \ \forall \omega \in \mathbb{R}, \tag{8.5}$$

where $W_a(s)$ is a stable transfer function that characterizes the frequency content of the uncertainty, confines the $P_\Delta(j\omega)$ to a disk centered at $P(j\omega)$ with radius $|W_a(j\omega)|$ at each frequency, as shown in Figure 8.6.

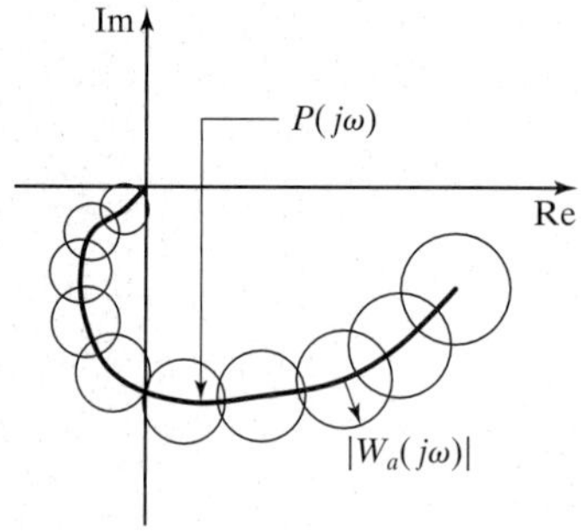

FIGURE 8.6: Nyquist plot of an additive uncertain model.

The representation does not imply a mechanism or structure that gives rise to $\Delta_a(s)$. The uncertainty may be caused by parameter changes, by neglected dynamics, or by any other unspecified effects.

An alternative representation to (8.5) is the so-called multiplicative form:

$$P_\Delta(s) = (1 + \Delta_m(s))P(s),\ |\Delta_m(j\omega)| \leq |W_m(j\omega)|,\ \forall\omega \in \mathbb{R}. \tag{8.6}$$

This representation confines $P_\Delta(s)$ to a normalized neighborhood of the nominal model $P(s)$ as shown in Figure 8.7. An advantage of (8.6) over (8.5) is that in (8.6) compensated transfer function, $P_\Delta(s)C(s)$ with any controller $C(s)$, has the same uncertainty representation as the uncompensated model $P_\Delta(s)$.

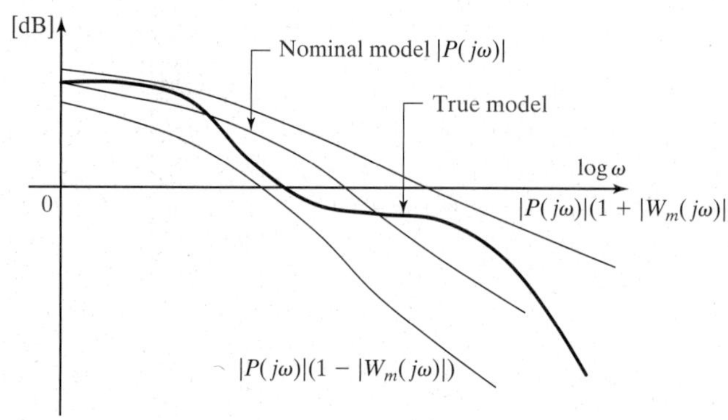

FIGURE 8.7: Typical behavior of multiplicative uncertainty: $P_\Delta(s) = [1 + \Delta_m(s)]P(s)$.

The best choice of uncertainty representation for a specific model depends, of course, on the errors the model makes. In practice, it is generally possible to represent some of these errors in a highly structured parameterized form. These are usually the low-frequency error components. There are always remaining higher frequency errors, however, which cannot be covered this way. These are caused by such effects as infinite-dimensional electromechanical resonance, time delays, diffusion processes, etc. Fortunately, the less structured representations, such as (8.5) and (8.6), are well suited to represent this latter class of errors. Consequently, (8.5) and (8.6) have become widely used "generic" uncertainty representations for finite-dimensional linear time-invariant models.

To see the effect of model uncertainty on the closed-loop stability of a feedback system, we shall first assume that the true plant can be modeled in an additive form

$$P_\Delta(s) = P(s) + \Delta_a(s), \quad |\Delta_a(j\omega)| \le |W_a(j\omega)|, \quad \forall\, \omega \in \mathbb{R}$$

where $P(s)$ is the nominal model, $\Delta_a(s)$ is the model uncertainty, and $W_a(s)$ is a stable proper transfer function.

In general, $\Delta_a(s)$ is not necessarily stable and, indeed, it would be too restrictive to assume the stability of the perturbation. For example, consider an uncertain system with a varying pole location

$$P_\Delta(s) = \frac{1}{s-a}, \quad a \in [2, 4].$$

This system can then be modeled as

$$P_\Delta(s) = \frac{1}{s-3} + \Delta_a(s), \qquad \Delta_a(s) := \frac{1}{s-a} - \frac{1}{s-3}$$

with $|\Delta_a(j\omega)| \le |W_a(j\omega)|$ and

$$W_a(s) = \frac{1}{(s+2)(s+3)}.$$

It is clear that $\Delta_a(s)$ is unstable but the number of unstable poles of the plant $P_\Delta(s)$ is always 1 for all allowable perturbation a.

Of course, the exact form of the model uncertainty $\Delta_a(s)$ is not explicitly known in many practical applications. What one might be able to get is an upper bound of the uncertainty, i.e., a representation of the bounding function $W_a(s)$. The question that a control engineer would be interested in is whether the closed-loop system can be stabilized with a given controller for all possible perturbation $\Delta_a(s)$ that is bounded by $W_a(s)$ or not. In this case, it is actually necessary to assume that $P_\Delta(s)$ has the same number of unstable poles for all admissible perturbation $\Delta_a(s)$. Without this restriction, it is not possible to guarantee the existence of a stabilizing controller for the uncertain systems described by this set.

EXAMPLE 8.14

Consider a set of uncertain systems described by

$$P_\Delta(s) = P(s) + \Delta_a(s)$$

with $P(s) = 0$ and

$$\Delta_a(s) = \left\{ \delta\frac{s+\epsilon}{s-\epsilon}, \quad \delta\frac{s-\epsilon}{s+\epsilon} \right\}.$$

It is then clear that $|\Delta_a(j\omega)| \le \delta$ but $P_\Delta(s)$ may have either no unstable pole or one unstable pole whenever $\epsilon \ne 0$.

It can be shown that there is no controller that can simultaneously stabilize the two plants

$$P_1(s) = \delta\frac{s+\epsilon}{s-\epsilon}, \quad P_2(s) = \delta\frac{s-\epsilon}{s+\epsilon}$$

for any given $\delta > 0$ and $\epsilon > 0$. In other words, there is no controller that will stabilize the uncertain family described by $P_\Delta(s)$.

It should be noted that the additive and multiplicative uncertain model sets given above can represent a rather large class of modes as illustrated by the following example.

EXAMPLE 8.15

A plant is described by the following multiplicative model set

$$P_\Delta(s) = \left\{ \frac{1}{s-1}(1+\Delta_m(s)) : \quad |\Delta_m(j\omega)| \le |W_m(j\omega)| \right\}$$

with

$$W_m(s) = \frac{\frac{1}{4}\left(\frac{1}{2}s+1\right)}{\frac{1}{32}s+1}$$

and $\Delta_m(s)$ can be any transfer function such that $\frac{1}{s-1}(1+\Delta_m(s))$ has exactly one unstable pole. This set of models then include all of the following transfer functions as special cases:

$$P_0(s) = \frac{1}{s-1},$$

$$P_1(s) = \frac{1}{s-1}\frac{6.1}{s+6.1}$$

$$P_2(s) = \frac{1.425}{s-1.425},$$

$$P_3(s) = \frac{0.67}{s-0.67}$$

$$P_4(s) = \frac{1}{s-1}\frac{-0.07s+1}{0.07s+1},$$

$$P_5(s) = \frac{1}{s-1}\frac{70^2}{s^2+2\times 0.15\times 70s+70^2},$$

$$P_6(s) = \frac{1}{s-1}\frac{70^2}{s^2+2\times 5.6\times 70s+70^2},$$

$$P_7(s) = \frac{1}{s-1}\left(\frac{50}{s+50}\right)^6$$

$$P_8(s) = \frac{1}{s-1}\frac{-2.9621(s-9.837)(s+0.76892)}{(s+32)(s+0.56119)},$$

$$P_9(s) = \frac{1}{s-1}\frac{s^2+3.6722s+34.848}{(s+7.2408)(s+32)}.$$

Hence, this set of models can represent variations of unstable pole locations, neglected high-frequency dynamics, high-frequency nonminimum phase zeros, etc. This also means that when such models are used to cover a specific type of uncertainty, the results will generally be conservative. This will be illustrated in Example 8.18.

Theorem 8.16. Assume that the true plant is in the form of $P_\Delta(s) = P(s) + \Delta_a(s)$ with $|\Delta_a(j\omega)| \leq |W_a(j\omega)|$ for all $\omega \in \mathbb{R}$ where $W_a(s)$ is a known stable transfer function. Furthermore, suppose $P(s)$ and $P_\Delta(s)$ have the same number of unstable poles for all allowable perturbations $\Delta_a(s)$, and suppose $C(s)$ is a stabilizing controller for the nominal plant $P(s)$. The controller $C(s)$ then also stabilizes the perturbed plant $P_\Delta(s)$ for all allowable perturbations $\Delta_a(s)$ if and only if

$$\|S(j\omega)C(j\omega)W_a(j\omega)\|_\infty = \left\| \frac{C(j\omega)W_a(j\omega)}{1 + P(j\omega)C(j\omega)} \right\|_\infty < 1. \tag{8.7}$$

Proof. To show that the system is stable for all allowable perturbations, it is sufficient to show that all zeros of $1 + P_\Delta(s)C(s)$ are stable. Note that

$$1 + P_\Delta(s)C(s) = (1 + P(s)C(s))\,[1 + S(s)C(s)\Delta_a(s)]$$

and the number of unstable poles of $1 + P_\Delta(s)C(s)$ is the same as the number of unstable poles of $1 + P(s)C(s)$ by assumption.

Hence, by Nyquist stability criterion, all zeros of $1 + P_\Delta(s)C(s)$ are stable if the number of encirclements of $1 + P_\Delta(s)C(s)$ around the origin is the same as the number of encirclements of $1 + P(s)C(s)$ around the origin since all zeros of $1 + P(s)C(s)$ are stable. Since the number of encirclements of $1 + P_\Delta(s)C(s)$ around the origin is the sum of the number of the encirclements of $1 + P(s)C(s)$ and that of $1 + S(s)C(s)\Delta_a(s)$ around the origin, it is now sufficient to show that the number of encirclements of $1 + S(s)C(s)\Delta_a(s)$ around the origin is zero for all allowable perturbations $\Delta_a(s)$ satisfying $|\Delta_a(j\omega)| \leq |W_a(j\omega)|$ for all $\omega \in \mathbb{R}$. Since

$$|S(j\omega)C(j\omega)\Delta_a(j\omega)| \leq |W_a(j\omega)S(j\omega)C(j\omega)| < 1$$

for all allowable $|\Delta_a(j\omega)| \leq |W_a(j\omega)|$ and for all ω, it is clear that

$$\mathrm{Re}\,(1 + S(j\omega)C(j\omega)\Delta_a(j\omega)) > 0$$

for all $\omega \in \mathbb{R}$, and hence the number of encirclements of $1+S(j\omega)C(j\omega)\Delta_a(j\omega)$ around the origin is zero.

To complete the proof, we must show that the condition given by the inequality (8.7) is necessary for stability, i.e., if the inequality is violated, then the system can be destabilized by an allowable perturbation. We shall leave the detailed proof to the readers. A key step for finding such destabilizing perturbation is given in Problem 8.7. □

Similarly, we can consider the case where the model uncertainty is represented in multiplicative form.

Theorem 8.17. Assume that the true plant is in the form of $P_\Delta(s) = P(s)(1+\Delta_m(s))$ with $|\Delta_m(j\omega)| \le |W_m(j\omega)|$ for all $\omega \in \mathbb{R}$ where $W_m(s)$ is a known stable transfer function. Furthermore, suppose $P(s)$ and $P_\Delta(s)$ have the same number of unstable poles for all allowable perturbations $\Delta_m(s)$, and suppose $C(s)$ is a stabilizing controller for the nominal plant $P(s)$. The controller $C(s)$ then also stabilizes the perturbed plant $P_\Delta(s)$ for all allowable perturbations $\Delta_m(s)$ if and only if

$$\|T(j\omega)W_m(j\omega)\|_\infty = \left\|\frac{P(j\omega)C(j\omega)W_m(j\omega)}{1+P(j\omega)C(j\omega)}\right\|_\infty < 1.$$

Proof. The same proof for Theorem 8.16 can be repeated here to prove this theorem. □

The above stability condition would, in general, amount to requiring that $|T(j\omega)|$ be small at those frequencies where $W_m(j\omega)$ is significant, typically in the high-frequency range, which, in turn, implies that the loop gain $|L(j\omega)|$ should be small at those frequencies.

Note that it is sometimes more convenient to write the uncertainty in a slightly different but equivalent form:

$$\{\Delta(s) : |\Delta(j\omega)| \le |W(j\omega)|\} = \{W(s)\Delta(s) : |\Delta(j\omega)| \le 1\}.$$

Given an uncertainty representation form and a controller, Theorems 8.16 and 8.17 can be used to determine the maximum size of allowable model uncertainties so that the closed-loop system remains stable.

EXAMPLE 8.18

Let a family of an uncertain system be described by

$$P_\Delta(s) = P(s) + W_a(s)\Delta(s)$$

with $W_a(s)$ stable and

$$|\Delta(j\omega)| \le \alpha, \ \forall\, \omega \in \mathbb{R}$$

such that $P_\Delta(s)$ and $P(s)$ have the same number of unstable poles. Let $C(s)$ be a controller that stabilizes the nominal model $P(s)$. Define

$$\alpha_{\max} := \left\|\frac{C(s)W_a(s)}{1+P(s)C(s)}\right\|_\infty^{-1}.$$

The closed-loop system then remains stable for any $\alpha < \alpha_{\max}$.

Now let

$$P(s) = \frac{1}{s-3}, \quad W_a(s) = \frac{1}{(s+2)(s+3)}, \quad C(s) = \frac{8(s+2)}{s}.$$

Then

$$\alpha_{\max} = \left\|\frac{C(s)W_a(s)}{1+P(s)C(s)}\right\|_\infty^{-1} = 1.9526.$$

If the true uncertain model is in the form of

$$P_\Delta(s) = \frac{1}{s-a}, \quad a > 0,$$

we can then find

$$\Delta(s) = \frac{(a-3)(s+2)(s+3)}{(s-a)(s-3)}.$$

It is not hard to calculate that

$$\|\Delta(s)\|_\infty = \begin{cases} |a-3|, & a \geq 2, \\ \dfrac{2(3-a)}{a}, & 0 < a < 2. \end{cases}$$

Setting $\|\Delta(s)\|_\infty < \alpha_{\max}$, we get

$$\frac{6}{2+\alpha_{\max}} < a < 3 + \alpha_{\max},$$

i.e.,

$$1.518 < a < 4.9526.$$

This estimated range of parameter variation is obviously conservative. Now apply the controller $C(s)$ directly to

$$P_\Delta(s) = \frac{1}{s-a}.$$

It is then easy to check that the closed-loop system is stable for all $a < 8$.

Two immediate questions from the above example are that given a plant uncertainty representation $P_\Delta(s)$,

- What is the largest possible $\alpha_{\max}$ over all possible stabilizing controllers?
- How do we find such controllers?

We shall consider these questions in the next chapter for some classes of uncertain systems. We shall demonstrate this through a simple example here.

EXAMPLE 8.19

Consider the uncertain plant given in Example 8.15:

$$P_\Delta(s) = \left\{ \frac{1}{s-1}(1 + W_m(s)\Delta(s)) : \ \|\Delta(s)\|_\infty \leq \alpha \right\}$$

with

$$W_m(s) = \frac{\frac{1}{4}\left(\frac{1}{2}s + 1\right)}{\frac{1}{32}s + 1}.$$

We shall find the largest α so that the uncertain system is robustly stable for all $\alpha < \alpha_{\max}$:

$$\alpha_{\max} := \left\{ \inf_{C(s)\in\mathcal{S}(P)} \left\| \frac{P(s)C(s)W_m(s)}{1+P(s)C(s)} \right\|_\infty \right\}^{-1}.$$

Let

$$C_0(s) = \frac{9s+15}{s} =: \frac{q(s)}{p(s)}$$

be a stabilizing controller for the nominal plant

$$P(s) = \frac{1}{s-1} =: \frac{b(s)}{a(s)}.$$

Then

$$a(s)p(s) + b(s)q(s) = (s+3)(s+5).$$

Let

$$M(s) = \frac{s-1}{s+3}, \quad N(s) = \frac{1}{s+3}, \quad X(s) = \frac{s}{s+5}, \quad Y(s) = \frac{9s+15}{s+5}.$$

Then $X(s)M(s)+N(s)Y(s) = 1$ and by Theorem 3.39, all controllers that stabilize $P(s)$ can be parameterized by

$$C(s) = \frac{Y(s)+M(s)Q(s)}{X(s)-N(s)Q(s)}$$

where $Q(s)$ is any stable transfer function.

Using this parametrization, we get

$$f(s) := \frac{P(s)C(s)W_m(s)}{1+P(s)C(s)} = W_m(s)N(s)(Y(s)+M(s)Q(s)).$$

Since

$$\|f(s)\|_\infty = \sup_{\operatorname{Re} s>0} |f(s)| \geq |f(1)| = W_m(1)N(1)Y(1) = \frac{4}{11} = 0.3636,$$

it is clear that

$$\alpha_{\max} \leq \frac{11}{4} = 2.75.$$

It turns out that we can get arbitrarily close to this value. To do this, we shall first allow $Q(s)$ to be improper and solve

$$W_m(s)N(s)(Y(s)+M(s)Q(s)) = f(1) = \frac{4}{11}.$$

Hence,

$$Q(s) = \frac{f(1) - W_m(s)N(s)Y(s)}{W_m(s)N(s)M(s)} = \frac{0.090909(s-60.48)(s+3)(s+2.48)}{(s+5)(s+2)}$$

and

$$C(s) = \frac{Y(s) + M(s)Q(s)}{X(s) - N(s)Q(s)} = 0.1(s + 32).$$

This is a PD controller. For practical implementation, we need to use a practical PD controller of the form

$$C(s) = \frac{0.1(s + 32)}{Ts + 1}$$

for a small $T > 0$. For example, let $T = 0.001$, then

$$C(s) = \frac{0.1(s + 32)}{0.001s + 1}.$$

Consequently,

$$\left\| \frac{P(s)C(s)W_m(s)}{1 + P(s)C(s)} \right\|_\infty = 0.3638$$

and the closed-loop system is robustly stable for all perturbations such that

$$\|\Delta(s)\|_\infty < 1/0.3638 = 2.7488.$$

Unfortunately, this solution technique does not work if $M(s)$ has more than one unstable zero (or equivalently $P(s)$ has more than one unstable pole). In that case, a more advanced algorithm such as the Nevanlinna–Pick interpolation algorithm has to be applied (see references at the end of the chapter).

8.4 CHORDAL AND SPHERICAL DISTANCES

In most situations, we measure the distance between two complex numbers c_1, $c_2 \in \mathbb{C}$ by

$$|c_1 - c_2|.$$

However, in some cases, this measure may not be the most convenient one. It may not be able to tell the relative difference between two numbers. For example, $|2 - 1| = |101 - 100| = 1$ whereas 2 has 100% difference from 1 but 101 has only 1% difference from 100. Its disadvantage becomes even more severe if we extend this distance to $\mathbb{C} \cup \{\infty\}$. In this case, the difference between ∞ and 0 is the same as that between ∞ and 10^{10}, though intuition tells that the difference between ∞ and 10^{10} should be small in a practical sense. In the following, we will define and use a distance measure in $\mathbb{C} \cup \{\infty\}$ that makes more practical sense in some applications including the robust stabilization.

Recall that $\phi(c_1)$ and $\phi(c_2)$ denote the stereographic projections of complex numbers c_1 and c_1 on the Riemann sphere $\mathbb{S}$. The **chordal distance** between c_1 and c_2, denoted by $\delta(c_1, c_2)$, is defined as the length of the chord connecting $\phi(c_1)$ and $\phi(c_2)$, i.e.,

$$\delta(c_1, c_2) = \text{Length of } \phi(c_1) - \phi(c_2).$$

Since we know that

$$\phi(c_i) = \left(\frac{\text{Re } c_i}{1+|c_i|^2}, \frac{\text{Im } c_i}{1+|c_i|^2}, \frac{|c_i|^2}{1+|c_i|^2} \right), \quad i = 1, 2,$$

it follows that

$$\delta(c_1, c_2) = \left[\left(\frac{\text{Re } c_1}{1+|c_1|^2} - \frac{\text{Re } c_2}{1+|c_2|^2} \right)^2 + \left(\frac{\text{Im } c_1}{1+|c_1|^2} - \frac{\text{Im } c_2}{1+|c_2|^2} \right)^2 + \left(\frac{|c_1|^2}{1+|c_1|^2} - \frac{|c_2|^2}{1+|c_2|^2} \right)^2 \right]^{1/2}$$

and

$$\delta(c_1, c_2) = \frac{|c_1 - c_2|}{\sqrt{1+|c_1|^2}\sqrt{1+|c_2|^2}}. \tag{8.8}$$

The equality above involves some nontrivial, but straightforward, derivation. We leave it to the readers to verify its validity. In the case when the complex numbers are represented by $c_1 = b_1/a_1$ and $c_2 = b_2/a_2$, we have

$$\delta\left(\frac{b_1}{a_1}, \frac{b_2}{a_2} \right) = \frac{|a_1 b_2 - a_2 b_1|}{\sqrt{|a_1|^2 + |b_1|^2}\sqrt{|a_2|^2 + |b_2|^2}}. \tag{8.9}$$

The **spherical distance** between c_1 and c_2, denoted by $\theta(c_1, c_2)$, is defined as the shortest length of an arc on the Riemann sphere connecting $\phi(c_1)$ and $\phi(c_2)$. It is easy to see that this shortest arc is the arc in the circle resulting from cutting the Riemann sphere with a plane containing $\phi(c_1)$, $\phi(c_2)$, and the center of the Riemann sphere as shown in Figure 8.8.

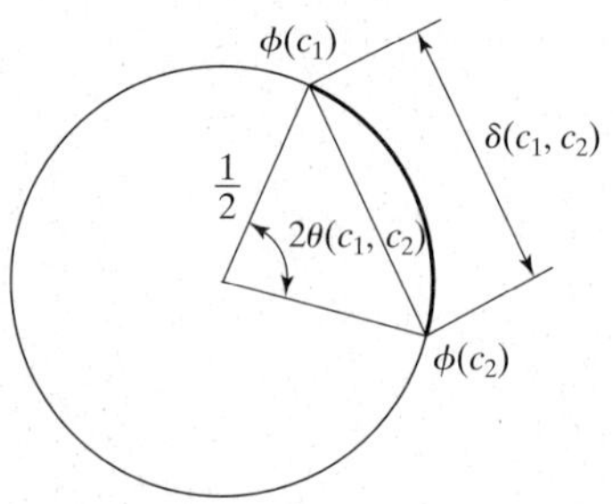

FIGURE 8.8: The relationship between $\delta(c_1, c_2)$ and $\theta(c_1, c_2)$.

Therefore, simple trigonometric arguments show that

$$\theta(c_1, c_2) = \arcsin \delta(c_1, c_2) = \arcsin \frac{|c_1 - c_2|}{\sqrt{1+|c_1|^2}\sqrt{1+|c_2|^2}} \tag{8.10}$$

and for complex numbers represented by $c_1 = b_1/a_1$ and $c_2 = b_2/a_2$,

$$\theta\left(\frac{b_1}{a_1}, \frac{b_2}{a_2}\right) = \arcsin \frac{|a_1 b_2 - a_2 b_1|}{\sqrt{|a_1|^2 + |b_1|^2}\sqrt{|a_2|^2 + |b_2|^2}}. \tag{8.11}$$

If we think of the Riemann sphere as our earth and the projections of two complex numbers as two points on earth, say Hong Kong and New York, then the chordal distance between the two is the straight line distance through the interior of the earth and the spherical distance is the shortest travel distance on the surface of the earth. They are different, but are closely related by (8.10).

Distance functions always satisfy the triangle inequality, i.e., for $c_1, c_2, c_3 \in \mathbb{C} \cup \{\infty\}$,

$$\delta(c_1, c_3) \leq \delta(c_1, c_2) + \delta(c_2, c_3) \tag{8.12}$$

$$\theta(c_1, c_3) \leq \theta(c_1, c_2) + \theta(c_2, c_3). \tag{8.13}$$

However, inequality (8.13) is tight in the sense that it is actually an equality in many special cases. For example, if $c_1 = 1, c_2 = j, c_3 = -1$, then

$$\theta(c_1, c_3) = \theta(c_1, c_2) + \theta(c_2, c_3).$$

However, inequality (8.12) is loose in the sense that it is always a strict inequality unless $c_1 = c_2$ or $c_2 = c_3$. As a matter of fact, inequality (8.13) implies inequality (8.12). This is because (8.13) is equivalent to

$$\begin{aligned}\delta(c_1, c_3) &\leq \sin[\theta(c_1, c_2) + \theta(c_2, c_3)] \\ &= \delta(c_1, c_2)\sqrt{1 - \delta^2(c_2, c_3)} + \sqrt{1 - \delta^2(c_1, c_2)}\delta(c_2, c_3).\end{aligned}$$

We often represent an uncertain real number by an interval. The middle point of the interval is often called the **nominal value** and the size of the uncertainty can then be given by the half length of the interval, called the **radius of uncertainty**. Similarly, an uncertain complex number can be represented by a disk in the complex plane with its center as the nominal value and its radius as the radius of uncertainty. Furthermore, an uncertain point in a three-dimensional Euclidean space can be represented by a ball whose center is called the **nominal position of the point** and whose radius is called the **radius of uncertainty**. All these uncertain objects have a common feature: each of them belongs to a certain set $\mathcal{S}$ with a distance measure d and can be represented by a subset of the following form:

$$\mathcal{B}(x_0, r) = \{x \in \mathcal{S} : d(x, x_0) \leq r\} \tag{8.14}$$

where x_0 is the nominal value of the uncertain object and r is the radius of uncertainty. Mathematically, a distance measure is called a **metric**, and a set of the form (8.14) is called a **ball** centered at x_0 with radius r. In the first two cases $d(x, x_0) = |x - x_0|$, and in the third case $d(x, x_0)$ is the Euclidean distance between

x and x_0, i.e., $d(x, x_0) = \sqrt{(x_1 - x_{01})^2 + (x_2 - x_{02})^2 + (x_3 - x_{03})^2}$. In the case of complex numbers (including infinity), if we change the distance from the usual distance $|c - c_0|$ to $\delta(c, c_0)$ or $\theta(c, c_0)$, we then get the sets

$$\mathcal{B}_\delta(c_0, r) = \{c \in \mathbb{C} \cup \{\infty\} : \delta(c, c_0) \leq r\}$$

and

$$\mathcal{B}_\theta(c_0, r) = \{c \in \mathbb{C} \cup \{\infty\} : \theta(c, c_0) \leq r\}.$$

These two balls are just hat-like discs in the Riemann sphere.

Figure 8.9 shows a circle in the complex plane centered at $(1, 0)$ with radius 0.5 and its stereographical projection to the Riemann sphere. The projection is also a circle with a chordal radius equal to 0.3630.

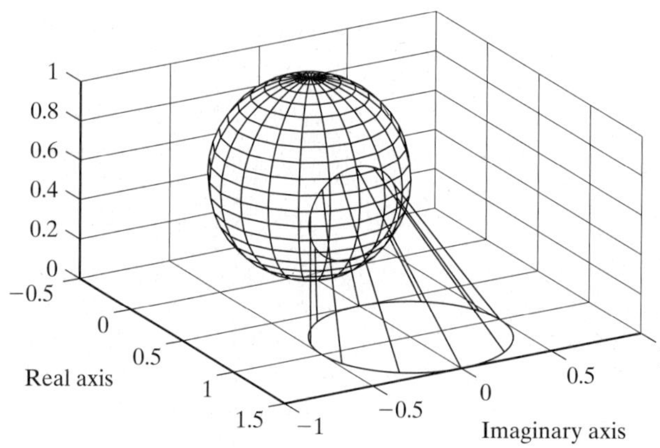

FIGURE 8.9: Projection of a disk in the complex plane onto the Riemann sphere.

8.5 DISTANCE BETWEEN SYSTEMS

To study robust stability, we need to see how to describe an uncertain system. This can easily be done using a distance function between systems. In the rest of this chapter, we assume the transfer functions are all proper.

For a polynomial $p(s)$, define its **inertia** to be three numbers

$$\nu[p(s)] = \{\nu_-[p(s)], \nu_0[p(s)], \nu_+[p(s)]\}$$

which are equal to the numbers of its roots with negative, zero, and positive real parts, respectively. For example,

$$\nu[s(s^2 - 1)] = \{1, 1, 1\}, \quad \nu[s^2 + s + 1] = \{2, 0, 0\}.$$

By the fundamental theorem of algebra, $p(s)$ is stable if and only if $\nu[p(s)] = \{\deg p(s), 0, 0\}$. It also follows from the fundamental theorem of algebra that $\nu_-[p(s)] + \nu_0[p(s)] + \nu_+[p(s)] = \deg p(s)$, i.e., if we know the degree of the polynomial, then only two numbers in its inertia are independent.

Let two transfer functions $G_i(s), i = 1, 2$, be given as

$$G_i(s) = \frac{b_i(s)}{a_i(s)}$$

where $a_i(s), b_i(s), i = 1, 2$, are polynomials. Assume that the orders of $G_i(s)$ are n_i, i.e., $\deg a_i(s) = n_i$.

Definition 8.20. Two systems $G_1(s)$ and $G_2(s)$ are said to be **comparable**, denoted by $G_1(s) \sim G_2(s)$, if

$$\nu[a_2(-s)a_1(s) + b_2(-s)b_1(s)] = \{n_1, 0, n_2\}.$$

It should be easy to see that if $G_1(s) = G_2(s)$, then $G_1(s) \sim G_2(s)$. Since the roots of $a_2(-s)a_1(s) + b_2(-s)b_1(s)$ are the mirror images of those of $a_1(-s)a_2(s) + b_1(-s)b_2(s)$, it follows that the condition

$$\nu[a_2(-s)a_1(s) + b_2(-s)b_1(s)] = \{n_1, 0, n_2\}$$

is equivalent to

$$\nu[a_1(-s)a_2(s) + b_1(-s)b_2(s)] = \{n_2, 0, n_1\}.$$

This shows that if $G_1(s) \sim G_2(s)$, then $G_2(s) \sim G_1(s)$.

Hence the relation "$\sim$" is reflexive and symmetric.

EXAMPLE 8.21

Let

$$G_1(s) = 0, \quad G_2(s) = \frac{1}{s + 0.5}, \quad G_3(s) = \frac{1}{s - 0.5}.$$

It is then natural to take

$$b_1(s) = 0, b_2(s) = b_3(s) = 1, a_1(s) = 1, a_2(s) = s + 0.5, a_3(s) = s - 0.5$$

and

$$n_1 = 0, n_2 = n_3 = 1.$$

Hence,

$$\nu[a_2(-s)a_1(s) + b_2(-s)b_1(s)] = \nu[(-s + 0.5)] = \{0, 0, 1\} = \{n_1, 0, n_2\}$$

$$\nu[a_3(-s)a_1(s) + b_3(-s)b_1(s)] = \nu[(-s - 0.5)] = \{1, 0, 0\} \neq \{n_1, 0, n_3\}$$

$$\nu[a_2(-s)a_3(s) + b_2(-s)b_3(s)] = \nu[(-s + 0.5)(s - 0.5) + 1] = \{1, 0, 1\} = \{n_3, 0, n_2\}.$$

Thus it is clear that

$$0 \sim \frac{1}{s + 0.5}, \quad 0 \not\sim \frac{1}{s - 0.5}, \quad \frac{1}{s + 0.5} \sim \frac{1}{s - 0.5}.$$

The example shows that even if $G_1(s) \sim G_2(s)$ and $G_2(s) \sim G_3(s)$, it may not be the case that $G_1(s) \sim G_3(s)$. This says that the relation "$\sim$" is not transitive.

Definition 8.22. The chordal distance between two systems $G_1(s)$ and $G_2(s)$ is defined as

$$\delta[G_1(s), G_2(s)] = \begin{cases} \max_{\omega\in\mathbb{R}} \delta[G_1(j\omega), G_2(j\omega)], & \text{if } G_1(s) \sim G_2(s) \\ 1, & \text{otherwise.} \end{cases}$$

Definition 8.23. The spherical distance between two systems $G_1(s)$ and $G_2(s)$ is defined as

$$\theta(G_1(s), G_2(s)) = \begin{cases} \max_{\omega\in\mathbb{R}} \theta[G_1(j\omega), G_2(j\omega)], & \text{if } G_1(s) \sim G_2(s) \\ \frac{\pi}{2}, & \text{otherwise.} \end{cases}$$

By expressions (8.9) and (8.11),

$$\delta\left(\frac{b_1(s)}{a_1(s)}, \frac{b_2(s)}{a_2(s)}\right) = \begin{cases} \max_{\omega\in\mathbb{R}} \dfrac{|a_1(j\omega)b_2(j\omega) - a_2(j\omega)b_1(j\omega)|}{\sqrt{|a_1(j\omega)|^2 + |b_1(j\omega)|^2}\sqrt{|a_2(j\omega)|^2 + |b_2(j\omega)|^2}}, & \text{if } G_1(s) \sim G_2(s) \\ 1, & \text{otherwise} \end{cases}$$

and

$$\theta\left(\frac{b_1(s)}{a_1(s)}, \frac{b_2(s)}{a_2(s)}\right) = \begin{cases} \max_{\omega\in\mathbb{R}} \arcsin \dfrac{|a_1(j\omega)b_2(j\omega) - a_2(j\omega)b_1(j\omega)|}{\sqrt{|a_1(j\omega)|^2 + |b_1(j\omega)|^2}\sqrt{|a_2(j\omega)|^2 + |b_2(j\omega)|^2}}, & \text{if } G_1(s) \sim G_2(s) \\ \frac{\pi}{2}, & \text{otherwise.} \end{cases}$$

Roughly speaking, $G_1(s)$ and $G_2(s)$ are considered to be close if their Riemann plots are close on the Riemann sphere, as long as the comparability condition is satisfied. The physical and geometric interpretations of the comparability condition are hard to explain using elementary language.

EXAMPLE 8.24

1. Let us first compute

$$\delta\left(\frac{1}{s+\epsilon}, \frac{1}{s}\right).$$

Since $\nu[(-s)(s+\epsilon)+1] = \{1,0,1\}$, it follows that the two systems concerned are comparable and

$$\delta\left(\frac{1}{s+\epsilon}, \frac{1}{s}\right) = \max_{\omega\in\mathbb{R}} \frac{|j\omega - (j\omega+\epsilon)|}{\sqrt{|j\omega|^2+1}\sqrt{|j\omega+\epsilon|^2+1}}$$

$$= \max_{\omega\in\mathbb{R}} \frac{|\epsilon|}{\sqrt{\omega^2+1}\sqrt{\omega^2+\epsilon^2+1}}.$$

Clearly, the maximum occurs at $\omega = 0$. Hence

$$\delta\left(\frac{1}{s+\epsilon}, \frac{1}{s}\right) = \frac{|\epsilon|}{\sqrt{1+\epsilon^2}}$$

no matter whether ϵ is positive or negative.

2. Let us then compute

$$\delta\left(\frac{1}{s^2+\omega_0^2}, \frac{1}{s^2+\omega_0^2+\epsilon^2}\right).$$

To determine whether the two systems concerned are comparable, we analyze the roots of polynomial

$$(s^2+\omega_0^2+\epsilon^2)(s^2+\omega_0^2)+1 = s^4+(2\omega_0^2+\epsilon^2)s^2+\omega_0^4+\omega_0^2\epsilon^2+1.$$

The roots of this polynomial are

$$\pm\sqrt{\frac{-(2\omega_0^2+\epsilon^2)\pm\sqrt{\epsilon^4-4}}{2}}. \tag{8.15}$$

When $\epsilon^2 < 2$, the small square root above is imaginary and so the fractions inside the large square root are complex. Hence, two of the roots are on the left-hand side of the complex plane and two are on the right. Therefore, in this case, the inertia of the polynomial is equal to $\{2,0,2\}$, which implies that the two systems concerned are comparable and

$$\delta\left(\frac{1}{s^2+\omega_0^2}, \frac{1}{s^2+\omega_0^2+\epsilon^2}\right) = \max_{\omega\in\mathbb{R}} \frac{|\epsilon^2|}{\sqrt{|\omega_0^2-\omega^2|^2+1}\sqrt{|\omega_0^2-\omega^2+\epsilon^2|^2+1}}.$$

The maximum can be obtained by considering the minimum of

$$[(\omega_0^2-\omega^2)^2+1][(\omega_0^2-\omega^2+\epsilon^2)^2+1].$$

The minimum occurs at $\omega^2 = \omega_0^2+\epsilon^2/2$ and the minimum is $(\epsilon^4/4+1)^2$. Therefore,

$$\delta\left(\frac{1}{s^2+\omega_0^2}, \frac{1}{s^2+\omega_0^2+\epsilon^2}\right) = \frac{4\epsilon^2}{\epsilon^4+4}.$$

When $\epsilon^2 \geq 2$, the small square root in (8.15) is real and the fractions inside the large square root are real and negative. Hence, the roots are all on the imaginary axis. Therefore, in this case, the inertia of the polynomial is equal to $\{0,4,0\}$, which implies that the two systems concerned are not comparable. In summary, we have

$$\delta\left(\frac{1}{s^2+\omega_0^2}, \frac{1}{s^2+\omega_0^2+\epsilon^2}\right) = \begin{cases} \dfrac{4\epsilon^2}{\epsilon^4+4}, & \text{if } \epsilon^2 < 2 \\ 1, & \text{otherwise.} \end{cases}$$

3. Let us now compute

$$\delta\left(0, \frac{\epsilon}{s-1}\right).$$

Notice that the system with transfer function 0 can be considered as a zeroth-order system whose transfer function has numerator zero and denominator 1. Since $\nu[(-s-1)] = \{1,0,0\} \neq \{0,0,1\}$, it follows that the two systems concerned are not comparable. Hence

$$\delta\left(0, \frac{\epsilon}{s-1}\right) = 1.$$

4. Let us finally compute

$$\delta\left(0, \frac{\epsilon}{s+1}\right).$$

Since $\nu[(-s+1)] = \{0,0,1\}$, the two systems concerned are comparable and

$$\delta\left(0, \frac{\epsilon}{s+1}\right) = \max_{\omega\in\mathbb{R}} \frac{|\epsilon|}{\sqrt{|j\omega+1|^2+\epsilon^2}} = \frac{|\epsilon|}{\sqrt{1+\epsilon^2}}.$$

Note that in the first two cases the differences in the usual distance between the frequency responses of the systems involved are equal to infinity. This shows that the usual distance function is not an appropriate measure of the difference between frequency responses of possibly unstable systems.

The spherical distance can be obtained from the chordal distance easily:

$$\theta\left(\frac{1}{s+\epsilon}, \frac{1}{s}\right) = \arcsin\frac{|\epsilon|}{\sqrt{1+\epsilon^2}}$$

$$\theta\left(\frac{1}{s^2+\omega_0^2}, \frac{1}{s^2+\omega_0^2+\epsilon^2}\right) = \begin{cases} \arcsin\dfrac{4\epsilon^2}{\epsilon^4+4}, & \text{if } \epsilon^2 < 2 \\ \dfrac{\pi}{2}, & \text{otherwise} \end{cases}$$

$$\theta\left(0, \frac{\epsilon}{s-1}\right) = \frac{\pi}{2}$$

$$\theta\left(0, \frac{\epsilon}{s+1}\right) = \arcsin\frac{|\epsilon|}{\sqrt{1+\epsilon^2}}.$$

For possibly high-order systems, closed-form expressions for the chordal or spherical distances are in general impossible. In the case when $G_1(s) \sim G_2(s)$, one possible way to compute the chordal distance or spherical distance is by plotting the function

$$\frac{|a_1(j\omega)b_2(j\omega) - a_2(j\omega)b_1(j\omega)|}{\sqrt{|a_1(j\omega)|^2 + |b_1(j\omega)|^2}\sqrt{|a_2(j\omega)|^2 + |b_2(j\omega)|^2}}$$

versus the frequency ω and read out the maximum. Another way is to convert it to the computation of a frequency-domain ∞-norm. Notice that

$$\begin{aligned}\delta[G_1(j\omega), G_2(j\omega)] &= \frac{|G_1(j\omega) - G_2(j\omega)|}{\sqrt{1 + G_1(-j\omega)G_1(j\omega)}\sqrt{1 + G_2(-j\omega)G_2(j\omega)}} \\ &= \left\{\frac{[1 + G_1(-j\omega)G_1(j\omega)][1 + G_2(-j\omega)G_2(j\omega)]}{[G_1(-j\omega) - G_2(-j\omega)][G_1(j\omega) - G_2(j\omega)]}\right\}^{-1/2}.\end{aligned}$$

The numerator inside the big brackets can be rewritten as

$$[G_1(-j\omega) - G_2(-j\omega)][G_1(j\omega) - G_2(j\omega)] \\ + [1 + G_1(-j\omega)G_2(j\omega)][1 + G_1(j\omega)G_2(-j\omega)].$$

Hence,

$$\begin{aligned}\delta[G_1(j\omega), G_2(j\omega)] &= \left\{1 + \frac{[1 + G_1(-j\omega)G_2(j\omega)][1 + G_1(j\omega)G_2(-j\omega)]}{[G_1(-j\omega) - G_2(-j\omega)][G_1(j\omega) - G_2(j\omega)]}\right\}^{-1/2} \\ &= \left\{1 + \left|\frac{G_1(j\omega) - G_2(j\omega)}{1 + G_1(j\omega)G_2(-j\omega)}\right|^{-2}\right\}^{-1/2}.\end{aligned}$$

Therefore, under the assumption that $G_1(s) \sim G_2(s)$, we have

$$\delta[G_1(s), G_2(s)] = \left(1 + \left\|\frac{G_1(s) - G_2(s)}{1 + G_1(s)G_2(-s)}\right\|_\infty^{-2}\right)^{-1/2}. \tag{8.16}$$

$$\theta[G_1(s), G_2(s)] = \arcsin\left(1 + \left\|\frac{G_1(s) - G_2(s)}{1 + G_1(s)G_2(-s)}\right\|_\infty^{-2}\right)^{-1/2}. \tag{8.17}$$

EXAMPLE 8.25

Let us compute

$$\delta\left(\frac{s-1}{s+1}, \frac{2s-1}{s+1}\right).$$

First we have

$$a_2(-s)a_1(s) + b_2(-s)b_1(s) = (-s+1)(s+1) + (-2s-1)(s-1) = -3s^2 + s + 2$$

whose roots are 1 and $-2/3$. Hence,

$$\nu[a_2(-s)a_1(s) + b_2(-s)b_1(s)] = \{1, 0, 1\},$$

which means that

$$\frac{s-1}{s+1} \sim \frac{2s-1}{s+1}.$$

Hence, the chordal distance can be computed in the following steps.

First,

$$\frac{G_1(s) - G_2(s)}{1 + G_1(s)G_2(-s)} = \frac{s^2 - s}{-3s^2 + s + 2} = -\frac{s}{3s+2}.$$

Then,

$$\left\| \frac{G_1(s) - G_2(s)}{1 + G_1(s)G_2(-s)} \right\|_\infty = \left\| -\frac{s}{3s+2} \right\|_\infty = \frac{1}{3}.$$

Finally,

$$\delta[G_1(s), G_2(s)] = \frac{1/3}{\sqrt{1 + (1/3)^2}} = \frac{1}{\sqrt{10}}.$$

The following propositions follow from the similar properties for constant complex numbers.

Proposition 8.26.

1. $\delta[G_1(s), G_2(s)] = 0$ if and only if $G_1(s) = G_2(s)$.
2. $\delta[G_1(s), G_2(s)] = \delta[G_2(s), G_1(s)]$.
3. $\delta[G_1(s), G_3(s)] \leq \sin\{\arcsin \delta[G_1(s), G_2(s)] + \arcsin \delta[G_2(s), G_3(s)]\}$.

Proposition 8.27.

1. $\theta[G_1(s), G_2(s)] = 0$ if and only if $G_1(s) = G_2(s)$.
2. $\theta[G_1(s), G_2(s)] = \theta[G_2(s), G_1(s)]$.
3. $\theta[G_1(s), G_3(s)] \leq \theta[G_1(s), G_2(s)] + \theta[G_2(s), G_3(s)]$.

8.6 UNCERTAINTY AND ROBUSTNESS

With the distance function δ between systems, we can describe an uncertain system such as a ball with center $G(s)$ and radius r:

$$\mathcal{B}_\delta[G(s), r] = \{\tilde{G}(s) : \delta[\tilde{G}(s), G(s)] \leq r\}.$$

The center $G(s)$ is called the **nominal system** and radius r is called the **radius of uncertainty**.

Alternatively, we can describe an uncertain system as a ball in terms of the distance function θ:

$$\mathcal{B}_\theta[G(s), r] = \{\tilde{G}(s) : \theta[\tilde{G}(s), G(s)] \le r\}.$$

Clearly.

$$\mathcal{B}_\theta[G(s), r] = \mathcal{B}_\delta[G(s), \sin r]$$

and

$$\mathcal{B}_\delta[G(s), r] = \mathcal{B}_\theta[G(s), \arcsin r].$$

EXAMPLE 8.28

Let

$$G(s) = \frac{10(s+1)}{s(s+2)(s+3)}.$$

The corresponding uncertain Nyquist diagram for $\delta[\tilde{G}(s), G(s)] \le 0.1$ with $\omega \in [1, 100]$ at each frequency is obtained from 8.16 and is shown in Figure 8.10 as a disk.

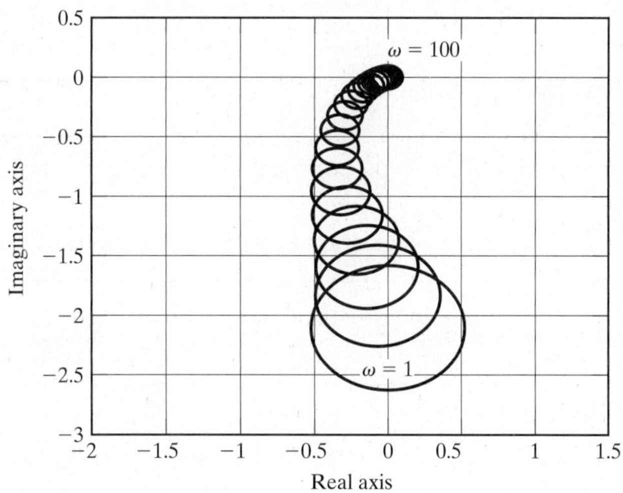

FIGURE 8.10: Uncertainty on the Nyquist diagram corresponding to the balls of uncertainty on the Riemann sphere centered at $G(s)$ with chordal radius 0.1.

Now consider the feedback system in Figure 8.11.

Definition 8.29. Assume that the feedback system in Figure 8.11 is stable. Its stability margin is defined as

$$b_{P,C} = \min_{\omega \in \mathbb{R}} \delta[-C^{-1}(j\omega), P(j\omega)].$$

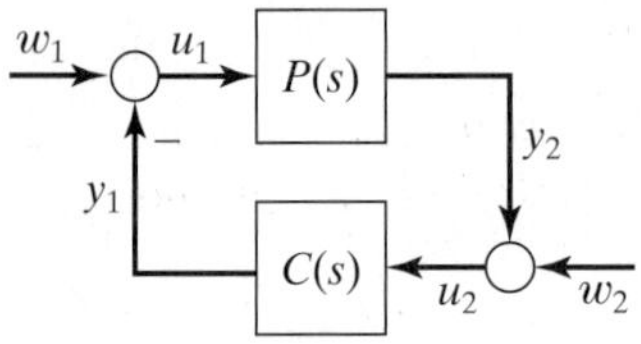

FIGURE 8.11: Feedback system for stabilization.

In other words, $b_{P,C}$ is the smallest chordal distance between $P(j\omega)$ and $-C^{-1}(j\omega)$.

The stability margin $b_{P,C}$ can also be related to the familiar closed-loop transfer functions. Since

$$\begin{aligned}\delta[-C^{-1}(j\omega), P(j\omega)] &= \frac{|P(j\omega)+C^{-1}(j\omega)|}{\sqrt{1+|P(j\omega)|^2}\sqrt{1+|C^{-1}(j\omega)|^2}} \\ &= \frac{|1+P(j\omega)C(j\omega)|}{\sqrt{1+|P(j\omega)|^2}\sqrt{1+|C(j\omega)|^2}} \\ &= \left[\sqrt{\frac{1+|P(j\omega)|^2+|C(j\omega)|^2+|P(j\omega)C(j\omega)|^2}{|1+P(j\omega)C(j\omega)|^2}}\right]^{-1}. \end{aligned} \tag{8.18}$$

Let us use the convention

$$S(s) = \frac{1}{1+C(s)P(s)}, \quad T(s) = \frac{C(s)P(s)}{1+C(s)P(s)}$$

for sensitivity and complementary sensitivity functions. Then

$$\boxed{b_{P,C} = \left[\max_{\omega\in\mathbb{R}} \sqrt{|S(j\omega)|^2+|C(j\omega)S(j\omega)|^2+|P(j\omega)S(j\omega)|^2+|T(j\omega)|^2}\right]^{-1}.}$$

We see that $b_{P,C}$ depends on all transfer functions in the gang of four.

Consider the feedback system shown in Figure 8.12. Here $\tilde{P}(s)$ is a perturbed version of $P(s)$ and $\tilde{C}(s)$ is a perturbed version of $C(s)$. If we know that the system shown in Figure 8.11 is stable, what can we say about the stability of the system shown in Figure 8.12?

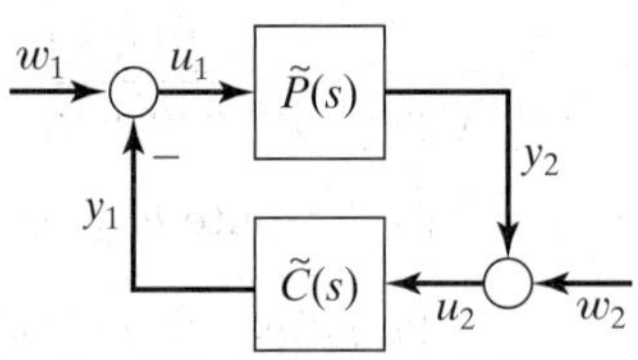

FIGURE 8.12: An uncertain feedback system.

Theorem 8.30.

1. The feedback system in Figure 8.12 is stable for all $\tilde{P}(s) \in \mathcal{B}_\delta[P(s), r_P]$ and $\tilde{C}(s) \in \mathcal{B}_\delta[C(s), r_C]$ if and only if

$$\arcsin r_P + \arcsin r_C < \arcsin b_{P,C}.$$

2. Let

$$\arcsin r_P + \arcsin r_C < \arcsin b_{P,C}.$$

Then,

$$\min_{\substack{\tilde{P}\in\mathcal{B}_\delta(P,r_P)\\ \tilde{C}\in\mathcal{B}_\delta(C,r_C)}} b_{\tilde{P},\tilde{C}} = \sin(\arcsin b_{P,C} - \arcsin r_P - \arcsin r_C).$$

The θ version of this theorem is as follows:

Theorem 8.31.

1. The feedback system in Figure 8.12 is stable for all $\tilde{P}(s) \in \mathcal{B}_\theta[P(s), r_P]$ and $\tilde{C}(s) \in \mathcal{B}_\theta[C(s), r_C]$ if and only if

$$r_P + r_C < \arcsin b_{P,C}.$$

2. Let

$$r_P + r_C < \arcsin b_{P,C}.$$

Then,

$$\min_{\substack{\tilde{P}\in\mathcal{B}_\theta(P,r_P)\\ \tilde{C}\in\mathcal{B}_\theta(C,r_C)}} b_{\tilde{P},\tilde{C}} = \sin(\arcsin b_{P,C} - r_P - r_C).$$

These theorems show that $b_{P,C}$ indeed gives a measure of the robustness of the feedback system. The next issue then is how we can compute $b_{P,C}$ from the data of plant $P(s)$ and controller $C(s)$ when $P(s)$ and $C(s)$ are both given.

Let

$$P(s) = \frac{b(s)}{a(s)}, \quad C(s) = \frac{q(s)}{p(s)}.$$

Then

$$\begin{aligned} b_{P,C} &= \min_{\omega\in\mathbb{R}} \frac{|b(j\omega)/a(j\omega) + p(j\omega)/q(j\omega)|}{\sqrt{1+|b(j\omega)/a(j\omega)|^2}\sqrt{|1+|p(j\omega)/q(j\omega)|^2}} \\ &= \min_{\omega\in\mathbb{R}} \frac{|a(j\omega)p(j\omega) + b(j\omega)q(j\omega)|}{\sqrt{|a(j\omega)|^2+|b(j\omega)|^2}\sqrt{|p(j\omega)|^2+|q(j\omega)|^2}}. \end{aligned}$$

Hence one possible method to compute $b_{P,C}$ is of course by plotting the function

$$\frac{|a(j\omega)p(j\omega) + b(j\omega)q(j\omega)|}{\sqrt{|a(j\omega)|^2+|b(j\omega)|^2}\sqrt{|p(j\omega)|^2+|q(j\omega)|^2}}$$

versus the frequency ω and reading out the minimum. Another way is to convert it to the computation of a frequency-domain ∞-norm. Start from the fraction inside (8.18),

$$\begin{aligned}&\frac{1+|P(j\omega)|^2+|C(j\omega)|^2+|P(j\omega)C(j\omega)|^2}{|1+P(j\omega)C(j\omega)|^2}\\ &\quad=\frac{1+P(-j\omega)P(j\omega)+C(-j\omega)C(j\omega)+C(-j\omega)P(-j\omega)P(j\omega)C(j\omega)}{[1+P(-j\omega)C(-j\omega)][1+P(j\omega)C(j\omega)]}.\end{aligned}$$

The numerator can be rewritten as

$$[1+P(-j\omega)C(-j\omega)][1+P(j\omega)C(j\omega)]+[P(j\omega)-C(-j\omega)][P(-j\omega)-C(j\omega)].$$

Hence

$$\begin{aligned}\delta[-C^{-1}(j\omega),P(j\omega)]&=\left[1+\frac{[P(j\omega)-C(-j\omega)][P(-j\omega)-C(j\omega)]}{[1+P(-j\omega)C(-j\omega)][1+P(j\omega)C(j\omega)]}\right]^{-1/2}\\ &=\left[1+\left|\frac{P(-j\omega)-C(j\omega)}{1+P(j\omega)C(j\omega)}\right|^2\right]^{-1/2}.\end{aligned}$$

Therefore

$$b_{P,C}=\left\{1+\left\|\frac{P(-s)-C(s)}{1+P(s)C(s)}\right\|_\infty^2\right\}^{-1/2}. \tag{8.19}$$

This relates $b_{P,C}$ with the frequency-domain ∞-norm of transfer function

$$\frac{P(-s)-C(s)}{1+P(s)C(s)}.$$

If the plant and the controller have proper transfer functions

$$P(s)=\frac{b(s)}{a(s)} \quad\text{and}\quad C(s)=\frac{q(s)}{p(s)}$$

with $\deg a(s)=n$ and $\deg p(s)=m$, then

$$\frac{P(-s)-C(s)}{1+P(s)C(s)}=\frac{b(-s)p(s)-a(-s)q(s)}{a(s)p(s)+b(s)q(s)}\frac{a(s)}{a(-s)}.$$

In the worst case, this transfer function has an order equal to $2n+m$. To make things worse, this transfer function is in general unstable even if the feedback system formed by $P(s)$ and $C(s)$ is internally stable, which is prone to trouble. Fortunately, there is a way out. Notice that $\dfrac{a(s)}{a(-s)}$ is all pass. It then follows that

$$\left\|\frac{P(-s)-C(s)}{1+P(s)C(s)}\right\|_\infty=\left\|\frac{b(-s)p(s)-a(-s)q(s)}{a(s)p(s)+b(s)q(s)}\right\|_\infty,$$

i.e.,

$$b_{P,C} = \left\{ 1 + \left\| \frac{b(-s)p(s) - a(-s)q(s)}{a(s)p(s) + b(s)q(s)} \right\|_\infty^2 \right\}^{-1/2}. \tag{8.20}$$

To acknowledge the importance of the norm on the right-hand side of (8.20), we denote it by $\gamma_{P,C}$. The transfer function now involved has the better property that its order is at most $n+m$ and it is stable as long as the feedback system formed by $P(s)$ and $C(s)$ is internally stable.

In summary, we have the following algorithm to compute the robustness $b_{P,C}$.

Algorithm 8.32 (Computation of stability robustness).

Step 1 (*Internal stability determination*) Check the internal stability of closed-loop system $(P(s), C(s))$. If it is unstable, set $b_{P,C} = 0$ and exit.

Step 2 (∞-*norm computation*) Compute

$$\gamma_{P,C} = \left\| \frac{b(-s)p(s) - a(-s)q(s)}{a(s)p(s) + b(s)q(s)} \right\|_\infty.$$

Step 3 (*Simple additional computation*) Finally,

$$b_{P,C} = (1 + \gamma_{P,C}^2)^{-1/2}.$$

EXAMPLE 8.33

Consider the double integrator $P(s) = 1/s^2$ stabilized by the controller

$$C(s) = \frac{2\sqrt{2}s + 1}{s^2 + 2\sqrt{2}s + 4}$$

achieving optimal transient response, as will be designed in Example 8.3. The closed-loop stability robustness $b_{P,C}$ can be computed as follows. First we compute the frequency domain ∞-norm

$$\begin{aligned}\gamma_{P,C} &= \left\| \frac{b(-s)p(s) - a(-s)q(s)}{a(s)p(s) + b(s)q(s)} \right\|_\infty \\ &= \left\| \frac{-2\sqrt{2}s^3 + 2\sqrt{2}s + 4}{(s^2 + \sqrt{2}s + 1)^2} \right\|_\infty = \left\| \frac{-2\sqrt{2}(s - \sqrt{2})}{s^2 + \sqrt{2}s + 1} \right\|_\infty.\end{aligned}$$

Careful hand computation gives $\gamma_{P,C} = 12\sqrt{2/17}$. Hence, $b_{P,C} = \sqrt{17/305}$.

We can also use MATLAB commands

```
>> sys=tf([-2*sqrt(2) 4],[1 sqrt(2) 1]);
>> gamma=norm(sys,inf);
```

give gamma $= 4.1160$. Hence, $b_{P,C} = 1/\sqrt{1 + 4.1160^2} = 0.2361$.

In the rest of this section, we show that $b_{P,C}$ gives an estimate of some classical stability margins.

Theorem 8.34. Let $C(s)$ be a stabilizing controller for $P(s)$. Then

$$\mathrm{GM} \geq \frac{1+b_{P,C}}{1-b_{P,C}}, \qquad \mathrm{PM} \geq 2\arcsin(b_{P,C}).$$

Proof. First of all, note that for any frequency ω we have

$$b_{P,C} \leq \frac{|1+P(j\omega)C(j\omega)|}{\sqrt{1+|P(j\omega)|^2}\,\sqrt{1+|C(j\omega)|^2}}.$$

So, at frequencies where $k := -C(j\omega)P(j\omega)$ is real,

$$\begin{aligned} b_{P,C} &\leq \frac{|1-k|}{\sqrt{(1+|P(j\omega)|^2)\left(1+\dfrac{k^2}{|P(j\omega)|^2}\right)}} \\ &\leq \frac{|1-k|}{\sqrt{\min\limits_{x}\left\{(1+x^2)\left(1+\dfrac{k^2}{x^2}\right)\right\}}} = \left|\frac{1-k}{1+k}\right|, \end{aligned}$$

which implies that

$$k \leq \frac{1-b_{P,C}}{1+b_{P,C}}, \quad \text{or} \quad k \geq \frac{1+b_{P,C}}{1-b_{P,C}}.$$

Thus, the system has at least the following gain margin:

$$\mathrm{GM} \geq \frac{1+b_{P,C}}{1-b_{P,C}}$$

Similarly, at frequencies where $P(j\omega)C(j\omega) = -e^{j\phi}$,

$$\begin{aligned} b_{P,C} &\leq \frac{|1-e^{j\phi}|}{\sqrt{(1+|P(j\omega)|^2)\left(1+\dfrac{1}{|P(j\omega)|^2}\right)}} \\ &\leq \frac{|2\sin\frac{\phi}{2}|}{\sqrt{\min\limits_{x}\left\{(1+x^2)\left(1+\dfrac{1}{x^2}\right)\right\}}} = \left|\sin\frac{\phi}{2}\right|, \end{aligned}$$

which implies that the system has at least the following phase margin:

$$\mathrm{PM} \geq 2\arcsin(b_{P,C}). \qquad \square$$

For example, $b_{P,C} = 1/2$ guarantees a gain margin of 3 and a phase margin of 60°.

EXAMPLE 8.35

Consider a closed-loop feedback system with

$$P(s) = \frac{1}{s^2 + s + 2}, \quad C(s) = \frac{s+1}{s}.$$

Then $b_{P,C} = 0.3371$, which gives the following estimates of the stability margins:

$$\text{GM} \geq \frac{1 + b_{P,C}}{1 - b_{P,C}} = 2.0169, \quad \text{PM} \geq 2\arcsin(b_{P,C}) = 0.6876[\text{rad}] = 39.4^\circ.$$

However, the actual stability margins are

$$\text{GM} \approx \infty, \quad \text{PM} = 90^\circ.$$

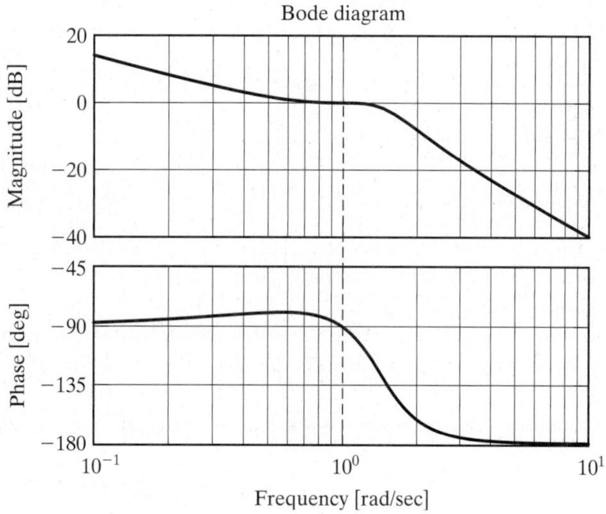

FIGURE 8.13: Bode diagrams of $L(s) = P(s)C(s)$.

Thus, the estimates on the classical stability margins obtained using $b_{P,C}$ can be very conservative.

The bigger the $b_{P,C}$, the more robust is the feedback system. One natural design problem then follows: Given $P(s)$, design $C(s)$ so that $b_{P,C}$ is maximized. This problem is called the **optimal robust stabilization problem**. Its solution will be given in the next chapter.

PROBLEMS

8.1. Let $G_1(s)$ be a stable strictly proper transfer function and $G_2(s)$ be an antistable, strictly proper transfer function. Show that $\langle G_1(s), G_2(s) \rangle = 0$.

8.2. On the basis of Problem 8.1, give a method to compute the 2-norm of a possibly unstable transfer function.

8.3. A stable system with transfer function $G(s)$ is said to be all pass if it satisfies $|G(j\omega)| = 1$ for all $\omega \in \mathbb{R}$. One example of an all-pass transfer function is $\dfrac{s-a}{s+a}$ for some $a > 0$.

1. Give another example of an all-pass transfer function.
2. Show that for an all-pass system the input $u(t)$ and output $y(t)$ satisfies $\|u(t)\|_2 = \|y(t)\|_2$.
3. Show that for an all-pass transfer function $G(s)$ and any other transfer function $F(s)$, we have $\|G(s)F(s)\|_2 = \|F(s)\|_2$ and $\|G(s)F(s)\|_\infty = \|F(s)\|_\infty$.

8.4. Compute the ∞-norm of $G(s) = \dfrac{s-1}{(s+1)(s^2+s+1)}$.

8.5. Let $G(s)$ be the Laplace transform of $g(t)$ and assume that $\|g(t)\|_1 < \infty$.

1. Prove that $\|G(s)\|_\infty \le \|g(t)\|_1$.
2. Prove that if $g(t) \ge 0$ for all $t \ge 0$ or if $g(t) \le 0$ for all $t \ge 0$, then $\|G(s)\|_\infty = \|g(t)\|_1$.

8.6. Let two plants be given by

$$P_1(s) = \delta\frac{s+\epsilon}{s-\epsilon}, \quad P_2(s) = \delta\frac{s-\epsilon}{s+\epsilon}.$$

Use the root-locus method to show that

$$P(s) := P_1(s) - P_2(s) = \frac{4\epsilon\delta s}{(s+\epsilon)(s-\epsilon)}$$

cannot be stabilized by any stable controller for any fixed $\epsilon > 0$ and $\delta > 0$. (It turns out that this also means that there is no controller, stable or not, that can stabilize simultaneously $P_1(s)$ and $P_2(s)$.)

8.7. Let $G(s)$ be a stable transfer function such that $|G(j\omega_0)| \ge 1$ for some ω_0. Find a stable transfer function $\Delta(s)$ such that $\|\Delta(s)\|_\infty \le 1$ and $1 + G(s)\Delta(s)$ has a zero at $j\omega_0$.

8.8. Consider the unity feedback system shown in Figure 8.14. Assume

$$\tilde{P}(s) = P(s)[1 + \Delta(s)]$$

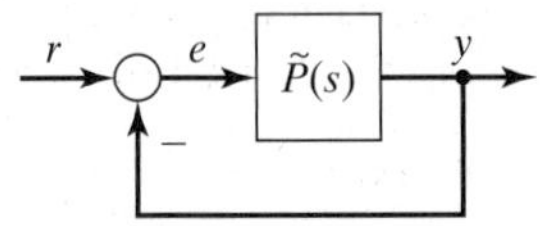

FIGURE 8.14: A unity feedback system.

where the nominal loop gain $P(s)$ is equal to $\dfrac{3(s+1)}{s^2(s+3)}$ and the uncertainty $\Delta(s)$ is stable and satisfies $|\Delta(j\omega)| < |0.5(j\omega + 0.01)|$. Is the feedback system stable for all $\tilde{P}(s)$?

8.9. Suppose that the plant transfer function is

$$\tilde{P}(s) = P(s)[1 + \Delta(s)]$$

where $P(s) = \dfrac{1}{s-1}$ and $\Delta(s)$ is stable and satisfies $|\Delta(j\omega)| < |(j\omega + 2)/10|$. Suppose that the controller is a pure gain $C(s) = K$. Determine the range of K such that the unity feedback system is stable for all possible $\tilde{P}(s)$.

8.10. Prove $\delta(c_1, c_2) = \delta(c_1^{-1}, c_2^{-1})$ for all $c_1, c_2 \in \mathbb{C} \cup \{\infty\}$.

8.11. **1.** Among the following two pairs of numbers, which pair is closer in the chordal distance: $(1, 2)$ or $(100, 200)$?

2. Compute the following chordal distances: $\delta\left(\frac{1}{s+0.1}, 0\right)$, $\delta\left(0, \frac{1}{s-0.1}\right)$, and $\delta\left(\frac{1}{s+0.1}, \frac{1}{s-0.1}\right)$.

8.12. Compute the chordal distance between the following two plants:

$$P_1(s) = \delta\frac{s+\epsilon}{s-\epsilon}, \quad P_2(s) = \delta\frac{s-\epsilon}{s+\epsilon}.$$

Explain how this distance measure can help you understand the simultaneous stabilization of these two plants in Problem 8.6.

8.13. Consider a unity feedback system with $P(s) = \dfrac{s+2}{(s+1)(s+3)}$ and $C(s) = K$. Show that the feedback system is internally stable for all $K \geq 0$. Plot $b_{P,C}$ as a function of $K \geq 0$. Choose the best K. There is a common perception that a feedback system is more robust if the closed-loop poles are further away from the imaginary axis. Use this example to explain that this perception is not correct.

8.14. We know that the plant $P(s) = \dfrac{1}{s}$ is stabilized by any proportional controller $C(s) = K, K > 0$. Plot $b_{P,C}$ as a function of K. What is the feedback gain that renders the most robust closed-loop system?

MATLAB PROBLEMS

8.15. Write a MATLAB program to compute the ∞-norm of a transfer function. Let us name it "norminf". Test the program on the systems given in Problem 4.27.

8.16. Write a MATLAB program to compute the chordal distance between two systems represented by two system variables. This should be a .m function file with two systems as input variables and their chordal distance as output variable. Use this program to compute the chordal distance between the following systems $G_1(s)$ and $G_2(s)$.

1. $G_1(s) = \dfrac{s+1.0001}{(s+1)(s+2)}$ and $G_2(s) = \dfrac{1}{s+2}$.

2. $G_1(s) = \dfrac{s-1.0001}{(s-1)(s+2)}$ and $G_2(s) = \dfrac{1}{s+2}$.

3. $G_1(s) = 0$ and $G_2(s) = \dfrac{(s+4)(s+5)}{(s+1)(s+2)(s+3)}$.

4. $G_1(s) = \dfrac{1}{s^2+1}$ and $G_2(s) = \dfrac{1}{s^2+a}$ for $a = 2$ and $a = 10$, respectively.

8.17. Write a MATLAB program to compute the stability margin $b_{P,C}$ for given plant $P(s)$ and controller $C(s)$. This should be a .m function file with plant and controller as input variables and the stability margin as output variable. Use this program to compute the stability margin of the following plant controller pairs.

$$P(s) = \frac{1}{s^2}, \quad C(s) = \frac{(1+\sqrt{2})s+1}{s+1+\sqrt{2}}.$$

NOTES AND REFERENCES

The simple proof of Theorem 8.8 is reported in

P. Fu, X. Zhao, and L. Qiu, "Solution to the Nehari problem using the Routh table," *Proc. of the 12th Mediterranean Conference on Control and Automation*, Kusadasi, Turkey, 2004.

This chapter also uses some elementary (but nontrivial) tools from complex analysis, such as the principle of arguments, the Riemann sphere, and the Cauchy integral theorem. A good reference on these materials is

R. V. Churchill, J. W. Brown, and R. F. Verhey, *Complex Variables and Applications*, 5th Edition, McGraw-Hill, New York, 1990.

Two plants $P_1(s)$ and $P_2(s)$ can be simultaneously stabilized by a controller if and only if $P(s) := P_1(s) - P_2(s)$ can be stabilized by a stable controller, i.e., $P(s)$ can be strongly stabilized. This result was first presented in

R. Saeks and J. Murray, "Fractional representation, algebraic geometry and simultaneous stabilization problem," *IEEE Transactions on Automatic Control*, vol. 27, pp. 895–903, 1982.

M. Vidyasagar and N. Viswanadham, "Algebraic design techniques for reliable stabilization," *IEEE Transactions on Automatic Control*, vol. 27, pp. 1085–1095, 1982.

Strong stabilization, simultaneous stabilization, Nevanlinna–Pick interpolation, and many other problems are discussed in detail in the following books:

M. Vidyasagar, *Control System Synthesis: a Factorization Approach*, The MIT Press, Cambridge, MA, 1985.

J. C. Doyle, B. Francis, and A. Tannenbaum, *Feedback Control Theory*, Macmillan College Division, New York, 1992.

The set of models in Example 8.15 were originally presented in

G. Balas, R. Chiang, A. Packard, and M. Safonov, *Robust Control Toolbox Users Guide Version 3*, The MathWorks, Inc., Natick, MA, 2006.

CHAPTER 9

Optimal and Robust Control

In this chapter, we study a few systematic methods to design a controller $C(s)$ for a given plant $P(s)$ so that the closed-loop system shown in Figure 9.1 is internally stable and has good performance and robustness. As we have already seen, and will again reinforce, the stabilizing controllers for a fixed plant are not unique. Choosing a good one among all the possibilities is a crucial issue. In this chapter, we will study several optimal control problems. More specifically, we will see how we can make a good choice among all the stabilizing controllers to achieve certain optimality in its performance and robustness. We will see that various optimal control problems can be converted into pole-placement problems, as covered in Chapter 3. The derivations of the optimal controller design algorithms will be given in star-marked sections for the investigation of research-oriented students. It is seen that the derivations are all based on making the best choices depending on various objectives in the set of all stabilizing controllers characterized in Section 3.7.

9.1 CONTROLLER WITH OPTIMAL TRANSIENT

In this section, we introduce a performance measure for a feedback system for stabilization and give an algorithm to design the best controller so that the performance measure is optimized. We assume that $P(s)$ is strictly proper in this section, i.e., it has the form

$$P(s) = \frac{b(s)}{a(s)} = \frac{b_1 s^{n-1} + \cdots + b_n}{a_0 s^n + a_1 s^{n-1} + \cdots + a_n}$$

where $a_0 \neq 0$.

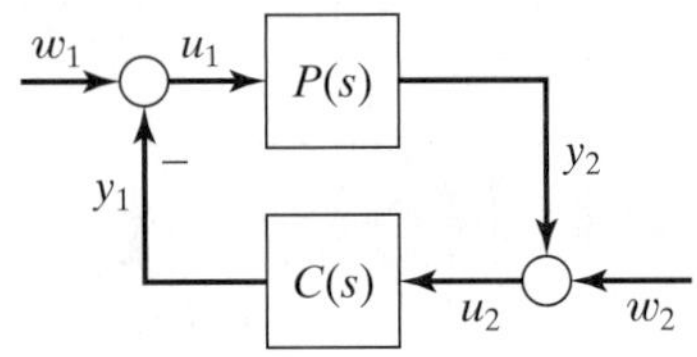

FIGURE 9.1: Feedback system for stabilization.

Consider Figure 9.1. If the closed-loop system is internally stable and the external inputs $w_i(t), i = 1, 2$, are impulse signals, then all the internal signals $u_j(t), y_j(t), j = 1, 2$, will eventually settle to zero when time goes to infinity. It is conceivable that the relatively more stable closed-loop systems will have less oscillations in the internal signals during the transient period. Because of this we can use the energy of $y_i(t)$ when $w_i(t), i = 1, 2$, are unit impulses as the performance criteria:

$$J = \left(\|y_1(t)\|_2^2 + \|y_2(t)\|_2^2\right)\Big|_{\substack{w_1(t)=\delta(t)\\ w_2(t)=0}} + \left(\|y_1(t)\|_2^2 + \|y_2(t)\|_2^2\right)\Big|_{\substack{w_1(t)=0\\ w_2(t)=\delta(t)}} . \tag{9.1}$$

When J is small, then we say the closed-loop system is more stable. It is then interesting to find the controller that minimizes J. The minimum value of J, denoted by J^*, when $C(s)$ is chosen among all stabilizing controllers yields the best performance that we can achieve. Hence, this number also reflects a measure of the difficulty in stabilizing $P(s)$. A small J^* means that $P(s)$ can easily be controlled. A large J^* means that $P(s)$ is hard to control.

Notice that if either $P(s)$ or $C(s)$ is not strictly proper, then the signal $y_1(t)$ or $y_2(t)$ would contain impulse functions if $w_1(t)$ or $w_2(t)$ is an impulse. Then the cost function J would be infinity. This is why we need to assume that $P(s)$ is strictly proper. The optimal controller $C(s)$ has to be strictly proper.

The transfer function from (w_1, w_2) to (y_1, y_2) is

$$\begin{bmatrix} \dfrac{P(s)C(s)}{1+P(s)C(s)} & \dfrac{C(s)}{1+P(s)C(s)} \\ \dfrac{P(s)}{1+P(s)C(s)} & \dfrac{-P(s)C(s)}{1+P(s)C(s)} \end{bmatrix}. \tag{9.2}$$

Since the transfer function is the Laplace transform of the impulse response of a system, by Parseval's identity,

$$\left(\|y_1(t)\|_2^2 + \|y_2(t)\|_2^2\right)\Big|_{\substack{w_1(t)=\delta(t)\\ w_2(t)=0}} = \left\|\frac{P(s)C(s)}{1+P(s)C(s)}\right\|_2^2 + \left\|\frac{P(s)}{1+P(s)C(s)}\right\|_2^2$$

and

$$\left(\|y_1(t)\|_2^2 + \|y_2(t)\|_2^2\right)\Big|_{\substack{w_1(t)=0\\ w_2(t)=\delta(t)}} = \left\|\frac{C(s)}{1+P(s)C(s)}\right\|_2^2 + \left\|\frac{-P(s)C(s)}{1+P(s)C(s)}\right\|_2^2 .$$

Hence, another expression for J is

$$J = \left\| \frac{P(s)C(s)}{1+P(s)C(s)} \right\|_2^2 + \left\| \frac{P(s)}{1+P(s)C(s)} \right\|_2^2 + \left\| \frac{C(s)}{1+P(s)C(s)} \right\|_2^2 + \left\| \frac{-P(s)C(s)}{1+P(s)C(s)} \right\|_2^2, \quad (9.3)$$

i.e., J is simply the squared sum of the 2-norms of all elements of matrix (9.2). The expression in (9.3) still looks cumbersome. Why don't we simply define the 2-norm of a matrix transfer function as the square root of the squared sum of the 2-norms of it elements? By doing this, we have

$$J = \left\| \begin{bmatrix} \dfrac{P(s)C(s)}{1+P(s)C(s)} & \dfrac{C(s)}{1+P(s)C(s)} \\ \dfrac{P(s)}{1+P(s)C(s)} & \dfrac{-P(s)C(s)}{1+P(s)C(s)} \end{bmatrix} \right\|_2^2,$$

a much more compact expression.

For a real polynomial $x(s)$, its **conjugate polynomial** is defined as $x(-s)$. It is easy to see that if z is a root of $x(s)$, then $-z$ is a root of $x(-s)$. Together with the fact that the roots of real polynomials are symmetric to the real axis, this implies that the roots of $x(-s)$ are mirror images of those of $x(s)$ with respect to the imaginary axis.

Given a strictly proper plant $P(s) = \dfrac{b(s)}{a(s)}$, where $a(s)$ and $b(s)$ are coprime polynomials, consider the polynomial

$$a(-s)a(s) + b(-s)b(s).$$

Since this polynomial is **self-conjugate**, i.e., its conjugate is itself, if z is a root of this polynomial, then so is $-z$. This implies that the roots of this polynomial are symmetric to the imaginary axis. This polynomial cannot have any roots on the imaginary axis since if it had a root at $j\omega$, i.e.,

$$a(-j\omega)a(j\omega) + b(-j\omega)b(j\omega) = |a(j\omega)|^2 + |b(j\omega)|^2 = 0,$$

then $a(j\omega) = b(j\omega) = 0$, which implies that $a(s)$ and $b(s)$ have a common root at $j\omega$, contradictory to the assumption that $a(s)$ and $b(s)$ are coprime. Therefore, there must exist a stable polynomial

$$d(s) = d_0 s^n + d_1 s^{n-1} + \cdots + d_n$$

such that

$$a(-s)a(s) + b(-s)b(s) = d(-s)d(s).$$

The polynomial $d(s)$ is called a **spectral factor** of $a(-s)a(s)+b(-s)b(s)$ and the process of finding $d(s)$ is called **spectral factorization**. The spectral factor is unique up to a factor of ±1. Also notice that because $\deg a(s) > \deg b(s)$, the leading coefficient of $d(s)$ has to be either the same as or the negative of that of $a(s)$. To remove this unpleasant nonuniqueness, we will always choose $d(s)$ in such a way that its leading coefficient is the same as that of $a(s)$, i.e., $d_0 = a_0$. Now let $C(s) = \dfrac{q(s)}{p(s)}$ be the unique nth-order strictly proper pole-placement controller such that

$$a(s)p(s) + b(s)q(s) = d^2(s).$$

This controller is then the unique optimal controller that minimizes J. The optimal value of the performance index can of course be obtained by substituting the plant and the optimal controller in (9.3). This gives

$$\boxed{J^* := \min J = 2\left\|\frac{b(s)q(s)}{d^2(s)}\right\|_2^2 + \left\|\frac{b(s)p(s)}{d^2(s)}\right\|_2^2 + \left\|\frac{a(s)q(s)}{d^2(s)}\right\|_2^2.}$$

This involves the computation of 2-norms of $2n$th-order transfer functions. Another expression involving the 2-norms of only nth-order systems, to be derived in Section 9.4, is given by

$$\boxed{J^* = \left\|\frac{d(s)-a(s)}{d(s)}\right\|_2^2 + \left\|\frac{b(s)}{d(s)}\right\|_2^2 + \left\|\frac{d(s)-p(s)}{d(s)}\right\|_2^2 + \left\|\frac{q(s)}{d(s)}\right\|_2^2.}$$

In summary, the optimal controller that minimizes J can be designed according to the following procedure:

Algorithm 9.1 (Design of the controller with optimal transient).

Step 1: Find a stable polynomial $d(s)$ such that

$$a(-s)a(s) + b(-s)b(s) = d(-s)d(s).$$

Step 2: The optimal controller $C(s) = \dfrac{q(s)}{p(s)}$ is the unique nth-order strictly proper pole-placement controller such that

$$a(s)p(s) + b(s)q(s) = d^2(s).$$

Step 3: The optimal performance index is given by

$$J^* = \left\|\frac{d(s)-a(s)}{d(s)}\right\|_2^2 + \left\|\frac{b(s)}{d(s)}\right\|_2^2 + \left\|\frac{d(s)-p(s)}{d(s)}\right\|_2^2 + \left\|\frac{q(s)}{d(s)}\right\|_2^2.$$

Algorithm 9.1 looks innocent but the reason why it gives the optimal control is not so simple. The detailed derivation of this algorithm, as a special case of a more general algorithm presented in the next section, is given in the star-marked Section 9.4.

Two examples are used to illustrate the design.

EXAMPLE 9.2

Let

$$P(s) = \frac{\beta}{s+\alpha}.$$

Design $C(s)$ such that J is minimized and find the optimal value of J.

Observe that

$$\begin{aligned} a(-s)a(s) + b(-s)b(s) &= -s^2 + \alpha^2 + \beta^2 \\ &= \left(-s + \sqrt{\alpha^2+\beta^2}\right)\left(s + \sqrt{\alpha^2+\beta^2}\right). \end{aligned}$$

Hence

$$d(s) = s + \sqrt{\alpha^2+\beta^2}.$$

Let

$$C(s) = \frac{q_1}{p_0 s + p_1}.$$

We then require

$$(s+\alpha)(p_0 s + p_1) + \beta q_1 = \left(s + \sqrt{\alpha^2+\beta^2}\right)^2 = s^2 + 2\sqrt{\alpha^2+\beta^2}s + \alpha^2 + \beta^2.$$

This gives

$$p_0 = 1, \quad p_1 = 2\sqrt{\alpha^2+\beta^2} - \alpha, \quad q_1 = \left(2\alpha^2 + \beta^2 - 2\alpha\sqrt{\alpha^2+\beta^2}\right)/\beta.$$

Therefore the optimal controller is

$$C(s) = \frac{\left(2\alpha^2 + \beta^2 - 2\alpha\sqrt{\alpha^2+\beta^2}\right)/\beta}{s + 2\sqrt{\alpha^2+\beta^2} - \alpha}.$$

The optimal value of the performance index J is

$$\begin{aligned} J^* = {} & \left\| \frac{\sqrt{\alpha^2+\beta^2} - \alpha}{s + \sqrt{\alpha^2+\beta^2}} \right\|_2^2 + \left\| \frac{\beta}{s + \sqrt{\alpha^2+\beta^2}} \right\|_2^2 \\ & + \left\| \frac{\alpha - \sqrt{\alpha^2+\beta^2}}{s + \sqrt{\alpha^2+\beta^2}} \right\|_2^2 + \left\| \frac{\left(2\alpha^2 + \beta^2 - 2\alpha\sqrt{\alpha^2+\beta^2}\right)/\beta}{s + \sqrt{\alpha^2+\beta^2}} \right\|_2^2. \end{aligned}$$

The formula in Section 4.9 gives

$$J^* = \frac{2\left(\sqrt{\alpha^2+\beta^2} - \alpha\right)^2 + \beta^2 + \left(2\alpha^2 + \beta^2 - 2\alpha\sqrt{\alpha^2+\beta^2}\right)^2/\beta^2}{2\sqrt{\alpha^2+\beta^2}}.$$

After simplification, we get

$$J^* = \frac{4(\alpha^2+\beta^2)\left(\sqrt{\alpha^2+\beta^2}-\alpha\right)}{\beta^2} - 2\sqrt{\alpha^2+\beta^2}.$$

The closed-loop transfer functions are

$$\begin{bmatrix} \dfrac{P(s)C(s)}{1+P(s)C(s)} & \dfrac{C(s)}{1+P(s)C(s)} \\ \dfrac{P(s)}{1+P(s)C(s)} & \dfrac{-P(s)C(s)}{1+P(s)C(s)} \end{bmatrix}$$

$$= \frac{\begin{bmatrix} 2\alpha^2+\beta^2-2\alpha\sqrt{\alpha^2+\beta^2} & (s+\alpha)\left(2\alpha^2+\beta^2-2\alpha\sqrt{\alpha^2+\beta^2}\right)/\beta \\ \beta\left(s+2\sqrt{\alpha^2+\beta^2}-\alpha\beta\right) & -2\alpha^2-\beta^2+2\alpha\sqrt{\alpha^2+\beta^2} \end{bmatrix}}{s^2+2\sqrt{\alpha^2+\beta^2}s+\alpha^2+\beta^2}.$$

In particular, if $\alpha = 0$ and $\beta = 1$, namely, if $P(s)$ is an integrator, then the optimal controller is

$$C(s) = \frac{1}{s+2}$$

and the optimal value of the performance index is $J^* = 2$. In this particular case, the optimal closed-loop transfer functions are

$$\begin{bmatrix} \dfrac{P(s)C(s)}{1+P(s)C(s)} & \dfrac{C(s)}{1+P(s)C(s)} \\ \dfrac{P(s)}{1+P(s)C(s)} & \dfrac{-P(s)C(s)}{1+P(s)C(s)} \end{bmatrix} = \frac{\begin{bmatrix} 1 & s \\ s+2 & -1 \end{bmatrix}}{s^2+2s+1}.$$

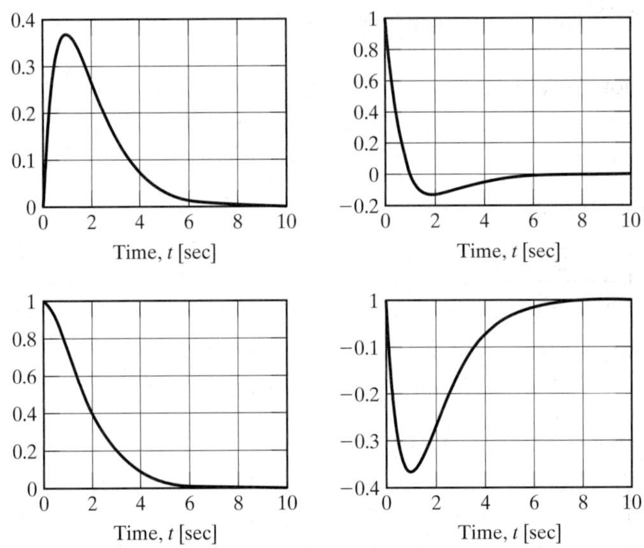

FIGURE 9.2: Impulse response of closed-loop transfer functions for Example 9.2.

The impulse responses of the closed-loop transfer functions are plotted in Figure 9.2. Here the position of the impulse response plot in the figure corresponds to the position of the transfer function in the 2×2 matrix above.

EXAMPLE 9.3

Let

$$P(s) = \frac{1}{s^2}.$$

Design $C(s)$ such that J is minimized and find the optimal value of J.

Observe that

$$a(-s)a(s) + b(-s)b(s) = s^4 + 1.$$

This polynomial has roots

$$s = \pm\frac{\sqrt{2}}{2} \pm j\frac{\sqrt{2}}{2}.$$

Hence a spectral factor of $s^4 + 1$ is

$$d(s) = \left(s + \frac{\sqrt{2}}{2} + j\frac{\sqrt{2}}{2}\right)\left(s + \frac{\sqrt{2}}{2} - j\frac{\sqrt{2}}{2}\right) = s^2 + \sqrt{2}s + 1.$$

Let

$$C(s) = \frac{q_1 s + q_2}{p_0 s^2 + p_1 s + p_2}.$$

We then require

$$s^2(p_0 s^2 + p_1 s + p_2) + q_1 s + q_2 = d(s)^2 = s^4 + 2\sqrt{2}s^3 + 4s^2 + 2\sqrt{2}s + 1.$$

This gives

$$p_0 = 1, \quad p_1 = 2\sqrt{2}, \quad p_2 = 4, \quad q_1 = 2\sqrt{2}, \quad q_2 = 1.$$

Therefore, the optimal controller is

$$C(s) = \frac{2\sqrt{2}s + 1}{s^2 + 2\sqrt{2}s + 4}.$$

The optimal value of J is

$$J^* = \left\|\frac{\sqrt{2}s + 1}{s^2 + \sqrt{2}s + 1}\right\|_2^2 + \left\|\frac{1}{s^2 + \sqrt{2}s + 1}\right\|_2^2 + \left\|\frac{-\sqrt{2}s - 3}{s^2 + \sqrt{2}s + 1}\right\|_2^2 + \left\|\frac{2\sqrt{2}s + 1}{s^2 + \sqrt{2}s + 1}\right\|_2^2.$$

By using the formula given in Section 4.9, we obtain

$$J^* = 6\sqrt{2}.$$

The closed-loop transfer functions are

$$\begin{bmatrix} \frac{P(s)C(s)}{1 + P(s)C(s)} & \frac{C(s)}{1 + P(s)C(s)} \\ \frac{P(s)}{1 + P(s)C(s)} & \frac{-P(s)C(s)}{1 + P(s)C(s)} \end{bmatrix} = \frac{\begin{bmatrix} 2\sqrt{2}s + 1 & 2\sqrt{2}s^3 + s^2 \\ s^2 + 2\sqrt{2}s + 4 & -2\sqrt{2}s - 1 \end{bmatrix}}{s^4 + 2\sqrt{2}s^3 + 4s^2 + 2\sqrt{2}s + 1}.$$

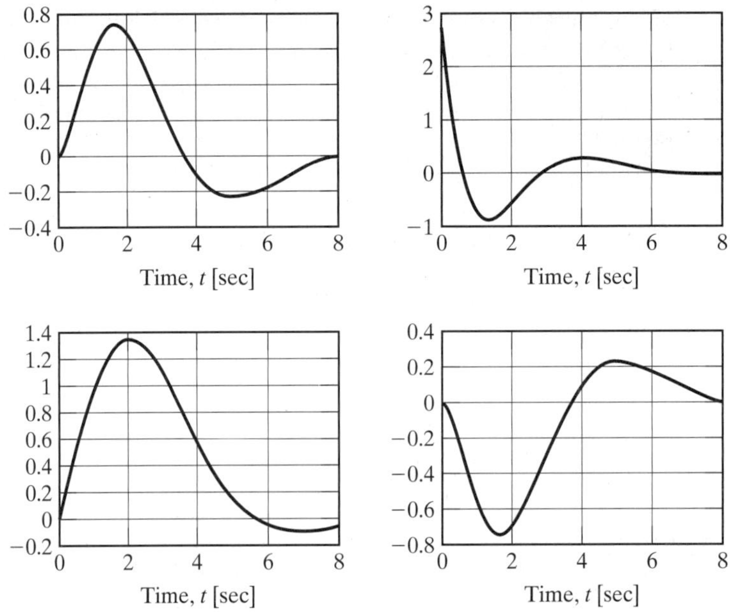

FIGURE 9.3: Impulse response of closed-loop transfer functions for Example 9.3.

The impulse responses of the closed-loop transfer functions are plotted in Figure 9.3.

9.2 CONTROLLER WITH WEIGHTED OPTIMAL TRANSIENT

Section 9.1 gave a way to design the stabilizing controller that gives optimal transient performance. In this section we will extend the method a bit further.

Again consider Figure 9.1. We now measure the performance of the closed-loop system by

$$J_{\rho,\mu} = \left(\|y_1(t)\|_2^2 + \rho^2\|y_2(t)\|_2^2\right)\Big|_{\substack{w_1(t)=\mu\delta(t)\\ w_2(t)=0}} + \left(\|y_1(t)\|_2^2 + \rho^2\|y_2(t)\|_2^2\right)\Big|_{\substack{w_1(t)=0\\ w_2(t)=\delta(t)}} .$$

Here, ρ is a positive number used to give a relative weight to $y_1(t)$ and $y_2(t)$, whereas μ is a positive number used to give a relative weight to $w_1(t)$ and $w_2(t)$. A small ρ means that we put more emphasis on $y_1(t)$, and a large ρ means that we put more emphasis on $y_2(t)$. Notice that for $t > 0$, $-y_1(t)$ is the input to the plant and $y_2(t)$ is the output of the plant. If $\rho \to 0$, then we put very little emphasis on plant output. This case is called **minimum-energy stabilization**, i.e., we wish to achieve stabilization using the least input energy. If $\rho \to \infty$, then we put very little emphasis on plant input. This case is called **cheap stabilization**, i.e., the input energy is cheap and it does not matter how much of it is spent in achieving stabilization. Similarly, a small μ means that we expect more severe noise in the plant output, i.e., high measurement noise, and a big μ means that we expect stronger noise in the plant input, i.e., high process noise.

In this case,

$$\left(\|y_1(t)\|_2^2+\rho^2\|y_2(t)\|_2^2\right)\Big|_{\substack{w_1(t)=\mu\delta(t)\\ w_2(t)=0}} = \left\|\frac{P(s)C(s)\mu}{1+P(s)C(s)}\right\|_2^2 + \left\|\frac{\rho P(s)\mu}{1+P(s)C(s)}\right\|_2^2$$

and

$$\left(\|y_1(t)\|_2^2+\rho^2\|y_2(t)\|_2^2\right)\Big|_{\substack{w_1(t)=0\\ w_2(t)=\delta(t)}} = \left\|\frac{C(s)}{1+P(s)C(s)}\right\|_2^2 + \left\|\frac{-\rho P(s)C(s)}{1+P(s)C(s)}\right\|_2^2.$$

Hence, another expression for $J_{\rho,\mu}$ is

$$J_{\rho,\mu} = \left\|\frac{P(s)C(s)\mu}{1+P(s)C(s)}\right\|_2^2 + \left\|\frac{\rho P(s)\mu}{1+P(s)C(s)}\right\|_2^2 + \left\|\frac{C(s)}{1+P(s)C(s)}\right\|_2^2 + \left\|\frac{-\rho P(s)C(s)}{1+P(s)C(s)}\right\|_2^2. \tag{9.4}$$

By using the 2-norms of matrix transfer functions, introduced in the last section, we have

$$J_{\rho,\mu} = \left\|\begin{bmatrix} \dfrac{P(s)C(s)\mu}{1+P(s)C(s)} & \dfrac{C(s)}{1+P(s)C(s)} \\ \dfrac{\rho P(s)\mu}{1+P(s)C(s)} & \dfrac{-\rho P(s)C(s)}{1+P(s)C(s)} \end{bmatrix}\right\|^2. \tag{9.5}$$

When $\rho=\mu=1$, then $J_{\rho,\mu}$ is the same as the performance measure J used in Section 9.1. It can also be seen that $J_{\rho,\mu}=J_{\mu,\rho}$.

The problem now is to design a stabilizing controller so that $J_{\rho,\mu}$ is minimized for a given strictly proper plant $P(s)=\dfrac{b(s)}{a(s)}$. The design procedure, whose derivation is given in Section 9.4, is as follows:

Algorithm 9.4 (Design for optimal weighted transient performance).

Step 1: Find a stable polynomial $d_\rho(s)$ such that

$$a(-s)a(s)+\rho^2 b(-s)b(s)=d_\rho(-s)d_\rho(s).$$

Step 2: Find a stable polynomial $d_\mu(s)$ such that

$$a(-s)a(s)+\mu^2 b(-s)b(s)=d_\mu(-s)d_\mu(s).$$

Step 3: The optimal controller $C(s) = \dfrac{q(s)}{p(s)}$ is the unique nth-order strictly proper pole-placement controller such that

$$a(s)p(s) + b(s)q(s) = d_\rho(s)d_\mu(s).$$

Step 4: The optimal performance index is given by

$$\boxed{J^*_{\rho,\mu} = \mu^2 \left\| \frac{d_\rho(s) - a(s)}{d_\rho(s)} \right\|_2^2 + \rho^2\mu^2 \left\| \frac{b(s)}{d_\rho(s)} \right\|_2^2 + \mu^2 \left\| \frac{d_\mu(s) - p(s)}{d_\mu(s)} \right\|_2^2 + \left\| \frac{q(s)}{d_\mu(s)} \right\|_2^2.}$$

The closed-loop poles are the roots of $d_\rho(s)$ and $d_\mu(s)$. When $\rho \to 0$, i.e., in the minimum-energy stabilization case,

$$d_\rho(-s)d_\rho(s) \to a(-s)a(s).$$

Since $d_\rho(s)$ is stable, the roots of $d_\rho(s)$ are near the stable roots of $a(s)$ and the mirror images of the unstable roots of $a(s)$ over the imaginary axis. This shows that if we only wish to minimize the input energy in the stabilization, we should not move the stable poles of the plant, which is reasonable, but should reflect the unstable poles of the plant over the imaginary axis, not just move them barely across the imaginary axis as one might expect.

When $\rho \to \infty$, i.e., in the cheap control case, $d_\rho(-s)d_\rho(s)$ is dominated by $\rho^2 b(-s)b(s)$. This shows that those roots of $d_\rho(-s)d_\rho(s)$ which are bounded as $\rho \to \infty$ will be near those of $b(-s)b(s)$. Hence the roots of $d_\rho(s)$ that are bounded as $\rho \to \infty$ are near the minimum phase zeros of the plant and the mirror images of the nonminimum phase zeros of the plant around the imaginary axis. Suppose that the relative degree of the plant is r. Then $2r$ roots of $d_\rho(-s)d_\rho(s)$ will tend to ∞. For large s, the polynomial $d_\rho(-s)d_\rho(s)$ can be approximated by

$$(-1)^n a_0^2 s^{2n} + \rho^2(-1)^{n-r} b_r^2 s^{2(n-r)}.$$

Therefore, the large roots will approximately be given by

$$s^{2r} = (-1)^{r+1}\frac{b_r^2}{a_0^2}\rho^2.$$

This shows that the $2r$ large roots of $d_\rho(-s)d_\rho(s)$ lie in a circle of radius $\left(\frac{b_r^2}{a_0^2}\rho^2\right)^{1/2r}$. Their phase angles are equal to those of $(-1)^{(r+1)/2r}$. This pattern is called a **Butterworth configuration**. The r large closed-loop poles, i.e., the large roots of $d_\rho(s)$, then tend to infinity following the Butterworth configuration on the left-hand side of the complex plane. Figure 9.4 gives examples of Butterworth configurations on the left-hand side of the complex plane.

For intermediate values of ρ, the roots of $a(-s)a(s) + \rho^2 b(-s)b(s)$ can be located by using the symmetric root locus. Let

$$P(s) = \frac{b(s)}{a(s)} = K\frac{(s-z_1)(s-z_2)\cdots(s-z_{n-r})}{(s-p_1)(s-p_2)\cdots(s-p_n)}.$$

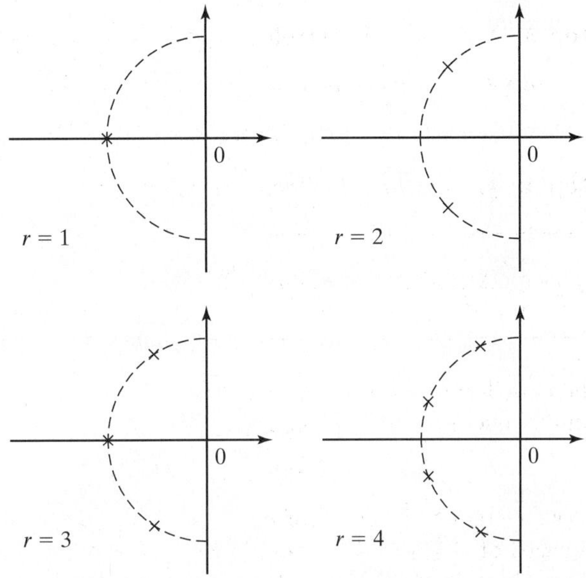

FIGURE 9.4: Butterworth configurations.

Then the roots of $a(-s)a(s) + \rho^2 b(-s)b(s)$ are also the solutions of

$$1 + (-1)^r \rho^2 K^2 \frac{(s - z_1)\cdots(s - z_{n-r})(s + z_1)\cdots(s + z_{n-r})}{(s - p_1)\cdots(s - p_n)(s + p_1)\cdots(s + p_n)} = 0.$$

Hence, if we know the pole/zero distribution of $P(s)$, flip them over the imaginary axis to get the complete symmetric pole/zero distribution of $P(-s)P(s)$ and then plot the complete root locus of these poles/zeros. This complete root locus will be symmetric to the imaginary axis since the poles and zeros of $P(-s)P(s)$ are symmetric to the imaginary axis and is hence called the **symmetric root locus**. Keep only the part of the root locus on the left-hand side of the complex plane. Then the root locations of $d_\rho(s)$ will be on this part of the root locus.

Identical discussions apply to the root locations of $d_\mu(s)$.

EXAMPLE 9.5

Let

$$P(s) = \frac{1}{s^2}.$$

Design $C(s)$ such that

$$J_{\rho,\mu} = \left(\|y_1(t)\|_2^2 + \rho^2\|y_2(t)\|_2^2\right)\Big|_{\substack{w_1(t)=\mu\delta(t)\\ w_2(t)=0}} + \left(\|y_1(t)\|_2^2 + \rho^2\|y_2(t)\|_2^2\right)\Big|_{\substack{w_1(t)=0\\ w_2(t)=\delta(t)}}$$

is minimized.

Observe that

$$a(-s)a(s) + \rho^2 b(-s)b(s) = s^4 + \rho^2.$$

This polynomial has roots

$$s = \pm\frac{\sqrt{2\rho}}{2} \pm j\frac{\sqrt{2\rho}}{2}.$$

Hence a spectral factor of $s^4 + \rho^2$ is

$$d_\rho(s) = \left(s + \frac{\sqrt{2\rho}}{2} + j\frac{\sqrt{2\rho}}{2}\right)\left(s + \frac{\sqrt{2\rho}}{2} - j\frac{\sqrt{2\rho}}{2}\right) = s^2 + \sqrt{2\rho}s + \rho.$$

Similarly, a spectral factor of $s^4 + \mu^2$ is

$$d_\mu(s) = s^2 + \sqrt{2\mu}s + \mu.$$

Let

$$C(s) = \frac{q_1 s + q_2}{p_0 s^2 + p_1 s + p_2}.$$

We then require

$$\begin{aligned} s^2(p_0 s^2 + p_1 s + p_2) + q_1 s + q_2 \\ &= d_\rho(s) d_\mu(s) \\ &= s^4 + (\sqrt{2\rho} + \sqrt{2\mu})s^3 + (\rho + \mu + 2\sqrt{\rho\mu})s^2 + (\mu\sqrt{2\rho} + \rho\sqrt{2\mu})s + \rho\mu. \end{aligned}$$

This gives

$$\begin{aligned} &p_0 = 1, \quad p_1 = \sqrt{2}(\sqrt{\rho} + \sqrt{\mu}), \quad p_2 = (\sqrt{\rho} + \sqrt{\mu})^2, \\ &q_1 = \sqrt{2\rho\mu}(\sqrt{\rho} + \sqrt{\mu}), \quad q_2 = \rho\mu. \end{aligned}$$

Therefore the optimal controller is

$$C(s) = \frac{\sqrt{2\rho\mu}(\sqrt{\rho} + \sqrt{\mu})s + \rho\mu}{s^2 + \sqrt{2}(\sqrt{\rho} + \sqrt{\mu})s + (\sqrt{\rho} + \sqrt{\mu})^2}.$$

The optimal value of the performance index is

$$\begin{aligned} J^*_{\rho,\mu} = \mu^2 \left\| \frac{\sqrt{2\rho}s + \rho}{s^2 + \sqrt{2\rho}s + \rho} \right\|_2^2 + \rho^2\mu^2 \left\| \frac{1}{s^2 + \sqrt{2\rho}s + \rho} \right\|_2^2 \\ + \mu^2 \left\| -\frac{\sqrt{2\rho}s + \rho + 2\sqrt{\rho\mu}}{s^2 + \sqrt{2\mu}s + \mu} \right\|_2^2 + \left\| \frac{\sqrt{2\rho\mu}(\sqrt{\rho} + \sqrt{\mu})s + \rho\mu}{s^2 + \sqrt{2\mu}s + \mu} \right\|_2^2. \end{aligned}$$

Detailed computation gives

$$J^*_{\rho,\mu} = \sqrt{2}[\mu^2\sqrt{\rho} + \rho^2\sqrt{\mu} + 2\rho\mu(\sqrt{\mu} + \sqrt{\rho})].$$

The symmetric root locus of $1/s^2$ is shown by the solid and dashed line in Figure 9.5. Its part on the left-hand side of the complex plane is shown by the solid line. We see that the roots of $d_\rho(s)$ change along the stable part of the symmetric root locus.

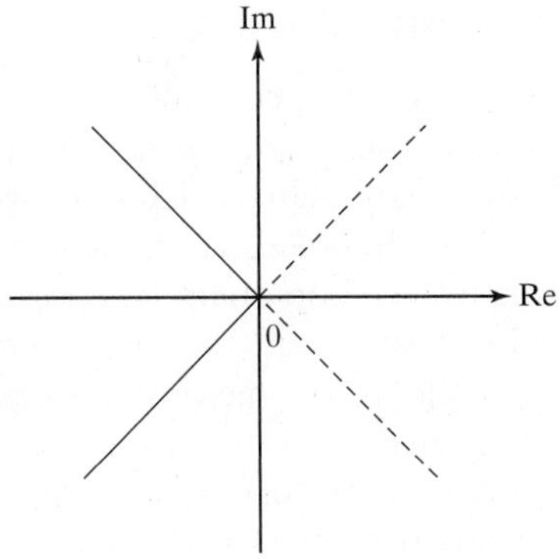

FIGURE 9.5: Symmetric root locus of $\frac{1}{s^2}$.

The closed-loop transfer functions are given by

$$\begin{bmatrix} \frac{P(s)C(s)}{1+P(s)C(s)} & \frac{C(s)}{1+P(s)C(s)} \\ \frac{P(s)}{1+P(s)C(s)} & \frac{-P(s)C(s)}{1+P(s)C(s)} \end{bmatrix}$$

$$= \frac{\begin{bmatrix} \sqrt{2\rho\mu}(\sqrt{\rho}+\sqrt{\mu})s+\rho\mu & \sqrt{2\rho\mu}(\sqrt{\rho}+\sqrt{\mu})s^3+\rho\mu s^2 \\ s^2+\sqrt{2}(\sqrt{\rho}+\sqrt{\mu})s+(\sqrt{\rho}+\sqrt{\mu})^2 & -\sqrt{2\rho\mu}(\sqrt{\rho}+\sqrt{\mu})s-\rho\mu \end{bmatrix}}{s^4+\sqrt{2}(\sqrt{\rho}+\sqrt{\mu})s^3+(\sqrt{\rho}+\sqrt{\mu})^2s^2+\sqrt{2\rho\mu}(\sqrt{\rho}+\sqrt{\mu})s+\rho\mu}.$$

The impulse responses of the closed-loop transfer functions are plotted in Figure 9.6. The solid lines are for $\rho = 0.1, \mu = 10$; the dashed lines are for $\rho = \mu = 10$; and the dash-dotted lines are for $\rho = \mu = 0.1$.

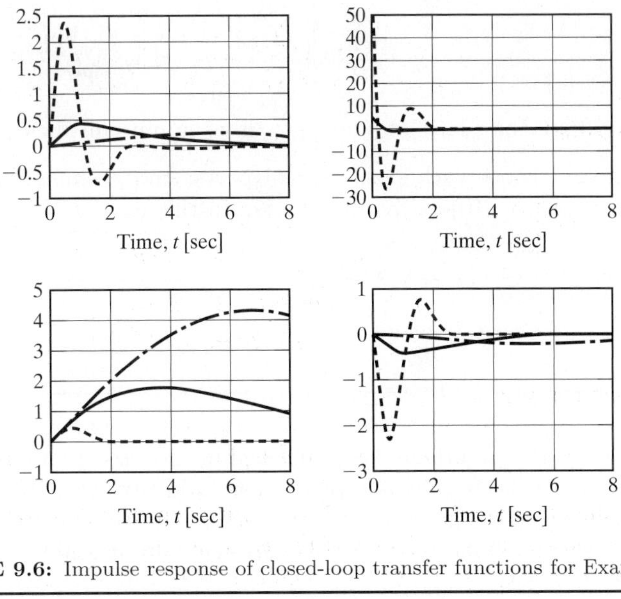

FIGURE 9.6: Impulse response of closed-loop transfer functions for Example 9.5.

9.3 MINIMUM-ENERGY STABILIZATION

In the real design of a stabilizing controller by optimizing a weighted optimal transient performance index, the weights ρ and μ should be chosen properly to get a good balance between various signals. Extreme values of ρ or μ will usually lead to unbalanced design, i.e., some signals are small but some other signals are excessively large. However, some closed-loop quantities designed under extreme values of ρ or μ are theoretically interesting and they may serve as performance limitations in stabilization. For example, it is sometimes interesting to know the smallest energy in the control effort needed to stabilize a system when the closed-loop system is only excited by a unit impulse in $w_1(t)$. Here by the control effort we mean the output $y_1(t)$ of the controller. Hence we are interested in the following quantity:

$$E^* = \inf_{C(s)\in\mathcal{S}(P)} \|y_1(t)\|_2^2 \Big|_{\substack{w_1(t)=\delta(t)\\ w_2(t)=0}}.$$

Trivially, if $P(s)$ is stable, we can set $C(s) = 0$ and then $y_1(t) = 0$, which gives $E^* = 0$. This is consistent with our intuition that no control effort is needed to stabilize a stable system if there is no other performance requirement. For general unstable systems, a moment of thought leads to

$$E^* = \lim_{\substack{\rho\to 0\\ \mu\to\infty}} \frac{1}{\mu^2} J^*_{\rho,\mu}.$$

This can also be seen from a transfer function point of view. The transfer function from w_1 to y_1 is

$$T(s) = \frac{P(s)C(s)}{1+P(s)C(s)}.$$

Hence, by using (9.4),

$$E^* = \inf_{C(s)\in\mathcal{S}(P)} \left\| \frac{P(s)C(s)}{1+P(s)C(s)} \right\|_2^2 = \lim_{\substack{\rho\to 0\\ \mu\to\infty}} \frac{1}{\mu^2} J^*_{\rho,\mu}.$$

Assume that the plant $P(s) = \dfrac{b(s)}{a(s)}$ is of minimum phase, i.e., $b(s)$ is a polynomial free of roots with positive real parts, and assume that $P(s)$ has m poles, $p_1, p_2, \ldots, p_m$, with positive real parts. In this case, E^* has a very simple expression as

$$E^* = 2\sum_{i=1}^{m} p_i. \tag{9.6}$$

Hence the minimum energy needed in stabilizing an unstable system depends only on the strictly unstable poles in an additive way. If the unstable poles are further away from the imaginary axis, more control effort will be needed to stabilize the system. One may wonder why critically unstable poles, i.e., poles on the imaginary axis, do not contribute to E^*. Does this means that no energy is needed to stabilize a critically unstable system? No, effort *is* needed to stabilize a critically unstable system but the effort can be made arbitrarily small.

If the plant is of nonminimum phase, the smallest control effort will in general be greater than the number given by (9.6) and will have a complicated dependence on the poles and zeros on the right-hand side of the complex plane. Let us gain some insight from the following example.

EXAMPLE 9.6

Consider a plant

$$P(s) = \frac{s-1-\epsilon}{(s+1+\epsilon)(s-1)}.$$

We are interested in finding

$$E^* = \lim_{\substack{\rho\to 0\\ \mu\to\infty}} \frac{1}{\mu^2} J^*_{\rho,\mu}.$$

Let us first find $J^*_{\rho,\mu}$ using the procedure in Section 9.2. The spectral factorization

$$\begin{aligned} a(-s)a(s) &+ \rho^2 b(-s)b(s) \\ &= (-s+1+\epsilon)(-s-1)(s+1+\epsilon)(s-1) + \rho^2(-s-1-\epsilon)(s-1-\epsilon) \\ &= d_\rho(-s)d_\rho(s) \end{aligned}$$

gives

$$d_\rho(s) = (s+1+\epsilon)\left(s+\sqrt{1+\rho^2}\right).$$

The spectral factorization

$$\begin{aligned} a(-s)a(s) &+ \mu^2 b(-s)b(s) \\ &= (-s+1+\epsilon)(-s-1)(s+1+\epsilon)(s-1) + \mu^2(-s-1-\epsilon)(s-1-\epsilon) \\ &= d_\mu(-s)d_\mu(s) \end{aligned}$$

gives

$$d_\mu(s) = (s+1+\epsilon)\left(s+\sqrt{1+\mu^2}\right).$$

Then the Diophantine equation becomes

$$a(s)p(s) + b(s)q(s) = (s+1+\epsilon)\left(s+\sqrt{1+\rho^2}\right)(s+1+\epsilon)\left(s+\sqrt{1+\mu^2}\right).$$

We omit the detailed computation. The final result is

$$E^* = \lim_{\substack{\rho\to 0\\ \mu\to\infty}} J^*_{\rho,\mu} = 2 + 8\left(1+\frac{1}{\epsilon}\right).$$

For small ϵ, this can be much greater than $2\sum_{i=1}^m p_i$, which is 2 in this case.

In this example, we see that the minimum control effort E^* is greater than twice of the sum of the unstable poles. As the nonminimum phase zero approaches the unstable pole, the minimum control effort E^* approaches infinity. These conclusions are true for general systems simultaneously having poles and zeros with positive real parts. This implies that nonminimum phase unstable systems, especially those with nonmimimum phase zeros close to the unstable poles, are hard to control. If a controller designer has influence in the process of designing the hardware, especially in locating the sensors and actuators for the control of a real physical process, he/she should aim at a minimum phase plant whenever possible for ease of control.

9.4 DERIVATION OF THE OPTIMAL CONTROLLER*

Since the unweighted optimal control problem in Section 9.1 is a special case of the weighted optimal control problem in Section 9.2, the derivation of the optimal controller for the weighted case also gives a derivation for the unweighted case.

We will need some technical tools in the derivation.

First, we need to extend the definition of 2-norm to possibly unstable systems. Let $G(s)$ be a system, stable or unstable. Define

$$\|G(s)\|_2 = \left[\frac{1}{2\pi}\int_{-\infty}^{\infty} |G(j\omega)|^2 d\omega\right]^{1/2}.$$

Lemma 9.7.

1. $\|G(s)\|_2 = \|G(-s)\|_2$.
2. If $G(s)$ is decomposed into $G(s) = G_1(s) + G_2(s)$ where $G_1(s)$ is stable and $G_2(-s)$ is stable, then

$$\|G(s)\|_2^2 = \|G_1(s)\|_2^2 + \|G_2(s)\|_2^2.$$

Secondly, for notational convenience, we will need the 2-norms of matrix transfer functions. They will be needed only in the following derivation, not in the computations of the optimal controller and the optimal performance index. Let $\boldsymbol{G}(s)$ be a matrix transfer function:

$$\boldsymbol{G}(s) = \begin{bmatrix} G_{11}(s) & \cdots & G_{1n}(s) \\ \vdots & & \vdots \\ G_{m1}(s) & \cdots & G_{mn}(s) \end{bmatrix}.$$

The 2-norm of $\boldsymbol{G}(s)$ is then defined as the square root of the squared sum of the 2-norms of all its elements, i.e.,

$$\|\boldsymbol{G}(s)\|_2 = \left(\sum_{i=1,j=1}^{m,n} \|G_{ij}(s)\|_2^2\right)^{1/2}.$$

As an example, an expression for the weighted transient performance index in Section 9.2 is

$$J_{\rho,\mu} = \left\| \begin{bmatrix} \dfrac{P(s)C(s)\mu}{1+P(s)C(s)} & \dfrac{C(s)}{1+P(s)C(s)} \\ \dfrac{\rho P(s)\mu}{1+P(s)C(s)} & \dfrac{-\rho P(s)C(s)}{1+P(s)C(s)} \end{bmatrix} \right\|_2^2 .$$

Thirdly, a technical lemma is required.

Lemma 9.8. Let $\boldsymbol{U}(s)$ be a matrix transfer function satisfying $\boldsymbol{U}^T(-s)\boldsymbol{U}(s) = \boldsymbol{I}$ and let $\boldsymbol{V}(s)$ be a matrix transfer function satisfying $\boldsymbol{V}(s)\boldsymbol{V}^T(-s) = \boldsymbol{I}$. Then

$$\|\boldsymbol{U}(s)\boldsymbol{G}(s)\boldsymbol{V}(s)\|_2 = \|\boldsymbol{G}(s)\|_2.$$

We are now ready to derive the optimal controller for the transient performance. It is obvious that the controller $C(s) = \dfrac{q(s)}{p(s)}$ given by the design procedure is a stabilizing controller since it is a pole-placement controller which assigns a stable, closed-loop characteristic polynomial $c(s) = d_\rho(s)d_\mu(s)$ to the closed-loop system. Let

$$M(s) = \frac{a(s)}{d_\rho(s)}, \quad N(s) = \frac{b(s)}{d_\rho(s)}, \quad X(s) = \frac{p(s)}{d_\mu(s)}, \quad Y(s) = \frac{q(s)}{d_\mu(s)}.$$

We then have

$$M(s)X(s) + N(s)Y(s) = 1 \tag{9.7}$$

and

$$M(-s)M(s) + \rho^2 N(-s)N(s) = 1. \tag{9.8}$$

According to the theory in Section 3.7, the set of all stabilizing controllers is given by

$$\mathcal{S}(P) = \left\{ C(s) = \frac{Y(s) + M(s)Q(s)}{X(s) - N(s)Q(s)} : Q(s) \text{ is stable} \right\}. \tag{9.9}$$

The idea in the rest of the derivation is to show that the optimal controller is obtained by $Q(s) = 0$. The optimal controller is then exactly

$$C(s) = \frac{Y(s)}{X(s)} = \frac{q(s)}{p(s)}.$$

It follows from (3.18) that under any controller of this form, we have

$$\begin{bmatrix} \dfrac{P(s)C(s)\mu}{1+P(s)C(s)} & \dfrac{C(s)}{1+P(s)C(s)} \\ \dfrac{\rho P(s)\mu}{1+P(s)C(s)} & \dfrac{-\rho P(s)C(s)}{1+P(s)C(s)} \end{bmatrix}$$
$$= \begin{bmatrix} N(s)Y(s)\mu & M(s)Y(s) \\ \rho N(s)X(s)\mu & -\rho N(s)Y(s) \end{bmatrix} + \begin{bmatrix} M(s) \\ -\rho N(s) \end{bmatrix} Q(s) \begin{bmatrix} N(s)\mu & M(s) \end{bmatrix}.$$

Observe from (9.7) that

$$\begin{bmatrix} N(s)Y(s)\mu & M(s)Y(s) \\ \rho N(s)X(s)\mu & -\rho N(s)Y(s) \end{bmatrix} = \begin{bmatrix} \mu - M(s)\mu & 0 \\ \rho N(s)\mu & 0 \end{bmatrix} + \begin{bmatrix} M(s) \\ -\rho N(s) \end{bmatrix} \begin{bmatrix} \mu - X(s)\mu & Y(s) \end{bmatrix}.$$

An expression of $J_{\rho,\mu}$ is then

$$J_{\rho,\mu} = \left\| \begin{bmatrix} \mu - M(s)\mu & 0 \\ \rho N(s)\mu & 0 \end{bmatrix} + \begin{bmatrix} M(s) \\ -\rho N(s) \end{bmatrix} \begin{bmatrix} \mu - X(s)\mu & Y(s) \end{bmatrix} + \begin{bmatrix} M(s) \\ -\rho N(s) \end{bmatrix} Q(s) \begin{bmatrix} N(s)\mu & M(s) \end{bmatrix} \right\|_2^2 . \tag{9.10}$$

Define

$$\boldsymbol{U}(s) = \begin{bmatrix} M(-s) & -N(-s)\rho \\ \rho N(s) & M(s) \end{bmatrix}.$$

Then by (9.8)

$$\boldsymbol{U}^T(-s)\boldsymbol{U}(s) = \begin{bmatrix} M(s) & N(-s)\rho \\ -\rho N(s) & M(-s) \end{bmatrix} \begin{bmatrix} M(-s) & -N(-s)\rho \\ \rho N(s) & M(s) \end{bmatrix} = \boldsymbol{I}.$$

Hence, by Lemma 9.8, we can multiply the matrix in (9.10) by $\boldsymbol{U}(s)$ from the left without changing its norm:

$$\begin{aligned} J_{\rho,\mu} &= \left\| \begin{bmatrix} M(-s)\mu - \mu & 0 \\ \rho N(s)\mu & 0 \end{bmatrix} + \begin{bmatrix} 1 \\ 0 \end{bmatrix} \begin{bmatrix} \mu - X(s)\mu & Y(s) \end{bmatrix} + \begin{bmatrix} 1 \\ 0 \end{bmatrix} Q(s) \begin{bmatrix} N(s)\mu & M(s) \end{bmatrix} \right\|_2^2 \\ &= \|M(-s)\mu - \mu + \mu - X(s)\mu + Q(s)N(s)\mu\|_2^2 \\ &\qquad + \|\rho N(s)\mu\|_2^2 + \|Y(s) + Q(s)M(s)\|_2^2 \\ &= \|M(-s)\mu - \mu\|_2^2 + \|\rho N(s)\mu\|_2^2 \\ &\qquad \|\mu - X(s)\mu + Q(s)N(s)\mu\|_2^2 + \|Y(s) + Q(s)M(s)\|_2^2 \\ &= \mu^2\|M(s) - 1\|_2^2 + \rho^2\mu^2\|N(s)\|_2^2 \\ &\qquad + \| \begin{bmatrix} \mu - X(s)\mu & Y(s) \end{bmatrix} + Q(s) \begin{bmatrix} N(s)\mu & M(s) \end{bmatrix} \|_2^2. \end{aligned} \tag{9.11}$$

Notice that Lemma 9.7 is used in getting the last equality above. Let us now define

$$\tilde{M}(s) = \frac{a(s)}{d_\mu(s)}, \quad \tilde{N}(s) = \frac{b(s)}{d_\mu(s)}, \quad \tilde{X}(s) = \frac{p(s)}{d_\rho(s)}, \quad \tilde{Y}(s) = \frac{q(s)}{d_\rho(s)}.$$

Then

$$\tilde{M}(s)\tilde{X}(s) + \tilde{N}(s)\tilde{Y}(s) = 1$$

$$\tilde{M}(-s)\tilde{M}(s) + \mu^2\tilde{N}(-s)\tilde{N}(s) = 1$$

$$\tilde{M}(-s)M(s) + \mu^2\tilde{N}(-s)\tilde{N}(s) = \frac{d_\mu(s)}{d_\rho(s)}.$$

Define

$$\boldsymbol{V}(s) = \begin{bmatrix} \mu\tilde{N}(-s) & \tilde{M}(s) \\ \tilde{M}(-s) & -\tilde{N}(s)\mu \end{bmatrix}.$$

Then

$$\boldsymbol{V}(s)\boldsymbol{V}^T(-s) = \begin{bmatrix} \mu\tilde{N}(-s) & \tilde{M}(s) \\ \tilde{M}(-s) & -\tilde{N}(s)\mu \end{bmatrix} \begin{bmatrix} \tilde{N}(s)\mu & \tilde{M}(s) \\ \tilde{M}(-s) & -\mu\tilde{N}(-s) \end{bmatrix} = \boldsymbol{I}.$$

Hence, again by Lemma 9.8, we can multiply $\boldsymbol{V}(s)$ to the right of the matrix in the third term of (9.11) without changing its norm. Then

$$\begin{aligned} &\| \begin{bmatrix} \mu - X(s)\mu & Y(s) \end{bmatrix} + Q(s) \begin{bmatrix} N(s)\mu & M(s) \end{bmatrix} \|_2^2 \\ &\qquad = \| \begin{bmatrix} W_1(s) & \mu W_2(s) \end{bmatrix} + Q(s) \begin{bmatrix} D(s) & 0 \end{bmatrix} \|_2^2 \\ &\qquad = \|W_1(s) + Q(s)D(s)\|_2^2 + \mu^2\|W_2(s)\|_2^2 \end{aligned}$$

where

$$\begin{aligned} W_1(s) &= \mu^2\tilde{N}(-s) - \mu^2 X(s)\tilde{N}(-s) + Y(s)\tilde{M}(-s) \qquad (9.12) \\ W_2(s) &= \tilde{M}(s) - X(s)\tilde{M}(s) - Y(s)\tilde{N}(s) \\ D(s) &= \frac{d_\mu(s)}{d_\rho(s)}. \end{aligned}$$

Now consider transfer function

$$\mu^2 X(s)\tilde{N}(-s) - Y(s)\tilde{M}(-s) = \frac{\mu^2 p(s)b(-s) - q(s)a(-s)}{d_\mu(s)d_\mu(-s)}.$$

Since polynomials $a(s)$ and $b(s)$ are coprime, there exist polynomials $f(s)$ and $g(s)$ such that

$$f(s)a(s) + g(s)b(s) = 1.$$

Now

$$\begin{aligned} &a(s)[\mu^2 p(s)b(-s) - q(s)a(-s)] \\ &= \mu^2[d_\rho(s)d_\mu(s) - q(s)b(s)]b(-s) - q(s)[d_\mu(-s)d_\mu(s) - \mu^2 b(-s)b(s)] \\ &= \mu^2[d_\rho(s)b(-s) - q(s)d_\mu(-s)]d_\mu(s) \end{aligned} \qquad (9.13)$$

and

$$\begin{aligned} &b(s)[\mu^2 p(s)b(-s) - q(s)a(-s)] \\ &\quad = p(s)[d_\mu(-s)d_\mu(s) - a(-s)a(s)] - [d_\rho(s)d_\mu(s) - a(s)p(s)]a(-s) \\ &\quad = [p(s)d_\mu(-s) - d_\rho(s)a(-s)]d_\mu(s). \end{aligned} \tag{9.14}$$

Multiplying (9.13) by $f(s)$ and (9.14) by $g(s)$ and then adding the two equations, we get

$$\begin{aligned} &\mu^2 p(s)b(-s) - q(s)a(-s) \\ &\quad = f(s)[b(-s)d_\rho(s) - q(s)d_\mu(-s)]d_\mu(s) + g(s)[p(s)d_\mu(-s) - a(-s)d_\rho(s)]d_\mu(s). \end{aligned}$$

Therefore

$$\begin{aligned} &\mu^2 X(s)\tilde{N}(-s) - Y(s)\tilde{M}(-s) \\ &\qquad = \frac{f(s)[b(-s)d_\rho(s) - q(s)d_\mu(-s)] + g(s)[p(s)d_\mu(-s) - a(-s)d_\rho(s)]}{d_\mu(-s)} \end{aligned}$$

which is now apparently antistable. Hence $W_1(s)$, as defined in (9.12), is antistable. Invoking Lemma 9.7, we obtain

$$\|W_1(s) + Q(s)D(s)\|_2^2 = \|W_1(s)\|_2^2 + \|Q(s)D(s)\|_2^2.$$

Summarizing the above, we get

$$\begin{aligned} J_{\rho,\mu} = {} & \|M(-s)\mu - \mu\|_2^2 + \|\rho N(s)\mu\|_2^2 + \mu^2\|\tilde{M}(s) - X(s)\tilde{M}(s) - Y(s)\tilde{N}(s)\|_2^2 \\ & + \|\tilde{N}(-s) - X(s)\tilde{N}(-s) + Y(s)\tilde{M}(-s)\|_2^2 + \|Q(s)D(s)\|_2^2. \end{aligned} \tag{9.15}$$

Clearly, to minimize $J_{\rho,\mu}$ we should choose $Q(s) = 0$.

The optimal value of $J_{\rho,\mu}$ can be computed by any of (9.10), (9.11), or (9.15) after plugging in $Q(s) = 0$. It turns out that (9.11) is the most convenient, which gives

$$\begin{aligned} J^*_{\rho,\mu} &= \mu^2\|M(s) - 1\|_2^2 + \rho^2\mu^2\|N(s)\|_2^2 + \mu^2\|1 - X(s)\|_2^2 + \|Y(s)\|_2^2 \\ &= \mu^2\left\|\frac{d_\rho(s) - a(s)}{d_\rho(s)}\right\|_2^2 + \rho^2\mu^2\left\|\frac{b(s)}{d_\rho(s)}\right\|_2^2 + \mu^2\left\|\frac{d_\mu(s) - p(s)}{d_\mu(s)}\right\|_2^2 + \left\|\frac{q(s)}{d_\mu(s)}\right\|_2^2. \end{aligned}$$

9.5 OPTIMAL ROBUST STABILIZATION

The purpose of this section is to design a stabilizing controller $C(s)$ for a given plant $P(s)$ so that the robustness of the feedback control system defined in Section 8.6

$$b_{P,C} = \left[\max_{\omega \in \mathbb{R}} \sqrt{|S(j\omega)|^2 + |C(j\omega)S(j\omega)|^2 + |P(j\omega)S(j\omega)|^2 + |T(j\omega)|^2}\right]^{-1}$$

is maximized. Here $S(s)$ is the sensitivity function and $T(s)$ is the complementary sensitivity function of the closed-loop system:

$$S(s) = \frac{1}{1+P(s)C(s)}, \quad T(s) = \frac{P(s)C(s)}{1+P(s)C(s)}, \quad S(s)+T(s) = 1.$$

At the same time, of course, we also wish to find the optimal robustness. Formally, we want to find the best possible robustness for a given plant

$$b^*(P) = \max_{C(s)\in\mathcal{S}(P)} b_{P,C}$$

and a controller $C(s)$ achieving

$$b^*(P) = b_{P,C}.$$

Since $b_{P,C}$ can also be expressed as

$$b_{P,C} = (1+\gamma_{P,C}^2)^{-1/2}$$

where

$$\gamma_{P,C} = \left\| \frac{P(-s) - C(s)}{1+P(s)C(s)} \right\|_\infty,$$

the problem of designing stabilizing controller $C(s)$ to maximize $b_{P,C}$ is then equivalent to the problem of designing stabilizing controller $C(s)$ to minimize $\gamma_{P,C}$. We also denote

$$\gamma^*(P) = \min_{C(s)\in\mathcal{S}(P)} \gamma_{P,C}.$$

For an nth-order proper plant, the optimal controller in this case is an $(n-1)$th-order proper pole-placement controller. The key issue in the design is to find the $2n-1$ optimal closed-loop poles. For first-order plants, this is extremely simple since the optimal controller is of zeroth order, i.e., a proportional controller. Let

$$P(s) = \frac{b(s)}{a(s)} = \frac{b_0 s + b_1}{a_0 s + a_1}$$

where $a_0 \neq 0$.

Algorithm 9.9.

Step 1: (*Spectral factorization*) Find a stable polynomial $d(s)$ such that

$$a(-s)a(s) + b(-s)b(s) = d(-s)d(s).$$

Step 2: (*Pole placement*) The optimal controller $C(s)$ is a zeroth order controller which places the closed-loop pole to the root of $d(s)$, i.e., $C(s) = \frac{q_0}{p_0}$ satisfying

$$a(s)p_0 + b(s)q_0 = d(s).$$

Step 3: (*Optimal robustness computation*) The optimal robustness is

$$b^*(P) = \frac{1}{\sqrt{p_0^2 + q_0^2}}.$$

EXAMPLE 9.10

Consider

$$P(s) = \frac{\beta}{s+\alpha}, \quad \beta \neq 0.$$

Design the controller $C(s)$ so that $b_{P,C}$ is maximized.

Step 1: Since

$$\begin{aligned} a(-s)a(s) + b(-s)b(s) &= (-s+\alpha)(s+\alpha) + \beta^2 \\ &= \left(-s + \sqrt{\alpha^2+\beta^2}\right)\left(s + \sqrt{\alpha^2+\beta^2}\right), \end{aligned}$$

it follows that $d(s) = s + \sqrt{\alpha^2+\beta^2}$.

Step 2: Let the optimal controller be $C(s) = \frac{q_0}{p_0}$. We then need

$$a(s)p_0 + b(s)q_0 = p_0(s+\alpha) + q_0\beta = s + \sqrt{\alpha^2+\beta^2}.$$

This gives $p_0 = 1$ and $q_0 = \frac{\sqrt{\alpha^2+\beta^2} - \alpha}{\beta}$. Therefore the optimal controller is

$$C(s) = \frac{\sqrt{\alpha^2+\beta^2} - \alpha}{\beta}.$$

Step 3:

$$b^*(P) = \frac{1}{\sqrt{p_0^2+q_0^2}} = \sqrt{\frac{\beta^2}{2\left(\alpha^2+\beta^2 - \alpha\sqrt{\alpha^2+\beta^2}\right)}}.$$

In particular, if $\alpha = 0$, i.e., $P(s) = \beta/s$, then

$$C(s) = \operatorname{sgn}\beta = \begin{cases} 1, & \beta > 0 \\ -1, & \beta < 0 \end{cases} \quad \text{and} \quad b^*(P) = \frac{1}{\sqrt{2}}.$$

The design procedure for high-order systems is a bit more complicated. Let a proper plant $P(s)$ be represented by

$$P(s) = \frac{b(s)}{a(s)} = \frac{b_0 s^n + b_1 s^{n-1} + \cdots + b_n}{a_0 s^n + a_1 s^{n-1} + \cdots + a_n}$$

where $a(s)$ and $b(s)$ are coprime and $a_0 \neq 0$. In the design procedure we need to use some matrix tools, such as eigenvalues, eigenvectors, evaluation of a polynomial at a matrix, and matrix inverse, which are covered in Appendix B.

Algorithm 9.11.

Step 1: (*Spectral factorization*) Find a stable polynomial

$$d(s) = d_0 s^n + d_1 s^{n-1} + \cdots + d_n$$

such that

$$a(-s)a(s) + b(-s)b(s) = d(-s)d(s).$$

Step 2: (*Matrix operation and construction*)

- Define a companion matrix and a diagonal sign matrix

$$\boldsymbol{A}_d = \begin{bmatrix} -\frac{d_1}{d_0} & 1 & \cdots & 0 \\ \vdots & \vdots & \ddots & \vdots \\ -\frac{d_{n-1}}{d_0} & 0 & \cdots & 1 \\ -\frac{d_n}{d_0} & 0 & \cdots & 0 \end{bmatrix}, \quad \boldsymbol{J} = \begin{bmatrix} (-1)^{n-1} & & & \\ & \ddots & & \\ & & -1 & \\ & & & 1 \end{bmatrix}.$$

- Form $2n \times n$ matrices

$$\boldsymbol{E} = \begin{bmatrix} -b(-\boldsymbol{A}_d) \\ a(-\boldsymbol{A}_d) \end{bmatrix}, \quad \boldsymbol{F} = \begin{bmatrix} a(\boldsymbol{A}_d) \\ b(\boldsymbol{A}_d) \end{bmatrix}$$

  by evaluating polynomials $a(s)$ and $b(s)$ at matrices $-\boldsymbol{A}_d$ and $\boldsymbol{A}_d$.

- Identify a nonsingular $n \times n$ submatrix $\boldsymbol{E}_0$ of $\boldsymbol{E}$. Let $\boldsymbol{F}_0$ be the corresponding $n \times n$ submatrix of $\boldsymbol{F}$. Namely, $\boldsymbol{F}_0$ takes the same position in $\boldsymbol{F}$ as $\boldsymbol{E}_0$ in $\boldsymbol{E}$.

- Set

$$\boldsymbol{H} = \boldsymbol{E}_0^{-1} \boldsymbol{F}_0 \boldsymbol{J}.$$

Step 3: (*Eigenvalue and eigenvector computation*) Compute the eigenvalues of $\boldsymbol{H}$. Let the one with the largest magnitude be γ^* and a corresponding eigenvector be $\boldsymbol{e}^*$. Define an $(n-1)$th-order polynomial

$$e^*(s) = \begin{bmatrix} s^{n-1} & \cdots & 1 \end{bmatrix} \boldsymbol{e}^*.$$

Step 4: (*Pole placement*) The optimal controller is the unique $(n-1)$th-order pole-placement controller $C(s) = \dfrac{q(s)}{p(s)}$ such that

$$a(s)p(s) + b(s)q(s) = d(s)e^*(s).$$

Step 5: (*Optimal robustness computation*)

$$b^*(P) = \frac{1}{\sqrt{1 + (\gamma^*)^2}}.$$

Algorithm 9.11 will have difficulty to continue if one or more of the following occurs:

1. The matrix $\boldsymbol{E}$ in Step 2 does not have an $n \times n$ nonsingular submatrix.
2. γ^* is not a real number, so $\boldsymbol{e}$ is not a real vector.
3. The first element of $\boldsymbol{e}^*$ vanishes, so $e^*(s)$ is actually an $(n-2)$th-order, or even lower order, polynomial.

It can be mathematically shown that none of these can ever occur. Therefore the algorithm can always be carried out to the end. It turns out that, most of the time, either $a(-\boldsymbol{A}_d)$ or $b(-\boldsymbol{A}_d)$ is nonsingular; so either of them can serve as $\boldsymbol{E}_0$ and the corresponding $\boldsymbol{F}_0$ can be taken as $a(\boldsymbol{A}_d)$ or $b(\boldsymbol{A}_d)$, respectively. One may wonder in case when the choice of nonsingular submatrix $\boldsymbol{E}_0$ is not unique whether or not the obtained matrix $\boldsymbol{H}$ is the same for different choices. Fortunately, the choice of $\boldsymbol{E}_0$ does not matter; any choice gives the same $\boldsymbol{H}$.

EXAMPLE 9.12

Consider

$$P(s) = \frac{1}{s^2}.$$

We wish to find the optimal $C(s)$ such that $b_{P,C}$ is maximized.

Step 1: (Spectral factorization)

$$s^4 + 1 = d(-s)d(s) \Rightarrow d(s) = s^2 + \sqrt{2}s + 1.$$

Step 2: (Matrix computation and construction)

- We have

$$\boldsymbol{A}_d = \begin{bmatrix} -\sqrt{2} & 1 \\ -1 & 0 \end{bmatrix}, \quad \boldsymbol{J} = \begin{bmatrix} -1 & 0 \\ 0 & 1 \end{bmatrix}.$$

- We then get

$$\boldsymbol{E} = \begin{bmatrix} -1 & 0 \\ 0 & -1 \\ 1 & -\sqrt{2} \\ \sqrt{2} & -1 \end{bmatrix}, \quad \boldsymbol{F} = \begin{bmatrix} 1 & -\sqrt{2} \\ \sqrt{2} & -1 \\ 1 & 0 \\ 0 & 1 \end{bmatrix}.$$

- We see that the top 2×2 submatrix of $\boldsymbol{E}$ is nonsingular, so we can take

$$\boldsymbol{E}_0 = \begin{bmatrix} -1 & 0 \\ 0 & -1 \end{bmatrix}, \quad \boldsymbol{F}_0 = \begin{bmatrix} 1 & -\sqrt{2} \\ \sqrt{2} & -1 \end{bmatrix}.$$

We also see that the bottom 2×2 submatrix of $\boldsymbol{E}$ is nonsingular, so we can also take

$$\boldsymbol{E}_0 = \begin{bmatrix} 1 & -\sqrt{2} \\ \sqrt{2} & -1 \end{bmatrix}, \quad \boldsymbol{F}_0 = \begin{bmatrix} 1 & 0 \\ 0 & 1 \end{bmatrix}.$$

- The first choice of $\boldsymbol{E}_0$ and $\boldsymbol{F}_0$ gives

$$\boldsymbol{H} = \begin{bmatrix} -1 & 0 \\ 0 & -1 \end{bmatrix}^{-1} \begin{bmatrix} 1 & -\sqrt{2} \\ \sqrt{2} & -1 \end{bmatrix} \boldsymbol{J} = \begin{bmatrix} 1 & \sqrt{2} \\ \sqrt{2} & 1 \end{bmatrix}.$$

The second choice of $\boldsymbol{E}_0$ and $\boldsymbol{F}_0$ gives

$$\boldsymbol{H} = \begin{bmatrix} 1 & -\sqrt{2} \\ \sqrt{2} & -1 \end{bmatrix}^{-1} \begin{bmatrix} 1 & 0 \\ 0 & 1 \end{bmatrix} \boldsymbol{J} = \begin{bmatrix} 1 & \sqrt{2} \\ \sqrt{2} & 1 \end{bmatrix}.$$

The two choices give the same answer. In the actual design, we need only to take one choice for the computation.

Step 3: (Eigenvalue and eigenvector computation) The eigenvalues of $\boldsymbol{H}$ are $1 \pm \sqrt{2}$. The one with the largest magnitude is $\gamma^* = 1 + \sqrt{2}$ and its corresponding eigenvector satisfies

$$\begin{bmatrix} 1 & \sqrt{2} \\ \sqrt{2} & 1 \end{bmatrix} \boldsymbol{e}^* = (1 + \sqrt{2})\boldsymbol{e}^*.$$

This gives $\boldsymbol{e}^* = \begin{bmatrix} 1 \\ 1 \end{bmatrix}$. Then

$$e^*(s) = \begin{bmatrix} s & 1 \end{bmatrix} \boldsymbol{e}^* = s + 1.$$

Step 4: (Pole placement) The Diophantine equation becomes

$$s^2 p(s) + q(s) = (s^2 + \sqrt{2}s + 1)(s + 1).$$

This gives $p(s) = s + 1 + \sqrt{2}$ and $q(s) = (1 + \sqrt{2})s + 1$. Therefore, the final optimal controller is

$$C(s) = \frac{(1 + \sqrt{2})s + 1}{s + 1 + \sqrt{2}}.$$

Step 5: (Optimal robustness) We have

$$b^*(P) = \frac{1}{\sqrt{1 + (\gamma^*)^2}} = \frac{1}{\sqrt{4 + 2\sqrt{2}}}.$$

In the example above, both $-b(-\boldsymbol{A}_d)$ and $a(-\boldsymbol{A}_d)$ are nonsingular, so either one can be taken as $\boldsymbol{E}_0$ and the other one need not be computed in the design procedure to save computational effort. However, there are cases in which neither $-b(-\boldsymbol{A}_d)$ nor $a(-\boldsymbol{A}_d)$ is invertible.

EXAMPLE 9.13

Consider

$$P(s) = \frac{b(s)}{a(s)} = \frac{(s+1)(s-2)}{(s-1)(s+2)} = \frac{s^2 - s - 2}{s^2 + s - 2}.$$

The spectral factorization

$$a(-s)a(s) + b(-s)b(s) = 2(-s+1)(-s+2)(s+1)(s+2) = d(-s)d(s)$$

gives

$$d(s) = \sqrt{2}(s+1)(s+2) = \sqrt{2}(s^2 + 3s + 2).$$

Then

$$\boldsymbol{A}_d = \begin{bmatrix} -3 & 1 \\ -2 & 0 \end{bmatrix}.$$

We can see that both

$$a(-\boldsymbol{A}_d) = \begin{bmatrix} 8 & -4 \\ 8 & -4 \end{bmatrix} \quad \text{and} \quad -b(-\boldsymbol{A}_d) = \begin{bmatrix} -2 & 2 \\ -4 & 4 \end{bmatrix}$$

are singular. The reason for such a situation is that $d(s)$ is not coprime with either $a(-s)$ or $b(-s)$.

We observe that the submatrix

$$\begin{bmatrix} -4 & 4 \\ 8 & -4 \end{bmatrix}$$

formed by rows 2 and 3 of

$$\boldsymbol{E} = \begin{bmatrix} -b(-\boldsymbol{A}_d) \\ a(-\boldsymbol{A}_d) \end{bmatrix} = \begin{bmatrix} -2 & 2 \\ -4 & 4 \\ 8 & -4 \\ 8 & -4 \end{bmatrix}$$

is nonsingular. The corresponding submatrix of

$$\begin{bmatrix} a(\boldsymbol{A}_d) \\ b(\boldsymbol{A}_d) \end{bmatrix} = \begin{bmatrix} 2 & -2 \\ 4 & -4 \\ 8 & -4 \\ 8 & -4 \end{bmatrix}$$

is

$$\begin{bmatrix} 4 & -4 \\ 8 & -4 \end{bmatrix}.$$

It follows that

$$\boldsymbol{H} = \begin{bmatrix} -4 & 4 \\ 8 & -4 \end{bmatrix}^{-1} \begin{bmatrix} 4 & -4 \\ 8 & -4 \end{bmatrix} \begin{bmatrix} -1 & 0 \\ 0 & 1 \end{bmatrix} = \begin{bmatrix} -3 & -2 \\ -4 & -3 \end{bmatrix}.$$

The eigenvalues of $\boldsymbol{H}$ are $-3 \pm 2\sqrt{2}$. Thus the eigenvalue with the largest magnitude is $\gamma^* = -3 - 2\sqrt{2}$ and the corresponding eigenvector is

$$e^* = \begin{bmatrix} 1 \\ \sqrt{2} \end{bmatrix}$$

since $\boldsymbol{H}e^* = \gamma^* e^*$. Now let

$$e^*(s) = \begin{bmatrix} s & 1 \end{bmatrix} e^* = s + \sqrt{2}$$

and solve the Diophantine equation

$$a(s)p(s) + b(s)q(s) = d(s)e^*(s)$$

with $p(s) = p_0 s + p_1$ and $q(s) = q_0 s + q_1$ to get

$$p(s) = 2(1+\sqrt{2})(s+1), \quad q(s) = -(2+\sqrt{2})(s+2).$$

Therefore, the optimal controller is

$$C(s) = -\frac{\sqrt{2}}{2}\frac{s+2}{s+1}$$

and the optimal robustness is given by

$$b^*(P) = \frac{1}{\sqrt{1+(\gamma^*)^2}} = \frac{1}{\sqrt{18+12\sqrt{2}}} = 0.1691.$$

The next example shows how the optimal robustness controller for one plant performs for a family of plants nearby.

EXAMPLE 9.14

Consider the following systems

$$P_1(s) = \frac{100}{s}, \; P_2(s) = \frac{100}{s+1}, \; P_3(s) = \frac{100}{s-1}, \; P_4(s) = \frac{100}{(s+1)^2}, \; P_5(s) = \frac{100}{s^2-1}.$$

Systems $P_2(s)$ and $P_3(s)$ have very different open-loop characteristics – one is stable, the other is unstable. However, Table 9.1 shows that their chordal distance is very small. Similarly, the chordal distance between $P_4(s)$ and $P_5(s)$ is also very small. On the other hand, the chordal distance between $P_2(s)$ and $P_4(s)$ is very large even though both $P_2(s)$ and $P_4(s)$ are stable. Thus, from the chordal distances among these plants we can conclude that $P_1(s)$, $P_2(s)$, and $P_3(s)$ are very close, $P_4(s)$ and $P_5(s)$ are also very close, while $P_1(s)$ and $P_4(s)$ (or $P_1(s)$ and $P_5(s)$, and so on) are quite far away. It is not surprising that any reasonable controller for $P_1(s)$ will do well for $P_2(s)$ and $P_3(s)$ but not necessarily for $P_4(s)$ and $P_5(s)$.

The optimal robust controller for $P_1(s)$, as designed in Example 9.10, is a proportional controller with the unit gain, i.e., $C_1^*(s) = 1$. The closed-loop step responses of unity feedback systems consisting of different plants $P_i(s)$, $i = 1, \ldots, 5$, and the same controller $C_1^*(s)$ are shown in Figure 9.7.

Table 9.1 shows that the corresponding stability margins for the closed-loop systems with $P_2(s)$ and $P_3(s)$ are almost optimal, while the stability margin for the closed-loop system with $P_4(s)$ is very small (in fact, $b_{P_4,C_1^*} = 0.099$). Thus it is expected that the closed-loop performance of this system is very poor. It is also noted that this controller $C_1^*(s)$ fails to stabilize $P_5(s)$. In fact, it is not hard to find a controller that will perform well for $P_1(s)$, $P_2(s)$, and $P_3(s)$ but will destabilize both $P_4(s)$ and $P_5(s)$.

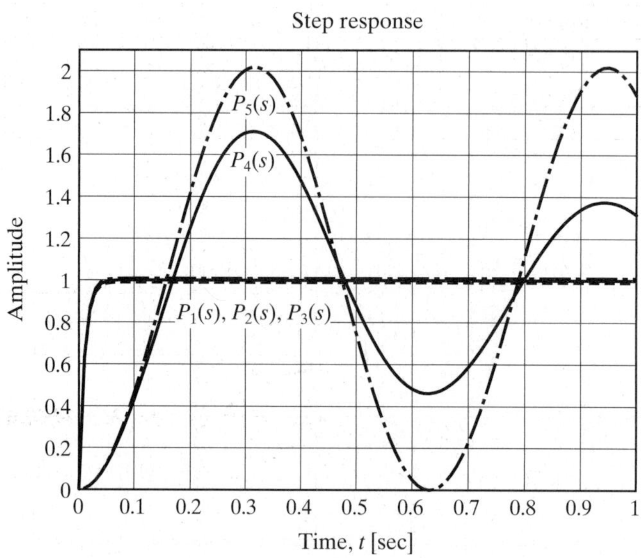

FIGURE 9.7: Closed-loop step responses with $C_1^*(s) = 1$.

Of course, this does not necessarily mean that all controllers performing reasonably well with $P_1(s)$, $P_2(s)$, and $P_3(s)$ will do badly with $P_4(s)$ and $P_5(s)$; it merely means that some do. It may be harder to find a controller that will perform reasonably well with all five plants; the optimal robust controllers of $P_4(s)$ and $P_5(s)$ are such controllers as can be seen from Table 9.1. The step responses under the optimal controller $C_5^*(s)$ are shown in Figure 9.8.

Since

$$\delta(C_1^*(s), C_4^*(s)) = 0.3431, \;\; \delta(C_1^*(s), C_5^*(s)) = 0.3849,$$

it is expected that $C_4^*(s)$ and $C_5^*(s)$ will perform far from optimal for the plant $P_1(s)$.

i	1	2	3	4	5
$\delta(P_1(s), P_i(s))$		0.01	0.01	0.956	0.959
$\delta(P_2(s), P_i(s))$			0.02	0.957	0.961
$\delta(P_3(s), P_i(s))$				0.954	0.957
$\delta(P_4(s), P_i(s))$					0.114
$b^*(P_i)$	0.707	0.711	0.703	0.431	0.380
$C_i^*(s)$	1	0.99	1.01	$\dfrac{2.1(s+5.16)}{s+23.26}$	$\dfrac{2.43(s+4.21)}{s+24.21}$
b_{P_i,C_1^*}	0.707	0.707	0.700	0.099	unstable
b_{P_i,C_2^*}	0.707	0.711	0.703	0.100	unstable
b_{P_i,C_3^*}	0.703	0.703	0.703	0.099	unstable
b_{P_i,C_4^*}	0.422	0.431	0.413	0.431	0.327
b_{P_i,C_5^*}	0.380	0.380	0.380	0.380	0.380
b_{P_i,C_6}	0.380	0.380	0.380	0.380	0.330

TABLE 9.1: Example 9.14.

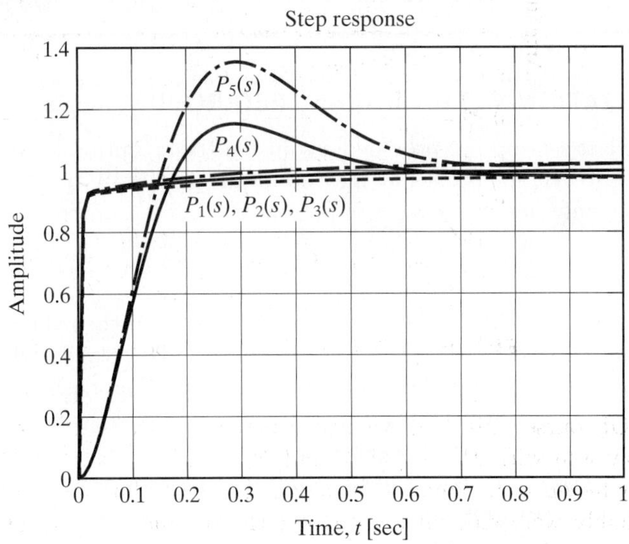

FIGURE 9.8: Closed-loop step responses with $C_5^*(s)$.

Since all optimal robust controllers do not take tracking into consideration, we shall connect a PI controller with $C_5^*(s)$ to obtain the following controller:

$$C_6(s) = \frac{s+1}{s} \frac{2.431(s+4.213)}{s+24.213}.$$

The step responses under this control law are shown in Figure 9.9.

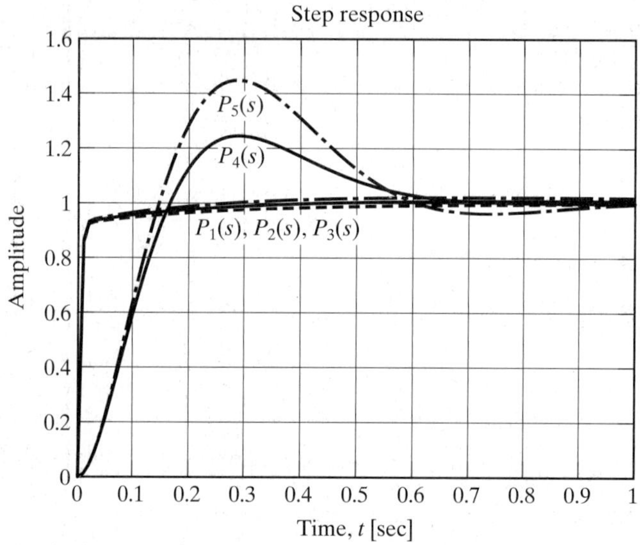

FIGURE 9.9: Closed-loop step responses with $C_6(s)$.

9.6 STABILIZATION WITH GUARANTEED ROBUSTNESS

In the last section, we provided a solution to the optimal robust stabilization problem. However, sometimes we may not need to optimize the robustness $b_{P,C}$. Instead, we may only need to meet a requirement on the robustness $b_{P,C}$ such as

$$b_{P,C} > \beta \tag{9.16}$$

for some prespecified β. Apparently, if $\beta \geq b^*(P)$, then there is no solution to this problem. If $\beta < b^*(P)$, then the problem can be solved. One solution is given by the optimal controller, and there are, in general, many other solutions. The freedom obtained by this can then be used to improve other performance objectives of the feedback system. Hence several questions can be asked. Firstly, one may wish to find a "reasonable" controller $C(s)$, if it exists, for a given $P(s)$ and $\beta > 0$ so that requirement (9.16) is satisfied. Secondly, one may wish to characterize the set of all controllers $C(s)$ satisfying (9.16) for a given $P(s)$ and $\beta > 0$. Thirdly, one may wish to select a $C(s)$, subject to constraint (9.16), to optimize some other performance objectives, for example the transient performance cost J defined in Section 9.1. We will give the answer to the first question. The answer to the second question can be found in the literature. The third question has not had a complete answer yet and is still attracting active research.

What is a "reasonable" controller? One way to define a reasonable controller is that it should not sacrifice too much of the other performance objectives when we pursue robustness. In this sense, the optimal robust controller is not quite "reasonable" since it is not strictly proper even when the plant is strictly proper and as a result the transient performance J is infinity when the optimal robust controller is used.

The procedure to design a particular controller $C(s)$, called the **central controller**, so that (9.16) is satisfied is given in the following algorithm. Here we assume that $P(s)$ is strictly proper in the form

$$P(s) = \frac{b(s)}{a(s)} = \frac{b_1 s^{n-1} + b_2 s^{n-2} + \cdots + b_n}{a_0 s^n + a_1 s^{n-1} + \cdots + a_n}, \qquad a_0 \neq 0.$$

Algorithm 9.15.

Step 1: (*Spectral factorization*) Find a stable polynomial

$$d(s) = d_0 s^n + d_1 s^{n-1} + \cdots + d_n$$

such that

$$a(-s)a(s) + b(-s)b(s) = d(-s)d(s).$$

Step 2: (*Matrix operation and construction*)

- Define a companion matrix and a diagonal sign matrix

$$\boldsymbol{A}_d = \begin{bmatrix} -\dfrac{d_1}{d_0} & 1 & \cdots & 0 \\ \vdots & \vdots & \ddots & \vdots \\ -\dfrac{d_{n-1}}{d_0} & 0 & \cdots & 1 \\ -\dfrac{d_n}{d_0} & 0 & \cdots & 0 \end{bmatrix}, \quad \boldsymbol{J} = \begin{bmatrix} (-1)^{n-1} & & & \\ & \ddots & & \\ & & -1 & \\ & & & 1 \end{bmatrix}.$$

- Form $2n \times n$ matrices

$$\boldsymbol{E} = \begin{bmatrix} -b(-\boldsymbol{A}_d) \\ a(-\boldsymbol{A}_d) \end{bmatrix}, \quad \boldsymbol{F} = \begin{bmatrix} a(\boldsymbol{A}_d) \\ b(\boldsymbol{A}_d) \end{bmatrix}$$

  by evaluating polynomials $a(s)$ and $b(s)$ at matrices $-\boldsymbol{A}_d$ and $\boldsymbol{A}_d$.

- Identify a nonsingular $n \times n$ submatrix $\boldsymbol{E}_0$ of $\boldsymbol{E}$. Let $\boldsymbol{F}_0$ be the corresponding $n \times n$ submatrix of $\boldsymbol{F}$. Namely, $\boldsymbol{F}_0$ takes the same position in $\boldsymbol{F}$ as $\boldsymbol{E}_0$ in $\boldsymbol{E}$.

- Set

$$\boldsymbol{H} = \boldsymbol{E}_0^{-1} \boldsymbol{F}_0 \boldsymbol{J}.$$

Step 3: (*Eigenvalue computation*) Let its eigenvalue with the largest magnitude be γ^*. If $\beta^2 \geq \frac{1}{1+(\gamma^*)^2}$, there is no solution and exit. Otherwise, define

$$e(s) = d(s) + 2\begin{bmatrix} s^{n-1} & s^{n-2} & \cdots 1 \end{bmatrix} \boldsymbol{H}^2[(\beta^{-2}-1)\boldsymbol{I} - \boldsymbol{H}^2]^{-1} \begin{bmatrix} d_1 \\ 0 \\ d_3 \\ \vdots \end{bmatrix}.$$

Step 4: (*Pole placement*) A stabilizing controller with guaranteed robustness is the unique nth-order strictly proper pole-placement controller $C(s) = \dfrac{q(s)}{p(s)}$ such that

$$a(s)p(s) + b(s)q(s) = d(s)e(s).$$

The controller $C(s)$ obtained depends on β. Hence the closed-loop stability robustness $b_{P,C}$ also depends on β. In general, $b_{P,C}$ is neither the optimal value $b^*(P)$ nor the boundary value β. It is rather somewhere in between.

When $\beta = 0$, i.e., there is no robustness requirement, we have $e(s) = d(s)$. This procedure gives a controller exactly the same as the optimal controller minimizing the transient cost J. This indicates the possibility that for other values of β the controller given by this procedure also yields a good J in addition to satisfying the robustness requirement (9.16). This turns out to be the case. The detailed justification is beyond the scope of this book. However, one can also show that, in general, the controller produced by this procedure is not the one with the best J among all controllers satisfying (9.16).

EXAMPLE 9.16

Consider

$$P(s) = \frac{1}{s^2}.$$

We wish to find the controller $C(s)$ such that $b_{P,C} \geq \beta$

Step 1: (Spectral factorization)

$$s^4 + 1 = d(-s)d(s) \Rightarrow d(s) = s^2 + \sqrt{2}s + 1.$$

Step 2: (Matrix computation and construction) This step is the same as in Example 9.12. We get

$$\boldsymbol{H} = \begin{bmatrix} 1 & \sqrt{2} \\ \sqrt{2} & 1 \end{bmatrix}.$$

Step 3: (Eigenvalue computation) The eigenvalue of $\boldsymbol{H}$ with the largest magnitude is $\gamma^* = 1 + \sqrt{2}$. If $\beta^2 < \frac{1}{1+(\gamma^*)^2} = \frac{1}{4+2\sqrt{2}}$, then we can proceed.

We have

$$2\boldsymbol{H}^2[(\beta^{-2}-1)\boldsymbol{I}-\boldsymbol{H}^2)]^{-1}\begin{bmatrix}\sqrt{2}\\0\end{bmatrix}=\frac{-\beta^2}{8\beta^4-8\beta^2+1}\begin{bmatrix}8\sqrt{2}\beta^2-6\sqrt{2}\\8\beta^2-8\end{bmatrix}.$$

This gives

$$\begin{aligned}e(s)&=d(s)-\frac{\beta^2}{8\beta^4-8\beta^2+1}[(8\sqrt{2}\beta^2-6\sqrt{2})s+8\beta^2-8]\\&=s^2+\frac{-2\sqrt{2}\beta^2+\sqrt{2}}{8\beta^4-8\beta^2+1}s+\frac{1}{8\beta^4-8\beta^2+1}.\end{aligned}$$

Step 4: (Pole placement) The Diophantine equation becomes

$$s^2p(s)+q(s)=(s^2+\sqrt{2}s+1)\left(s^2+\frac{-2\sqrt{2}\beta^2+\sqrt{2}}{8\beta^4-8\beta^2+1}s+\frac{1}{8\beta^4-8\beta^2+1}\right)$$

$$=d(s)s^2+\frac{(-2\beta^2+1)\sqrt{2}s^3+(-4\beta^2+3)s^2+(-2\beta^2+2)\sqrt{2}s+1}{8\beta^4-8\beta^2+1}.$$

This gives

$$\begin{aligned}p(s)&=d(s)+\frac{(-2\beta^2+1)\sqrt{2}s-4\beta^2+3}{8\beta^4-8\beta^2+1}\\q(s)&=\frac{(-2\beta^2+2)\sqrt{2}s+1}{8\beta^4-8\beta^2+1}.\end{aligned}$$

Therefore, a guaranteed robust controller is given by

$$C(s)=\frac{(-2\beta^2+2)\sqrt{2}s+1}{(8\beta^4-8\beta^2+1)d(s)+(-2\beta^2+1)\sqrt{2}s+(-4\beta^2+3)}.$$

We see that when $\beta^2=\frac{1}{4+2\sqrt{2}}$, i.e., when β is equal to the optimal value, the controller becomes

$$C(s)=\frac{(1+\sqrt{2})s+1}{s+1+\sqrt{2}},$$

which is the optimal robust stabilizing controller obtained in Example 9.12. When $\beta=0$, i.e., when there is no robust requirement, the controller becomes

$$C(s)=\frac{2\sqrt{2}s+1}{s^2+2\sqrt{2}s+4},$$

which is the controller minimizing J as obtained in Example 9.3, Section 9.1.

The closed-loop robustness $b_{P,C}$ for different β is plotted in the first plot of Figure 9.10. The closed-loop transient performance J is plotted in the second plot of Figure 9.10.

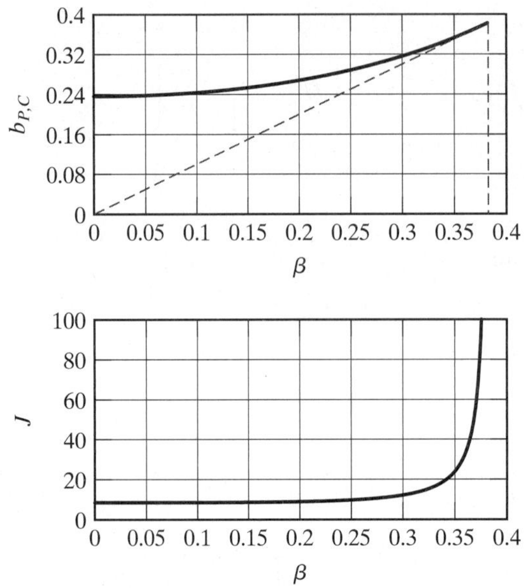

FIGURE 9.10: Values of the robustness margin $b_{P,C}$ and the transient performance index J for the guaranteed robustness controllers designed for different β.

In particular, if $\beta = 1/\sqrt{10}$, then

$$C(s) = \frac{2.5456s + 1}{0.28s^2 + 1.5274s + 2.88}.$$

In this case, $b_{P,C} = 0.3267$ and $J = 13.89$.

PROBLEMS

9.1. Find the optimal controller $C(s)$ minimizing the transient cost J for the following plants.

1. $P(s) = \dfrac{1}{s+1}$;
2. $P(s) = \dfrac{1}{s-1}$;
3. $P(s) = \dfrac{1}{s(s+1)}$;
4. $P(s) = \dfrac{s-1}{s(s+1)}$;
5. $P(s) = \dfrac{1}{s^3}$.

9.2. Let $P(s) = \dfrac{1/2}{s(s+1)}$.

1. Design a stabilizing controller so that the closed-loop poles are $-1, -2, -3$.

2. Design a stabilizing controller to minimize

$$J = \left(\|y_1(t)\|_2^2 + \|y_2(t)\|_2^2\right)\Big|_{\substack{w_1(t)=\delta(t)\\ w_2(t)=0}} + \left(\|y_1(t)\|_2^2 + \|y_2(t)\|_2^2\right)\Big|_{\substack{w_1(t)=0\\ w_2(t)=\delta(t)}}$$

and find the optimal cost J^*.

9.3. Let $P(s) = \dfrac{4}{s^2}$.

1. Design a proper controller so that the closed-loop system has poles at $-1 \pm j$ and -2.

2. Design a strictly proper controller so that the closed-loop system has poles at $-1 \pm j$ and $-2 \pm 2j$.

3. Design a controller so that the cost

$$J = \left(\|y_1(t)\|_2^2 + \|y_2(t)\|_2^2\right)\Big|_{\substack{w_1(t)=\delta(t)\\ w_2(t)=0}} + \left(\|y_1(t)\|_2^2 + \|y_2(t)\|_2^2\right)\Big|_{\substack{w_1(t)=0\\ w_2(t)=\delta(t)}}$$

is minimized. Find J^*.

4. Design a controller so that the cost

$$J = \left(4\|y_1(t)\|_2^2 + 4\|y_2(t)\|_2^2\right)\Big|_{\substack{w_1(t)=\delta(t)\\ w_2(t)=0}} + \left(\|y_1(t)\|_2^2 + \|y_2(t)\|_2^2\right)\Big|_{\substack{w_1(t)=0\\ w_2(t)=2\delta(t)}}$$

is minimized. Find J^*.

9.4. Design a stabilizing controller for $P(s) = \dfrac{1}{s-12}$ such that

$$J = \left(512\|y_1(t)\|_2^2 + 41472\|y_2(t)\|_2^2\right)\Big|_{\substack{w_1(t)=\delta(t)\\ w_2(t)=0}} + \left(2\|y_1(t)\|_2^2 + 162\|y_2(t)\|_2^2\right)\Big|_{\substack{w_1(t)=0\\ w_2(t)=\delta(t)}}$$

is minimized. What is the optimal value of J?

9.5. Design a stabilizing controller for $P(s) = \dfrac{1}{s-1}$ such that

$$J = \left(2\|y_1(t)\|_2^2 + 32\|y_2(t)\|_2^2\right)\Big|_{\substack{w_1(t)=\delta(t)\\ w_2(t)=0}} + \left(\frac{1}{8}\|y_1(t)\|_2^2 + 2\|y_2(t)\|_2^2\right)\Big|_{\substack{w_1(t)=0\\ w_2(t)=\delta(t)}}$$

is minimized. What is the optimal value of J?

9.6. Find the optimal robust controller $C(s)$ and the optimal robustness $b^*(P)$ for the following plants.

1. $P(s) = \dfrac{1}{s+1}$;

2. $P(s) = \dfrac{1}{s-1}$;

3. $P(s) = \dfrac{1}{s(s+1)}$;

4. $P(s) = \dfrac{s-1}{s(s+1)}$;

5. $P(s) = \dfrac{1}{s^3}$.

MATLAB PROBLEMS

9.7. Write a MATLAB program "sfac" to find the spectral factorization of $a(-s)a(s) + b(-s)b(s)$ for given coprime polynomials $a(s)$ and $b(s)$.

9.8. Write a MATLAB program "opt-tran" to compute the optimal controller for the weighted transient performance. Test the program on the plants in Problem 9.1.

9.9. Write a MATLAB program "opt-robust" to implement the design procedure in Algorithm 9.11. Test the program on the plants in Problem 9.6.

EXTRA CREDIT PROBLEMS

9.10. Prove Lemma 9.7.

9.11. Prove Lemma 9.8.

NOTES AND REFERENCES

The optimal control problem minimizing $J_{\rho,\mu}$ considered in Sections 9.1 and 9.2 is usually called a **linear quadratic Gaussian (LQG) optimal control problem**, which was the climax of the so-called modern control theory, developed from 1950s to 1970s. The particular case when $\rho = \mu = 1$ is called a **normalized LQG control problem**. The problem is usually formulated for systems described by state space models and subject to Gaussian stochastic external signals. This explains how the name LQG came about. Good books covering the LQG optimal control problem are

H. Kwakernaak and R. Sivan, *Linear Optimal Control Systems*, Wiley-Interscience, New York, 1972.

B. D. O. Anderson and J. B. Moore, *Optimal Control: Linear Quaqratic Method*, Prentice Hall, Englewood Cliffs, New Jersy, 1989.

The robust stabilization problems considered in Sections 9.5 and 9.6 are special cases of the $\mathcal{H}_\infty$ control problem that were the main focus of the so-called postmodern control theory, developed in the 1980s. The representative books covering the $\mathcal{H}_\infty$ control problem are

K. Zhou with J. C. Doyle and K. Glover, *Robust and Optimal Control*, Prentice Hall, Upper Saddle River, New Jersy, 1996.

M. Green and D. J. N. Limebeer, *Linear Robust Control*, Prentice Hall, Englewood Cliffs, New Jersy, 1995.

The particular robust stabilization problems discussed in Sections 9.5 and 9.6 are thoroughly studied in

G. Vinnicombe, *Uncertainty and Feedback: $\mathcal{H}_\infty$ Loop-Shaping and the ν-Gap Metric*, Imperial Collage Press, London, 2001.

APPENDIX A

Laplace Transform

A.1 DEFINITION

Let $x(t)$ be a signal, i.e., a real-valued function defined on the real line $(-\infty, \infty)$. Its Laplace transform, denoted by $X(s)$, is a complex function and is defined in this book as

$$X(s) = \mathcal{L}[x(t)] = \int_{-\infty}^{\infty} x(t)e^{-st}dt. \tag{A.1}$$

This Laplace transform is sometimes called the **two-sided or bilateral Laplace transform** in some other books to emphasize that the integral in the definition is from $-\infty$ to ∞, and to distinguish it from the one-sided or unilateral Laplace transform, which is defined by the same integral but from 0 to ∞, even for two-sided or bilateral signals. We will not use the one-sided Laplace transform in this book.

Most of the time, we deal with one-sided or unilateral signals, i.e., signals $x(t)$ with $x(t) = 0$ for $t < 0$. If $x(t)$ is one-sided, then its Laplace transform is equal to

$$X(s) = \mathcal{L}[x(t)] = \int_{0^-}^{\infty} x(t)e^{-st}dt. \tag{A.2}$$

We wish to emphasize that in (A.2) a one-sided signal is considered a special two-sided signal whose value at the whole negative time axis is equal to zero and the two-sided Laplace transform is then applied. In other words, we shall use two-sided

Laplace transform for both two-sided and one-sided signals but will never use one-sided Laplace transform for either two-sided or one-sided signals, though most of the time the signals considered are actually one-sided signals.

Notice that the lower limit of the integral in (A.2) is 0^-, i.e., infinitesimally left of 0. If the signal $x(t)$ is not impulsive at time $t = 0$, whether the lower limit is 0^- or 0^+ is not critical, but if $x(t)$ is impulsive at time $t = 0$, for example $x(t) = \delta(t)$, it does make a big difference. Such technicality can be avoided if definition (A.1) is always used.

The integral on the right-hand side of (A.1) may not be meaningful for every complex number s. The set of s that makes the integral meaningful is called the **region of convergence** (ROC). In other words, (A.1) and (A.2) are valid only when s is in the ROC.

EXAMPLE A.1

The Laplace transform of the unit impulse signal $\delta(t)$ can be computed as

$$\Delta(s) = \mathcal{L}[\delta(t)] = \int_{0^-}^{0^+} \delta(t)e^{-st}dt = 1,$$

and that of the unit step signal $\sigma(t)$ can be computed as

$$\Sigma(s) = \mathcal{L}[\sigma(t)] = \int_{0^-}^{\infty} e^{-st}dt = \begin{cases} \dfrac{1}{s}, & \operatorname{Re} s > 0; \\ \text{undefined}, & \operatorname{Re} s \le 0. \end{cases}$$

In this case, it is more common to say

$$\Sigma(s) = \mathcal{L}[\sigma(t)] = \frac{1}{s}$$

with an ROC $\operatorname{Re} s > 0$.

For a one-sided signal, the ROC of its Laplace transform has the following three possibilities: the entire complex plane; a half plane to the right of a vertical line in the complex plane, i.e., a set like $\{s \in \mathbb{C} : \operatorname{Re} s > \rho\}$ for some real number ρ; or empty. In the third case, we simply say that the Laplace transform does not exist. For example, the ROC of $\Delta(s)$ is the entire complex plane, that of $\Sigma(s)$ is the half of the complex plane to the right of the imaginary axis, and that of the Laplace transform of $e^{t^2}\sigma(t)$ is empty.

The Laplace transforms, as well as their ROCs, of some commonly used one-sided functions are listed in Table A.1.

A.2 PROPERTIES

Some useful properties of the Laplace transform are listed in Table A.2.

Let us take a closer look at the derivative property and the integration property. Let us first prove why they are true. The proofs below are valid for two-sided

Function name	$x(t)$	$X(s)$	ROC
Unit impulse	$\delta(t)$	1	$\mathbb{C}$
Unit step	$\sigma(t)$	$\frac{1}{s}$	$\mathrm{Re}\, s > 0$
Unit ramp	$t\sigma(t)$	$\frac{1}{s^2}$	$\mathrm{Re}\, s > 0$
Unit acceleration	$\frac{t^2}{2}\sigma(t)$	$\frac{1}{s^3}$	$\mathrm{Re}\, s > 0$
nth-order ramp	$\frac{t^n}{n!}\sigma(t)$	$\frac{1}{s^{n+1}}$	$\mathrm{Re}\, s > 0$
Exponential	$e^{-\alpha t}\sigma(t)$	$\frac{1}{s+\alpha}$	$\mathrm{Re}\, s > -\alpha$
nth-order exponential	$\frac{t^n}{n!}e^{-\alpha t}\sigma(t)$	$\frac{1}{(s+\alpha)^{n+1}}$	$\mathrm{Re}\, s > -\alpha$
Sine	$(\sin \omega t)\sigma(t)$	$\frac{\omega}{s^2+\omega^2}$	$\mathrm{Re}\, s > 0$
Cosine	$(\cos \omega t)\sigma(t)$	$\frac{s}{s^2+\omega^2}$	$\mathrm{Re}\, s > 0$
Damped sine	$(e^{-\alpha t}\sin \omega t)\sigma(t)$	$\frac{\omega}{(s+\alpha)^2+\omega^2}$	$\mathrm{Re}\, s > -\alpha$
Damped cosine	$(e^{-\alpha t}\cos \omega t)\sigma(t)$	$\frac{s+\alpha}{(s+\alpha)^2+\omega^2}$	$\mathrm{Re}\, s > -\alpha$

TABLE A.1: Laplace transform table.

signals and hence also for one-sided signals. To prove the derivative property, we use integration by parts:

$$\mathcal{L}[\dot{x}(t)] = \int_{-\infty}^{\infty} \dot{x}(t)e^{-st}dt = \int_{-\infty}^{\infty} e^{-st}dx(t) = e^{-st}x(t)\Big|_{-\infty}^{\infty} - \int_{-\infty}^{\infty} x(t)de^{-st}.$$

The first term vanishes when s is in the ROC of $\mathcal{L}[x(t)]$. Hence

$$\mathcal{L}[\dot{x}(t)] = s\int_{-\infty}^{\infty} x(t)e^{-st}dt = s\mathcal{L}[x(t)].$$

Number	Properties	Comments
1	$\mathcal{L}[x(t)] = X(s)$, $\mathcal{L}[y(t)] = Y(s)$	Notation
2	$\mathcal{L}[\alpha x(t) + \beta y(t)] = \alpha X(s) + \beta Y(s)$	Linearity
3	$\mathcal{L}[e^{-\alpha t}x(t)] = X(s+\alpha)$	Frequency shift
4	$\mathcal{L}[x(t-T)] = e^{-Ts}X(s)$	Time delay
5	$\mathcal{L}[x(\alpha t)] = \frac{1}{\alpha}X\left(\frac{s}{\alpha}\right)$	Scaling
6	$\mathcal{L}[\dot{x}(t)] = sX(s)$	Derivative
7	$\mathcal{L}\left[x^{(n)}(t)\right] = s^n X(s)$	High-order derivative
8	$\mathcal{L}\left[\int_{-\infty}^{t} x(\tau)d\tau\right] = \frac{X(s)}{s}$	Integration
9	$\mathcal{L}[x(t) * y(t)] = X(s)Y(s)$	Convolution

TABLE A.2: Properties of Laplace transform.

To prove the integration property, we let $y(t) = \int_{-\infty}^{t} x(\tau)d\tau$. Then

$$\mathcal{L}[y(t)] = \int_{-\infty}^{\infty} y(t)e^{-st}dt = -\frac{1}{s}\int_{-\infty}^{\infty} y(t)de^{-st}.$$

Using integration by parts, we see

$$\mathcal{L}[y(t)] = -\frac{1}{s}y(t)e^{-st}\Big|_{-\infty}^{\infty} + \frac{1}{s}\int_{-\infty}^{\infty} e^{-st}dy(t).$$

The first term vanishes when s is in the ROC of $\mathcal{L}[y(t)]$. Hence

$$\mathcal{L}[y(t)] = \frac{1}{s}\int_{-\infty}^{\infty} x(t)e^{-st}dt = \frac{1}{s}X(s).$$

EXAMPLE A.2

When the derivative of a one-sided signal is taken, it is very easy to result in an impulse at time 0, which is easily overlooked. For example, since $\dot{\sigma}(t) = \delta(t)$, it follows that

$$s\Sigma(s) = s\frac{1}{s} = 1 = \Delta(s).$$

For another example, let $x(t) = (\cos\omega t)\sigma(t)$, then $\dot{x}(t) = \delta(t) - \omega(\sin\omega t)\sigma(t) \neq \omega(\sin\omega t)\sigma(t)$. Therefore

$$s\mathcal{L}[x(t)] = \frac{s^2}{s^2+\omega^2} = 1 - \frac{\omega^2}{s^2+\omega^2} = \mathcal{L}[\delta(t) - \omega(\sin\omega t)\sigma(t)].$$

By using the properties of the Laplace transform listed in Table A.2, together with the Laplace transforms of simple functions listed in Table A.1, we can obtain the Laplace transforms of a wide class of functions, obtainable from the simple ones by scaling, addition, differentiation, integration, time shifting, etc.

In the application of the Laplace transform, the following two theorems are of great importance.

Theorem A.3 (Initial Value Theorem). Let $X(s) = \mathcal{L}[x(t)]$. Then

$$\lim_{t\to 0^+} x(t) = \lim_{s\to\infty} sX(s) \tag{A.3}$$

in the case when the two limits exist and are finite.

When $X(s)$ is a strictly proper rational function, $x(t)$ does not have an impulsive mode. In this case, the two limits in (A.3) exist, are finite, and are hence equal.

Theorem A.4 (Final Value Theorem). Let $X(s) = \mathcal{L}[x(t)]$. Then

$$\lim_{t\to\infty} x(t) = \lim_{s\to 0} sX(s) \tag{A.4}$$

in the case when the two limits exist and are finite.

When $X(s)$ is a rational function with poles located in the open left complex plane except possibly a simple one at the origin, $x(t)$ has a finite steady-state value. In this case, the two limits in (A.4) exist, are finite, and are hence equal.

EXAMPLE A.5

The impulse function $\delta(t)$ is often a trouble maker. Since $\Delta(s) = 1$ and $\lim_{s\to\infty} s\Delta(s) = \infty$, the initial value theorem does not apply. On the other hand, the final value theorem applies since $\lim_{t\to\infty}\delta(t) = 0$ and $\lim_{s\to 0} s\Delta(s) = 0$.

For sinusoidal functions, on the other hand, the final value theorem does not apply but the initial value theorem applies. To check, we see $\lim_{t\to\infty}(\sin\omega t)\sigma(t)$ does not exist, but $\lim_{s\to 0} s\mathcal{L}[(\sin\omega t)\sigma(t)] = 0$, whereas both $\lim_{t\to 0}(\sin\omega t)\sigma(t)$ and $\lim_{s\to\infty} s\mathcal{L}[(\sin\omega t)\sigma(t)]$ are zero.

A.3 INVERSE LAPLACE TRANSFORM

Quite often we also need to find the signal $x(t)$ from its Laplace transform $X(s)$. In this case, we say that $x(t)$ is the inverse Laplace transform of $X(s)$. Let $\{\rho + j\omega : \omega \in \mathbb{R}\}$

be a vertical line in the ROC of $X(s)$. Then the formula for the inverse Laplace transform is

$$x(t) = \mathcal{L}^{-1}[X(s)] = \frac{1}{2\pi j}\int_{\rho-j\infty}^{\rho-j\infty} X(s)e^{st}ds. \tag{A.5}$$

However, this formula is almost useless in actual computation of the inverse Laplace transform.

EXAMPLE A.6

Since the ROC of $\Delta(s)$ is the entire complex plane, it follows that for any $\rho \in \mathbb{R}$,

$$\mathcal{L}^{-1}[\Delta(s)] = \frac{1}{2\pi j}\int_{\rho-j\infty}^{\rho+j\infty} e^{st}ds.$$

In particular, we can choose $\rho = 0$. Then

$$\mathcal{L}^{-1}[\Delta(s)] = \frac{1}{2\pi}\int_{-\infty}^{\infty} e^{j\omega t}d\omega.$$

The integral on the right does not exist in the usual sense. How to make sense of the integral goes much beyond the scope of this book, though we know that $\mathcal{L}^{-1}[\Delta(s)] = \delta(t)$.

Also, since the ROC of $\Sigma(s)$ is $\mathbb{C}^+ = \{s : \operatorname{Re} s > 0\}$, it follows that for any $\rho > 0$,

$$\mathcal{L}^{-1}[\Sigma(s)] = \frac{1}{2\pi j}\int_{\rho-j\infty}^{\rho+j\infty} \frac{1}{s}e^{st}ds = \frac{1}{2\pi}\int_{-\infty}^{\infty} \frac{1}{\rho + j\omega}e^{\rho t}e^{j\omega t}d\omega.$$

How to find the integral on the right is not an easy issue.

It is seen that even for simple functions like $\Delta(s)$ and $\Sigma(s)$, finding their inverse Laplace transforms using integral (A.5) is almost impossible. We usually do not resort to integral (A.5) to find the inverse Laplace transform of a rational function. The common ways to find the inverse Laplace transform of a rational function is via the partial fractional expansion, together with Tables A.1 and A.2.

Let $G(s)$ be a strictly proper rational function

$$G(s) = \frac{b(s)}{a(s)} = \frac{b_1 s^{n-1} + b_2 s^{n-2} + \cdots b_n}{a_0 s^n + a_1 s^{n-1} + \cdots a_n}$$

where $a_0 \neq 0$. Let the denominator polynomial $a(s)$ have roots $p_1, p_2, \ldots, p_n$, i.e.,

$$a(s) = a_0(s - p_1)(s - p_2)\cdots(s - p_n).$$

Let us also assume that $p_1, p_2, \ldots, p_n$ are all distinct. We allow them to be complex, but we assume that the original coefficients $a_0, \ldots, a_n$ and $b_1, \ldots, b_n$ are all real. Hence, the complex poles among $p_1, p_2, \ldots, p_n$ appear in conjugate pairs. Therefore, $G(s)$ can be expanded as

$$G(s) = \frac{c_1}{s - p_1} + \frac{c_2}{s - p_2} + \cdots + \frac{c_n}{s - p_n} \tag{A.6}$$

where c_i can be obtained as

$$c_i = \lim_{s \to p_i} (s - p_i)G(s),$$

or by solving polynomial equation

$$b(s) = a(s)\left[\frac{c_1}{s-p_1} + \frac{c_2}{s-p_2} + \cdots + \frac{c_n}{s-p_n}\right]. \tag{A.7}$$

Notice that the right-hand side of (A.7) is indeed a polynomial because of cancellations. The standard way of solving the polynomial equation (A.7) is to convert it to a set of n linear equations by comparing the coefficients of both sides of the polynomial equation.

In the case when p_i and p_j are complex conjugates to each other, then c_i and c_j are also complex conjugates to each other.

After the partial fractional expansion is obtained, the inverse Laplace transform can be obtained by using Table A.1 as

$$\mathcal{L}^{-1}[G(s)] = (c_1 e^{p_1 t} + c_2 e^{p_2 t} + \cdots + c_n e^{p_n t})\sigma(t).$$

In this expression, some p_i and c_i might be complex, which might be undesirable. Since the complex parameters always appear in conjugate pairs, the imaginary part of the above expression is completely cancelled and the time function on the right-hand side is a real-valued function. More specifically, if $p_j = \bar{p}_i$ and $c_j = \bar{c}_i$, then the sum of the corresponding two terms can be rewritten as

$$c_i e^{p_i t} + c_j e^{p_j t} = 2e^{\mathrm{Re}\, p_i t}\left[(\mathrm{Re}\, c_i)\cos(\mathrm{Im}\, p_i t) - (\mathrm{Im}\, c_i)\sin(\mathrm{Im}\, p_i t)\right].$$

EXAMPLE A.7

Let

$$G(s) = \frac{2}{s(s^2 + 2s + 2)}.$$

We wish to find its inverse Laplace transform. First notice

$$G(s) = \frac{2}{s(s+1-j)(s+1+j)}.$$

Then

$$G(s) = \frac{c_1}{s} + \frac{c_2}{s+1-j} + \frac{c_3}{s+1+j}$$

where

$$c_1 = \lim_{s \to 0} sG(s) = 1$$

$$c_2 = \lim_{s \to -1+j} (s+1-j)G(s) = \frac{1}{j(-1+j)}$$

$$c_3 = \lim_{s \to -1-j} (s+1+j)G(s) = \frac{1}{-j(-1-j)}.$$

The determination of c_3 can also be simply done as $c_3 = \bar{c}_2$.

Another way to obtain c_i is by comparing the coefficients of

$$2 = s(s+1-j)(s+1+j)\left(\frac{c_1}{s} + \frac{c_2}{s+1-j} + \frac{c_3}{s+1+j},\right),$$

which gives

$$2 = c_1(s+1-j)(s+1+j) + c_2 s(s+1+j) + c_3 s(s+1-j).$$

Comparing the coefficients, we get

$$\begin{aligned} c_1 + c_2 + c_3 &= 0 \\ 2c_1 + (1+j)c_2 + (1-j)c_3 &= 0 \\ 2c_1 &= 2. \end{aligned}$$

This gives $c_1 = 1, c_2 = -\frac{1}{2} + \frac{1}{2}j, c_3 = -\frac{1}{2} - \frac{1}{2}j$, which are the same as what we obtained using the first method, though in different forms. Therefore

$$\begin{aligned} \mathcal{L}^{-1}[G(s)] &= \left[1 + \left(-\frac{1}{2} + \frac{1}{2}j\right)e^{(-1+j)t} + \left(-\frac{1}{2} - \frac{1}{2}j\right)e^{(-1-j)t}\right]\sigma(t) \\ &= \left[1 + 2e^{-t}\left(-\frac{1}{2}\cos t - \frac{1}{2}\sin t\right)\right]\sigma(t) \\ &= \left[1 - e^{-t}(\cos t + \sin t)\right]\sigma(t). \end{aligned}$$

If $G(s)$ has multiple poles, the situation is a bit more complicated. In this case, the denominator $a(s)$ can be rewritten as

$$a(s) = a_0(s-p_1)^{n_1}(s-p_2)^{n_2}\cdots(s-p_r)^{n_r}.$$

where $p_1, p_2, \ldots, p_r$ are all distinct and $n_1 + n_2 + \cdots + n_r = n$. Then the partial fractional expansion is of the form

$$\begin{aligned} G(s) = &\frac{c_{11}}{s-p_1} + \frac{c_{12}}{(s-p_1)^2} + \cdots + \frac{c_{1n_1}}{(s-p_1)^{n_1}} \\ &+ \frac{c_{21}}{s-p_2} + \frac{c_{22}}{(s-p_2)^2} + \cdots + \frac{c_{2n_2}}{(s-p_2)^{n_2}} \\ &+ \cdots + \frac{c_{r1}}{s-p_r} + \frac{c_{r2}}{(s-p_r)^2} + \cdots + \frac{c_{rn_r}}{(s-p_r)^{n_r}}. \end{aligned}$$

where

$$\begin{aligned} c_{i1} &= \frac{1}{(n_i-1)!}\lim_{s\to p_i}\frac{d^{n_i-1}}{ds^{n_i-1}}[(s-p_i)^{n_i}G(s)] \\ c_{i2} &= \frac{1}{(n_i-2)!}\lim_{s\to p_i}\frac{d^{n_i-2}}{ds^{n_i-2}}[(s-p_i)^{n_i}G(s)] \\ &\ \vdots \\ c_{in_i} &= \lim_{s\to p_i}(s-p_i)^{n_i}G(s) \end{aligned}$$

If one finds these formulas hard to remember, one can also obtain the coefficients by solving the polynomial equation

$$b(s) = a(s)\left[\frac{c_{11}}{s-p_1} + \cdots + \frac{c_{1n_1}}{(s-p_1)^{n_1}} + \cdots + \frac{c_{r1}}{s-p_r} + \cdots + \frac{c_{rn_r}}{(s-p_r)^{n_r}}\right].$$

Notice that the right-hand side of the above equation is indeed a polynomial since all the denominators are cancelled by $a(s)$. The standard way to solve such a polynomial equation is to convert it to a set of linear equations by comparing the coefficients.

EXAMPLE A.8

Let

$$G(s) = \frac{1}{(s+1)^2 s^3}$$

We wish to find its inverse Laplace transform. First let us expand $G(s)$ as

$$G(s) = \frac{c_{11}}{s+1} + \frac{c_{12}}{(s+1)^2} + \frac{c_{21}}{s} + \frac{c_{22}}{s^2} + \frac{c_{23}}{s^3}$$

where

$$\begin{aligned}
c_{11} &= \frac{1}{1!}\lim_{s\to -1}\frac{d}{ds}[(s+1)^2 G(s)] = -3 \\
c_{12} &= \lim_{s\to -1}(s+1)^2 G(s) = -1 \\
c_{21} &= \frac{1}{2!}\lim_{s\to 0}\frac{d^2}{ds^2}s^3 G(s) = 3 \\
c_{22} &= \frac{1}{1!}\lim_{s\to 0}\frac{d}{ds}s^3 G(s) = -2 \\
c_{23} &= \lim_{s\to 0} s^3 G(s) = 1.
\end{aligned}$$

Another way to find these coefficients is by solving the polynomial equation

$$1 = c_{11}(s+1)s^3 + c_{12}s^3 + c_{21}(s+1)^2 s^2 + c_{22}(s+1)^2 s + c_{23}(s+1)^2.$$

By comparing the coefficients, we obtain a set of linear equations

$$\begin{aligned}
c_{11} + c_{21} &= 0 \\
c_{11} + c_{12} + 2c_{21} + c_{22} &= 0 \\
c_{21} + 2c_{22} + c_{23} &= 0 \\
c_{23} &= 1.
\end{aligned}$$

The solution is

$$c_{11} = -3, \quad c_{12} = -1, \quad c_{21} = 3, \quad c_{22} = -2, \quad c_{23} = 1.$$

Both methods give the same answer. The partial fractional expansion of $G(s)$ is

$$G(s) = \frac{-3}{s+1} + \frac{-1}{(s+1)^2} + \frac{3}{s} + \frac{-2}{s^2} + \frac{1}{s^3}.$$

Finally, the inverse Laplace transform can then be obtained from Table A.1 and the linearity property:

$$\mathcal{L}^{-1}[G(s)] = \left[-(3+t)e^{-t} + 3 - 2t + \frac{1}{2}t^2\right]\sigma(t).$$

PROBLEMS

A.1. Derive $\mathcal{L}[(e^{-\alpha t}\cos\omega t)\sigma(t)]$.

A.2. Derive $\mathcal{L}[(e^{-\alpha t}\sin\omega t)\sigma(t)]$ using the derivative property and the result in Problem A.1

A.3. Prove Properties 1–5 as listed in Table A.2.

NOTES AND REFERENCES

The best reference for Laplace transform and other background materials on signals and systems is

A. V. Oppenheim and A. S. Willsky with S. H. Newab, *Signals & Systems*, 2nd Edition, Prentice Hall, Upper Saddle River, New Jersey, 1997.

Another good reference is

H. Kwakernaak and R. Sivan, *Modern Signals and Systems*, Prentice Hall, Englewood Cliffs, New Jersey, 1991.

In these books, both two-sided and one-sided Laplace transforms are introduced. Both books emphasize the two-sided Laplace transform, showing that the two-sided Laplace transform is more convenient in signal and system analysis. In history, however, the one-sided Laplace transform appeared first in mathematical literature and was first used in solving linear differential equations. The one-sided Laplace transform, usually defined as

$$\mathcal{L}_{+}[x(t)] = \int_0^\infty x(t)e^{st}dt,$$

for a signal $x(t)$ defined on the positive real axis $[0,\infty)$, has difficulty in dealing with singularity or even discontinuity at time 0. There is a confusion on whether the lower limit 0 of the integral should be 0^- or 0^+. There is also a confusion on whether functions that are discontinuous or singular at time 0 should be allowed. The differentiation property for the one-sided Laplace transform takes a more complicated form.

An interesting article rectifying the long-lasting and widespread confusions surrounding the one-sided Laplace transform is

> K. H. Lundberg, H. R. Miller, and D. L. Trumper, "Initial conditions, generalized functions, and the Laplace transform: Trouble at the origin," *IEEE Control Systems Magazine*, vol. 27, pp. 22–35, 2007.

APPENDIX B

Matrices and Polynomials

B.1 MATRICES

We assume that the readers are familiar with basic matrix operations. One needs to pay special attention to the compatibility of the matrices involved in various operations. For example, the matrix sum $\boldsymbol{A}+\boldsymbol{B}$ requires matrices $\boldsymbol{A}$ and $\boldsymbol{B}$ to have the same size; the matrix multiplication $\boldsymbol{AB}$ requires the width (number of columns) of $\boldsymbol{A}$ to be the same as the height (number of rows) of $\boldsymbol{B}$, and the determinant $\det \boldsymbol{A}$ of matrix $\boldsymbol{A}$ requires $\boldsymbol{A}$ to be square.

Let us now consider an $n \times n$ square matrix

$$\boldsymbol{A} = \begin{bmatrix} a_{11} & a_{12} & \cdots & a_{1n} \\ a_{21} & a_{22} & \cdots & a_{2n} \\ \vdots & \vdots & & \vdots \\ a_{n1} & a_{n2} & \cdots & a_{nn} \end{bmatrix}.$$

This matrix is said to be **nonsingular** if its determinant is nonzero, i.e.,

$$\det \boldsymbol{A} \neq 0.$$

The matrix is said to be **invertible** if there is an $n \times n$ matrix $\boldsymbol{B}$ such that

$$\boldsymbol{AB} = I$$

or

$$\boldsymbol{BA} = I.$$

Such a matrix $\boldsymbol{B}$ is called the **inverse** of $\boldsymbol{A}$ and is denoted by $\boldsymbol{A}^{-1}$. The inverse of $\boldsymbol{A}$ can be obtained from $\boldsymbol{A}$ using the following formula:

$$\boldsymbol{A}^{-1} = \frac{\mathrm{adj}\boldsymbol{A}}{\det \boldsymbol{A}}$$

where adj$\boldsymbol{A}$ is the **adjugate** of the matrix $\boldsymbol{A}$ defined as

$$\mathrm{adj}\boldsymbol{A} = \begin{bmatrix} A_{11} & A_{21} & \cdots & A_{n1} \\ A_{12} & A_{22} & \cdots & A_{n2} \\ \vdots & \vdots & & \vdots \\ A_{1n} & A_{2n} & \cdots & A_{nn} \end{bmatrix}$$

where A_{ij} is the so-called **algebraic cofactor** of a_{ij}:

$$A_{ij} = (-1)^{i+j} \det \begin{bmatrix} a_{11} & \cdots & a_{1j} & \cdots & a_{1n} \\ \vdots & \vdots & \vdots & & \vdots \\ a_{i1} & \cdots & a_{ij} & \cdots & a_{in} \\ \vdots & \vdots & \vdots & & \vdots \\ a_{n1} & \cdots & a_{nj} & \cdots & a_{nn} \end{bmatrix}.$$

Here the matrix on the right is an $(n-1) \times (n-1)$ matrix obtained by deleting the ith row and jth column. We see that a matrix is invertible if and only if it is nonsingular.

For an $n \times n$ matrix $\boldsymbol{A}$, the polynomial

$$\det(s\boldsymbol{I} - \boldsymbol{A})$$

is called the **characteristic polynomial** of $\boldsymbol{A}$ and is denoted by $\chi_{\boldsymbol{A}}(s)$. The roots of $\chi_{\boldsymbol{A}}(s)$ are called the **eigenvalues** of $\boldsymbol{A}$. Let λ be an eigenvalue of $\boldsymbol{A}$. Then $\lambda\boldsymbol{I} - \boldsymbol{A}$ is singular and there must exist a nonzero vector $\boldsymbol{x}$ such that $(\lambda\boldsymbol{I} - \boldsymbol{A})\boldsymbol{x} = 0$ or, equivalently, $\boldsymbol{A}\boldsymbol{x} = \lambda\boldsymbol{x}$. Such a vector is called an **eigenvector** of $\boldsymbol{A}$ corresponding to eigenvalue λ.

We often need to evaluate a polynomial

$$p(s) = p_0 s^m + p_1 s^{m-1} + \cdots + p_{m-1} s + p_m$$

at a square matrix $\boldsymbol{A}$. This can be simply done as

$$p(\boldsymbol{A}) = p_0 \boldsymbol{A}^m + p_1 \boldsymbol{A}^{m-1} + \cdots + p_{m-1}\boldsymbol{A} + p_m \boldsymbol{I}.$$

A very important theorem in linear algebra is the Cayley–Hamilton theorem.

Theorem B.1 (Cayley–Hamilton theorem).

$$\chi_{\boldsymbol{A}}(\boldsymbol{A}) = 0.$$

MATLAB provides a host of commands for matrix computation. For example, the meanings of

```
>> A + B
```

```
>> A * B
```

```
>> det(A)
>> inv(A)
>> eig(A)
```

are self-explanatory. MATLAB even has a command for the evaluation of a polynomial $p(s)$ at a square matrix $\boldsymbol{A}$

```
>> polyvalm(p,A)
```

where p is a vector containing the coefficients of the polynomial $p(s)$.

B.2 POLYNOMIALS

In feedback control theory, we have to deal with a large number of polynomials. A polynomial is a function of the form

$$a(s) = a_0 s^n + a_1 s^{n-1} + \dots a_n.$$

Here we assume that $a_0 \neq 0$ so $\deg a(s) = n$.

The **companion matrix** of polynomial $a(s)$ is defined as

$$\begin{bmatrix} -\frac{a_1}{a_0} & 1 & \cdots & 0 \\ \vdots & \vdots & \ddots & \vdots \\ -\frac{a_{n-1}}{a_0} & 0 & \cdots & 1 \\ -\frac{a_n}{a_0} & 0 & \cdots & 0 \end{bmatrix}$$

and is denoted by $\mathbf{\Gamma}_a$. One can easily check that the characteristic polynomial of $\mathbf{\Gamma}_a$ is $a(s)/a_0$, i.e., $\chi_{\mathbf{\Gamma}_a}(s) = a(s)/a_0$. An immediate consequence of this is that the roots of $a(s)$ are the eigenvalues of $\mathbf{\Gamma}_a$. It is probably surprising that although theoretically the eigenvalues of a matrix are defined as the roots of its characteristic polynomial, things are turned around in numerical computation. The numerical computation of the roots of a polynomial in most of the standard scientific computation software, such as MATLAB, is carried out by computing the eigenvalues of the companion matrix of the polynomial.

Let $b(s)$ be another polynomial,

$$b(s) = b_0 s^m + b_1 s^{m-1} + \dots + b_m.$$

Here we do not assume $b_0 \neq 0$ so $\deg b(s) \leq m$. We then say that $a(s)$ divides $b(s)$, or $b(s)$ can be divided by $a(s)$, if $a(s)$ is a factor of $b(s)$, i.e., there exists another polynomial $q(s)$ such that $b(s) = a(s)q(s)$. Equivalently, $a(s)$ divides $b(s)$ if all roots of $a(s)$ are also the roots of $b(s)$.

Theorem B.2. $a(s)$ divides $b(s)$ if and only if $b(\mathbf{\Gamma}_a) = 0$.

Polynomials $a(s)$ and $b(s)$ are said to be **coprime** if they do not have common roots. There are many ways to test if $a(s)$ and $b(s)$ are coprime from their coefficients.

Define an $(n+m) \times m$ matrix from $a(s)$

$$
\boldsymbol{T}(a(s), m) = \overbrace{\begin{bmatrix} a_0 & 0 & \cdots & 0 \\ a_1 & a_0 & \ddots & \vdots \\ \vdots & a_1 & \ddots & 0 \\ a_n & \vdots & \ddots & a_0 \\ 0 & a_n & \ddots & a_1 \\ \vdots & \vdots & \ddots & \vdots \\ 0 & 0 & \cdots & a_n \end{bmatrix}}^{m \text{ columns}}
$$

and an $(n+m) \times n$ matrix from $b(s)$

$$
\boldsymbol{T}(b(s), n) = \overbrace{\begin{bmatrix} b_0 & 0 & \cdots & 0 \\ b_1 & b_0 & \ddots & \vdots \\ \vdots & b_1 & \ddots & 0 \\ b_m & \vdots & \ddots & b_0 \\ 0 & b_m & \ddots & b_1 \\ \vdots & \vdots & \ddots & \vdots \\ 0 & 0 & \cdots & b_m \end{bmatrix}}^{n \text{ columns}}.
$$

These matrices have the property that elements in each of their diagonals are equal. Such matrices are called **Toeplitz matrices**.

Further, define

$$
\boldsymbol{S}(a(s), b(s)) = \begin{bmatrix} \boldsymbol{T}(a(s), m) & \boldsymbol{T}(b(s), n) \end{bmatrix}.
$$

This matrix is called a **Sylvester's resultant matrix**. Notice that $\boldsymbol{S}(a(s), b(s))$ is an $(n+m) \times (n+m)$ matrix.

Theorem B.3. The following statements are equivalent:

1. $a(s)$ and $b(s)$ are coprime.
2. $\det \boldsymbol{S}(a(s), b(s)) \neq 0$.
3. $\det b(\boldsymbol{\Gamma}_a) \neq 0$.
4. There exist unique polynomials $x(s)$ and $y(s)$ with $\deg x(s) < m$ and $\deg y(x) < n$ such that

$$
a(s)x(s) + b(s)y(s) = 1. \tag{B.1}
$$

The determinant in Theorem B.3.2 is known as the **Sylvester resultant**, and that in Theorem B.3.3 is known as the **MacDuffee's resultant**. Equation (B.1) in Theorem B.3 is called a **Bezout equation**.

One way to solve a Bezout equation is to turn it into a set of $n+m$ linear equations with $n+m$ unknowns by comparing the coefficients. Let

$$x(s) = x_1 s^{m-1} + \cdots + x_m$$

$$y(s) = y_1 s^{n-1} + \cdots + y_n.$$

Then the linear equation is exactly

$$\boldsymbol{S}(a(s), b(s)) \begin{bmatrix} x_1 \\ \vdots \\ x_m \\ y_1 \\ \vdots \\ y_n \end{bmatrix} = \begin{bmatrix} 0 \\ \vdots \\ 0 \\ 1 \end{bmatrix}.$$

It is now easy to see why statement 2 in Theorem B.3 is equivalent to statement 4.

EXAMPLE B.4

Let us first consider the polynomials

$$a(s) = s^2 - 2s, \quad b(s) = s - 1.$$

They are apparently coprime. We have

$$\boldsymbol{S}(a(s), b(s)) = \begin{bmatrix} 1 & 1 & 0 \\ -2 & -1 & 1 \\ 0 & 0 & -1 \end{bmatrix}$$

and

$$\boldsymbol{\Gamma}_a = \begin{bmatrix} 2 & 1 \\ 0 & 0 \end{bmatrix}.$$

Clearly,

$$\det \boldsymbol{S}(a(s), b(s)) = -1$$

and

$$\det b(\boldsymbol{\Gamma}_a) = \det \begin{bmatrix} 1 & 1 \\ 0 & -1 \end{bmatrix} = -1$$

which are nonzero. The unique solution to the Bezout equation (B.1) is given by

$$x(s) = -1, \quad y(s) = s - 1.$$

Let us next consider the polynomials

$$a(s) = s^2 - s, \quad b(s) = s - 1.$$

They are apparently not coprime. We have

$$\boldsymbol{S}(a(s), b(s)) = \begin{bmatrix} 1 & 1 & 0 \\ -1 & -1 & 1 \\ 0 & 0 & -1 \end{bmatrix}$$

and

$$\boldsymbol{\Gamma}_a = \begin{bmatrix} 1 & 1 \\ 0 & 0 \end{bmatrix}.$$

Clearly,

$$\det \boldsymbol{S}(a(s), b(s)) = 0$$

and

$$\det b(\boldsymbol{\Gamma}_a) = \det \begin{bmatrix} 0 & 1 \\ 0 & -1 \end{bmatrix} = 0.$$

The Bezout equation (B.1) does not have a solution since no matter what $x(s)$ and $y(s)$ are, the left-hand side has a factor $s - 1$, whereas the right-hand side does not have any nontrivial factor.

In several places in this book we need to deal with the limiting roots of

$$c(s) = a(s) + Kb(s)$$

as K goes to ∞ or 0. Here we assume that $\deg a(s) = n$, $\deg b(s) = m$, and $m < n$.

First notice that the degree of $c(s)$ is n, so it has n roots. When $K \to 0$, then $c(s) \to a(s)$. Because of continuity, the roots of $c(s)$ converge to the roots of $a(s)$. When K goes to infinity, then $c(s)$ is dominated by $Kb(s)$. It is sensible to believe that the roots of $c(s)$ converge to the roots of $b(s)$. This is partially correct. The trouble is that the number of roots of $b(s)$ is m while that of $c(s)$ is n. What happens to the $n - m$ extra roots of $c(s)$? It is probably easier to investigate the roots of

$$\frac{1}{K}c(s) = \frac{1}{K}a(s) + b(s).$$

They are the same as those of $c(s)$. When K is very big, the roots of normal magnitudes have to be close to the roots of $b(s)$. The rest of the $n - m$ roots then have to be very big. When these roots are plugged into the polynomials $a(s)$ and $b(s)$, the polynomials are dominated by their leading terms

$$a(s) \approx a_0 s^n + a_1 s^{n-1}$$
$$b(s) \approx b_0 s^m + b_1 s^{m-1}.$$

Consequently,

$$c(s) \approx a_0 s^n + a_1 s^{n-1} + Kb_0 s^m + b_1 s^{m-1}.$$

Hence the big roots of $c(s)$ approach the following $n - m$ numbers:

$$\frac{b_1}{b_0} - \frac{a_1}{a_0} + \left(K\frac{b_0}{a_0}e^{(2k+1)\pi j}\right)^{1/(n-m)}, \qquad k = 0, 1, \ldots, n - m - 1,$$

as K approaches infinity. The locus of these numbers, as K changes from 0 to ∞, are called the asymptotes of the big roots of $c(s)$. Also notice

$$-\frac{a_1}{a_0} = \sum \text{ all roots of } a(s)$$

$$-\frac{b_1}{a_0} = \sum \text{ all roots of } b(s).$$

Hence the big roots of $c(s)$ converge to

$$\sum \text{ all roots of } a(s) - \sum \text{ all roots of } b(s) + \left(K\frac{b_0}{a_0}e^{(2k+1)\pi j}\right)^{1/(n-m)}$$

for $k = 0, 1, \ldots, n-m-1$.

PROBLEMS

B.1. Let

$$A = \begin{bmatrix} 1 & 2 & 3 \\ 4 & 5 & 6 \\ 7 & 8 & 9 \end{bmatrix}$$

Verify that the Cayley–Hamilton theorem holds for this matrix.

B.2. Without computing the roots, determine whether the two polynomials

$$a(s) = s^3 + 3s^2 + 3s + 1 \quad \text{and} \quad b(s) = s^2 + 3s + 2$$

are coprime.

B.3. A polynomial is said to be simple if it does not have repeated roots. Show that a polynomial $p(s)$ is simple if and only if $p(s)$ and $\dot{p}(s)$ are coprime. Use Theorem B.3 to derive conditions for $p(s)$ being simple.

EXTRA CREDIT PROBLEMS

B.4. Prove the Cayley–Hamilton theorem.

B.5. Complete the proof of Theorem B.3.

NOTES AND REFERENCES

There are many elementary textbooks on linear algebra, which deal with matrices. A particularly good one is

G. Strang, *Linear Algebra and Its Applications*, Third Edition, Harcourt Brace Jovanovich Publishers, San Diego, 1988.

A good book combining materials on matrices, polynomials, and even something on control theory is

P. A. Fuhrmann, *A Polynomial Approach to Linear Algebra*, Springer-Verlag, New York, 1996.

APPENDIX C

Answers to Selected Problems

C.1 CHAPTER 1

PROBLEMS

1.4. The derivatives are

$$\frac{d(\sin \omega t)\sigma(t)}{dt} = \omega(\cos \omega t)\sigma(t)$$

$$\frac{d(\cos \omega t)\sigma(t)}{dt} = \delta(t) - \omega(\sin \omega t)\sigma(t)$$

$$\frac{de^{\lambda t}\sigma(t)}{dt} = \delta(t) + \lambda e^{\lambda t}\sigma(t).$$

C.2 CHAPTER 2

PROBLEMS

2.1. The transfer function of the system is

$$G(s) = \frac{V_o(s)}{V_i(s)} = \frac{(L_1 s + R_2)C_1 s}{a(s)}$$

where

$$\begin{aligned} a(s) =& C_1C_2L_1(R_1+R_3)s^3 + [C_1C_2(R_1R_2+R_1R_3+R_2R_3)+(C_1+C_2)L_1]s^2 \\ &+ (C_1R_1+C_2R_2+C_1R_2+C_2R_3)s+1. \end{aligned}$$

2.3. **1.** The differential equation model is

$$\left(\frac{1}{12}M_b+\frac{1}{4}M_l\right)L\ddot{\theta}(t)=\frac{1}{2}[f(t)-M_lg]\cos\theta(t).$$

2. A state space description of the system with $u(t)=f(t)$, $x_1(t)=\theta(t)$, $x_2(t)=\dot{\theta}(t)$, and $y(t)=\theta(t)$ is

$$\begin{bmatrix}\dot{x}_1(t)\\ \dot{x}_2(t)\end{bmatrix}=\begin{bmatrix}x_2(t)\\ \dfrac{6[u(t)-M_lg]\cos x_1(t)}{(M_b+3M_l)L}\end{bmatrix}$$

$$y(t)=x_1(t).$$

3. The system is not linear. The operating point with $\theta_0=0$ is $(u_0,\boldsymbol{x}_0,y_0)=(M_lg,\mathbf{0},0)$. Let $\tilde{u}(t)=u(t)-M_lg$, $\tilde{\boldsymbol{x}}(t)=\boldsymbol{x}(t)$, and $\tilde{y}(t)=y(t)$. The linearized system around this operating point is:

$$\begin{bmatrix}\dot{\tilde{x}}_1(t)\\ \dot{\tilde{x}}_2(t)\end{bmatrix}=\begin{bmatrix}0&1\\0&0\end{bmatrix}\begin{bmatrix}\tilde{x}_1(t)\\ \tilde{x}_2(t)\end{bmatrix}+\begin{bmatrix}0\\ \dfrac{6}{(M_b+3M_l)L}\end{bmatrix}\tilde{u}(t),$$

$$\tilde{y}(t)=\begin{bmatrix}1&0\end{bmatrix}\begin{bmatrix}\tilde{x}_1(t)\\ \tilde{x}_2(t)\end{bmatrix}.$$

3. The transfer function is $G(s)=\dfrac{\tilde{Y}(s)}{\tilde{U}(s)}=\dfrac{6}{(M_bL+3M_lL)s^2}$.

2.6. **1.** Let $x(t)=h(t)$, $u(t)=f_i(t)$, $y(t)=x(t)$, then the state space model is

$$\dot{x}(t)=\frac{16}{\pi}\left(\frac{u(t)}{x(t)^2}-\frac{\sqrt{3x(t)}}{x(t)^2}\right),$$

$$y(t)=x(t).$$

2. The system is not linear. The operating point with $x_0=3$ is $(u_0,x_0,y_0)=(3,3,3)$. Let $\tilde{u}(t)=u(t)-u_0$, $\tilde{x}(t)=x(t)-x_0$ and $\tilde{y}(t)=y(t)-y_0$. The linearized system around the operating point is:

$$\dot{\tilde{x}}(t)=-\frac{8}{9\pi}\tilde{x}(t)+\frac{16}{9\pi}\tilde{u}(t),$$

$$\tilde{y}(t)=\tilde{x}(t).$$

3. The transfer function of the linearized system is $G(s) = \dfrac{\tilde{Y}(s)}{\tilde{U}(s)} = \dfrac{16}{9\pi s + 8}$.

2.9. The closed-loop transfer function from r to z is $G(s) = \dfrac{P(s)F(s) + P(s)C(s)}{1 + P(s)C(s)}$.

If $F(s) = \dfrac{1}{P(s)}$, then $G(s) = 1$.

2.10. 1. The transfer function from r to y is

$$G(s) = \frac{P(s)(H_1(s) + H_2(s)H_3(s))}{1 + P(s)(H_5(s) + H_3(s)H_4(s))}.$$

2. The equivalent 2DOF controller is

$$\boldsymbol{C}(s) = \begin{bmatrix} H_1(s) + H_3(s)H_2(s) & H_5(s) + H_3(s)H_4(s) \end{bmatrix}.$$

2.11. The transfer function from r to y is $G(s) = \dfrac{H(s) + P(s)}{1 + P(s)F(s)}$. If $H(s) = 1 - P(s) + P(s)F(s)$, then $G(s) = 1$.

2.12. The transfer function from r to y is

$$\frac{G_4(s) + G_3(s)G_2(s)G_1(s)}{1 + G_4(s) + G_2(s)G_1(s) + G_3(s)G_2(s) - G_4(s)G_2(s) + G_3(s)G_2(s)G_1(s)}.$$

MATLAB PROBLEMS

2.14. 1. `>> sys=ss(A,B,C,D)`

2. `>> systf=tf(sys)`

```
Transfer function:
```

$$\frac{0.002416s^3 - 1.634s^2 + 0.2485s + 43.98}{s^4 + 0.1741s^3 + 27.92s^2 + 0.02582s + 0.002138}$$

3. `>> syszpk=zpk(sys)`

```
Zero/pole/gain:
```

$$\frac{0.0024161(s - 676.1)(s - 5.286)(s + 5.094)}{(s^2 + 0.0009242s + 7.659e - 005)(s^2 + 0.1732s + 27.92)}$$

The DC gain is $43.98/0.002138 = 2.0571 \times 10^4$. Note that the DC gain is not the gain obtained from the command `zpk`.

4. The controller form is

$$\dot{x}(t) = \begin{bmatrix} -0.1741 & -27.92 & -0.02582 & -0.002138 \\ 1 & 0 & 0 & 0 \\ 0 & 1 & 0 & 0 \\ 0 & 0 & 1 & 0 \end{bmatrix} x(t) + \begin{bmatrix} 1 \\ 0 \\ 0 \\ 0 \end{bmatrix} u(t)$$

$$y(t) = \begin{bmatrix} 0.002416 & -1.634 & 0.2485 & 43.98 \end{bmatrix} x(t),$$

and the observer form is

$$\dot{x}(t) = \begin{bmatrix} -0.1741 & 1 & 0 & 0 \\ -27.92 & 0 & 1 & 0 \\ -0.02582 & 0 & 0 & 1 \\ -0.002138 & 0 & 0 & 0 \end{bmatrix} x(t) + \begin{bmatrix} 0.002416 \\ -1.634 \\ 0.2485 \\ 43.98 \end{bmatrix} u(t)$$

$$y(t) = \begin{bmatrix} 1 & 0 & 0 & 0 \end{bmatrix} x(t).$$

5. `>> sysss=ss(systf)`

$a =$

	$x1$	$x2$	$x3$	$x4$
$x1$	-0.1741	-6.98	-0.05163	-0.06843
$x2$	4	0	0	0
$x3$	0	0.125	0	0
$x4$	0	0	0.0625	0

$b =$

	$u1$
$x1$	2
$x2$	0
$x3$	0
$x4$	0

$c =$

	$x1$	$x2$	$x3$	$x4$
$y1$	$7.55e-0.05$	-0.01277	0.01553	43.98

$d =$

	$u1$
$y1$	0

It is not the same as the original state space equation and neither the same as one of state space models obtained in part 4.

C.3 CHAPTER 3

PROBLEMS

3.2. Select input $u(t) = (\sin \omega t^2)\sigma(t)$.

3.3. **1.** Stable; **2.** Unstable; **3.** Stable if and only if $K > \frac{\sqrt{129}-3}{12}$.

3.4. $K > 2$.

3.7. The two special polynomials are $a_3(s) = s^3 + \underline{a}_1 s^2 + \underline{a}_2 s + \bar{a}_3$ and $a_4(s) = s^3 + \bar{a}_1 s^2 + \underline{a}_2 s + \underline{a}_3$.

3.8. **1.** Yes; **2.** Yes; **3.** No; **4.** Yes; **5.** Yes.

3.9. **1.** No; **2.** Yes.

3.11. **1.** No; **2.** No; **3.** No.

3.12. **1.** No; **2.** No; **3.** Yes.

3.13. Yes.

3.15. $-\frac{1}{3} < K < 1.$

3.17. No.

3.18. $C(s) = \frac{6s+4}{s+4}.$

3.19. **1.** Impossible.

2. $C(s) = \frac{-22s+10}{s+29}.$

3.20. $C(s) = \frac{567s+567}{s^2+25s+207}.$

3.21. $C(s) = \frac{12s+12}{s+5}.$

3.22. $C(s) = \frac{74s+106}{s+16}.$

MATLAB PROBLEMS

3.24. **1.** Unstable; **2.** Unstable; **3.** Stable; **4.** Unstable.

3.26. **1.** Stable; **2.** Stable; **3.** Unstable; **4.** Stable; **5.** Stable.

C.4 CHAPTER 4

PROBLEMS

4.1. $5G(0)$.

4.2. $G(0)$.

4.3. The closed-loop transfer function is

$$G(s) = \frac{Y(s)}{R(s)} = \frac{K_1}{s^2 + (1 + K_1K_2)s + K_1}.$$

The desired gains are $K_1 = 100$ and $K_2 = 0.19$.

4.4. The estimated rise time t_r, settling time t_s, and the percentage overshoot PO from the reduced order system are

$$t_r = 0.6655,\ t_s = 3.3744,\ \text{PO} = 28.14\%.$$

The simulation of the full order system is shown in Figure C.1. The actual performances are

$$t_r \approx 0.7,\ t_s \approx 3.4,\ \text{PO} \approx 28\%.$$

4.5. **1.** For $T = 0.25$, the estimates are $t_r = 0.2986$, $t_s = 1.295$, PO $= 23.56\%$.

2. For $T = 0.025$, the estimates are $t_r = 0.4092$, $t_s = 5.4$, PO $= 57.91\%$.

4.6. $1.2 \le K < 6$.

4.9. **1.** $L(s) = \frac{Ks+b}{s^2+(a-K)s};$

2. If $a = K$, $L(s)$ is type 2, or $L(s)$ is type 1.

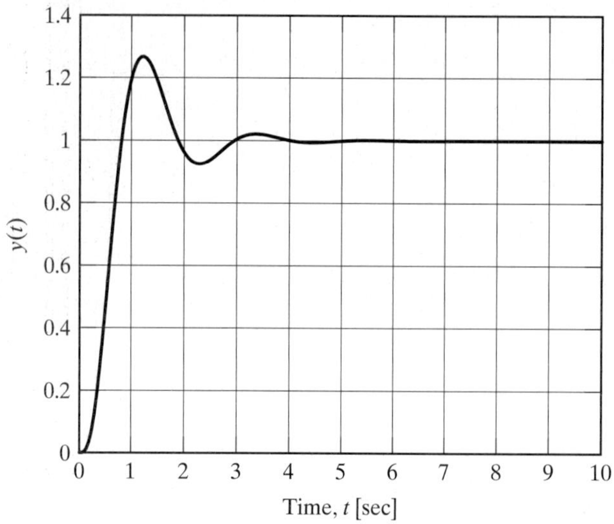

FIGURE C.1: Step response of $G(s)$.

3. If $a = K$, $K_p = \infty$, $K_v = \infty$ and $K_a = b$, or $K_p = \infty$, $K_v = \dfrac{b}{a-K}$ and $K_a = 0$.

4.11. **1.** Yes; **2.** Yes; **3.** Type 2, $K_p = \infty$, $K_v = \infty$, $K_a = 1$; **4.** 1/3.

4.12. **1.** $K = 10$; **2.** $K = 1$; **3.** $K = 2$; **4.** $K = 2$.

4.16. Its step response has a type B undershoot.

4.21. **1.** Impossible.

2. The 2DOF controller is $C(s) = \dfrac{9}{s+9}[s+3 \quad 3(s+1)]$.

3. No overshoot.

4.23. The required controller is $C(s) = \dfrac{36s^2 + 60s + 24}{-s^2 - 10s}$. The velocity error constant of the closed-loop system is $K_v = -2.4$.

4.24. $K = 1/2$.

4.27. **1.** $\sqrt{41}/(4\sqrt{2})$; **2.** $\sqrt{341}/32$; **3.** 2; **4.** $\sqrt{20}/\sqrt{27}$.

4.29. Let

$$\alpha_1 = \frac{a_0}{a_1}, \qquad \alpha_2 = \frac{a_1^2}{a_1a_2 - a_0a_3},$$

$$\alpha_3 = \frac{(a_1a_2 - a_0a_3)^2}{a_1^2a_2a_3 - a_0a_1a_3^2 - a_1^3a_4}, \qquad \alpha_4 = \frac{a_1a_2a_3 - a_0a_3^2 - a_1^2a_4}{a_1a_2a_4 - a_0a_3a_4}$$

$$\beta_1 = \frac{b_1}{a_1}, \qquad \beta_2 = \frac{a_1b_2}{a_1a_2 - a_0a_3},$$

$$\beta_3 = \frac{(a_1a_2 - a_0a_3)(a_1b_2 - a_3b_1)}{a_1^2a_2a_3 - a_0a_1a_3^2 - a_1^3a_4}, \qquad \beta_4 = \frac{a_1a_2b_4 - a_0a_3b_4 - a_1a_4b_2}{a_1a_2a_4 - a_0a_3a_4}.$$

Then

$$\left\| \frac{b_1 s^3 + b_2 s^2 + b_3 s + b_4}{a_0 s^4 + a_1 s^3 + a_2 s^2 + a_3 s + a_4} \right\|_2^2 = \frac{\beta_1^2}{2\alpha_1} + \frac{\beta_2^2}{2\alpha_2} + \frac{\beta_3^2}{2\alpha_3} + \frac{\beta_4^2}{2\alpha_4}.$$

C.5 CHAPTER 5

PROBLEMS

5.1. **2.** Draw a root locus for

$$\hat{L}(s) = \frac{10Ts}{s^3 + 2s^2 + 10}.$$

3. Draw a root locus for

$$\hat{L}(s) = \frac{a(s+10)(s+20)}{s^3 + 35s^2 + 200s - 5}.$$

5.2. All roots can be parameterized as

$$s_1 = -a,\ s_2 = -2 - \frac{2-a}{2} + j\frac{\sqrt{3}|a-2|}{2},\ s_3 = -2 - \frac{2-a}{2} - j\frac{\sqrt{3}|a-2|}{2}$$

with $K = a^3 - 6a^2 + 12a$ and $a \geq 0$.

5.3. The range of K for stability is $0 < K < 80$. The persistent oscillation occurs at $K = 80$ with frequency $\omega = 2$.

The roots of this problem can be parameterized as follows: 1. when $0 \leq K \leq 16 (0 \leq a \leq 2)$

$$s_1 = -a,\ s_2 = -4 + a,\ s_3 = -2 + j(2-a),\ s_4 = -2 - j(2-a),$$

$$K = a(4 - a)(8 - 4a + a^2)$$

2. when $K > 16$

$$s_{1,2} = -2 - b \pm jb,\ s_{3,4} = -2 + b \pm jb,\ K = (4 - 4b + 2b^2)(4 + 4b + 2b^2), b > 0.$$

5.4. $K = 1$ and the corresponding closed-loop poles are $-1 \pm j$.

5.6. **1.** A simple PD controller will do. For example, $C(s) = 2s + 1$.

5.7. There are infinitely many choices. For example, $a = 3, b = 0$, and $K = 2$.

5.8. A lead controller

$$C(s) = \frac{K(s+a)}{s+b}$$

must satisfy

$$\angle C(-1 + j\sqrt{3}) = 60^\circ$$

There are possibly many choices. For example, $K = 8, a = 4$, and $b = 1$.

5.9. $b < 0$ and $a > 0$.

5.10. The PI controller is given in the form of

$$C(s) = \frac{K(s+a)}{s}$$

So the velocity error constant K_v is given by

$$K_v = -Ka = -5.$$

Then

$$K = 1 + 2\sqrt{5}, \quad a = \frac{5}{1 + 2\sqrt{5}}$$

and the closed-loop poles are at $-\sqrt{5}, -\sqrt{5}$.

5.11.

$$C(s) = 6 + \frac{4}{s} + 4s.$$

5.12. A stable stabilizing controller is

$$C(s) = \frac{a(s+4)}{K(s+5)}$$

for any $5 < a < 6$.

C.6 CHAPTER 6

PROBLEMS

6.1. The steady-state response is approximately $y_{ss}(t) = 100 - 5\sin(t - \pi/2)$.

6.3. $K > 1$.

6.7. $K = 60$.

6.10.

1. PM $= 59.3^\circ$, GM $= \infty$. 2. PM $= 99.6^\circ$, GM $= \infty$.

3. PM $= -59.7^\circ$, GM $= -27.1$ [dB]. 4. PM $= 8.5^\circ$, GM $= 0.08$ [dB].

5. PM $= 40.4^\circ$, GM $= -6$ [dB]. 6. PM $= 84.3^\circ$, GM $= \infty$.

6.14. See Figures C.2 and C.3.

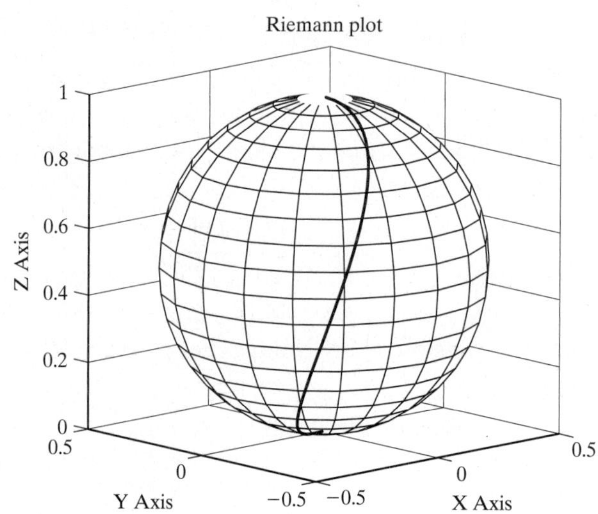

FIGURE C.2: The Riemann plot of $G(s) = \dfrac{1}{s(s+1)}$.

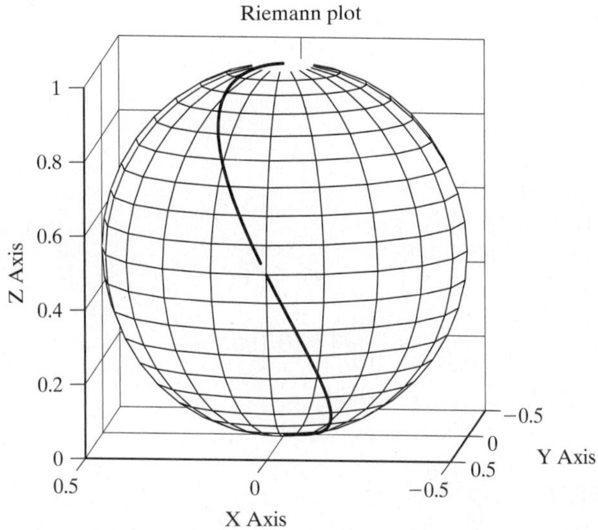

FIGURE C.3: The Riemann plot of $G(s) = \dfrac{s+1}{s^2(s-1)}$.

C.7 CHAPTER 7

Answers to most of the problems in this chapter are not unique.

PROBLEMS

7.1. $C(s) = \dfrac{2(s/0.28+1)}{s/0.0287+1}.$

7.2. $C(s) = \dfrac{2 \times 0.028(s/0.28+1)}{s}.$

7.3. $C(s) = \dfrac{2.5(s+1)}{s/0.6+1}.$

7.4. $C(s) = \dfrac{5(s/0.06+1)}{s/0.0042+1}.$

7.5. $C(s) = \dfrac{20(s/10.08+1)}{s/37.31+1}.$

7.6. $C(s) = \dfrac{4.12(s/1.24+1)}{s/7.24+1}.$

7.7. $C(s) = \dfrac{0.9629(s/0.83+1)}{s/4.83+1}.$

7.8. $C(s) = \dfrac{10(s+1)}{s}.$

7.9. $C(s) = C_{\text{lead}}(s)C_{\text{lag}}(s) = \dfrac{s/0.728+1}{s/5.5+1}\dfrac{5(s/0.2+1)}{s/0.0339+1}.$

7.10. $K = 40$ and $\omega = 2$.

C.8 CHAPTER 8

PROBLEMS

8.4. $\|G(s)\|_\infty = 2/\sqrt{3}$.

8.8. Yes.

8.9. $K \in (5/4, 10)$.

8.11. **1.** $(100, 200)$ is closer.

2. $\delta\left(\frac{1}{s+0.1}, 0\right) = 0.995$, $\delta\left(0, \frac{1}{s-0.1}\right) = 1$,
$\delta\left(\frac{1}{s+0.1}, \frac{1}{s-0.1}\right) = 0.198$.

8.13. The best K is $K = 0.303$.

8.15. The best K is $K = 1$.

C.9 CHAPTER 9

PROBLEMS

9.1. **1.** $C(s) = \frac{3-2\sqrt{2}}{s+2\sqrt{2}-1}$ and $J^* = 6\sqrt{2} - 8$.

2. $C(s) = \frac{3+2\sqrt{2}}{s+2\sqrt{2}+1}$ and $J^* = 6\sqrt{2} + 8$.

3. $C(s) = \frac{(4\sqrt{3}-6)s+1}{s2+(2\sqrt{3}-1)s+6-2\sqrt{3}}$ and $J^* = 22\sqrt{3} - 36$.

4. $C(s) = -\frac{s+1}{s^2+3s+4}$ and $J^* = 4$.

5. $C(s) = \frac{8s^2+4s+1}{s^3+4s^2+8s+10}$ and $J^* = 52$.

9.2. **1.** $C(s) = \frac{12(s+1)}{s+5}$.

2. $C(s) = \frac{0.4852s+0.5}{s^2+1.8284s+1.1716}$ and $J^* = 1.0538$.

9.3. **1.** $C(s) = \frac{1.5s+1}{s+4}$.

2. $C(s) = \frac{6s+4}{s^2+4s+18}$.

3. $C(s) = \frac{4\sqrt{2}s+4}{s^2+4\sqrt{2}s+16}$ and $J^* = 12\sqrt{2}$.

4. $C(s) = \frac{4\sqrt{2}s+4}{s^2+4\sqrt{2}s+16}$ and $J^* = 48\sqrt{2}$.

9.4. $C(s) = \frac{864}{s+47}$ and $J^* = 60480$.

9.5. $C(s) = \frac{2\sqrt{15}+16}{s+\sqrt{15}+1}$ and $J^* = 25.4930$.

9.6. **1.** $C(s) = \sqrt{2} - 1$ and $b^*(P) = \dfrac{1}{\sqrt{4 - 2\sqrt{2}}}$.

2. $C(s) = \sqrt{2} + 1$ and $b^*(P) = \dfrac{1}{\sqrt{4 + 2\sqrt{2}}}$.

3. $C(s) = \dfrac{1.452s + 1.618}{s + 2.35}$ and $b^*(P) = 0.5671$.

4. $C(s) = -\dfrac{(1 + \sqrt{2})(s + 1)}{s + 3 + 2\sqrt{2}}$ and $b^*(P) = \dfrac{1}{\sqrt{4 + 2\sqrt{2}}}$.

5. $C(s) = \dfrac{(2\sqrt{2} + 3)s^2 + (\sqrt{2} + 2)s + 1}{s^2 + (\sqrt{2} + 2)s + 2\sqrt{2} + 3}$ and $b^*(P) = \dfrac{1}{\sqrt{18 + 12\sqrt{2}}}$.

Bibliography

1. L. V. Ahlfors, *Complex Analysis*, 3rd Edition, McGraw-Hill, New York, 1979.
2. K. J. Åström, *Introduction to Stochastic Control Theory*, Academic Press, New York, 1970.
3. K. J. Åström and T. Hägglund, *Advanced PID Control*, The Instrumentation, Systems, and Automation Society, Research Triangle Park, NC, 2006
4. K. J. Åström and R. Murray, *Feedback Systems: An Introduction for Scientists and Engineers*, Princeton University Press, Princeton and Oxford, 2008.
5. B. D. O. Anderson and J. B. Moore,*Optimal Control: Linear Quadratic Methods*, Prentice Hall, Englewood Cliffs, NJ, 1989.
6. G. Balas, R. Chiang, A. Packard, and M. Safonov, *Robust Control Toolbox Users Guide Version 3*, The MathWorks, Inc., Natick, MA, 2006.
7. B. R. Barmish, *New Tools for Robustness of Linear Systems*, Macmillan, New York, 1994.
8. A. Bedford and W. Fowler, *Engineering Mechanics*, 3rd Edition, Prentice Hall, Upper Saddle River, NJ, 2002.
9. S. P. Bhattacharyya, H. Chapellat, and L. H. Keel, *Robust Control, the Parametric Approach*, Prentice Hall, Upper Saddle River, NJ, 1995.
10. H. W. Bode, "Relations between attenuation and phase in feedback amplifier design," *Bell System Technical Journal*, vol. 19, pp. 421–454, 1940.
11. H. W. Bode, *Network Analysis and Feedback Amplifier Design*, Van Nostrand Reinhold, New York, 1945.
12. N. K. Bose, "A system theoretic approach to stability of sets of polynomials," *Contemporary Mathematics*, vol. 47, pp. 25–34, 1985.

13. W. L. Brogan, *Modern Control Theory*, 3rd Edition, Prentice Hall, Englewood Cliffs, NJ, 1991.

14. H. Chapellat and S. P. Bhattacharyya, "An alternative proof of Kharitonov's theorem," *IEEE Transactions on Automatic Control*, vol. 34, pp. 448–450, 1989.

15. H. Chapellat, M. Mansour, and S. P. Bhattacharyya, "Elementary proofs of some classical stability criteria," *IEEE Transactions on Education*, vol. 33, pp. 232–239, 1990.

16. C. T. Chen, *Linear System Theory and Design*, 3rd Edition, Oxford University Press, New York, 1998.

17. R. V. Churchill, J. W. Brown, and R. F. Verhey, *Complex Variables and Applications*, 5th Edition, McGraw-Hill, New York, 1990.

18. S. Darbha and S. P. Bhattacharyya, "On the synthesis of controllers for a nonovershooting step response," *IEEE Transactions on Automatic Control*, vol. 48, pp. 797–799, 2003.

19. E. J. Davison, "The robust control of a servomechanism problem for linear time-invariant multivariable systems," *IEEE Transactions on Automatic Control*, vol. AC-21, pp. 25–34, 1976.

20. C. A. Desoer and E. S. Kuh, *Basic Circuit Theory*, McGraw-Hill International, Auckland, 1969.

21. C. A. Desoer and C. L. Gustafson, "Algebraic theory of linear multivariable feedback systems," *IEEE Transactions on Automatic Control*, vol. AC-29, pp. 909–917, 1984.

22. R. C. Dorf and R. H. Bishop, *Modern Control System*, 11th Edition, Pearson Prentice Hall, Upper Saddle River, NJ, 2008.

23. J. C. Doyle, B. A. Francis, and A. R. Tannenbaum, *Feedback Control Theory*, Macmillan Publishing Company, New York, 1992.

24. A. K. El-Sakkary, "Estimating robustness on the Riemann sphere," *International Journal of Control*, vol. 42, pp. 561–567, 1989.

25. W. R. Evans, "Graphical analysis of control systems," *Transactions of the American Institute of Electrical Engineers*, vol. 67, pp. 547–551, 1948,

26. W. R. Evans, "Control system synthesis by root locus method," *Transactions of the American Institute of Electrical Engineers*, vol. 69, pp. 66–69, 1950.

27. W. R. Evans, *Control-System Dynamics*, McGraw-Hill, New York, 1954.

28. A. Ferrante, A. Lepschy, and U. Viaro, "A simple proof of the Routh test," *IEEE Transactions on Automatic Control*, vol. 44, pp. 1306–1309, 1999.

29. B. F. Francis and W. M. Wonham, "The internal model principle of control theory," *Automatica*, vol. 12, pp. 457–465, 1976.

30. G. F Franklin, J. D. Powell, and A. Emami-Naeini, *Feedback Control of Dynamic Systems*, 5th Edition, Pearson Prentice Hall, Upper Saddle River, NJ, 2006.

31. J. S. Freudenberg and D. P. Looze, *Frequency Domain Properties of Scalar and Multivariable Feedback Systems*, Lecture Notes in Contr. and Info. Science, Vol. 104, Springer-Verlag, Berlin, 1988.

32. P. Fu, X. Zhao, and L. Qiu, "Solution to the Nehari problem using the Routh table," *Proceedings of the 12th Mediterranean Conference on Control and Automation*, Kusadasi, Turkey, 2004.

33. P. A. Fuhrmann, *A Polynomial Approach to Linear Algebra*, Springer, New York, 1996.

34. F. R. Gantmacher, *Theory of Matrices*, Vol. 2, Clelsea, New York, 1960.

35. J. Garloff and D. G. Wagner, "Hadamard products of stable polynomials are stable," *Journal of Mathematical Analysis and Applications*, vol. 202, pp. 797–809, 1996.

36. G. C. Goodwin, S. F. Graebe, and M. E. Salgado, *Control System Design*, Prentice Hall, Upper Saddle River, NJ, 2001.

37. M. Green and D. J. N. Limebeer, *Linear Robust Control*, Prentice Hall, Englewood Cliffs, NJ, 1995.

38. J. W. Helton and O. Merino, *Classical Control Using H^{∞} Methods: Theory, Optimization, and Design*, SIAM, Philadelphia, 1998.

39. R. A. Horn and C. R. Johnson, *Matrix Analysis*, Cambridge University Press, Cambridge, 1990.

40. I. M. Horowitz, *Synthesis of Feedback Systems*, Academic Press, London, 1963.

41. A. Hurwitz, "Über die Bedingungen unter welchen eine Gleichung nur Wurzeln mit negativen reellen Eeilen besitzt," *Mathematische Annalen*, vol. 46, pp. 273–284, 1895.

42. H. M. James, N. B. Nichols, and R. S. Philips, *Theory of Servomechanisms*, McGraw-Hill, New York, 1947.

43. J. L. Jensen,"Sur un nouvel et important thorme de la thorie des fonctions," *Acta Mathematica*, vol. 22, 359–364, 1899.

44. T. Kailath, *Linear Systems*, Prentice Hall, Englewood Cliffs, NJ, 1980.

45. R. E. Kalman, "When is a control system optimal?" *Transactions of the ASME Series D: Journal of Basic Engineering*, vol. 86, pp. 1–10, 1964.

46. R. E. Kalman and R. S. Bucy, "New results in linear filtering and prediction theory," *Transactions of the ASME Series D: Journal of Basic Engineering*, vol. 83, pp. 95–108, 1960.

47. V. Kharitonov, "Asymptotic stability an equilibrium position of a family of systems of differential equations," *Differential Uravnen*, vol. 14, pp. 2086–2088, 1978.

48. V. Kucĕra, "Stability of discrete linear systems," *Preprints of the 6th IFAC World Congress*, Boston, MA, 1975.

49. B. C. Kuo and F. Golnaraghi, *Automatic Control Systems*, 8th Edition, Prentice Hall, Englewood Cliffs, NJ, 2002.

50. H. Kwakernaak and R. Sivan, *Linear Optimal Control Systems*, Wiley-Interscience, New York, 1972.

51. H. Kwakernaak and R. Sivan, *Modern Signals and Systems*, Prentice Hall, Englewood Cliffs, NJ, 1991.

52. V. B. Larin, K. I. Naumenko, and V. N. Suntsev, *Spectral Methods for Synthesis of Linear Systems with Feedback* (in Russian), Naukova Dumka, Kiev, USSR, 1971.

53. E. A. Lee and P. Varaiya, *Structure and Interpretation of Signals and Systems*, Addison Wesley, Boston, 2003.

54. A. Liénard and M. H. Chipart, "Sur la signe de la partie réelle des racines d'une équations algébrique," *Journal de Mathematiques Pures et Appliquees*, vol. 10, pp. 291–346, 1914.

55. K. H. Lundberg, H. R. Miller, and D. L. Trumper, "Initial conditions, generalized functions, and the Laplace transform: Trouble at the origin," *IEEE Control Systems Magazine*, vol. 27, pp. 22–35, 2007.

56. M. Margaliot and G. Langholz, "The Routh-Hurwitz array and realization of characteristic polynomials," *IEEE Transactions on Automatic Control*, vol. 45, pp. 2424–2426, 2000.

57. J. C. Maxwell, "On governors," *Proceedings of the Royal Society of London*, vol. 16, pp. 270–283, 1868.

58. D. C. McFarlane and K. Glover, *Robust Controller Design Using Normalized Coprime Factor Plant Descriptions*, Lecture Notes in Contr. and Info. Science, Vol. 138, Springer-Verlag, Berlin, 1990.

59. G. Meinsma, "Elementary proof of the Routh-Hurwitz test," *Systems & Control Letters*, vol. 25, pp. 227–242, 1995.

60. R. J. Minnichelli, J. J. Anagnost, and C. Desoer, "An elementary proof of Kharitonov's stability theorem with extensions," *IEEE Transactions on Automatic Control*, vol. 34, pp. 995–998, 1989.

61. T. Mita and H. Yoshida, "Undershooting phenomenon and its control in linear multivariable servomechanisms," *IEEE Transactions on Automatic Control*, vol. AC-26, pp. 402–407, 1981.

62. M. Morari, and E. Zafiriou, *Robust Process Control*. Prentice Hall, Englewood Cliffs, NJ, 1989.

63. G. C. Newton, L. A. Gould, and J. F. Kaiser, *Analytical Design of Linear Feedback Control*, John Wiley & Sons, Inc., New York, 1957.

64. T. Norimatsu and M. Ito, "On the zero non-regular control system," *Journal of Institute of Electrical Engineering of Japan*, vol. 81, pp. 566–575, 1961.

65. H. Nyquist, "Regeneration theory," *Bell System Technical Journal*, vol. 11, pp. 126–147, 1932.

66. K. Ogata, *Discrete-Time Control Systems*, 2nd Edition, Prentice Hall, Englewood Cliffs, NJ, 1994.

67. K. Ogata, *Modern Control Engineering*, 5th Edition, Pearson Prentice Hall, Upper Saddle River, NJ, 2008.

68. A. V. Oppenheim and A. S. Willsky, *Signals & Systems*, Prentice Hall, Upper Saddle River, NJ, 1997.

69. P. C. Parks, "A new proof of the Routh-Hurwitz stability criterion using the second method of Lyapunov," *Proceedings of the Cambridge Philosophical Society*, vol. 58, pp. 694–702, 1962.

70. C. L. Phillips and R. D. Harbor, *Feedback Control Systems*, 3rd Edition, Prentice Hall, Upper Saddle River, NJ, 2000.

71. L. Qiu and E. J. Davison, "Feedback stability under simultaneous gap metric uncertainties in plant and controller," *Systems and Control Letters*, vol. 18, pp. 9–22, 1992.

72. L. Qiu and E. J. Davison, "Pointwise gap metrics on transfer matrices," *IEEE Transactions on Automatic Control*, vol. 37 (6), pp. 741–758, 1992.

73. E. J. Routh, *A Treatise on the Stability of a Given State of Motion*, Macmillan, London, 1877.

74. E. J. Routh, *The Advanced Part of a Treatise on the Dynamics of A Rigid Body*, 6th Edition, Macmillan, London, 1905; reprint, Dover, New York:1959.

75. R. Saeks and J. Murray, "Fractional representation, algebraic geometry and simultaneous stabilization problem," *IEEE Transactions on Automatic Control*, vol. 27, pp. 895–903, 1982

76. M. M. Seron, J. H. Braslavsky, and G. C. Goodwin, *Fundamental Limitations in Filtering and Control*, Springer, London, 1997.

77. S. Skogestad and I. Postlethwaite, *Multivariable Feedback Control*, John Wiley & Sons, New York, 1996.

78. R. J. Smith and R. C. Dorf, *Circuits, Devices, and Systems*, 5th Edition, John Wiley & Sons, New York, 1992.

79. G. Stein, "Respect the unstable," *IEEE Control Systems Magazine*, vol. 23, no. 4, pp. 12–25, 2003.

80. G. Strang, *Linear Algebra and Its Applications*, 3rd Edition, Harcourt Brace Jovanovich Publishers, San Diego, 1988.

81. C. P. Therapos, "Balanced minimal realization of SISO systems," *Electronics Letters*, vol. 19, pp. 424–426, 1983.

82. M. Vidyasagar, "The graph metric for unstable plants and robustness estimates for feedback stability," *IEEE Transactions on Automatic Control*, vol. 29, pp. 403–417, 1984.

83. M. Vidyasagar, "On undershoot and nonminimum phase zeros," *IEEE Transactions on Automatic Control*, vol. AC-31, p. 440, 1985.

84. M. Vidyasagar, *Control System Synthesis: A Factorization Approach*, The MIT Press, Cambridge, MA, 1985.

85. M. Vidyasagar, "Normalized coprime factorizations for non-strictly proper systems," *IEEE Transactions on Automatic Control*, vol. 33, pp. 300–301, 1988.

86. M. Vidyasagar and N. Viswanadham, "Algebraic design techniques for reliable stabilization," *IEEE Transactions on Automatic Control*, vol. 27, pp. 1085–1095, 1982.

87. G. Vinnicombe, "On the frequency response interpretation on an indexed $\mathcal{L}_2$-gap metric," *Proceedings of the American Control Conference*, Chicago, Illinois, pp. 1133–1137, 1992.

88. G. Vinnicombe, "Robust Design in the graph topology: a benchmark example," *Proceedings of the American Control Conference*, Chicago, Illinois, pp. 2063–2064, 1992.

89. G. Vinnicombe, "Frequency domain uncertainty and the graph topology," *IEEE Transactions on Automatic Control*, vol. 38, pp. 1371–1383, 1993.

90. G. Vinnicombe, *Measuring the Robustness of Feedback Systems,* Ph.D. dissertation, Cambridge University, Cambridge, 1993.

91. G. Vinnicombe, *Uncertainty and Feedback: $\mathcal{H}_\infty$ Loop-Shaping and the ν-Gap Metric*, Imperial Collage Press, London, 2001.

92. W. A. Wolovich, *Automatic Control Systems*, Sauders College Publishing, Fort Worth, 1994.

93. K. S. Yeung and S. S. Wang, "A simple proof of Kharitonov's theorem," *IEEE Transactions on Automatic Control*, vol. 32, pp. 822–823, 1987.

94. D. C. Youla and J. J. Bongiorno Jr., "A feedback theory of two-degree-of-freedom optimal Wiener-Hopf design," *IEEE Transactions on Automatic Control*, vol. AC-30, pp. 652–665, 1985.

95. D. C. Youla, H. A. Jabr, and J. J. Bongiorno, "Modern Wiener-Hopf design of optimal controllers: part I," *IEEE Transactions on Automatic Control*, vol. AC-21, pp. 3–13, 1976.

96. D. C. Youla, H. A. Jabr, and J. J. Bongiorno, "Modern Wiener-Hopf design of optimal controllers: part II," *IEEE Transactions on Automatic Control*, vol. AC-21, pp. 319–338, 1976.

97. D. C. Youla, H. A. Jabr, and C. N. Lu, "Single-loop feedback stabilization of linear multivariable dynamical plants," *Automatica*, vol. 10, pp. 159–173, 1974.

98. G. Zames, "Feedback and optimal sensitivity: model reference transformations, multiplicative seminorms, and approximate inverses," *IEEE Transactions on Automatic Control*, vol. AC-26, pp. 301–320, 1981.

99. K. Zhou and J. C. Doyle, *Essentials of Robust Control*, Prentice Hall, Upper Saddle River, NJ, 1998.

100. K. Zhou, J. C. Doyle, and K. Glover, *Robust and Optimal Control*, Prentice Hall, Upper Saddle River, NJ, 1996.

101. J. G. Ziegler and N. B. Nichols, "Optimum setting for automatic controllers," *Transactions of the ASME*, vol. 64, pp. 759–768, 1942,

102. J. G. Ziegler, N. B. Nichols, and N. Y. Rochester, "Process lags in automatic control circuits," *Transactions of the ASME*, vol. 65, pp. 433–444, 1943.

Index